PRINCIPLES OF ELECTRICAL ENGINEERING MATERIALS AND DEVICES

S. O. Kasap

Professor of Electrical Engineering
University of Saskatchewan
Canada

Boston, Massachusetts   Burr Ridge, Illinois   Dubuque, Iowa
Madison, Wisconsin   New York, New York   San Francisco, California   St. Louis, Missouri

To *Güler,* my mother; *Nicolette,* my wife; and *Alp,* my dad

McGraw-Hill

A Division of The McGraw·Hill Companies

PRINCIPLES OF ELECTRICAL ENGINEERING MATERIALS AND DEVICES

2 3 4 5 6 7 8 9 0 DON DON 9 0 9 8 7

ISBN 0-256-16173-9

Publisher: *Tom Casson*
Senior sponsoring editor: *Scott Isenberg*
Director of marketing: *Kurt L. Strand*
Project supervisor: *Karen M. Smith*
Production supervisor: *Bob Lange*
Cover designer: *Randy Scott*
Director, prepress purchasing: *Kimberly Meriwether David*
Compositor: *Interactive Composition Corporation*
Typeface: *10/12 Times Roman*
Printer: *R. R. Donnelley & Sons Company*

Library of Congress Cataloging-in-Publication Data

Kasap, S. O. (Safa O.)
   Principles of electrical engineering materials and devices / S.O.
Kasap.
      p.   cm.
   Includes index.
   ISBN 0-256-16173-9
   1. Electric engineering—Materials.   2. Electric apparatus and
appliances.   I. Title.
TK453.K26   1997
621.3—dc21   96-47581

http://www.mhcollege.com

ABOUT THE AUTHOR

Safa Kasap received his Bachelor of Science (Engineering), Master of Science, and Ph.D. degrees from the Department of Electrical Engineering, Imperial College of Science and Technology, University of London, in 1976, 1978, and 1983, specializing in amorphous semiconductor materials and devices. From 1982 to 1983 he was a faculty member in the Department of Electrical Engineering, South Bank University (London) and from 1983 to 1986 he was a Gestetner Postdoctoral Research Fellow at Imperial College working on electrophotography.

Professor Kasap is currently with the Department of Electrical Engineering, University of Saskatchewan (Canada), where, since 1986, he has been teaching electrical engineering materials and devices, optoelectronics, physical electronics, and solid state devices at the undergraduate and postgraduate levels. His research interests cover electrical engineering materials (which includes polymers and glasses), electrophotography, photodetectors, X-ray image detectors, characterization of electrical noise in semiconductor devices, and measurement science and technology, with more than 70 journal papers in these fields.

He has recently been awarded the Doctor of Science degree from the University of London for his contributions to materials science in electrical engineering and electrophotography. He is a Fellow of the Institution of Electrical Engineers, the Institute of Physics, and the Institute of Materials; and he is a registered professional engineer in Canada and the European Union.

Concepts form the basis for any science. These are ideas, usually somewhat vague (especially when first encountered), which often defy really adequate definition. The meaning of a new concept can seldom be grasped from reading a one-paragraph discussion. There must be time to become accustomed to the concept, to investigate it with prior knowledge, and to associate it with personal experience. Inability to work with details of a new subject can often be traced to inadequate understanding of its basic concepts.

William C. Reynolds, *Thermodynamics*
(New York: McGraw-Hill, 1968, now out of print)

PREFACE

Over the last two decades, there have been dramatic advances in semiconductors; and many Electrical Engineering (EE), Materials Science, and Engineering Physics departments have introduced extensive courses on semiconductors, particularly on the fabrication of microelectronic devices. There are now many excellent texts on the subject. The first course in materials is generally taught by non-electrical engineers in other departments. In all these courses, the mechanical properties, fracture, phase diagrams, phase transformations, metals, alloys, steels and cast irons, ceramics, glasses, plastics, and composites are greatly overemphasized, and usually the students do not satisfactorily cover electrical engineering materials and applications of materials in the EE discipline. The first materials course is generally followed by an EE course in semiconductor materials and devices that invariably includes extensive semiconductor physics. The consequence has been an overemphasis in mechanical properties and semiconductors. In many cases, dielectric and magnetic properties have been either underrepresented or totally bypassed.

We are now realizing the importance of having a first course that covers EE materials in a general way that also includes dielectric and magnetic properties. This was the rationale in preparing the current text. Over the next five to ten years, many Electrical Engineering, Materials Science, and Physics departments will be looking for a materials text at the junior level that covers a broad spectrum of electrical materials, includes dielectric and magnetic materials, has applications and extensive problems, has elementary quantum mechanics, and has

an overall goal of satisfying various accreditation requirements across international boarders. This textbook should answer these needs.

Organization and Features

The text represents a *first* course in electrical engineering materials and devices suitable for one- or two-semester courses at the second- or third-year level of a four-year curriculum. It can either follow a short elementary materials science course or can be used on its own as a starting text. This is not a specialized text in electronic materials emphasizing semiconductor physics and technology. There are many books in the market in this field.

The organization of the text allows it to be used for both one or two semester courses. Some chapters have additional topics to allow a more detailed treatment, usually including quantum mechanics or more mathematics. The text may be used in short or extended (two-semester) courses according to the instructor's choice of chapters and sections. Cross referencing has been avoided as much as possible to reduce "student irritation" to an acceptable level without too much repetition and to allow various sections to be skipped as desired by the reader.

The majority of the problems have been solved on software math packages (Theorist and Mathcad) that will be available to the instructor. I chose to divide the material into eight chapters based on my personal experience with the students. They felt more comfortable in covering something in every chapter rather than leaving out whole chapters in a course.

When a whole chapter is skipped, they tend to feel that they missed out on an important topic and feel a loss of continuity. I currently use the text in a two-semester EE materials course.

I tried to keep the general treatment and various proofs at a semiquantitative level without going into detailed physics. On the one hand, we are required to cover as much as possible and, on the other hand, professional engineering accreditation requires students to solve numerical problems and carry out "design calculations." In preparing the text, I tried to satisfy B.Sc. Engineering accreditation requirements in as much breadth as possible. Many of the problems have been set to satisfy engineering accreditation requirements. Obviously one cannot solve numerical problems, carry out design calculations, and at the same time derive each equation in a text without exploding the size of the text to an unacceptable level.

Some important features:

- The principles are developed with the minimum of mathematics and with the emphasis on physical ideas. Quantum mechanics is part of the course but without its difficult mathematical formalism.

- There are more than 120 worked examples, most of which have a significance in electrical engineering. Students learn by way of examples, however simple, and to that end more than 130 problems have been provided.

- Even the simplest concepts have examples to endow a feeling of fulfillment in the student.

- Most students would like to have clear diagrams to help them visualize the explanations and understand concepts. The text includes numerous illustrations (427 original diagrams, figures, and graphs) that have been computer drafted to reflect the concepts and aid the explanations in the text.

- The end-of-chapter questions and problems are graded so that they start with easy concepts and eventually lead to more sophisticated concepts. Difficult problems are identified with an asterisk (*). Many practical applications with diagrams have been included. All of these problems have been classroom tested and have been solved on Theorist and/or Mathcad. There is a detailed *Solutions Manual* for the instructor.

- Some of the end-of-chapter problems also instill the idea of concurrent engineering, which requires the student to apply several concepts simultaneously from various diverse fields even though they may not have covered some of the topics in detail.

- There is a glossary, "Important Terms," at the end of each chapter that defines some of the concepts and terms used, not only within the text but also in the problems.

- The end of each chapter includes a section on "Additional Topics" to further develop important concepts, to introduce interesting applications, or to prove a theorem. These topics are intended for the keen student and can be used as part of the text for a two-semester course.

This book is supported by the author's Web site (http://Kasap3.Usask.Ca), which contains additional worked examples, additional topics, color photographs, more illustrations, an extended glossary of *Defining Terms in Electronic Materials and Devices,* links to other materials science Web sites, a list of interesting reading material (helpful for term papers), tables of useful materials data, a corrigendum page, and an extensive reading list with comments. No undergraduate book can claim originality, and I have benefitted from many excellent books. These are listed in Web site with the author's comments.

Prerequisites

The level of sophistication has been kept at the junior undergraduate level where the students have not seen quantum mechanics or any elementary materials science. Indeed, this would be their first exposure to quantum mechanics. They would have been exposed to first-year physics and chemistry courses covering only classical concepts. Their mathematical level is assumed to include ordinary differentiation and integration but to exclude partial differentiation. It is quite likely that partial differentiation is covered in a math class in the same semester. Partial differentiation occurs only in the Schrödinger equation in Chapter 3, where it is simply treated as if it were ordinary differentiation but with respect to one of the variables. For those students who have taken an elementary class in materials science (van Vlack, *Elements of Materials Science,* or a similar text), a suitable starting point is Chapter 2.

Acknowledgments

I have many to thank, directly and indirectly, who have helped me in one way or another eventually make it to the editing and production departments of Irwin, not least the Irwin team themselves. My gracious thanks go to my past and present graduate students and post-doctoral research fellows, especially Randy Thakur (Micron Technology, Boise), Brad Polischuk (Noranda Advanced Materials, Montreal), Vish Aiyah (AECL), Don Scansen (NRC, Ottawa), Chris Haugen (completing his Ph.D.), Mark Nesdoly (Ph.D. student), and Reza Tanha (completing his M.Sc.), who have kept me on my toes and read various sections of this book. I owe a lot to Ron Fleming, who was the head of the department when I first joined the faculty in 1986. He understood the time-consuming and arduous nature of running experimental research laboratories at the same time as having a heavy teaching load and, throughout his headship, kept a tight lid on my administrative duties (i.e., almost none). With his frequent visits to my office and research labs, Ron kept on encouraging me and thereby unwittingly initiated the whole project.

Various reviewers at one time or another read various portions of the manuscript and provided extensive comments. I incorporated the majority of the suggestions, which I believe make this a better book. I'd like to personally thank them all for their invaluable critiques, some of whom include:

Professor Çetin Aktik, University of Sherbrooke, Quebec, Canada

Professor Richard Bube, Stanford University, California

Dr. K. W. Cheah, Hong Kong Baptist University, Hong Kong

Professor Bruce A. Ferguson, University of Portland, Oregon

Dr. Z. Ghassemlooy, Sheffield-Hallam University, Sheffield, England

Professor Furrukh Khan, Ohio State University, Ohio

Professor Michael Kozicki, Arizona State University, Arizona

Professor Hilary Lackritz, Purdue University, Indiana

Professor Aaron Peled, Center for Technological Education, Holon, Israel

Dr. Charbel Tannous, University of Brest, France

Dr. M. Vaezi-Nejad, Greenwich University, London, England

Professor Linda Vanasupa, California Polytechnic State University, California

S. O. Kasap

BRIEF CONTENTS

CONTENTS

x   Contents

Chapter 8

Magnetic Properties and Superconductivity   599

1

Elementary Materials Science Concepts[1]

Understanding the basic building blocks of matter has been one of the most intriguing endeavors of mankind. Our understanding of interatomic interactions has now reached a point where we can quite comfortably explain the macroscopic properties of matter, based on quantum mechanics and electrostatic interactions between electrons and ionic nuclei in the material. There are many properties of materials that can be explained by a classical treatment of the subject. In this chapter, as well as the next, we treat the interactions in a material from a classical perspective and introduce a number of elementary concepts. These concepts do not invoke any quantum mechanics, which is a subject of modern physics and is introduced in Chapter 3. Although many useful engineering properties of materials can be treated with hardly any quantum mechanics, it is impossible to develop the science of electronic materials and devices without modern physics.

1.1 ATOMIC STRUCTURE

The model of the atom that we must use to understand its general behavior involves quantum mechanics, a topic we will study in detail in Chapter 3. For the present, we will simply accept the following facts about a simplified, but intuitively satisfactory, atomic model called the **shell model,** based on the Bohr model (1913).

The mass of the atom is concentrated at the nucleus, which contains protons and neutrons. Protons are positively charged particles, whereas neutrons are neutral particles, and both have about the same mass. Although there is a coulombic repulsion between the protons, all the protons and neutrons are held together in the nucleus by the **strong force,** which is a powerful, fundamental, natural force

[1] This chapter may be skipped by readers who have already been exposed to an elementary course in materials science.

between particles. This force has a very short range of influence, typically less than 10^{-15} m. When the protons and neutrons are brought together very closely, the strong force overcomes the electrostatic repulsion between the protons and keeps the nucleus intact. The number of protons in the nucleus is the atomic number Z of the element.

The electrons are assumed to be orbiting the nucleus at very large distances compared to the size of the nucleus. There are as many orbiting electrons as there are protons in the nucleus. An important assumption in the Bohr model is that only certain orbits with fixed radii are stable around the nucleus. For example, the closest orbit of the electron in the hydrogen atom can only have a radius of 0.053 nm. Since the electron is constantly moving around an orbit with a given radius, over a long time period (perhaps $\sim 10^{-12}$ seconds on the atomic time scale), the electron would appear as a spherical negative-charge cloud around the nucleus and not as a single dot representing a finite particle. We can therefore view the electron as a charge contained within a spherical **shell** of a given radius.

Due to the requirement of stable orbits, the electrons therefore do not randomly occupy the whole region around the nucleus. Instead, they occupy various well-defined spherical regions. They are distributed in various shells and **subshells** within the shells, obeying certain occupation (or seating) rules.[2] The example for the carbon atom is shown in Figure 1.1.

The shells and subshells that define the whereabouts of the electrons are labeled using two sets of integers, n and ℓ. These integers are called the **principal** and **orbital angular momentum quantum numbers,** respectively. (The meanings of these names are not critical at this point.) The integers n and ℓ have the values $n = 1, 2, 3, \ldots$, and $\ell = 0, 1, 2, \ldots n - 1$, and $\ell < n$. For each choice of n, there are n values of ℓ, so higher-order shells contain more subshells. The shells

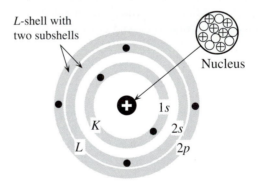

L-shell with two subshells

Nucleus

K

L

$1s$

$2s$

$2p$

$1s^2 2s^2 2p^2$ or $[He]2s^2 2p^2$

Figure 1.1   The shell model of the carbon atom, in which the electrons are confined to certain shells and subshells within shells.

[2] In Chapter 3, in which we discuss the quantum mechanical model of the atom, we will see that these shells and subshells are spatial regions around the nucleus where the electrons are most likely to be found.

corresponding to $n = 1, 2, 3, 4, \ldots$ are labeled by the capital letters $K, L, M,$ $N, \ldots$, and the subshells denoted by $\ell = 0, 1, 2, 3, \ldots$ are labeled $s, p, d, f \ldots$. The subshell with $\ell = 1$ in the $n = 2$ shell is thus labeled the $2p$ subshell, based on the standard notation $n\ell$.

There is a definite rule to filling up the subshells with electrons; we cannot simply put all the electrons in one subshell. The number of electrons a given subshell can take is fixed by nature to be[3] $2(2\ell + 1)$. For the s subshell ($\ell = 0$), there are 2 electrons, whereas for the p subshell, there are 6 electrons and so on. Table 1.1 summarizes the most number of electrons that can be put into various subshells and shells of an atom. Obviously, the larger the shell, the more electrons it can take, simply because it contains more subshells.

The number of electrons in a subshell is indicated by a superscript on the subshell symbol, so the electronic structure, or configuration, of the carbon atom (atomic number 6) shown in Figure 1.1 becomes $1s^2 2s^2 2p^2$. The K shell has only one subshell, which is full with two electrons. This is the structure of the inert element He. We can therefore write the electronic configuration more simply as $[\text{He}]2s^2 2p^2$. The general rule is put the nearest previous inert element, in this case He, in square brackets and write the subshells thereafter.

The electrons occupying the outer subshells are the farthest away from the nucleus and have the most important role in atomic interactions, as in chemical reactions, because these electrons are the first to interact with outer electrons on neighboring atoms. The outermost electrons are called **valence electrons** and they determine the valency of the atom. Figure 1.1 shows that carbon has four valence electrons in the L-shell.

When a subshell is full of electrons, it cannot accept any more electrons and it is said to have acquired a stable configuration. This is the case with the inert elements at the right-hand side of the Periodic Table, all of which have completely filled subshells and are rarely involved in chemical reactions. The majority of such elements are gases inasmuch as the atoms do not bond together easily to form a liquid or solid. They are sometimes used to provide an inert atmosphere instead of air for certain reactive materials.

Table 1.1   Maximum possible number of electrons in the shells and subshells of an atom

		Subshell			
		$\ell = 0$	1	2	3
n	**Shell**	s	p	d	f
1	K	2			
2	L	2	6		
3	M	2	6	10	
4	N	2	6	10	14

| [3] We will actually show this in Chapter 3 using quantum mechanics.

1.2   BONDING AND TYPES OF SOLIDS

1.2.1   Molecules and General Bonding Principles

When two atoms are brought together, the valence electrons interact with each other and with the neighbor's positively charged nucleus. The result of this interaction is often the formation of a bond between the two atoms, producing a molecule. The formation of a bond means that the energy of the system of two atoms together must be less than that of the two atoms separated, so that the molecule formation is energetically favorable, that is, more stable. The general principle of molecule formation is illustrated in Figure 1.2a, showing two atoms brought together from infinity. As the two atoms approach each other, the atoms exert attractive and repulsive forces on each other as a result of mutual electrostatic interactions. Initially, the attractive force F_A dominates over the repulsive force F_R. The net force F_N is the sum of the two,

$$F_N = F_A + F_R$$

and this is initially attractive, as indicated in Figure 1.2a.

The potential energy $E(r)$ of the two atoms can be found from[4]

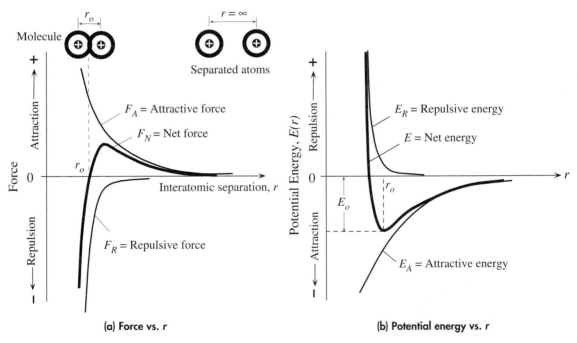

(a) **Force vs. r**   (b) **Potential energy vs. r**

Figure 1.2   (a) Force versus interatomic separation and (b) potential energy versus interatomic separation.

| [4]Remember that the change dE in the PE is the work done against the force, $dE = Fdr$.

$$F_N = \frac{dE}{dr}$$

by integrating the net force F_N. Figures 1.2a and b show the variation of the net force $F_N(r)$ and the overall potential energy $E(r)$ with the interatomic separation r as the two atoms are brought together from infinity. The lowering of energy corresponds to an attractive interaction between the two atoms.

The variations of F_A and F_R with distance are different. Force F_A varies slowly, whereas F_R varies strongly with separation and is strongest when the two atoms are very close. When the atoms are so close that the individual electron shells overlap, there is a very strong electron-to-electron shell repulsion and F_R dominates. An equilibrium will be reached when the attractive force just balances the repulsive force and the net force is zero, or

$$F_N = F_A + F_R = 0 \qquad \text{[1.1]}$$

Net force in bonding between atoms

In this state of equilibrium, the atoms are separated by a certain distance r_o, as shown in Figure 1.2. This distance is called the equilibrium separation and is effectively the **bond length.** On the energy diagram, $F_N = 0$ means $dE/dr = 0$, which means that the equilibrium of two atoms corresponds to the potential energy of the system acquiring its minimum value. Consequently, the molecule will only be formed if the energy of the two atoms as they approach each other can attain a minimum. This minimum energy also defines the bond energy of the molecule, as depicted in Figure 1.2b. An energy of E_o is required to separate the two atoms, and this represents the **bond energy.**

Although we considered only two atoms, similar arguments also apply to bonding between many atoms, or between millions of atoms as in a typical solid. Although the actual details of F_A and F_R will change from material to material, the general principle that there is a bonding energy E_o per atom and an equilibrium interatomic separation r_o will still be valid. Even in a solid in the presence of many interacting atoms, we can still identify a general potential energy curve $E(r)$ per atom similar to the type shown in Figure 1.2b. We can also use the curve to understand the properties of the solid, such as the thermal expansion coefficient, elastic and bulk moduli, and so forth.

1.2.2 Covalently Bonded Solids: Diamond

Two atoms can form a bond with each other by sharing some or all of the their valence electrons and thereby reducing the overall potential energy of the combination. The covalent bond results from the sharing of valence electrons to complete the subshells of each atom. Figure 1.3 shows the formation of a covalent bond between two hydrogen atoms as they come together to form the H_2 molecule. When the $1s$ subshells overlap, the electrons are shared by both atoms and each atom now has a complete subshell. As illustrated in Figure 1.3, electrons 1 and 2 must now orbit both atoms; they therefore cross the overlap region more frequently, indeed twice as often. Thus, electron sharing results in a greater concentration of negative

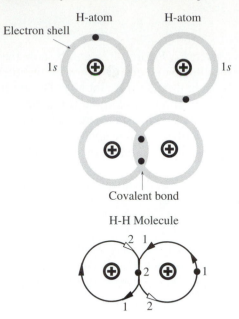

Figure 1.3   Formation of a covalent bond between two H atoms, leading to the H_2 molecule.

Electrons spend a majority of their time between the two nuclei, which results in a net attraction between the electrons and the two nuclei, which is the origin of the covalent bond.

charge in the region between the two nuclei, which keeps the two nuclei bonded to each other. Furthermore, by synchronizing their motions, electrons 1 and 2 can avoid crossing the overlap region at the same time. For example, when electron 1 is at the far right (or left), electron 2 is in the ovelap region; later, the situation is reversed.

The electronic structure of the carbon atom is $[He]2s^2 2p^2$ with four empty seats in the $2p$ subshell. The $2s$ and $2p$ subshells, however, are quite close. When other atoms are in the vicinity, as a result of interatomic interactions the two subshells become indistinguishable and we can consider only the shell itself, which is the L shell with a capacity of eight electrons. It is clear that the C atom with four vacancies in the L shell can readily share electrons with four H atoms, as depicted in Figure 1.4 whereby the C atom and each of the H atoms attain complete shells. This is the CH_4 molecule, which is the gas methane. The repulsion between the electrons in one bond and the electrons in a neighboring bond causes the bonds to spread as far out from each other as possible, so that in three dimensions, the H atoms occupy the corners of an imaginary tetrahedron and the CH bonds are at an angle of 109.5 degrees to each other, as sketched in Figure 1.4.

The C atom can also share electrons with other C atoms, as shown in Figure 1.5. Each neighboring C atom can share electrons with other C atoms, leading to a three-dimensional network of a covalently bonded structure. This is the

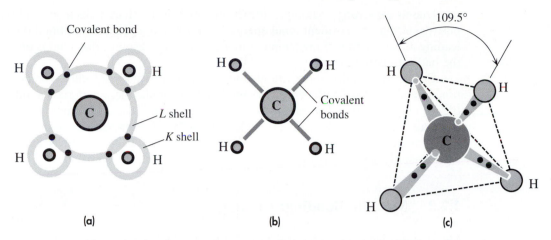

Covalent bond

H   O • • O H

_L_ shell

C

_K_ shell

H   O • • O H

(a)

H O   O H

C

Covalent
bonds

H O   O H

(b)

109.5°

H   H

C

H

H

(c)

Figure 1.4   **(a)** Covalent bonding in methane, CH_4, which involves four hydrogen atoms sharing electrons with one carbon atom. Each covalent bond has two shared electrons. The four bonds are identical and repel each other. **(b)** Schematic sketch of CH_4 on paper. **(c)** In three dimensions, due to symmetry, the bonds are directed toward the corners of a tetrahedron.

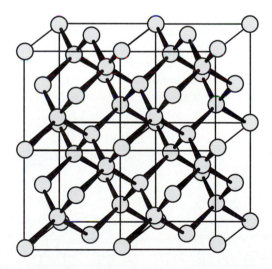

Figure 1.5   The diamond crystal is a covalently bonded network of carbon atoms.

Each carbon atom is bonded covalently to four neighbors, forming a regular three-dimensional pattern of atoms that constitutes the diamond crystal.

structure of the precious diamond crystal, in which all the carbon atoms are covalently bonded to each other, as depicted in the figure. The **coordination number (CN)** is the number of nearest neighbors for a given atom in the solid. As is apparent in Figure 1.5, the coordination number for a carbon atom in the diamond crystal structure is four.

Due to the strong coulombic attraction between the shared electrons and the positive nuclei, the covalent bond energy is usually the highest for all bond types, leading to very high melting temperatures and very hard solids: diamond is one of the hardest known materials.

Covalently bonded solids are also insoluble in nearly all solvents. The directional nature and strength of the covalent bond also make these materials nonductile (or nonmalleable). Under a strong force, they exhibit brittle fracture. Further, since all the valence electrons are locked in the bonds between the atoms, these electrons are not free to drift in the crystal when an electric field is applied. Consequently, the electrical conductivity of such materials is very poor.

1.2.3   Metallic Bonding: Copper

Metal atoms have only a few valence electrons, which can be readily removed by a small amount of energy. When many metal atoms are brought together to form a solid, these valence electrons are lost from individual atoms and become collectively shared by all the ions. The valence electrons therefore become **delocalized** and form an **electron gas** or **electron cloud,** permeating the space between the ions, as depicted in Figure 1.6. The attraction between the negative charge of this electron gas and the metal ions more than compensates for the energy initially required to remove the valence electrons from the individual atoms. Thus, the bonding in a metal is essentially due to the attraction between the stationary metal ions and the freely wandering electrons between the ions.

The bond is a collective sharing of electrons and is therefore nondirectional. Consequently, the metal ions try to get as close as possible, which leads to close-

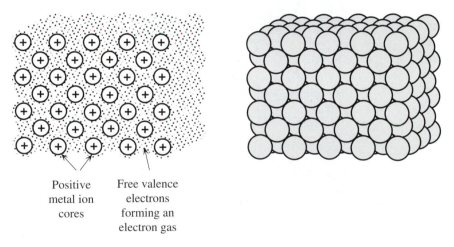

Positive
metal ion
cores

Free valence
electrons
forming an
electron gas

Figure 1.6   In metallic bonding, the valence electrons from the metal atoms form a "cloud of electrons," which fills the space between the metal ions and "glues" the ions together through coulombic attraction between the electron gas and the positive metal ions.

packed crystal structures with high coordination numbers, compared to covalently bonded solids. In the particular example shown in Figure 1.6, Cu^+ ions are packed as closely as possible by the gluing effect of the electrons between the ions, forming a crystal structure called the **face centered cubic (FCC).** Any smallest square area in the crystal that has Cu^+ ions at its corners also has a Cu^+ at its center, as is apparent from the figure.

The results of this type of bonding are dramatic. First, the nondirectional nature of the bond means that under an applied force, metal ions are able to move with respect to each other, especially in the presence of certain crystal defects (such as dislocations). Thus, metals tend to be ductile. Most importantly, however, the "free" valence electrons in the electron gas can respond readily to an applied electric field and drift along the force of the field, which is the reason for the high electrical conductivity of metals. Furthermore, if there is a temperature gradient along a metal bar, the free electrons can also contribute to the energy transfer from the hot to the cold regions, since they frequently collide with the metal ions and thereby transfer energy. Metals therefore also have good thermal conductivities.

1.2.4   Ionically Bonded Solids: Salt

Common table salt, NaCl, is a classic example of a solid in which the atoms are held together by ionic bonding. Ionic bonding is frequently found in materials that normally have a metal and a nonmetal as the constituent elements. Sodium (Na) is an alkaline metal with only one valence electron that can easily be removed to form an Na^+ ion with complete subshells. The ion Na^+ looks like the inert element Ne, but with a positive charge. Chlorine has five electrons in its $3p$ subshell and can readily accept one more electron to close this subshell. By taking the electron given up by the Na atom, the Cl atom becomes negatively charged and looks like the inert element Ar with a net negative charge. Transferring the valence electron of Na to Cl thus results in two oppositely charged ions, Na^+ and Cl^-, which are called the **cation** and **anion,** respectively, as shown in Figure 1.7. As a result of the Coulombic force, the two ions pull each other until the attractive force is just balanced by the repulsive force between the closed electron shells. Initially, energy is needed to remove the electron from the Na atom, this is the **energy of ionization.** However, this is more than compensated by the energy of Coulombic attraction between the two resulting oppositely-charged ions, and the net effect is a lowering of the potential energy of the Na^+ and Cl^- ion pair.

When many Na and Cl atoms are ionized and brought together, the resulting collection of ions is held together by the Coulombic attraction between the Na^+ and Cl^- ions. The solid thus consists of Na^+ cations and Cl^- anions holding each other through the Coulombic force, as depicted in Figure 1.8. The Coulombic force around a charge is nondirectional; also, it can be attractive or repulsive, depending on the polarity of the interacting ions. There are also repulsive Coulombic forces between the Na^+ ions themselves and between the Cl^- ions themselves. For the solid to be stable, each Na^+ ion must therefore have Cl^- ions as neighbors and vice versa so that like-ions are not close to each other.

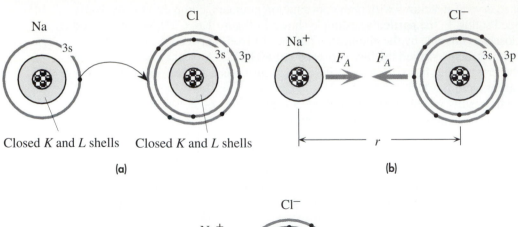

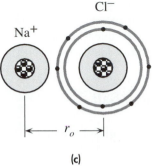

Figure 1.7   The formation of an ionic bond between Na and Cl atoms in NaCl. The attraction is due to Coulombic forces.

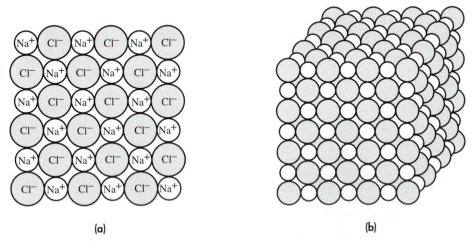

Figure 1.8   **(a)** A schematic illustration of a cross section from solid NaCl. Solid NaCl is made of Cl^- and Na^+ ions arranged alternatingly, so the oppositely charged ions are closest to each other and attract each other. There are also repulsive forces between the like-ions. In equilibrium, the net force acting on any ions is zero.
(b) Solid NaCl.

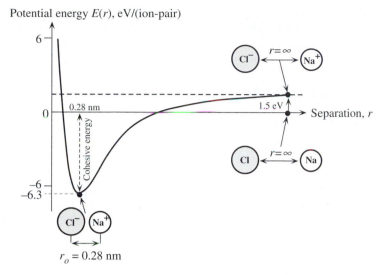

Figure 1.9   Sketch of the potential energy per ion pair in solid NaCl. Zero energy corresponds to neutral Na and Cl atoms infinitely separated.

The ions are in equilibrium and the solid is stable when the net potential energy is minimum, or $dE/dr = 0$. Figure 1.9 illustrates the variation of the net potential energy for a pair of ions as the interatomic distance r is reduced from infinity to less than the equilibrium separation, that is, as the ions are brought together from infinity. Zero energy corresponds to separated Na and Cl atoms. Initially, about 1.5 eV is required to transfer the electron from the Na to Cl atom and thereby form Na^+ and Cl^- ions. Then, as the ions come together, the energy is lowered, until it reaches a minimum at about 6.31 eV below the energy of the separated Na and Cl atoms. When $r = 0.28$ nm, the energy is minimum and the ions are in equilibrium. The bonding energy per ion for solid NaCl is thus 3.15 eV, as is apparent in Figure 1.9. The energy required to take solid NaCl apart into individual Na and Cl atoms is the **cohesive energy** of the solid, which is 3.15 eV per atom.

In solid NaCl, the Na^+ and Cl^- ions are thus arranged with each one having oppositely charged ions as its neighbors, to attain a minimum of potential energy. Since there is a size difference between the ions and since we must avoid like ions getting close to each other means, to achieve a stable structure, each ion can have only six oppositely charged ions as nearest neighbors. Figure 1.8b shows the packing of Na^+ and Cl^- ions in the solid. As is apparent in this figure, the coordination number for both cations and anions in the NaCl crystal is 6.

A number of solids consisting of metal–nonmetal elements follow the Nacl example and have ionic bonding. They are called **ionic crystals** and, by virtue of their ionic bonding characteristics, they share many physical properties. For example, LiF, MgO (magnesia), CsCl, and ZnS are all ionic crystals. They are strong, brittle materials with high melting temperatures compared to metals. Most become soluble in polar liquids such as water. Since all the electrons are within the rigidly positioned ions, there are no free or loose electrons to wander around in the crystal as in metals. Therefore, ionic solids are typically electrical insulators. Compared to metals and covalently bonded solids, ionically bonded solids have poor thermal conductivity since ions cannot readily pass vibrational kinetic energy to their neighbors.

1.2.5   Secondary Bonding

Since the atoms of inert elements have full shells and therefore cannot accept any electrons nor share any electrons, you might think that no bonding is possible between them. However, a solid form of argon exists at low temperatures, below $-189\ °C$, which means there must be some bonding mechanism between the Ar atoms. The magnitude of this bond cannot be strong, since above $-189\ °C$ solid argon melts.

   A weak type of attraction exists between all atoms and molecules, the so-called **van der Waals–London force,** which is due to a net electrostatic attraction between the electron distribution of one atom and the positive nucleus of the other. To understand this concept, we first must introduce the notion of an **electric dipole moment,** which is also useful in the studies of dielectric materials and signal radiation from antennas.

   An electric dipole moment occurs whenever there is a separation between a negative and a positive charge of equal magnitude Q, as shown in Figure 1.10. This moment is defined as a vector

Definition of the electric dipole moment

$$\mathbf{p} = Q\mathbf{x} \qquad\qquad [1.2]$$

where $\mathbf{x}$ is a distance vector from the negative to the positive charge.

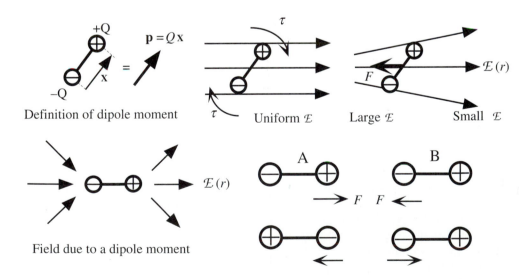

Definition of dipole moment   Uniform $\mathcal{E}$ Large $\mathcal{E}$ Small $\mathcal{E}$

Field due to a dipole moment

Figure 1.10   Definition of the electrical dipole moment and its properties.
In a uniform electric field, the dipole experiences a torque, whereas in a nonuniform electric field, it also experiences a net force. There is an electric field $\mathcal{E}(r)$ around a dipole moment. Two dipoles can attract each other, or repel each other, depending on their orientations.

When an electric dipole is placed in an external electric field $\mathcal{E}(r)$, the dipole will experience both a torque and a force, as depicted in Figure 1.10. In a uniform field, the torque τ will simply try to orient the dipole to line up with the field, because the charges $+Q$ and $-Q$ experience forces in opposite directions. In a nonuniform field, the dipole will also experience a net force F, which tries to move the dipole toward larger fields. The force will depend on both the orientation of the dipole and the spatial gradient of the field. The magnitude of the force is given by

$$F = p\frac{d\mathcal{E}}{dr}$$

Force on a dipole

A dipole moment creates an electric field $\mathcal{E}$ around itself, just as a single charge has an associated field. By calculating the net (total) electric field $\mathcal{E}$ from two charges $+Q$ and $-Q$, we can show that a dipole moment creates an electric field for which the magnitude at a distance r directly ahead of the dipole is given by

$$\mathcal{E}(r) = \frac{2p}{4\pi\varepsilon_o r^3}$$ [1.3]

Electric field from a dipole

an equation that is reminiscent of Coulomb's Law for the field of a single charge of Q, with $2p$ instead of Q and r^3 instead of r^2. Therefore, two dipoles, such as A and B in Figure 1.10, will exert forces on each other because each dipole will be in the field of the other. This is the origin of the van der Waals–London force.

A dipole moment can be either permanent or induced by an electric field. A permanent dipole moment is the result of a fixed separation of negative and positive charge distributions in a molecule, as in the HCl molecule depicted in Figure 1.11. An atom in an external electric field will develop an induced dipole moment; that is, it will become polarized. A molecule is said to be **polarized** if it possesses an effective dipole moment, that is, if there is a separation between the negative and positive charge distributions. The electrons, being much lighter than the positive nucleus, are easily displaced by an electric field and the result is an induced dipole moment, as shown in Figure 1.11b. In general, this induced dipole moment

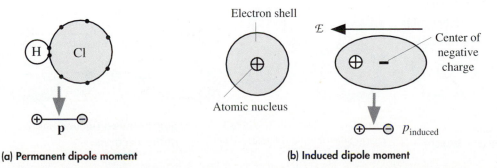

(a) Permanent dipole moment   (b) Induced dipole moment

Figure 1.11   Origin of the permanent dipole moment in polarized molecules (HCl) and the induced dipole moment in an atom.

depends on the electric field causing it. We can simply express the moment as being proportional to $\mathcal{E}$ or

Definition of polarizability

$$p_{induced} = \alpha \mathcal{E}$$

where α is a constant of proportionality called the **polarizability** of the atom. The value of α tells us how easily the atom can be polarized. Since the electrons in an atom are not rigidly fixed, each atom will possess a certain degree of polarizability.

To understand the physical origin of the van der Waals–London force, we consider two Ne atoms approaching each other to form a **van der Waals bond.** Each has closed (or full) electron shells. The center of mass of the electrons in the closed shells, when averaged over time, coincides with the location of the positive nucleus. At any one instant, however, the center of mass is displaced from the nucleus, due to various motions of the individual electrons around the nucleus, as depicted in Figure 1.12. In fact, the center of mass of all the electrons fluctuates about the nucleus. Consequently, the electron charge distribution is not static around the nucleus, but fluctuates asymmetrically, giving rise to an instantaneous dipole moment.

When two Ne atoms, A and B, approach each other, the rapidly fluctuating negative charge distribution on one affects the motion of the negative charge distribution on the other. A lower energy configuration (i.e., attraction) is produced when the fluctuations are synchronized such that the negative charge distribution on A gets closer to the nucleus of B while the negative distribution on B at that instant stays away from the negative charge on A, as shown in Figure 1.12. The strongest interaction arises from the closest charges, which are the displaced electrons in A and the nucleus in B. This means there will be a net attraction between the two atoms and hence a lowering of the net energy, which in turn leads to bonding.

This type of attraction between two atoms is due to induced synchronization of the electron motions around the nuclei. We refer to this as induced dipole–dipole attraction. The instantaneous dipole on A puts a field at B, which causes an induced

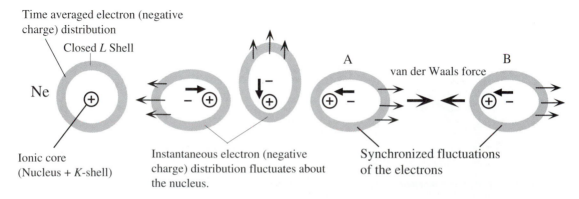

Time averaged electron (negative charge) distribution

Closed *L* Shell

Ne

Ionic core (Nucleus + *K*-shell)

Instantaneous electron (negative charge) distribution fluctuates about the nucleus.

A

van der Waals force

B

Synchronized fluctuations of the electrons

Figure 1.12   Induced dipole–induced dipole interaction and the resulting van der Waals force.

dipole in B. This dipole interacts with the dipole on A. The field at B due to the dipole at A is given by Equation 1.3, and the induced dipole at B is

$$p_{induced} = \alpha' E \propto \frac{\alpha}{r^3}$$

Similarly, B puts a field at A, which induces a dipole in A also given by Equation 1.3. The force pulling B toward A is therefore

$$F = p_{induced}\left[\frac{dE}{dr}\right] \propto \frac{\alpha}{r^3}\left[\frac{d}{dr}\left(\frac{1}{r^3}\right)\right] \propto \frac{\alpha}{r^7}$$

[1.4]   *Induced dipole–dipole force*

The magnitude of this force drops sharply with distance.

The bond energy of this type of induced dipole–dipole interaction is at least an order of magnitude lower than those of ionic, covalent, and metallic bonding. This is why the inert elements Ne and Ar solidify at temperatures below 25 K ($-248\,^{\circ}$C) and 84 K ($-189\,^{\circ}$C).

It is also possible to have an attraction and bonding between molecules that already possess a permanent dipole moment, such as H_2O. As shown in the simplistic view in Figure 1.13 each hydrogen atom shares all its electrons with the oxygen atom, whereas the oxygen atom has four unshared electrons. Consequently, the negative charge distribution is shifted more toward the oxygen atom, and the positive hydrogen proton is relatively exposed. This means that the H_2O molecule has a net dipole moment $\mathbf{p}_o$.

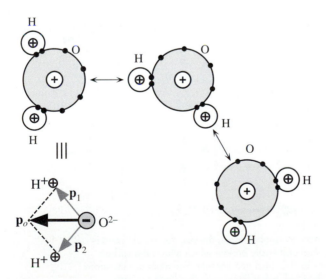

Figure 1.13   The origin of van der Waals bonding between water molecules.

The H_2O molecule has a net permanent dipole moment. Attractions between the various dipole moments in water give rise to van der Waals bonding.

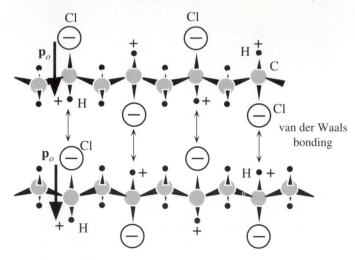

(a) van der Waals bonding in PVC (polyvinylchloride). Cl atom has a net negative charge whereas the hydrogen atom at that location has a net positive charge which gives rise to a net permanent dipole moment. Attraction between the dipoles results in van der Waals bonding between the chains.

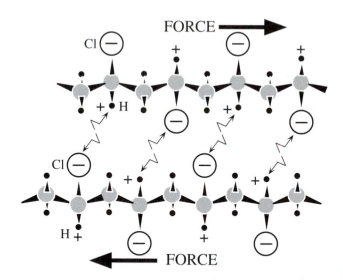

(b) Under an applied force, the van der Waals bond is stretched and easily broken which allows the carbon chains to slide past one another. Polymers are therefore ductile; much softer than solids that have primary bonding.

Figure 1.14   Secondary bonding in PVC.

When two water molecules are close to each other, the unshared electrons on the oxygen will be atttracted to the exposed proton of the hydrogen in the neighboring molecule, leading to what is known as **hydrogen bonding,** depicted in Figure 1.13. The true attraction is actually between two permanent dipoles. Dipole $\mathbf{p}_o$ is pulled in by the field of the other dipole $\mathbf{p}_o$, which means that the attraction is

$$F = p_o \frac{dE}{dr} \propto p_o \frac{d}{dr}\left(\frac{p_o}{r^3}\right) \propto \frac{p_o^2}{r^4} \qquad [1.5]$$

Permanent dipole–dipole interaction

The attraction between permanent dipoles is stronger than that between induced dipoles. This is borne out by experiments, which show that water has a melting temperature well above those of solid Ne or Ar. In the case of water, the H_2O dipoles cluster together and some of the molecules can even dissociate, as we know from elementary chemistry.

Van der Waals bonding is responsible for holding the carbon chains in polymers together. Although the C-to-C bond in a chain is due to covalent bonding, the interaction between the chains arises from van der Walls forces, and the interchain bonding is therefore secondary in nature, as indicated in Figure 1.14 for polyvinyl chloride (PVC). In PVC, the Cl atom has five unshared electrons and therefore has a slightly more negative charge distribution than the hydrogen atom, in which the proton is relatively exposed. Consequently, a hydrogen bond develops between the Cl and the H of the neighboring chain. These bonds are weak and can easily be stretched or broken. Polymers, therefore, have lower elastic moduli and melting temperatures than metals and ceramics.

Table 1.2 compares the energies involved in the five types of bonding found in materials and lists some important properties of these materials, showing the correlation with the bond type and its associated energy. The greater the bond energy, for example, the higher the melting temperature. Similarly, strong bond energies lead to greater elastic moduli and smaller expansion coefficients. Metals generally have the greatest electrical conductivity, since only this type of bonding allows a very large number of free charges (valence electrons) to wander in the solid and thereby contribute to electrical conduction. Electrical conduction in other types of solid may involve the motions of ions or charged defects from one fixed location to another.

1.2.6   Mixed Bonding

In many solids, the bonding between atoms is generally not just one type; rather, it is a mixture of bond types. We know that bonding in the silicon crystal is totally covalent, because the shared electrons in the bonds are equally attracted by the neighboring positive ion cores and are therefore equally shared. When there is a covalent type bond between two different atoms, the electrons become unequally shared, because the two neighboring ion cores are different and hence have different electron-attracting abilities. The bond is no longer purely covalent; it has some ionic character, because the shared electrons spend more time close to one of the ion cores. Covalent bonds that have an ionic character, due to an unequal sharing

Table 1.2   Comparison of bond types and typical properties (general trends)

	Typical Solids	Bond Energy (eV/atom)	Melt. Temp. (°C)	Elastic Modulus (GPa)	Density (g cm^{-3})	Typical Properties
Ionic	NaCl (rock salt)	3.2	801	40	2.17	Generally electrical insulators. May become conductive at high temperatures.
	MgO (magnesia)	10	2852	250	3.58	High elastic modulus. Hard and brittle but cleavable. Thermal conductivity between covalent and metallic solids.
Metallic	Cu	3.1	1083	120	8.96	Electrical conductor.
	Mg	1.1	650	44	1.74	Good thermal conduction. High elastic modulus. Generally ductile. Can be shaped.
Covalent	Si	4	1410	190	2.33	Large elastic modulus. Hard and brittle.
	C (diamond)	7.4	3550	827	3.52	Diamond is the hardest material. Good electrical insulator. Moderate thermal conduction, though diamond has exceptionally high thermal conductivity.
van der Waals: hydrogen bonding	PVC (polymer)		212	4	1.3	Low elastic modulus. Some ductility.
	H$_2$O (ice)	0.52	0	9.1	0.917	Electrical insulator. Poor thermal conductor. Large thermal expansion coefficient.
van der Waals: induced dipole	Crystalline argon	0.01	−189	8	1.8	Low elastic modulus. Electrical insulator. Poor thermal conductor. Large thermal expansion coefficient.

of electrons, are generally called **polar bonds.** Many technologically important semiconductor materials, such as III–V compounds (e.g., GaAs), have polar covalent bonds. In GaAs, for example, the electrons in a covalent bond spend slightly more time around the As^{5+} ion core than the Ga^{3+} ion core.

Ceramic materials are compounds that generally contain metallic and non-metallic elements. They are well known for their brittle mechanical properties, hardness, high melting temperatures, and electrical insulating properties. The type of bonding in a ceramic material may be covalent, ionic, or a mixture of the two, in which the bond between the atoms involves some electron sharing and, to some extent, the partial formation of cations and anions; the shared electrons spend more time with one type of atom, which then becomes a partial anion while the other becomes a partial cation. Silicon nitride (Si_3N_4), magnesia (MgO), and alumina (Al_2O_3) are all ceramics, but they have different types of bonding: Si_3N_4 has covalent, MgO has ionic, and Al_2O_3 has a mixture of ionic and covalent bonding. All three are brittle, have high melting temperatures, and are electrical insulators.

ENERGY OF SECONDARY BONDING   Consider the van der Waals bonding in solid argon. The | **Example 1.1**
potential energy as a function of interatomic separation can generally be modelled by the
Lennard–Jones 6–12 potential energy curve, that is,

$$E(r) = -Ar^{-6} + Br^{-12}$$

where A and B are constants. Given that $A = 1.037 \times 10^{-77}$ J m^6 and $B = 1.616 \times 10^{-134}$ J m^{12}, calculate the bond length and bond energy (in eV) for solid argon.

Solution

Bonding occurs when the potential energy is minimum. We therefore differentiate the Lennard–Jones potential $E(r)$ and set it to zero at $r = a$, the interatomic equilibrium separation or

$$\frac{dE}{dr} = 6Ar^{-7} - 12Br^{-13} = 0 \quad \text{at} \quad r = a$$

that is,

$$a^6 = \frac{2B}{A}$$

or

$$a = \left[\frac{2B}{A}\right]^{1/6}$$

Substituting $A = 1.037 \times 10^{-77}$ and $B = 1.616 \times 10^{-134}$ and solving for a we find

$$a = 3.82 \times 10^{-10} \text{ m} \quad \text{or} \quad 0.382 \text{ nm}$$

When $r = a = 3.82 \times 10^{-10}$ m, the potential energy is minimum and corresponds to $-E_{bond}$ so that

$$E_{bond} = \left| -Aa^{-6} + Ba^{-12} \right| = \left| -\frac{1.037 \times 10^{-77}}{(3.82 \times 10^{-10})^6} + \frac{1.616 \times 10^{-134}}{(3.82 \times 10^{-10})^{12}} \right|$$

that is,

$$E_{bond} = 1.66 \times 10^{-21} \text{ J} \quad \text{or} \quad 0.0104 \text{ eV}$$

Notice how small this energy is compared to primary bonding.

1.3   KINETIC MOLECULAR THEORY

1.3.1   Mean Kinetic Energy and Temperature

The kinetic molecular theory of matter is a classical theory that can explain such seemingly diverse topics as the pressure of a gas, the heat capacity of metals, the average speed of electrons in a semiconductor and electrical noise in resistors, among many interesting phenomena. We start with the kinetic molecular theory of gases, which considers a collection of gas molecules in a container and applies the classical equations of motion from elementary mechanics to these molecules. We assume that the collisions between the gas molecules and the walls of the container result in the gas pressure P. Newton's second law, $dp/dt = $ force, where $p = mv$ is the momentum, is used to relate the pressure P (force per unit area) to the mean square velocity $\overline{v^2}$, and the number of molecules per unit volume N/V. The result can be stated simply as

Kinetic molecular theory for gases

$$PV = \frac{1}{3}Nm\overline{v^2} \qquad\qquad \text{[1.6]}$$

where m is the mass of the gas molecule. Comparing this theoretical derivation with the experimental observation that

$$PV = \left(\frac{N}{N_A}\right)RT$$

where N_A is **Avogadro's number** and R is the gas constant, we can relate the mean kinetic energy of the molecules to the temperature. Our objective is to derive Equation 1.6; to do so, we make the following assumptions:

1. The molecules are in constant random motion. Since we are considering a large number of molecules, perhaps 10^{20} m^{-3}, there are as many molecules travelling in one direction as in any other direction, so the center of mass of the gas is at rest.

2. The range of intermolecular forces is short compared to the average separation of the gas molecules. Consequently:
   a. Intermolecular forces are negligible, except during a collision.
   b. The volume of the gas molecules (all together) is negligible compared to the volume occupied by the gas (that is, the container).

3. The duration of a collision is negligible compared to the time spent in free motion between collisions.

4. Each molecule moves with uniform velocity between collisions, and the acceleration due to the gravitational force or other external forces is neglected.

5. On average, the collisions of the molecules with one another and with the walls of the container are perfectly elastic. Collisions between molecules result in exchanges of kinetic energy.

6. Newtonian mechanics can be applied to describe the motion of the molecules.

We consider a collection of N gas molecules within a cubic container of side a. We focus our attention on one of the molecules moving toward one of the walls. The velocity can be decomposed into two components, one directly toward the wall v_x, and the other parallel to the wall v_y, as shown in Figure 1.15. Clearly, the collision of the molecule, which is perfectly elastic, does not change the component v_y along the wall, but reverses the perpendicular component v_x. The change in the momentum of the molecule following its collision with the wall is

$$\Delta p = 2mv_x$$

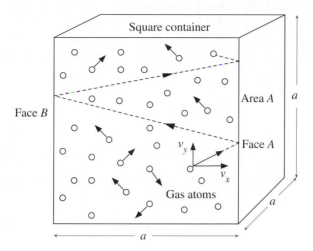

Figure 1.15   The gas molecules in the container are in random motion.

where m is the mass of the molecule. Following its collision, the molecule travels back across the box, collides with the opposite face B, and returns to hit face A again. The time interval Δt is the time to traverse twice the length of the box, or $\Delta t = 2a/v_x$. Thus, every Δt seconds, the molecule collides with face A and changes its momentum by $2mv_x$. To find the force F exerted by this molecule on face A, we need the rate of change of momentum, or

$$F = \frac{\Delta p}{\Delta t} = \frac{2mv_x}{\left(\dfrac{2a}{v_x}\right)} = \frac{mv_x^2}{a}$$

The total pressure P exerted by N molecules on face A, of area a^2, is due to the sum of all individual forces F, or

$$P = \frac{\text{Total force}}{a^2} = \frac{mv_{x1}^2 + mv_{x2}^2 + \cdots + mv_{xN}^2}{a^3}$$

$$= \frac{m}{a^3}(v_{x1}^2 + v_{x2}^2 + \cdots + v_{xN}^2)$$

that is,

$$P = \frac{mN\overline{v_x^2}}{V}$$

where $\overline{v_x^2}$ is the average of v_x^2 for all the molecules and is called the *mean square velocity,* and V is the volume a^3.

Since the molecules are in random motion and collide randomly with each other, thereby exchanging kinetic energy, the mean square velocity in the x direction is the same as those in the y and z directions, or

$$\overline{v_x^2} = \overline{v_y^2} = \overline{v_z^2}$$

For any molecule, the velocity v is given by

$$\overline{v^2} = \overline{v_x^2} + \overline{v_y^2} + \overline{v_z^2} = 3\overline{v_x^2}$$

The relationship between the pressure P and the mean square velocity of the molecules is therefore

Gas pressure in the kinetic theory

$$P = \frac{Nm\overline{v^2}}{3V} = \frac{1}{3}\rho\overline{v^2} \qquad [1.7]$$

where ρ is the density of the gas, or Nm/V. By using elementary mechanical concepts, we have now related the pressure exerted by the gas to the number of molecules per unit volume and to the mean square of the molecular velocity.

Equation 1.7 can be written explicitly to show the dependence of PV on the mean kinetic energy of the molecule. Rearranging Equation 1.7, we obtain

$$PV = \frac{2}{3}N\left(\frac{1}{2}m\overline{v_x^2}\right)$$

where $(1/2)\,m\overline{v_x^2}$ is the average kinetic energy $\overline{KE}$ per molecule. If we consider one mole of gas, then N is simply N_A, Avogadro's number.

Experiments on gases lead to the empirical gas equation

$$PV = \eta RT$$

where η is the number of moles of gas ($\eta = N/N_A$) and R is the universal gas constant. Comparing this equation with the kinetic theory equation shows that the average kinetic energy per molecule must be proportional to the temperature.

Mean kinetic energy per atom

$$\overline{KE} = \frac{1}{2}m\overline{v^2} = \frac{3}{2}kT \qquad [1.8]$$

where $k = R/N_A$ is called the **Boltzmann constant.** Thus, the mean square velocity is proportional to the absolute temperature. This is a major conclusion from the kinetic theory and we will use it frequently.

When heat is added to a gas, its internal energy and, by virtue of Equation 1.8, its temperature both increase. The rise in the internal energy per unit temperature is called the **heat capacity.** If we consider 1 mole of gas, then the heat capacity is called the **molar heat capacity** C_m. The total internal energy U of 1 mole of monatomic gas (i.e., a gas with only one atom in each molecule) at constant volume is

$$U = N_A\left(\frac{1}{2}m\overline{v^2}\right) = \frac{3}{2}N_AkT \qquad [1.9]$$

so that, from the definition of C_m, we have

Molar heat capacity

$$C_m = \frac{dU}{dT} = \frac{3}{2}N_Ak = \frac{3}{2}R \qquad [1.10]$$

Thus, the heat capacity per mole of a monatomic gas is simply $(3/2)R$. By comparison, we will see later that the heat capacity of metals is twice this amount.

There is a useful theorem called **Maxwell's principle of equipartition of energy,** which assigns an average of $(1/2)kT$ to each independent energy term in the expression for the total energy of the system. A monatomic molecule can only have translational kinetic energy, which is the sum of kinetic energies in the x, y, and z directions. The total energy is therefore

$$E = \frac{1}{2}mv_x^2 + \frac{1}{2}mv_y^2 + \frac{1}{2}mv_z^2$$

Each of these terms represents an independent way in which the molecule can be made to absorb energy. Each method by which a system can absorb energy is called a **degree of freedom.** A monatomic molecule has only three degrees of freedom. According to Maxwell's principle, for a collection of molecules in thermal equilibrium, each degree of freedom has an average energy of $(1/2)kT$, so the average kinetic energy of the monatomic molecule is $3[(1/2)kT]$.

A rigid diatomic molecule (such as an O_2 molecule) can acquire energy as translational motion and rotational motion, as depicted in Figure 1.16. Assuming the moment of inertia, I_x, about the molecular axis (along x) is negligible, the energy of the molecule is

$$E = \frac{1}{2}mv_x^2 + \frac{1}{2}mv_y^2 + \frac{1}{2}mv_z^2 + \frac{1}{2}I_y\omega_y^2 + \frac{1}{2}I_z\omega_z^2$$

where I_y and I_z are moments of inertia about the y and z axes and ω_y and ω_z are angular velocities about the y and z axes (Figure 1.16).

This molecule has five degrees of freedom and hence an average energy of $5[(1/2)kT]$. Its molar heat capacity is therefore $(5/2)R$.

The atoms in the molecule will also vibrate by stretching or bending the bond, which behaves like a "spring." At room temperature, the addition of heat only

TRANSLATIONAL MOTION   ROTATIONAL MOTION

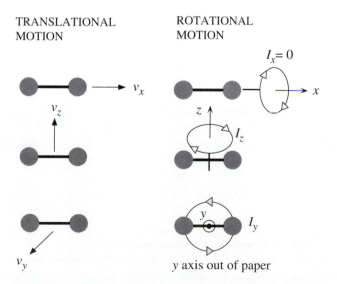

Figure 1.16   Possible translational and rotational motions of a diatomic molecule. Vibrational motions are neglected.

results in the translational and rotational motions becoming more energetic (excited), whereas the molecular vibrations remain the same and therefore do not absorb energy. This occurs because the vibrational energy of the molecule can only change in finite steps; in other words, the vibrational energy is quantized. For many molecules, the energy required to excite a more energetic vibration is much more than the energy possessed by the majority of molecules. Therefore, energy exchanges via molecular collisions cannot readily excite more energetic vibrations; consequently, the contribution of molecular vibrations to the heat capacity is negligible.

In a solid, the atoms are bonded to each other and can only move by vibrating about their equilibrium positions. In the simplest view, a typical atom in a solid is joined to its neighbors by "springs" that represent the bonds, as depicted in Figure 1.17. If we consider a given atom, its potential energy as a function of displacement from the equilibrium position is such that if it is displaced slightly in any direction, it will experience a restoring force proportional to the displacement. Thus, this atom can acquire energy by vibrations in three directions. The energy associated with the x direction, for example, is the kinetic energy of vibration plus the potential energy of the "spring," or $(1/2)mv_x^2 + (1/2)K_x x^2$, where v_x is the velocity, x is the extension of the spring, and K_x is the spring constant, all along the x direction. Clearly, there are similar energy terms in the y and z directions, so that there are six energy terms in the total energy equation:

$$E = \frac{1}{2}mv_x^2 + \frac{1}{2}mv_y^2 + \frac{1}{2}mv_z^2 + \frac{1}{2}K_x x^2 + \frac{1}{2}K_y y^2 + \frac{1}{2}K_z z^2$$

We know that for simple harmonic motion, the average KE is equal to the average PE. Since, by virtue of the equipartition of energy principle, each average KE term has an energy of $(1/2)kT$, the average total energy per atom is $6[(1/2)kT]$.

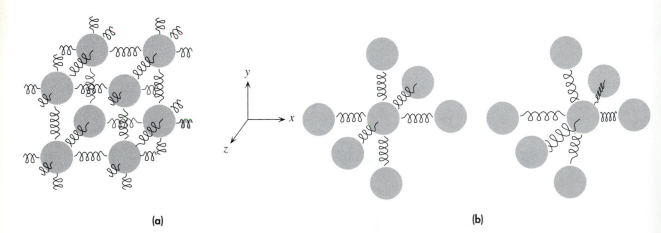

(a)   (b)

Figure 1.17   (a) The ball-and-spring model of solids, in which the springs represent the interatomic bonds. Each ball (atom) is linked to its neighbors by springs. Atomic vibrations in a solid involve three dimensions.
(b) An atom vibrating about its equilibrium position. The atom stretches and compresses its springs to its neighbors and has both kinetic and potential energy.

The internal energy U per mole is

$$U = N_A 6\left(\frac{1}{2}kT\right) = 3RT$$

The molar heat capacity then becomes

$$C_m = \frac{dU}{dT} = 3R = 25 \text{ J K}^{-1} \text{ mol}^{-1}$$

This is the **Dulong–Petit rule.**

   The kinetic molecular theory of matter is one of the successes of classical physics, with a beautiful simplicity in its equations and predictions. Its failures, however, are numerous. For example, the theory fails to predict that, at low temperatures, the heat capacity increases as T^3 and that the resistivity of a metal increases linearly with the absolute temperature. We will explain the origins of these phenomena in later chapters.

SPEED OF SOUND IN AIR   Calculate the root mean square (rms) velocity of nitrogen molecules in atmospheric air at 27 °C. Also calculate the root mean square velocity in one direction (v_{rmsx}). Compare the speed of propagation of sound waves in air, 350 m s^{-1}, with v_x and explain the difference. | **Example 1.2**

Solution

From the kinetic theory

$$\frac{1}{2}mv_{rms}^2 = \frac{3}{2}kT$$

so that

$$v_{rms} = \sqrt{\left[\frac{3kT}{m}\right]}$$

where m is the mass of the nitrogen molecule N_2. The atomic mass of nitrogen is $M_{at} = 14$ g mol^{-1}, so that in kilograms

$$m = \frac{2M_{at}(10^{-3})}{N_A}$$

Thus

$$v_{rms} = \left[\frac{3kN_A T}{2M_{at}(10^{-3})}\right]^{1/2} = \left[\frac{3RT}{2M_{at}(10^{-3})}\right]^{1/2}$$

$$= \left[\frac{3(8.314 \text{ J mol}^{-1} \text{ K}^{-1})(300 \text{ K})}{2(14 \times 10^{-3} \text{ kg mol}^{-1})}\right]^{1/2} = 517 \text{ m s}^{-1}$$

Consider an rms velocity in one direction. Then

$$v_{rmsx} = \sqrt{\overline{v_x^2}} = \sqrt{\frac{1}{3}\overline{v^2}} = \frac{1}{\sqrt{3}}v_{rms} = 298 \text{ m s}^{-1}$$

which is slightly less than the velocity of sound in air (350 m s^{-1}). The difference is due to the fact that the propagation of a sound wave involves rapid compressions and rarefactions of air, and the result is that the propagation is not isothermal. Note that accounting for oxygen in air lowers $v_{rmsx.}$ (Why?)

Example 1.3 | **SPECIFIC HEAT CAPACITY**   Estimate the heat capacity of copper per unit gram, given that its atomic mass is 63.6.

Solution

From the Dulong–Petit rule, $C_m = 3R$ for N_A atoms. But N_A atoms have a mass of M_{at} grams, so the heat capacity per gram (specific heat) is

$$C_{gram} = \frac{3R}{M_{at}} = \frac{(24.9 \text{ J mol}^{-1} \text{ K}^{-1})}{(63.6 \text{ g mol}^{-1})}$$

$$\approx 0.39 \text{ J g}^{-1} \text{ K}^{-1} \qquad \text{(The experimental value is 0.38 J g}^{-1}\text{ K}^{-1}\text{.)}$$

1.3.2   Molecular Velocity and Energy Distribution

Although the kinetic theory allows us to determine the root mean square velocity of the gas molecules, it says nothing about the distribution of velocities. Due to random collisions between the molecules and the walls of the container and between the molecules themselves, the molecules do not all have the same velocity. The velocity distribution of molecules can be determined experimentally by the simple scheme illustrated in Figure 1.18. Gas molecules are allowed to escape from a small aperture of a hot oven in which the substance is vaporized. Two blocking slits allow only those molecules that are moving along the line through the two slits to pass through, which results in a **collimated beam.** This beam is directed toward two rotating disks, which have slightly displaced slits. The molecules that pass through the first slit can only pass through the second if they have a certain speed; that is, the exact speed at which the second slit lines up with the first slit. Thus, the

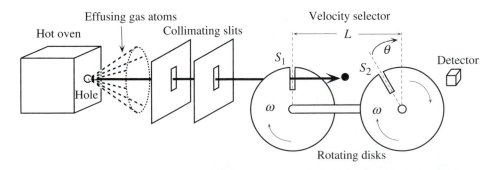

Figure 1.18   Schematic diagram of a Stern-type experiment for determining the distribution of molecular speeds.

two disks act as a speed selector. The speed of rotation of the disks determines which molecular speeds are allowed to go through. The experiment therefore measures the number of molecules ΔN with speeds in the range v to $(v + \Delta v)$.

It is generally convenient to describe the number of molecules dN with speeds in a certain range v to $(v + dv)$ by defining a **velocity density function** n_v as follows:

$$dN = n_v \, dv$$

where n_v is the number of molecules per unit velocity that have velocities in the range v to $(v + dv)$. This number represents the velocity distribution among the molecules and is a function of the molecular velocity, $n_v = n_v(v)$. From the experiment, we can easily obtain n_v by $n_v = \Delta N / \Delta v$ at various velocities. Figure 1.19 shows the velocity density function n_v of nitrogen gas at two temperatures. The average (v_{av}), most probable (v^*), and rms (v_{rms}) speeds are marked to show their relative positions. As expected, these speeds all increase with increasing temperature. From various experiments of the type shown in Figure 1.18, the velocity distribution function n_v has been widely studied and found to obey the following equation:

$$n_v = 4\pi N \left(\frac{m}{2\pi kT} \right)^{3/2} v^2 \exp\left(-\frac{mv^2}{2kT} \right) \qquad \text{[1.11]}$$

Maxwell–Boltzmann distribution for molecular speeds

where N is the total number of molecules and m is the molecular mass. This is the **Maxwell–Boltzmann distribution function,** which describes the statistics of particle velocities in thermal equilibrium. The function assumes that the particles do not interact with each other while in motion and that all the collisions are elastic in the sense that they involve an exchange of kinetic energy. Figure 1.19 clearly shows that molecules move around randomly, with a variety of velocities ranging from nearly zero to almost infinity. The kinetic theory speaks of their rms value only.

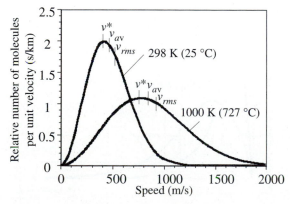

Figure 1.19   Maxwell–Boltzmann distribution of molecular speeds in nitrogen gas at two temperatures.

The ordinate is $dN/(Ndv)$, the fractional number of molecules per unit speed interval in $(km/s)^{-1}$.

What is the energy distribution of molecules in a gas? In the case of a monatomic gas, the total energy E is purely translational kinetic energy, so we can use $E = (1/2)mv^2$. To relate an energy range dE to a velocity range dv, we have $dE = mv\, dv$. Suppose that n_E is the number of atoms per unit volume per unit energy at an energy E. Then $n_E\, dE$ is the number of atoms with energies in the range E to $(E + dE)$. These are also the atoms with velocities in the range v to $(v + dv)$, because an atom with a velocity v has an energy E. Thus,

$$n_E\, dE = n_v\, dv$$

i.e.,

$$n_E = n_v \left(\frac{dv}{dE}\right)$$

If we substitute for n_v and (dv/dE), we obtain the expression for n_E as a function of E:

Maxwell–Boltzmann distribution for translational kinetic energies

$$n_E = \frac{2}{\pi^{1/2}} N \left(\frac{1}{kT}\right)^{3/2} E^{1/2} \exp\left(-\frac{E}{kT}\right) \qquad [1.12]$$

Thus, the total internal energy is distributed among the atoms according to the Maxwell–Boltzmann distribution in Equation 1.12. The exponential factor $\exp(-E/kT)$ is called the **Boltzmann factor.** Atoms have widely differing kinetic energies, but a mean energy of $(3/2)\,kT$. Figure 1.20 shows the Maxwell–Boltzmann energy distribution among the gas atoms in a tank at two temperatures. As the temperature increases, the distribution extends to higher energies. The area under the curve is the total number of molecules, which remains the same for a closed container.

Equation 1.12 represents the energy distribution among the N gas atoms at any time. Since the atoms are continually colliding and exchanging energies, the energy

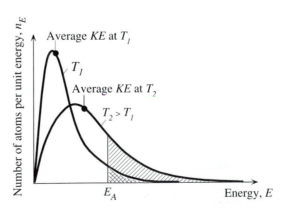

Figure 1.20   Energy distribution of gas molecules at two different temperatures.

The shaded area shows the number of molecules that have energies greater than E_A. This area depends strongly on the temperature as $\exp(-E_A/kT)$.

of one atom will sometimes be small and sometimes be large, but averaged over a long time, this energy will be $(3/2)\,kT$ as long as all the gas atoms are in thermal equilibrium (i.e., the temperature is the same everywhere in the gas). Thus, we can also use Equation 1.12 to represent all possible energies an atom can acquire over a long period. There are a total of N atoms, and $n_E dE$ of them have energies in the range E to $(E + dE)$. Thus,

$$\text{Probability of energy being in } E \text{ to } (E + dE) = \frac{n_E dE}{N} \qquad \text{[1.13]}$$

When the probability in Equation 1.13 is integrated (i.e., summed) for all energies ($E = 0$ to ∞), the result is unity, because the atom must have an energy somewhere in the range of zero to infinity.

What happens to the Maxwell–Boltzmann energy distribution law in Equation 1.12 when the total energy is not simply translational kinetic energy? What happens when we do not have a montatomic gas? Suppose that the total energy of a molecule (which may simply be an atom) in a system of N molecules has vibrational and rotational kinetic energy contributions, as well as potential energy due to intermolecular interactions. In all cases, the number of molecules per unit energy n_E turns out to contain the Boltzmann factor, and the energy distribution obeys what is called the **Boltzmann energy distribution:**

$$\frac{n_E}{N} = C \exp\!\left(-\frac{E}{kT}\right) \qquad \text{[1.14]}$$

Boltzmann energy distribution

where E is the total energy ($KE + PE$), N is the total number of molecules in the system, and C is a constant that relates to the specific system (e.g., a monatomic gas, or a liquid, etc.). The constant C may depend on the energy E, as in Equation 1.12, but not as strongly as the exponential term. Equation 1.14 is the **probability per unit energy** that a molecule in a given system has an energy E.

MEAN AND RMS SPEEDS OF MOLECULES   Given the Maxwell–Boltzmann distribution law for the velocities of molecules in a gas, derive expressions for the mean speed (v_{av}), most probable speed (v^*), and rms velocity (v_{rms}) of the molecules and calculate the corresponding values for a gas of noninteracting electrons. | **Example 1.4**

Solution

The number of molecules with speeds in the range v to $(v + dv)$ is

$$dN = n_v\, dv = 4\pi N\!\left(\frac{m}{2\pi kT}\right)^{3/2} v^2 \exp\!\left(\frac{mv^2}{2kT}\right) dv$$

By definition, then, the mean speed is given by

$$v_{av} = \frac{\int v\, dN}{\int dN} = \frac{\int v n_v\, dv}{\int n_v\, dv} = \sqrt{\frac{8kT}{\pi m}}$$

where the integration is over all speeds ($v = 0$ to ∞). The mean square velocity is given by

$$\bar{v}^2 = \frac{\int v^2\, dN}{\int dN} = \frac{\int v^2 n_v\, dv}{\int n_v\, dv} = \frac{3kT}{m}$$

so that

$$v_{rms} = \sqrt{\frac{3kT}{m}}$$

where n_v/N is the probability per unit speed that a molecule has a speed in the range v to $(v + dv)$. Differentiating n_v with respect to v and setting this to zero, $dn_v/dv = 0$, gives the position of the peak of n_v versus v, and thus the most probable speed v^*,

$$v^* = \left[\frac{2kT}{m}\right]^{1/2}$$

Substituting $m = 9.1 \times 10^{-31}$ kg for electrons and using $T = 300$ K, we find $v^* = 95.3$ km s^{-1}, $v_{av} = 108$ km s^{-1}, and $v_{rms} = 117$ km s^{-1}, all of which are close in value. We often use the term **thermal velocity** to describe the mean speed of the electrons due to their thermal random motion. Also, the integrations shown are not trivial and they involve substitution and integration by parts.

1.4   HEAT, THERMAL FLUCTUATIONS, AND NOISE

Generally, thermal equilibrium between two objects implies that they have the same temperature, where temperature (from the kinetic theory) is a measure of the mean kinetic energy of the molecules. Consider a solid in a monatomic gas atmosphere such as He gas, as depicted in Figure 1.21. Both the gas and the solid are at the same temperature. The gas molecules move around randomly, with a mean kinetic energy given by $(1/2)\,m\overline{v^2} = (3/2)\,kT$, where m is the mass of the gas molecule. We also know that the atoms in the solid vibrate with a mean kinetic energy given by $(1/2)\,M\overline{V^2} = (3/2)\,kT$, where M is the mass of the solid atom and V

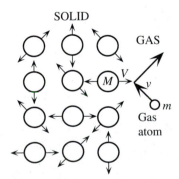

Figure 1.21   Solid in equilibrium in air.

During collisions between the gas and solid atoms, kinetic energy is exchanged.

is the velocity of vibration. The gas molecules will collide with the atoms on the surface of the solid and will thus exchange energy with those solid atoms. Since both are at the same temperature, the solid atoms and gas molecules have the same mean kinetic energy, which means that over a long time, there will be no net transfer of energy from one to the other. This is basically what we mean by **thermal equilibrium.**

If, on the other hand, the solid is hotter than the gas, $T_{solid} > T_{gas}$, and thus $(1/2)\, M\overline{V^2} > (1/2)\, m\overline{v^2}$, then when an average gas molecule and an average solid atom collide, energy will be transferred from the solid atom to the gas molecule. As many more gas molecules collide with solid atoms, more and more energy will be transferred, until the mean kinetic energy of atoms in each substance is the same and they reach the same temperature: the bodies have **equilibrated.** The amount of energy transferred from the kinetic energy of the atoms in the hot solid to the kinetic energy of the gas molecules is called **heat.** Heat represents the energy transfer from the hot body to the cold body by virtue of the *random* motions and collisions of the atoms and molecules.

Although, over a long time, the energy transferred between two systems in thermal equilibrium is certainly zero, this does not preclude a net energy transfer from one to the other at one instant. For example, at any one instant, an average solid atom may be hit by a fast gas molecule with speed at the far end of the Maxwell–Boltzmann distribution. There will then be a transfer of energy from the gas molecule to the solid atom. At another instant, a slow gas molecule hits the solid, and the reverse is true. Thus, although the mean energy transferred from one atom to the other is zero, the instantaneous value of this energy is not zero and varies randomly about zero.

As an example, consider a small mass attached to a spring, as illustrated in Figure 1.22. The gas or air molecules will bombard and exchange energy with the

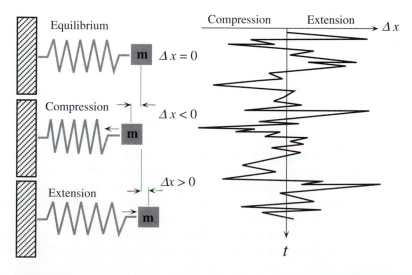

Figure 1.22   Fluctuations of a mass attached to a spring, due to random bombardment by air molecules.

solid atoms. Some air molecules will be fast and some will be slow, which means that there will an instantaneous exchange of energy. Consequently, the spring will be compressed when the bombarding air molecules are fast (more energetic) and extended when they are less energetic. This leads to a mechanical fluctuation of the mass about its equilibrium position, as depicted in Figure 1.22. These fluctuations make the measurement of the exact position of the mass uncertain, and it is futile to try to measure the position more accurately than these fluctuations permit.

If the mass m compresses the spring by Δx, then at time t, the energy stored as potential energy in the spring is

$$PE(t) = \frac{1}{2}K(\Delta x)^2 \qquad [1.15]$$

where K is the spring constant. At a later instant, this energy will be returned to the gas by the spring. The spring will continue to fluctuate because of the fluctuations in the velocity of the bombarding air molecules. Over a long period, the average value of PE will be the same as KE and, by virtue of the Maxwell equipartition of energy theorem, it will be given by

$$\overline{\frac{1}{2}K(\Delta x)^2} = \frac{1}{2}kT \qquad [1.16]$$

Thus, the root mean square (rms) value of the fluctuations of the mass about its equilibrium position is

Rms fluctuations of a body attached to a spring of stiffness K

$$(\Delta x)_{rms} = \sqrt{\frac{kT}{K}} \qquad [1.17]$$

To understand the origin of electrical noise, for example, we consider the thermal fluctuations in the instantaneous local electron concentration in a conductor, such as that shown in Figure 1.23. Due to fluctuations in the electron concentration at any one instant, end A of the conductor can become more negative with respect to end B, which will give rise to a voltage across the conductor. This fluctuation in the electron concentration is due to more electrons at that instant moving toward end A than toward B. At a later instant, the situation reverses and more electrons move toward B than toward A, resulting in end B becoming more negative and leading to a reversal of the voltage between A and B. Clearly, there will therefore be voltage fluctuations across the conductor, even though the mean voltage across it over a long period is always zero. If the conductor is connected to an amplifier, these voltage fluctuations will be amplified and recorded as noise at the output. This noise corrupts the actual signal at the amplifier input and is obviously undesirable. As engineers, we have to know how to calculate the magnitude of this noise. Although the mean voltage due to thermal fluctuations is zero, the root mean square (rms) value is not. The average voltage from a power outlet is zero, but the rms value is 110V. We use the rms value to calculate the amount of average power available.

Consider a conductor of resistance R and a capacitor C, connected in parallel and in thermal equilibrium; that is, both are at the same temperature, as depicted

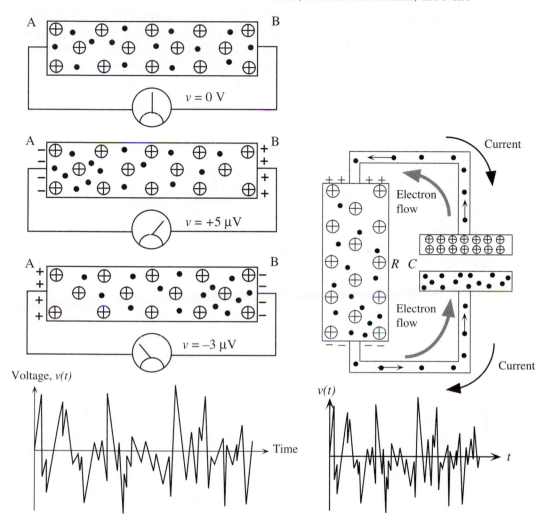

Figure 1.23   Random motion of conduction electrons in a conductor, resulting in electrical noise.

Figure 1.24   Charging and discharging of a capacitor by a conductor, due to the random thermal motions of the conduction electrons.

in Figure 1.24. Due to thermal fluctuations, the instantaneous voltage across the conductor may be positive or negative, which charges and discharges the capacitor C positively and negatively. The energy transferred to or from the capacitor at instant t when the voltage across is $v(t)$ will be

$$E(t) = \frac{1}{2} C v(t)^2$$

Over a long period, this has a mean value of

$$\bar{E} = \frac{1}{2} C \overline{v(t)^2}$$

This is the mean energy stored in C and removed from C due to the thermal fluctuations experienced by the electrons in the conductor. By virtue of Maxwell's equipartition of energy theorem, this mean energy must be $(1/2)\,kT$, because the two bodies are at the same temperature. Thus,

$$\frac{1}{2}\,C\overline{v(t)^2} = \frac{1}{2}kT$$

which means that the mean square voltage across the capacitor is given by

$$\overline{v(t)^2} = \frac{kT}{C} \qquad\qquad [1.18]$$

Interestingly, the rms noise voltage across an RC network seems to be independent of the resistance. However, the origin of the noise voltage arises from the electron fluctuations in the conductor and we must somehow re-express Equation 1.18 to reflect this fact; that is, we must relate the electrical fluctuations to R.

The voltage fluctuations across the network will have many sinusoidal components, but only those below the cutoff frequency of the RC network will contribute to the mean square voltage (that is, we effectively have a low-pass filter). If B is the bandwidth of the RC network,[5] then $B = 1/(2\pi RC)$ and we can eliminate C in Equation 1.18 to obtain

$$\overline{v(t)^2} = 2\pi kTRB$$

This is the key equation for calculating the mean square noise voltage from a resistor over a bandwidth B. A more rigorous derivation makes the numerical factor 4 rather than 2π. For a network with a bandwidth B, the **rms noise voltage** is therefore

Rms noise voltage across a resistance

$$v_{rms} = \sqrt{4kTRB} \qquad\qquad [1.19]$$

Equation 1.19 is known as the **Johnson resistor noise equation** and it sets the lower limit of the magnitude of small signals that can be amplified. Note that Equation 1.19 basically tells us the rms value of the voltage fluctuations within a given bandwidth (B) and not the origin and spectrum (noise voltage vs. frequency) of the noise. The origin of noise is attributed to the random motion of electrons in the conductor (resistor), and Equation 1.19 is the fundamental description of electrical fluctuations; that is, the fluctuations in the conductor's instantaneous local electron concentration that charges and discharges the capacitor. To determine the rms noise voltage across a network with an impedance $Z(j\omega)$, all we have to do is find the real part of Z, which represents the resistive part, and use this for R in Equation 1.19.

---

[5] A low-pass filter allows all signal frequencies up to the cutoff frequency B to pass. B is $1/[2\pi RC]$.

NOISE IN AN *RLC* CIRCUIT   Most radio receivers have a tuned parallel-resonant circuit, | **Example 1.5**
which consists of an inductor L, capacitor C, and resistance R in parallel. Suppose L is
100 μH, C is 100 pF, and R, the equivalent resistance due to the input resistance of the
amplifier and to the loss in the coil (coil resistance plus ferrite losses), is about 200 kΩ. What
is the minimum rms radio signal that can be detected?

Solution

Consider the bandwidth of this tuned *RLC*, which can be found in any electrical engineering
textbook:

$$B = \frac{f_o}{Q}$$

where $f_o = 1/[2\pi \sqrt{(LC)}]$ is the resonant frequency and $Q = 2\pi f_o CR$ is the quality factor.
Substituting for L, C, and R, we get, $f_o = 10^7/2\pi = 1.6 \times 10^6$ Hz and $Q = 200$, which
gives $B = 10^7/[2\pi(200)]$ Hz, or 8 kHz. The rms noise voltage is

$$v_{rms} = [4kTRB]^{1/2} = [4(1.38 \times 10^{-23} \text{ J K}^{-1})(300 \text{ K})(200 \times 10^3 \ \Omega)(8 \times 10^3 \text{ Hz})]^{1/2}$$

$$= 5.1 \times 10^{-6} \text{ V or } 5.1 \ \mu\text{V}$$

This rms voltage is within a bandwidth of 8 kHz centered at 1.6 MHz. This last
information is totally absent in Equation 1.19. If we attempt to use

$$v_{rms} = \left[\frac{kT}{C}\right]^{1/2}$$

we get

$$v_{rms} = \left[\frac{(1.38 \times 10^{-23} \text{ J K}^{-1})(300 \text{ K})}{100 \times 10^{-12} \text{ F}}\right]^{1/2} = 6.4 \ \mu\text{V}$$

However, Equation 1.18 was derived using the *RC* circuit in Figure 1.23, whereas we
now have an *LCR* circuit. The correct approach uses Equation 1.19, which is generally
valid, and the appropriate bandwidth B.

1.5   THERMALLY ACTIVATED PROCESSES

Many physical and chemical processes strongly depend on temperature and exhibit
what is called an **Arrhenius type behavior,** in which the rate of change is proportional to $\exp(-E_A/kT)$, where E_A is a characteristic energy parameter applicable to
the particular process. For example, when we store food in the refrigerator, we are
effectively using the Arrhenius rate equation: cooling the food diminishes the rate
of decay. Processes that exhibit an Arrhenius type temperature dependence are
referred to as **thermally activated.**

For an intuitive understanding of a thermally activated process, consider a
vertical filing cabinet that stands in equilibrium, with its center of mass at A, as
sketched in Figure 1.25. Tilting the cabinet left or right increases the potential
energy *PE* and requires external work. If we could supply this energy, we could

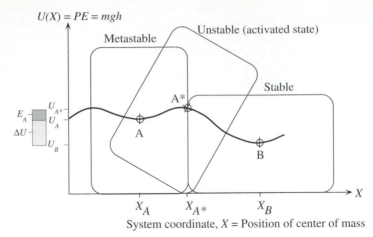

$U(X) = PE = mgh$

Figure 1.25   Tilting a filing cabinet from state A to its edge in state A* requires an energy E_A.

After reaching A*, the cabinet spontaneously drops to the stable position B.
The *PE* of state B is lower than A, and therefore state B is more stable than A.

move the cabinet over its edge and lay it flat, where its *PE* would be lower than at A. Clearly, since the *PE* at B is lower, this is a more stable position than A. Further, in going from A to B, we had to overcome a **potential energy barrier** of amount E_A, which corresponds to the cabinet standing on its edge with the center of mass at the highest point at A*. To topple the cabinet, we must first provide energy equal to E_A to take the center of mass to A*, from which point the cabinet, with the slightest encouragement, will fall spontaneously to B to attain the lowest *PE*. During the whole tilting process, the internal energy change for the cabinet, ΔU, is due to the change in the *PE* ($= mgh$) from A to B, which in turn comes from external work.[6]

Suppose, for example, a person with an average energy less than E_A tries to topple the cabinet. Like everyone else, that person experiences energy fluctuations as a result of interactions with the environment (e.g., what type of day the person had). During one of those high-energy periods, he can be topple the cabinet, even though most of the time he cannot do so because his average energy is less then E_A. The rate at which the cabinet is toppled depends on the number of times (frequency) the person tries and the probability that he possesses energy greater than E_A.

As an example of a thermally activated process, consider the diffusion of impurity atoms in a solid, one of which is depicted in Figure 1.26. In this example, the impurity atom is at an interatomic void A in the crystal, called an **interstitial site.** For the impurity atom to move from A to a neighboring void B, the atom must push the host neighbors apart as it moves across. This requires energy in much the same way as toppling the filing cabinet. There is a potential energy barrier E_A to the motion of this atom from A to B.

---

[6] According to the conservation of energy principle, the increase in the *PE* from A to A* must come from the external work.

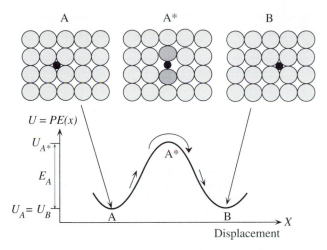

Figure 1.26 Diffusion of an interstitial impurity atom in a crystal from one void to a neighboring void.

The impurity atom at position A must posses an energy E_A to push the host atoms away and move into the neighboring void at B.

Both the host and the impurity atoms in the solid vibrate about their equilibrium positions, with a distribution of energies, and they also continually exchange energies, which leads to energy fluctuations. In thermal equilibrium, at any instant, we can expect the energy distribution of the atoms to obey the Boltzmann distribution law, (see Equation 1.14). The average kinetic energy per atom is vibrational and is $(3/2) kT$, which will not allow the impurity simply to overcome the PE barrier E_A, because typically $E_A \gg (3/2) kT$.

The rate of jump, called the **diffusion,** of the impurity from A to B depends on two factors. The first is the number of times the atom tries to go over the potential barrier, which is the vibrational frequency v_o, in the AB direction. The second factor is the probability that the atom has sufficient energy to overcome the PE barrier. Only during those times when the atom has an energy greater than the potential energy barrier $E_A = U_{A*} - U_A$ will it jump across from A to B. During this diffusion process, the atom attains an **activated state,** labelled A* in Figure 1.26, with an energy E_A above U_A, so the crystal internal energy is higher than U_A. E_A is called the activation energy.

Suppose there are N impurity atoms. At any instant, according to the Boltzmann distribution, $n_E dE$ of these will have kinetic energies in the range E to $(E + dE)$, so the probability that an impurity atom has an energy E greater than E_A is

$$\text{Probability} \ (E > E_A) = \frac{\text{Number of impurities with } E > E_A}{\text{Total number of impurities}}$$

$$\text{Probability} \ (E > E_A) = \frac{\int_{E_A}^{\infty} n_E \, dE}{N} = A \exp\left(-\frac{E_A}{kT}\right)$$

where A is a dimensionless constant that has only a weak temperature dependence. The rate of jumps, jumps per seconds, or simply the **frequency of jumps** ϑ from void to void is

$$\vartheta = \text{(Frequency of attempt along AB) (Probability of } E > E_A)$$

Rate for a thermally activated process

$$\vartheta = Av_o \exp\left(-\frac{E_A}{kT}\right) \qquad E_A = U_{A*} - U_A \qquad \text{[1.20]}$$

Equation 1.20 describes the rate of thermally activated process, for which increasing the temperature causes more atoms to be energetic and hence results in more jumps over the potential barrier. Equation 1.20 is the well-known **Arrhenius rate equation** and is generally valid for a vast number of transformations, both chemical and physical.

Example 1.6 | **ATOMIC DIFFUSION AND THE BOLTZMANN FACTOR**   Consider the motion of the impurity atom in Figure 1.26. For simplicity, assume a two-dimensional crystal in the plane of the paper, as in Figure 1.27. The impurity atom has four neighboring voids into which it can jump. If θ is the angle with respect to the x axis, then these voids are at directions $\theta = 0°$, $90°$, $180°$ and $270°$, as depicted in Figure 1.27. Each jump is in a random direction along one of these four angles. As the impurity atom jumps from void to void, it leaves its original location at O, and after N jumps, after time t, it has been displaced from O to O'.

Let a be the closest void-to-void separation. Each jump results in a displacement along x which is equal to $a\cos\theta$, with $\theta = 0°$, $90°$, $180°$, or $270°$. Thus, each jump results in a displacement along x which can be a, 0, $-a$, or 0, corresponding to the four possibilities. After N jumps, the mean displacement along x will be close to zero, just as the mean voltage of the ac voltage from a power outlet is zero, even though it has an rms value of 110 V. We therefore consider the square of the displacements. The total square displacement, denoted X^2, is

$$X^2 = a^2 \cos^2 \theta_1 + a^2 \cos^2 \theta_2 + \cdots + a^2 \cos^2 \theta_N$$

Clearly, $\theta = 90°$ and $270°$ give $\cos^2 \theta = 0$. Of all N jumps, $(1/2)N$ are $\theta = 0$ and $180°$, each of which gives $\cos^2 \theta = 1$. Thus,

$$X^2 = \frac{1}{2} a^2 N$$

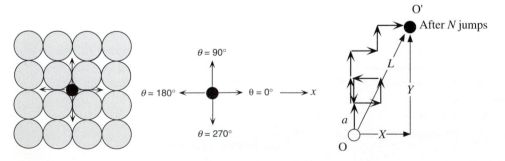

Figure 1.27   An impurity atom has four site choices for diffusion to a neighboring interstitial vacancy.
After N jumps, the impurity atom would have been displaced from the original position at O.

There will be a similar expression for Y^2, which means that after N jumps, the total square distance L^2 from O to O′ in Figure 1.26 is

$$L^2 = X^2 + Y^2 = a^2 N$$

The rate of jumping (frequency of jumps) is given by

$$\vartheta = v_o A \, \exp\left(-\frac{E_A}{kT}\right)$$

so the time per jump is $1/\vartheta$. Time t for N jumps is N/ϑ. Thus, $N = \vartheta t$ and

$$L^2 = a^2 \vartheta t = 2Dt \qquad\qquad \textbf{[1.21]}$$

where, by definition, $D = (1/2)\, a^2 \vartheta$, which is a constant that depends on the diffusion process, as well as the temperature, by virtue of ϑ. This constant is generally called the **diffusion coefficient.** Substituting for ϑ, we find

$$D = \frac{1}{2} a^2 v_o A \, \exp\left(-\frac{E_A}{kT}\right)$$

or

$$D = D_o \, \exp\left(-\frac{E_A}{kT}\right) \qquad\qquad \textbf{[1.22]} \qquad \textit{Diffusion}$$
$$\textit{coefficient}$$

where D_o is a constant. The root square displacement L in time t, from Equation 1.21, is given by $L = [2Dt]^{1/2}$. In a more rigorous analysis, we would consider a number of impurity atoms in the crystal and evaluate the mean square displacement per impurity atom, so that L becomes the **root mean square displacement.** The final result is the same as that in Equation 1.21.

1.6   THE CRYSTALLINE STATE

1.6.1   Types of Crystals

A **crystalline solid** is a solid in which the atoms bond with each other in a regular pattern to form a periodic collection (or array) of atoms, as shown for the copper crystal in Figure 1.28. The most important property of a crystal is **periodicity,** which leads to what is termed **long-range order.** In a crystal, the local bonding geometry is repeated many times at regular intervals, to produce a periodic array of atoms that constitutes the crystal structure. The location of each atom is well known by virtue of periodicity. There is therefore a long-range order, since we can always predict the atomic arrangement anywhere in the crystal. Nearly all metals, many ceramics and semiconductors, and various polymers are crystalline solids in the sense that the atoms or molecules are positioned on a **periodic array of points in space.**

This periodic array on which the atoms are located is called the **crystal lattice.** When the atoms or molecules are positioned at the points of the lattice, they form the crystal structure. In Figure 1.28a, the Cu atoms are positioned at points that

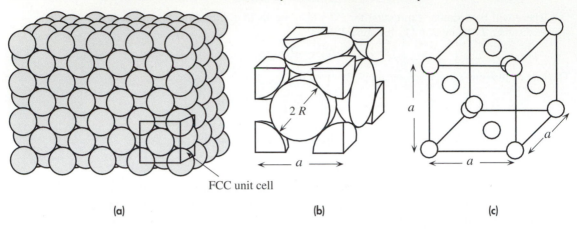

FCC unit cell

(a)   (b)   (c)

Figure 1.28   **(a)** The crystal structure of copper which is face-centered cubic (FCC).
The atoms are positioned at well-defined sites arranged periodically, and there is a long-range order in the crystal.
(b) An FCC unit cell with close-packed spheres.
(c) Reduced-sphere representation of the FCC unit cell. Examples: Ag, Al, Au, Ca, Cu, γ-Fe (> 912 °C), Ni, Pd, Pt, Rh.

form a lattice of cubic symmetry, which is only one of the many possible periodic arrays.

Since the crystal is essentially a periodic repetition of a small volume (or cell) of atoms in three dimensions, it is useful to identify the repeating unit so that the crystal properties can be described through this unit. The **unit cell** is the most convenient small cell in the crystal structure that carries the properties of the crystal. The repetition of the unit cell in three dimensions generates the whole crystal structure, as is apparent in Figure 1.28a for the copper crystal.

The unit cell of the copper crystal is cubic with Cu atoms at its corners and one Cu atom at the center of each face, as indicated in Figure 1.28b. The unit cell of Cu is thus said to have a **face-centered cubic (FCC)** structure. The Cu atoms are shared with neighboring unit cells. Effectively, then, only one-eighth of a corner atom is in the unit cell and one-half of the face-centered atom belongs to the unit cell, as shown in Figure 1.28b. This means there are effectively four atoms in the unit cell. The length of the cubic unit cell is termed the **lattice parameter** a of the crystal structure. For Cu, for example, a is 0.362 nm, whereas the radius R of the Cu atom in the crystal is 0.128 nm. Assuming the Cu atoms are spheres that touch each other, we can geometrically relate a and R. For clarity, it is often more convenient to draw the unit cell with the spheres reduced, as in Figure 1.28c.

The FCC crystal structure of Cu is known as a **close-packed crystal structure** because the Cu atoms are packed as closely as possible, as is apparent in Figures 1.28a and b. The volume of the FCC unit cell is 74 percent full of atoms, which is the maximum packing possible with identical spheres. By comparison, iron has a **body-centered cubic (BCC)** crystal structure and its unit cell is shown in Figure 1.29a. The BCC unit cell has Fe atoms at its corners and one Fe atom at the center of the cell. The volume of the BCC unit cell is 68 percent full of atoms, which is lower than the maximum possible packing.

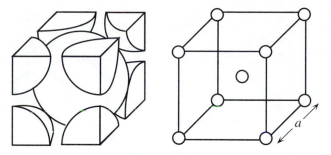

Figure 1.29   Body-centered cubic (BCC) crystal structure.

(a) A BCC unit cell with closely-packed hard spheres representing the Fe atoms.

(b) A reduced-sphere unit cell.

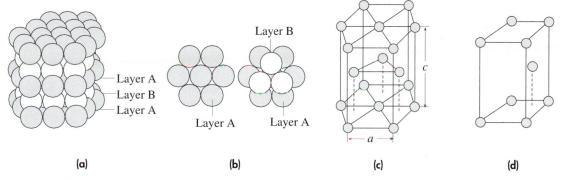

Figure 1.30   The hexagonal close-packed (HCP) crystal structure.

(a) The hexagonal close-packed (HCP) structure. A collection of many Zn atoms. Color difference distinguishes layers (stacks).

(b) The stacking sequence of closely packed layers is ABAB.

(c) A unit cell with reduced spheres.

(d) The smallest unit cell with reduced spheres.

The FCC crystal structure is only one way to pack the atoms as closely as possible. For example, in zinc, the atoms are arranged as closely as possible in a hexagonal symmetry, to form the **hexagonal close-packed (HCP) structure** shown in Figure 1.30a. This structure corresponds to packing spheres as closely as possible first as one layer A, as shown in Figure 1.30b. You can visualize this by arranging six pennies as closely as possible on a table top. On top of layer A we can place an identical layer B, with the spheres taking up the voids on layer A, as depicted in Figure 1.30b. The third layer can be placed on top of B and lined up with layer A. The stacking sequence is therefore *ABAB* . . . . A unit cell for the HCP structure is shown in Figure 1.30c, which shows that this is not a cubic structure. The unit cell shown, although convenient, is not the smallest unit cell. The smallest unit cell for the HCP structure is shown in Figure 1.30d and is called the **hexagonal unit cell.** The repetition of this unit cell will generate the whole HCP structure. The atomic packing density in the HCP crystal structure is 74 percent, which is the same as that in the FCC structure.

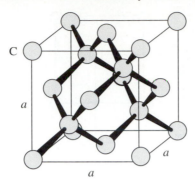

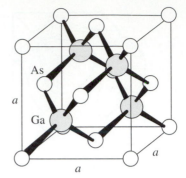

Figure 1.31   The diamond unit cell which is cubic. The cell has eight atoms.

Gray Sn (α-Sn) and the elemental semiconductors Ge and Si have this crystal structure.

Figure 1.32   The zinc blende (ZnS) cubic crystal structure.

Many important compound crystals have the zinc blende structure. Examples: AlAs, GaAs, GaP, GaSb, InAs, InP, InSb, ZnS, ZnZe.

Covalently bonded solids, such as silicon and germanium, have a diamond crystal structure brought about by the directional nature of the covalent bond, as shown in Figure 1.31 (see also Figure 1.5). The rigid local bonding geometry of four Si−Si bonds in the tetrahedral configuration forces the atoms to form what is called the **diamond cubic crystal structure.** The unit cell in this case can be identified with the cubic structure. Although there are atoms at each corner and at the center of each face, indicating an FCC structure, there are four atoms within the cell as well. Thus, the structure is not a true FCC and there are eight atoms in the unit cell.

In the GaAs crystal, as in the silicon crystal, each atom forms four directional bonds with its neighbors. The unit cell is a diamond cubic, as indicated in Figure 1.32 but with the Ga and As atoms alternating positions. This unit cell is termed the **zinc blende** structure after ZnS, which has this type of unit cell. Many important compound semiconductors have this crystal structure, GaAs being the most commonly known.

In ionic solids, the cations (e.g., Na$^+$) and the anions (Cl$^-$) attract each other nondirectionally. The crystal structure depends on how closely the opposite ions can be brought together and how the same ions can best avoid each other while maintaining long-range order, or maintaining symmetry. These depend on the relative charge and size per ion. A typical crystal structure for an ionic solid, taking NaCl as an example, is shown in Figure 1.8b.

To demonstrate the importance of the size effect in two dimensions, consider identical coins, say pennies (1-cent coins). At most, we can make six pennies touch one penny, as shown in Figure 1.33. On the other hand, if we use quarters[7] (25-cent coins) to touch one penny, at most only five quarters can do so. However, this arrangement cannot be extended to the construction of a two-dimensional crystal

---

[7] Although many are familiar with the United States coinage, any two coins with a size ratio of about 0.75 would work out the same.

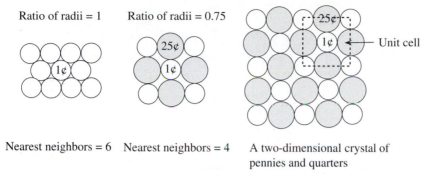

Ratio of radii = 1   Ratio of radii = 0.75

Nearest neighbors = 6   Nearest neighbors = 4   A two-dimensional crystal of
                                                   pennies and quarters

Figure 1.33   Packing of coins on a table top to build a two-dimensional crystal.

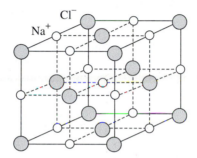

Figure 1.34   A possible reduced-sphere unit cell for the NaCl (rock salt) crystal.

An alternative unit cell may have Na^+ and Cl^- interchanged. Examples: AgCl, CaO, CsF, LiF, LiCl, NaF, NaCl, KF, KCl, MgO.

with periodicity. To fulfill the long-range symmetry requirements for crystals, we can only use four quarters to touch the penny and thereby build a two-dimensional "penny–quarter" crystal, which is shown in the figure. In the two-dimensional crystal, a penny has four quarters as nearest neighbors; similarly, a quarter has four pennies as nearest neighbors. A convenient unit cell is a square cell with one-quarter of a penny at each corner and a full penny at the center (as shown in the figure).

The three-dimensional equivalent of the unit cell of the penny–quarter crystal is the **NaCl unit cell** shown in Figure 1.34. The Na^+ ion is about half the size of the Cl^- ion, which permits six nearest neighbors while maintaining long-range order. The repetition of this unit cell in three dimensions generates the whole NaCl crystal, which was depicted in Figure 1.8b.

A similar unit cell with Na^+ and Cl^- interchanged is also possible and equally convenient. We can therefore describe the whole crystal with two interpenetrating FCC unit cells, each having oppositely charged ions at the corners and face centers. Many ionic solids have the rock salt (NaCl) crystal structure.

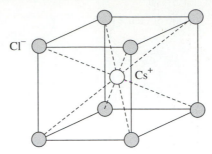

Figure 1.35   A possible reduced-sphere unit cell for the CsCl crystal.

An alternative unit cell may have Cs$^+$ and Cl$^-$ interchanged. Examples: CsCl, CsBr, CsI, TlCl, TlBr, TlI.

When the cation and anions have equal charges and are about the same size, as in the CsCl crystal, the unit cell is called the **CsCl structure,** which is shown in Figure 1.35. Each cation is surrounded by eight anions (and vice versa), which are at the corners of a cube. This is not a true BCC unit cell because the atoms at various BCC lattice points are different.

Table 1.3 summarizes some of the important properties of the main crystal structures considered in this section.

Table 1.3   Properties of some important crystal structures

Crystal Structure	a and R (R is the Radius of the Atom)	Coordination Number (CN)	Number of Atoms per Unit Cell	Atomic Packing Factor	Examples
Simple cubic	$a = 2R$	6	1	0.52	None
BCC	$a = \dfrac{4R}{\sqrt{3}}$	8	2	0.68	Many metals: α–Fe, Cr, Mo, W
FCC	$a = \dfrac{4R}{\sqrt{2}}$	12	4	0.74	Many metals Ag, Au, Cu, Pt
HCP	$a = 2R$ $c = 1.633a$	12	2	0.74	Many metals: Co, Mg, Ti, Zn
Diamond	$a = \dfrac{8R}{\sqrt{3}}$	4	8	0.34	Covalent solids: Diamond, Ge, Si, a-Sn
Zinc blende		4	8	0.34	Many covalent and ionic solids. Many compound semiconductors. ZnS, GaAs, GaSb, InAs, InSb
NaCl		6	4 cations	0.67	Ionic solids such as NaCl, AgCl, LiF MgO, CaO
			4 anions	(NaCl)	Ionic packing factor depends on relative sizes of ions.
CsCl		8	1 cation 1 anion		Ionic solids such as CsCl, CsBr, CsI

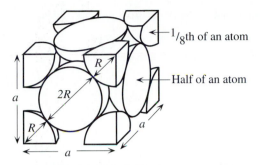

Figure 1.36   The FCC unit cell.
The atomic radius is R and the lattice parameter is a.

THE COPPER (FCC) CRYSTAL   Consider the FCC unit cell of the copper crystal shown in Figure 1.36. | **Example 1.7**

a.   How many atoms are there per unit cell?

b.   If R is the radius of the Cu atom, show that the lattice parameter a is given by $a = R2\sqrt{2}$.

c.   Calculate the atomic packing factor (APF) defined by

$$APF = \frac{\text{Volume of atoms in unit cell}}{\text{Volume of unit cell}}$$

d.   Calculate the density of Cu, given its atomic mass is 63.55 and the radius of the copper atom is 0.128 nm.

Solution

a.   There are four atoms per unit cell. The Cu atom at each corner is shared with eight other adjoining unit cells. Each Cu atom at the face center is shared with the neighboring unit cell. Thus, the number of atoms in the unit cell = 8 corners (1/8 atom) + 6 faces (1/2 atom) = 4 atoms.

b.   Consider the unit cell shown in Figure 1.36 and one of the cubic faces. The face is a square of side a and the diagonal is $\sqrt{[a^2 + a^2]}$ or $a\sqrt{2}$. The diagonal has one atom at the center of diameter $2R$, which touches two atoms centered at the corners. The diagonal, going from corner to corner, is therefore $R + 2R + R$. Thus, $4R = a\sqrt{2}$ and $a = 4R/\sqrt{2} = R2\sqrt{2}$. Therefore, $a = 0.3620$ nm.

c.
$$APF = \frac{(\text{Number of atoms in unit cell}) \times (\text{Volume of atom})}{\text{Volume of unit cell}}$$

$$APF = \frac{4 \times \frac{4}{3}\pi R^3}{a^3} = \frac{\frac{4^2}{3}\pi R^3}{(R2\sqrt{2})^3} = \frac{4^2 \pi}{3(2\sqrt{2})^3} = 0.74$$

d.   To calculate the density ρ, we need the mass of the atoms in the unit cell and the volume of the unit cell. Let n be the number of atoms per unit cell, M_{at} be the atomic mass (amu, or grams per mole), and N_A be Avogadro's number. Then,

$$\text{Density} = \frac{\text{Mass of atoms in unit cell}}{\text{Volume of unit cell}}$$

$$= \frac{(\text{Number of atoms in unit cell}) \times (\text{Mass of the atom})}{\text{Volume of unit cell}}$$

That is,

$$\rho = \frac{n(M_{at}10^{-3})}{a^3 N_A} = \frac{4(64.55 \times 10^{-3} \text{ kg mol}^{-1})}{(0.3620 \times 10^{-9} \text{ m})^3(6.022 \times 10^{23} \text{ mol}^{-1})}$$

or

$$\rho = 9038 \text{ kg m}^{-3} \text{ or } 9.0 \text{ g cm}^{-3}$$

1.6.2   Bravais Lattices

There can be a number of possibilities for choosing a unit cell for a given crystal structure, as is apparent in Figure 1.30c and d for the HCP crystal. As a convention, we generally represent the **geometry of the unit cell** as a parallelepiped with sides a, b, and c and angles α, β, and γ, as depicted in Figure 1.37. The sides a, b, and c and angles α, β, and γ are referred to as the **lattice parameters.** To establish a reference frame and to apply three-dimensional geometry, we insert an xyz coordinate system. The x, y, and z axes follow the edges of the parallelepiped and the origin is at the lower-left rear corner of the cell. The unit cell extends along the x axis from 0 to a, along y from 0 to b, and along z from 0 to c.

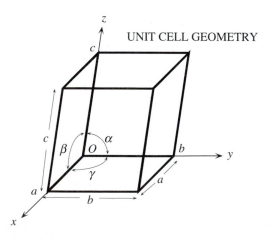

Figure 1.37   We choose a parallelepiped as a general description of the geometry of a unit cell. We line the x, y, and z axes with the edges of the parallelepiped taking lower-left rear corner as the origin.

For Cu and Fe, the unit-cell geometry has $a = b = c$, $\alpha = \beta = \gamma = 90°$, and cubic symmetry. For Zn, the unit cell has hexagonal geometry, with $a = b \neq c$, $\alpha = \beta = 90°$, and $\gamma = 120°$, as shown in Figure 1.30d. Based on different lattice parameters, there are only seven possible distinct unit-cell geometries, which we call **crystal systems,** each with a particular distinct symmetry. The seven crystal

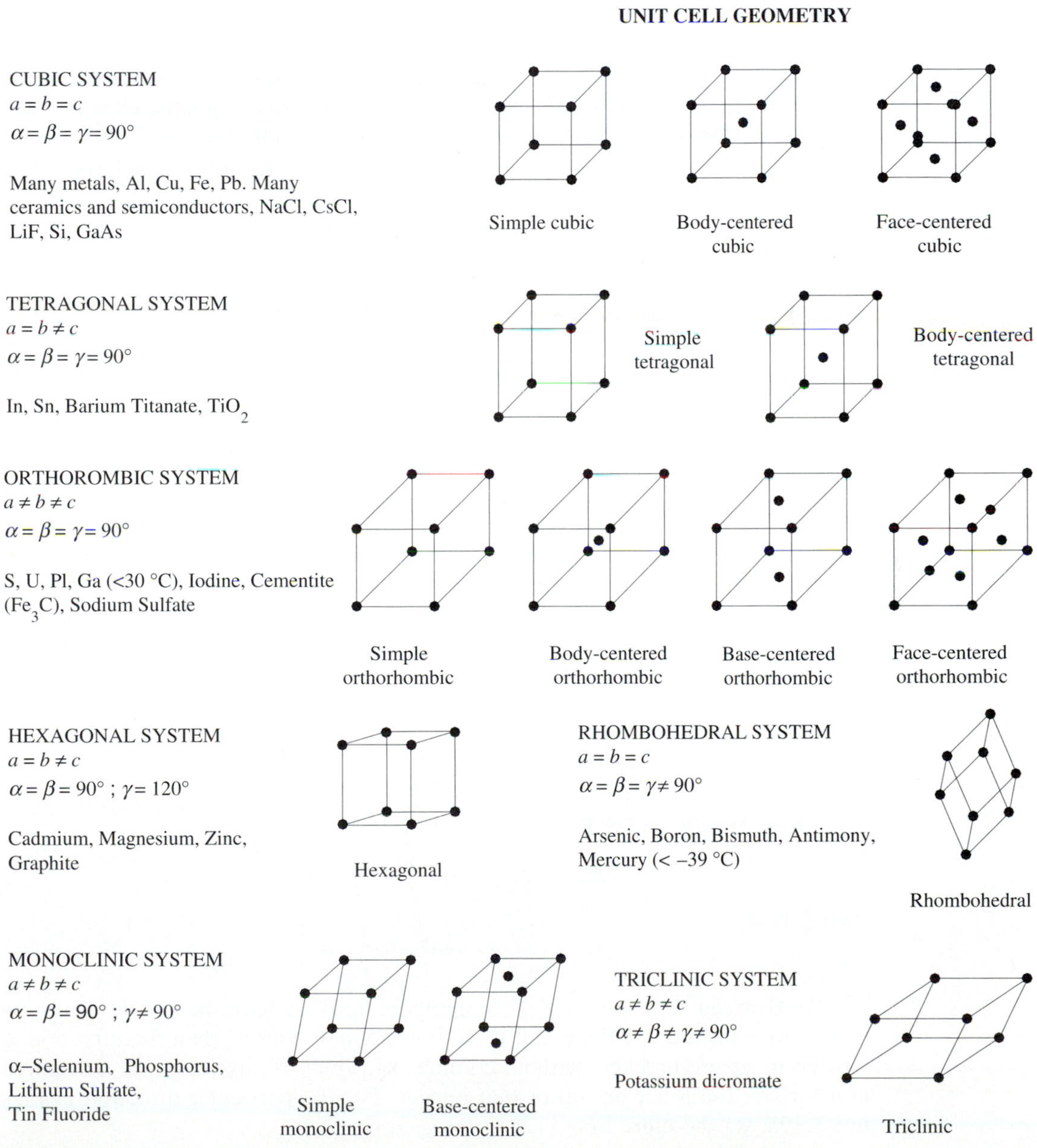

UNIT CELL GEOMETRY

CUBIC SYSTEM
$a = b = c$
$\alpha = \beta = \gamma = 90°$

Many metals, Al, Cu, Fe, Pb. Many ceramics and semiconductors, NaCl, CsCl, LiF, Si, GaAs

Simple cubic   Body-centered cubic   Face-centered cubic

TETRAGONAL SYSTEM
$a = b \neq c$
$\alpha = \beta = \gamma = 90°$

In, Sn, Barium Titanate, TiO_2

Simple tetragonal   Body-centered tetragonal

ORTHOROMBIC SYSTEM
$a \neq b \neq c$
$\alpha = \beta = \gamma = 90°$

S, U, Pl, Ga (<30 °C), Iodine, Cementite (Fe_3C), Sodium Sulfate

Simple orthorhombic   Body-centered orthorhombic   Base-centered orthorhombic   Face-centered orthorhombic

HEXAGONAL SYSTEM
$a = b \neq c$
$\alpha = \beta = 90°$; $\gamma = 120°$

Cadmium, Magnesium, Zinc, Graphite

Hexagonal

RHOMBOHEDRAL SYSTEM
$a = b = c$
$\alpha = \beta = \gamma \neq 90°$

Arsenic, Boron, Bismuth, Antimony, Mercury (< –39 °C)

Rhombohedral

MONOCLINIC SYSTEM
$a \neq b \neq c$
$\alpha = \beta = 90°$; $\gamma \neq 90°$

α–Selenium, Phosphorus, Lithium Sulfate, Tin Fluoride

Simple monoclinic   Base-centered monoclinic

TRICLINIC SYSTEM
$a \neq b \neq c$
$\alpha \neq \beta \neq \gamma \neq 90°$

Potassium dicromate

Triclinic

Figure 1.38   The seven crystal systems (unit cell geometries) and fourteen Bravais lattices.

systems are depicted in Figure 1.38, with typical examples. We are already familiar with the cubic and hexagonal systems.

The seven crystal systems only categorize the unit cells based on geometry; they do not show where the atoms may be placed within the cell. Do not confuse the unit cell geometry with the lattice, which is a periodic array of points at which the atoms are placed. In the cubic system, for example, there are three possible distinct lattices, corresponding to simple cubic (SC), BCC, and FCC, as shown in Figure 1.38. All three have the same cubic geometry: $a = b = c$ and $\alpha = \beta = \gamma = 90°$. However, when each unit cell is repeated in three dimensions, the resulting crystal structures are different, as is the case for iron and copper.

Many distinctly different lattices, or distinct patterns of points, exist in three dimensions. There are 14 distinct lattices whose unit cells have one of the seven geometries indicated in Figure 1.38. Each of these is called a **Bravais lattice.** The copper crystal, for example has the FCC Bravais lattice, and the arsenic, antimony, and bismuth crystals have the rhombohedral Bravais lattice. Tin's unit cell belongs to the tetragonal crystal system and its crystal lattice is a **body-centered tetragonal (BCT).**

1.6.3   Miller Indices: Crystal Directions and Planes

Crystallographic Directions   In explaining crystal properties, we must frequently specify a direction in a crystal, or a particular plane of atoms. Many properties, for example, the elastic modulus, electrical resistivity, magnetic susceptibility, etc., are directional within the crystal. We use the convention described here for labelling crystal directions based on three-dimensional geometry.

All parallel vectors have the same indices. Therefore, the direction to be labelled can be moved to pass through the origin of the unit cell. As an example, Figure 1.39a shows a direction whose indices are to be determined. A point P on the vector can be expressed by the coordinates x_o, y_o, z_o, where $x_o, y_o,$ and z_o are projections from point P onto the $x, y,$ and z axes, as shown in Figure 1.39a. It is generally convenient to place P where the line cuts a surface (though this is not necessary). We can express these coordinates in terms of the lattice parameters $a, b,$ and c, respectively. We then have three coordinates, say $x_1, y_1,$ and z_1, for point P in terms of $a, b,$ and c. For example, if

$$x_o, y_o, z_o \quad \text{are} \quad \tfrac{1}{2}a, b, \tfrac{1}{2}c$$

then P is at

$$x_1, y_1, z_1 \quad \text{and} \quad \tfrac{1}{2}, 1, \tfrac{1}{2}$$

We then multiply or divide these numbers until we have the smallest integers (which may include 0). If we call these integers $u, v,$ and w, then the direction is written in square brackets without commas as $[uvw]$. If any integer is a negative number, we use a bar on top of that integer. For the particular direction in Figure 1.39a, we therefore have [121].

Some of the important directions in a cubic lattice are shown in Figure 1.39b. For example, the $x, y,$ and z directions in the cube are [100], [010], and [001], as

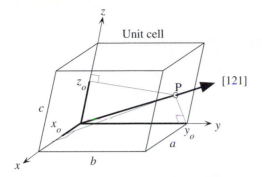

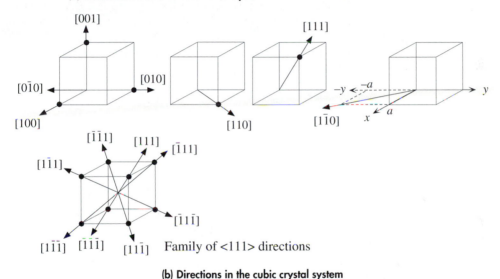

(a) Identification of a direction in a crystal

Family of <111> directions

(b) Directions in the cubic crystal system

Figure 1.39   Labelling of crystal directions and various directions in the cubic lattice.

shown. Reversing a direction simply changes the sign of each index. The negative x, y, and z directions are $[\bar{1}00]$, $[0\,\bar{1}\,0]$, and $[00\,\bar{1}]$, respectively.

Certain directions in the crystal are equivalent because the differences between them are based only on our arbitrary decision for labelling x, y, and z directions. For example, $[100]$ and $[010]$ are different simply because of the way in which we labelled the x and y axes. Indeed, directional properties of a material (e.g., elastic modulus, dielectric susceptibility, etc.) along the edge of the cube $[100]$, are invariably the same as along the other edges, for example, $[010]$, $[001]$, etc. All of these directions along the edges of the cube constitute a **family of directions,** which is any set of directions considered to be equivalent. We label a family of directions, for example, $[100]$, $[010]$, $[001]$, etc., by using a common notation, triangular brackets. Thus, $<100>$ represents the family of six directions, $[100]$, $[010]$, $[001]$, $[\bar{1}00]$, $[0\,\bar{1}\,0]$, and $[00\,\bar{1}]$. Similarly, the family of diagonal directions in the cube, shown in Figure 1.39b, is denoted $<111>$.

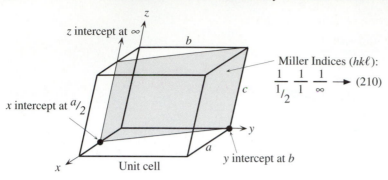

(a) Identification of a plane in a crystal

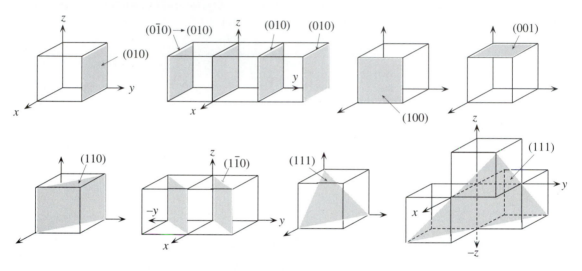

(b) Various planes in the cubic lattice.

Figure 1.40   Labelling of crystal planes and typical examples in the cubic lattice.

Crystallographic Planes   We also frequently need to describe a particular plane in a crystal. Figure 1.40 shows a general unit cell with a plane to be labelled. We use the following convention, called the **Miller indices** of a plane, for this purpose.

We take the intercepts x_o, y_o, and z_o of the plane on the x, y, and z axes, respectively. If the plane passes through the origin, we can use another convenient parallel plane, or simply shift the origin to another point. All planes that have been shifted by a lattice parameter have identical Miller indices.

We express the intercepts x_o, y_o, and z_o in terms of the lattice parameters a, b, and c, to obtain x_1, y_1, and z_1. We then invert these numbers. Taking the reciprocals, we obtain

$$\frac{1}{x_1}, \frac{1}{y_1}, \frac{1}{z_1}$$

We then clear all fractions, without reducing to lowest integers, to obtain a set of integers, say h, k, and ℓ. We then put these integers into parentheses, without commas, that is, $(hk\ell)$. For the plane in Figure 1.40a, we have:

Intercepts x_o, y_o, and z_o are $(1/2)\,a$, $1b$, and $\infty\,c$.

Intercepts x_1, y_1, and z_1, in terms of a, b, and c,   are $1/2$, 1, and ∞.

Reciprocals $1/x_1$, $1/y_1$, and $1/z_1$ are $1/\tfrac{1}{2}$, $1/1$, $1/\infty = 2, 1, 0$.

This set of numbers does not have fractions, so it is not necessary to clear fractions. Hence, the Miller indices $(hk\ell)$ are (210).

If there is a negative integer due to a negative intercept, a bar is placed across the top of the integer. Also, if parallel planes differ only by a shift that involves a multiple number of lattice parameters, then these planes may be assigned the same Miller indices. For example, the plane $(0\,\bar{1}\,0)$ is the xz plane that cuts the y axis at $-b$. If we shift the plane along y by two lattice parameters $(2b)$, it will cut the y axis at b and the Miller indices will become (010). In terms of the unit cell, the $(0\,\bar{1}\,0)$ plane is the same as the (010) plane, as shown in Figure 1.40b. Note that not all parallel planes are identical. Planes can have the same Miller indices *only* if they are separated by a multiple of the lattice parameter. For example, the (010) plane is not identical to the (020) plane, even though they are geometrically parallel. In terms of the unit cell, plane (010) is a face of the unit cell cutting the y axis at b, whereas (020) is a plane that is halfway inside the unit cell, cutting the y axis at $(1/2)\,b$. The planes contain different numbers of atoms. The (020) plane cannot be shifted by the lattice parameter b to coincide with plane (010).

It is apparent from Figure 1.40b that in the case of the cubic crystal, the $[hk\ell]$ direction is always perpendicular to the $(hk\ell)$ plane.

Certain planes in the crystal belong to a **family of planes** because their indices differ only as a consequence of the arbitrary choice of axis labels. For example, the indices of the (100) plane become (010) if we switch the x and y axes. All the (100), (010), and (001) planes, and hence the parallel $(\bar{1}\,00)$, $(0\,\bar{1}\,0)$, $(00\,\bar{1})$ planes, form a family of planes, conveniently denoted by curly brackets as {100}.

MILLER INDICES   Consider the plane shown in Figure 1.41, which passes through one side of a face and the center of an opposite face in the FCC lattice. What are the Miller indices of this plane?   **Example 1.8**

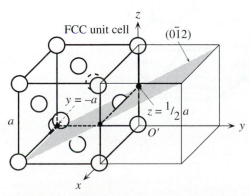

Figure 1.41   The plane that passes through one side of a face and the center of an opposite face in the FCC lattice.

Solution

The plane passes through the origin at the lower-left rear corner. We therefore shift the origin to point O′ at the lower-right rear corner of the unit cell. In terms of a, the plane cuts the x, y, and z axis at ∞, -1, and $1/2$. We take the reciprocals to obtain, 0, -1, and 2. Therefore, the Miller indices are $(0\ \bar{1}\ 2)$.

1.6.4   Allotropy and the Three Phases of Carbon

Certain substances can have more than one crystal structure, iron being one of the best known examples. This characteristic is termed **polymorphism** or **allotropy.** Below 912 °C, iron has the BCC structure and is called α-Fe. Between 912 °C and 1400 °C, iron has the FCC structure and is called γ-Fe. Above 1400 °C, iron again

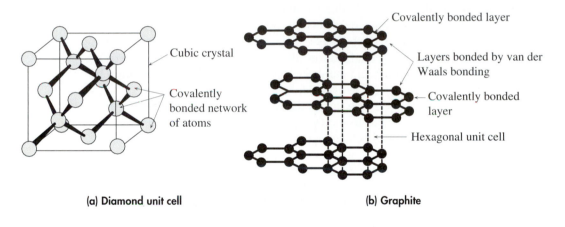

Cubic crystal

Covalently bonded network of atoms

Covalently bonded layer

Layers bonded by van der Waals bonding

Covalently bonded layer

Hexagonal unit cell

(a) Diamond unit cell

(b) Graphite

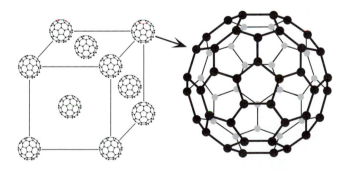

The FCC unit cell of the Buckminsterfullerene crystal. Each lattice point has a C_{60} molecule

Buckminsterfullerene (C_{60}) molecule (the "buckyball" molecule)

(c) Buckminsterfullerene

Figure 1.42   The three allotropes of carbon.

has the BCC structure and is called δ-Fe. Since iron has more than one crystal structure, it is called **polymorphic.** Each iron crystal structure is an allotrope or a polymorph.

The allotropes of iron are all metals. Furthermore, one allotrope changes to another at a well-defined temperature called a **transition temperature,** which in this case is 912 °C.

Many substances have allotropes that exhibit diversely different properties. Moreover, for some polymorphic substances, the transformation from one allotrope to another cannot be achieved by a change of temperature, but requires the application of pressure, as in the transformation of graphite to diamond.

Carbon has three crystalline allotropes: diamond, graphite, and the newly discovered **buckminsterfullerene.** These crystal structures are shown in Figure 1.42a, b and c, respectively, and their properties are summarized in Table 1.4. Graphite is the carbon form that is stable at room temperature. Diamond is the stable form at very high pressures. Once formed, diamond continues to exist at atmospheric pressures and below about 900 °C, because the transformation rate of diamond to graphite is virtually zero under these conditions. Graphite and diamond have widely differing properties, which lead to diverse applications. For example, graphite is an electrical conductor, whereas diamond is an insulator. Diamond is

Table 1.4   Crystalline allotropes of carbon (ρ is the density and Y is the elastic modulus or Young's modulus)

	Graphite	Diamond	Buckminsterfullerene Crystal
Structure	Covalent bonding within layers. Van der Waals bonding between layers. Hexagonal unit cell.	Covalently bonded network. Diamond crystal structure.	Covalently bonded C_{60} spheroidal molecules held in an FCC crystal structure by van der Waals bonding.
Electrical and Thermal Properties	Good electrical conductor. Thermal conductivity comparable to metals.	Very good electrical insulator. Excellent thermal conductor, about five times more than silver or copper.	Semiconductor. Compounds with alkali metals (e.g., K_3C_{60}) exhibit superconductivity.
Mechanical Properties	Lubricating agent. Machinable. Bulk Graphite: $Y \approx 27$ GPa $\rho = 2.25$ g cm^{-3}	The hardest material. $Y = 827$ GPa $\rho = 3.25$ g cm^{-3}	Mechanically soft. $Y \approx 18$ GPa $\rho = 1.65$ g cm^{-3}
Comment	Stable allotrope at atmospheric pressure.	High pressure allotrope.	Laboratory synthesized. Occurs in the soot of partial combustion.
Uses, Potential Uses	Metallurgical crucibles, welding electrodes, heating elements, electrical contacts, refractory applications.	Cutting tool applications. Diamond anvils. Diamond film coated drills, blades, bearings etc. Jewelry. Heat conductor for ICs. Possible thin-film semiconductor devices, as the charge carrier motilities are large.	Possible future semiconductor or superconductivity applications.

the hardest substance known. On the other hand, the carbon layers in graphite can readily slide over each other under shear stresses, because the layers are only held together by weak secondary bonds (van der Waals bonds). This is the reason for graphite's lubricating properties.

Buckminsterfullerene is another polymorph of carbon. In the buckminsterfullerene molecule (called the "buckyball"), 60 carbon atoms bond with each other to form a perfect soccer ball-type molecule. The C_{60} molecule has 12 pentagons and 20 hexagons joined together to form a spherical molecule, with each C atom at a corner, as depicted in Figure 1.42c. The molecules are produced in the laboratory by a carbon arc in a partial atmosphere of an inert gas (He); they are also found in the soot of partial combustion. The crystal form of buckminsterfullerene has the FCC structure, with each C_{60} molecule occupying a lattice point and being held together by van der Waals forces, as shown in Figure 1.42c. The Buckminsterfullerene crystal is a semiconductor, and its compounds with alkali metals, such as K_3C_{60}, exhibit superconductivity at low temperatures (below 18 K). Mechanically, it is a soft material.

1.7   CRYSTALLINE DEFECTS AND THEIR SIGNIFICANCE

By bringing all the atoms together to try to form a perfect crystal, we lower the total potential energy of the atoms as much as possible for that particular structure. What happens when the crystal is grown from a liquid or vapor; do you always get a perfect crystal? What happens when the temperature is raised? What happens when impurities are added to the solid?

There is no such thing as a perfect crystal. We must therefore understand the types of defects that can exist in a given crystal structure. Quite often, key mechanical and electrical properties are controlled by these defects.

1.7.1   Point Defects: Vacancies and Impurities

Above the absolute zero temperature, all crystals have atomic vacancies or atoms missing from lattice sites in the crystal structure. The vacancies exist as a requirement of thermal equilibrium and are called **thermodynamic defects.** Vacancies introduce disorder into the crystal by upsetting the perfect periodicity of atomic arrangements.

We know from the kinetic molecular theory that all the atoms in a crystal vibrate about their equilibrium positions with a distribution of energies, a distribution that closely resembles the Boltzmann distribution. At some instant, there may be one atom with sufficient energy to break its bonds and jump to an adjoining site on the surface, as depicted in Figure 1.43. This leaves a vacancy behind, just below the surface. This vacancy can then diffuse into the bulk of the crystal, because a neighboring atom can diffuse into it.

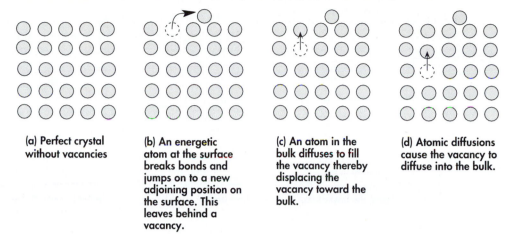

(a) Perfect crystal without vacancies

(b) An energetic atom at the surface breaks bonds and jumps on to a new adjoining position on the surface. This leaves behind a vacancy.

(c) An atom in the bulk diffuses to fill the vacancy thereby displacing the vacancy toward the bulk.

(d) Atomic diffusions cause the vacancy to diffuse into the bulk.

Figure 1.43 Generation of a vacancy by the diffusion of an atom to the surface and the subsequent diffusion of the vacancy into the bulk.

This latter process of vacancy creation has been shown to be a sequence of events in Figure 1.43. Suppose that E_v is the average energy required to create such a vacancy. Then only a fraction, $\exp(-E_v/kT)$, of all the atoms in the crystal can have sufficient energy to create vacancies. If the number of atoms per unit volume in the crystal is N, then the vacancy concentration n_v is given by

$$n_v = N \exp\left(-\frac{E_v}{kT}\right) \qquad \textbf{[1.23]}$$

Equilibrium concentration of vacancies

At all temperatures above absolute zero, there will always be an equilibrium concentration of vacancies, as dictated by Equation 1.23.[8] Although we considered only one possible vacancy creation process in Figure 1.43 there are other processes that also create vacancies. Furthermore, we have shown the vacancy to be the same size in the lattice as the missing atom, which is not entirely true. The neighboring atoms around a vacancy close in to take up some of the slack, as shown in Figure 1.44a. This means that the crystal lattice around the vacancy is distorted from the perfect arrangement over a few atomic dimensions. The vacancy volume is therefore smaller than the volume of the missing atom.

Vacancies are only one type of **point defect** in a crystal structure. Point defects generally involve lattice changes or distortions of a few atomic distances, as depicted in Figure 1.44. The crystal structure may contain impurities, either naturally or as a consequence of intentional addition, as in the case of silicon crystals grown for microelectronics. If the impurity atom substitutes directly for the host atom, the result is called a **substitutional impurity** and the resulting crystal structure is that of a **substitutional solid solution,** as shown in Figure 1.44b and c. When an Si

[8] Equation 1.23 can be rigorously derived from thermodynamics and Gibbs free energy concepts, which are beyond the scope of this book.

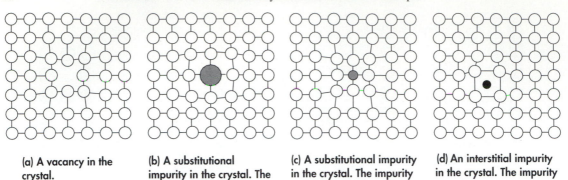

(a) A vacancy in the crystal.

(b) A substitutional impurity in the crystal. The impurity atom is larger than the host atom.

(c) A substitutional impurity in the crystal. The impurity atom is smaller than the host atom.

(d) An interstitial impurity in the crystal. The impurity occupies an empty space between host atoms.

Figure 1.44   Point defects in the crystal structure.
The regions around the point defect become distorted; the lattice becomes strained.

crystal is "doped" with small amounts of arsenic (As) atoms, the As atoms substitute directly for the Si atoms in the Si crystal; that is, the arsenic atoms are substitutional impurities. The impurity atom can also place itself in an interstitial site, that is, in a void between the host atoms, as carbon does in the BCC iron crystal. In that case, the impurity is called an **interstitial impurity,** as shown in Figure 1.44d.

In general, the impurity atom will have both a different valency and a different size. It will therefore distort the lattice around it. For example, if a substitutional impurity atom is larger than the host atom, the neighboring host atoms will be pushed away, as in Figure 1.44b. The crystal region around an impurity is therefore distorted from the perfect periodicity and the lattice is said to be **strained around a point defect.** A smaller substitutional impurity atom will pull in the neighboring atoms, as in Figure 1.44c. Typically, interstitial impurities tend to be small atoms compared to the host atoms, a typical example being the small carbon atom in the BCC iron crystal.

In an ionic crystal, such as NaCl, which consists of anions (Cl^-) and cations (Na^+), one common type of defect is called a **Schottky defect.** This involves a missing cation–anion pair (which may have migrated to the surface), so the neutrality is maintained, as indicated in Figure 1.45a. These Schottky defects are responsible for the major optical and electrical properties of alkali halide crystals. Another type of defect in the ionic crystal is the **Frenkel defect,** which occurs when a host ion is displaced into an interstitial position, leaving a vacancy at its original site. The interstitial ion and the vacancy pair constitute the Frenkel defect, as identified in Figure 1.45a. For the AgCl crystal, which has predominantly Frenkel defects, an Ag^+ is in an interstitial position. The concentration of such Frenkel defects is given by Equation 1.23, with an appropriate defect creation energy E_{defect} instead of E_v.

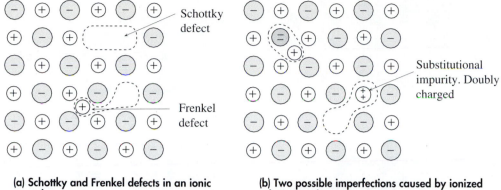

(a) Schottky and Frenkel defects in an ionic crystal.

(b) Two possible imperfections caused by ionized substitutional impurity atoms in an ionic crystal.

Figure 1.45   Point defects in ionic crystals.

Ionic crystals can also have substitutional and interstitial impurities that become ionized in the lattice. Overall, the ionic crystal must be neutral. Suppose that an Mg^{2+} ion substitutes for an Na^+ ion in the NaCl crystal, as depicted in Figure 1.45b. Since the overall crystal must be neutral, either one Na^+ ion is missing somewhere in the crystal, or an additional Cl^- ion exists in the crystal. Similarly, when a doubly-charged negative ion, such as O^{2-}, substitutes for Cl^-, there must either be an additional cation (usually in an interstitial site) or a missing Cl^- somewhere in order to maintain charge neutrality in the crystal. The most likely type of defect depends on the composition of the ionic solid and the relative sizes and charges of the ions.

1.7.2   Line Defects: Edge and Screw Dislocations

A line defect is formed in a crystal when an atomic plane terminates within the crystal instead of passing all the way to the end of the crystal, as depicted in Figure 1.46a. The edge of this short plane of atoms is therefore like a line running inside the crystal. The planes neighboring (i.e., above) this short plane are dislocated (displaced) with respect to those below the line. We therefore call this type of defect an **edge dislocation** and use an inverted T symbol. The vertical line corresponds to the half-plane of atoms in the crystal, as illustrated in Figure 1.46a. It is clear that the atoms around the dislocation line have been effectively displaced from their perfect-crystal equilibrium positions, which results in atoms being out of registry above and below the dislocation. The atoms above the dislocation line are pushed together, whereas those below it are pulled apart, so there are regions of compression and tension above and below the dislocation line, respectively, as depicted by the shaded region around the dislocation line in Figure 1.45b. Therefore, around a dislocation line, we have a **strain field** due to the stretching or compressing of bonds.

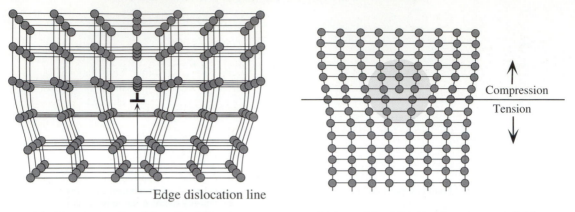

(a) Dislocation is a line defect. The dislocation shown runs into the paper.

(b) Around the dislocation there is a strain field as the atomic bonds have been compressed above and stretched below the dislocation line.

Figure 1.46   Dislocation in a crystal. This is a line defect, which is accompanied by lattice distortion and hence a lattice strain around it.

The energy required to create a dislocation is typically in the order of 100 eV per nm of dislocation line. On the other hand, it takes only few eV to form a point defect, which is a few nanometers in dimension. In other words, forming a number of point defects is energetically more favorable than forming a dislocation. Dislocations are not equilibrium defects. They normally arise when the crystal is deformed by stress, or when the crystal is actually being grown.

Another type of dislocation is the **screw dislocation,** which is essentially a shearing of one portion of the crystal with respect to another, by one atomic distance, as illustrated in Figure 1.47a. The displacement occurs on either side of the **screw dislocation line.** The circular arrow around the line symbolizes the screw dislocation. As we move away from the dislocation line, the atoms in the upper portion become more out of registry with those below; at the edge of the crystal, this displacement is one atomic distance, as illustrated in Figure 1.47b.

Both edge and screw dislocations are generally created by stresses resulting from thermal and mechanical processing. A line defect is not necessarily either a pure edge or a pure screw dislocation; it can be a mixture, as depicted in Figure 1.48. Screw dislocations frequently occur during crystal growth, which involves atomic stacking on the surface of a crystal. Such dislocations aid crystallization by providing an additional "edge" to which the incoming atoms can attach, as illustrated in Figure 1.49. To explain, if an atom arrives at the surface of a perfect crystal, it can only attach to one atom in the plane below. However, if there is a screw dislocation, the incoming atom can attach to an edge and thereby form more bonds; hence, it can lower its potential energy more than anywhere else on the surface. With incoming atoms attaching to the edges, the growth occurs spirally around the screw dislocation, and the final crystal surface reflects this spiral growth geometry.

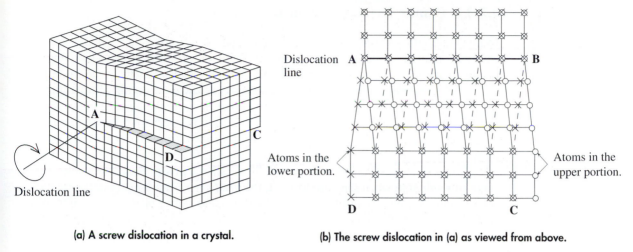

Dislocation line

Dislocation line A ——————— B

Atoms in the lower portion.

Atoms in the upper portion.

D   C

(a) A screw dislocation in a crystal.

(b) The screw dislocation in (a) as viewed from above.

Figure 1.47   A screw dislocation, which involves shearing one portion of a perfect crystal with respect to another, on one side of a line (AB).

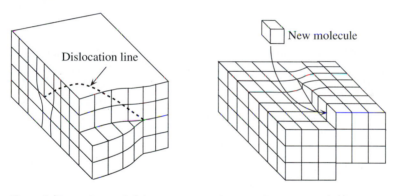

Dislocation line

New molecule

Figure 1.48   A mixed dislocation.

Figure 1.49   Screw dislocation aids crystal growth because the newly arriving atom can attach to two or three atoms instead of one atom and thereby form more bonds.

The phenomenon of **plastic** or **permanent deformation** of a metal depends totally on the presence and motions of dislocations, as discussed in elementary books on the mechanical properties of materials. In the case of electrical properties of metals, we will see later in this book that dislocations increase the resistivity of materials, cause significant leakage current in a pn junction, and give rise to unwanted noise in various semiconductor devices. Fortunately, the occurrence of dislocations in semiconductor crystals can be controlled and nearly eliminated. In a metal interconnection line on a chip, there may be on average of 10^4-10^5 dislocation lines per mm^2 of crystal, whereas a silicon crystal wafer that is carefully grown may typically have only 1 dislocation line per mm^2 of crystal.

1.7.3   Planar Defects: Grain Boundaries

Many materials are polycrystalline, that is, they are composed of many small crystals oriented in different directions. In fact, the growth of a flawless single crystal from what is called the **melt** (liquid) requires special skills, in addition to scientific knowledge. When a liquid is cooled to below its freezing temperature, solidification does not occur at every point, rather, it occurs at certain sites called **nuclei,** which are small crystal-like structures containing perhaps 50 to 100 atoms. Figures 1.50a to c depict a typical solidification process from the melt. The liquid atoms adjacent to a nucleus diffuse into the nucleus, thereby causing it to grow in size to become a small crystal, or a crystallite, called a **grain.** Since the nuclei are randomly oriented when they are formed, the grains have random crystallographic orientations during crystallite growth. As the liquid between the grains is consumed, some grains meet and obstruct each other. At the end of solidification, therefore, the whole structure has grains with irregular shapes and orientations, as shown in Figure 1.50c.

It is apparent from Figure 1.50c that in contrast to a single crystal, a polycrystalline material has grain boundaries where differently oriented crystals meet. As indicated in Figure 1.51, the atoms at the grain boundaries obviously cannot follow their natural bonding habits, because the crystal orientation suddenly changes across the boundary. Therefore, there are both voids at the grain boundary and stretched and broken bonds. In addition, in this region, there are misplaced atoms that do not follow the crystalline pattern on either side of the boundary. Consequently, the grain boundary represents a high-energy region per atom with respect to the energy per atom within the bulk of the grains themselves. The atoms can diffuse more easily along a grain boundary because: (a) less bonds are broken due to the presence of voids; and (b) the bonds are strained and easily broken anyway. In many polycrystalline materials, impurities therefore tend to congregate in the grain boundary region. We generally refer to the atomic arrangement in the grain

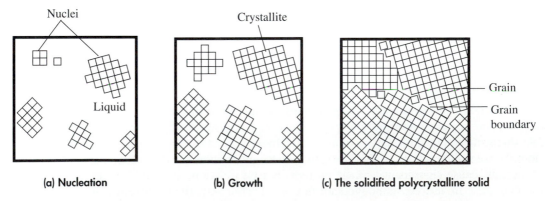

(a) **Nucleation**   (b) **Growth**   (c) **The solidified polycrystalline solid**

Figure 1.50   Solidification of a polycrystalline solid from the melt. For simplicity, cubes represent atoms.

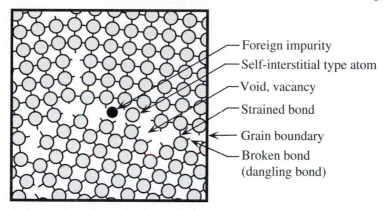

— Foreign impurity

— Self-interstitial type atom

— Void, vacancy

— Strained bond

← Grain boundary

— Broken bond
(dangling bond)

Figure 1.51   The grain boundaries have broken bonds, voids, vacancies, strained bonds, and interstitial type atoms.

The structure of the grain boundary is disordered, and the atoms in the grain boundaries have higher energies than those within the grains.

boundary region as being **disordered** due to the presence of the voids and misplaced atoms.

Since the energy of an atom at the grain boundary is greater than that of an atom within the grain, these grain boundaries are nonequilibrium defects; consequently, they try to reduce in size to give the whole structure a lower potential energy. At or around room temperature, the atomic diffusion process is slow; thus, the reduction in the grain boundary is insignificant. At elevated temperatures, however, atomic diffusion allows big grains to grow, at the expense of small grains, which leads to **grain coarsening (grain growth)** and hence to a reduction in the grain boundary area.

Mechanical engineers have learned to control the grain size, and hence the mechanical properties of metals to suit their needs, through various thermal treatment cycles. For electrical engineers, the grain boundaries become important when designing electronic devices based on polysilicon or any polycrystalline semiconductor. For example, in highly polycrystalline materials, particularly thin-film semiconductors (e.g., polysilicon), the resistivity is invariably determined by polycrystallinity, or grain size, of the material, as discussed in Chapter 2.

1.7.4   Crystal Surfaces and Surface Properties

In describing crystal structures, we assume that the periodicity extends to infinity, which means that the regular array of atoms is not interrupted anywhere by the presence of the real surfaces of the material. In practice, we know that all substances have real surfaces. When the crystal lattice is abruptly terminated by a surface, the atoms at the surface cannot fulfill their bonding requirements, as illustrated in Figure 1.52. For simplicity, the figure shows an Si crystal schematically sketched in two dimensions, where each atom in the bulk of the crystal has

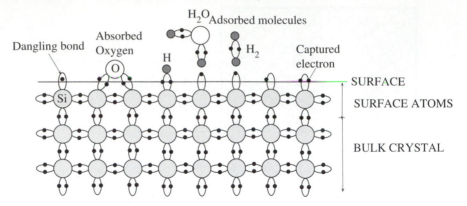

Figure 1.52   At the surface of a hypothetical two-dimensional Si crystal, the Si atoms cannot fulfill their bonding requirements and therefore have dangling bonds.

These bonds capture atoms and molecules from the environment, as well as any free electron.

four covalent bonds, each covalent bond having two electrons. The Si atoms at the surface are left with **dangling bonds,** that is, bonds that are half full, having only one electron. These dangling bonds look for atoms to which they can bond.

Atoms from the environment can therefore easily bond with the Si atoms on the surface. For example, a hydrogen atom can be captured by a dangling bond and hence become **absorbed.** A foreign atom, or molecule, is absorbed if it chemically bonds with the atoms on the surface. The H atom in Figure 1.52 forms a covalent bond with an Si atom and hence becomes absorbed. However, an H_2 molecule cannot form a covalent bond, but due to hydrogen bonding, it can form a secondary bond with a surface Si atom and become **adsorbed.** Molecules that only form secondary bonds with surface atoms are said to be adsorbed. Similarly, a water molecule in the air can readily become adsorbed at the surface of a crystal.

Also, if there are any free electrons, these can be captured by dangling bonds. When an Si crystal and a metal are in contact, there is a grain boundary between the two crystals. Any dangling bond at the interface between the Si crystal and the metal will capture electrons from the metal.

When left unprocessed, the surface of a crystal will have absorbed and adsorbed atoms and molecules from the environment. This is one of reasons that the surface of an Si wafer in microelectronics technology is first etched and then oxidized to form an SiO_2 **passivating layer** on the crystal surface. Many substances have a natural oxide layer on the surface. For example, aluminum surfaces always have a thin aluminum oxide layer. In addition, the oxide surface often has adsorbed organic atoms, usually from machining and handling.

Figure 1.52 shows only the dangling bonds at the surface of a crystal. The surface structure generally depends greatly on the mode of surface formation, which invariably involves thermal and mechanical processing, as well as on previous environmental history. One visualization of a crystal surface is based on the **terrace-ledge-kink model,** which is the so-called **Kossel model,** illustrated in

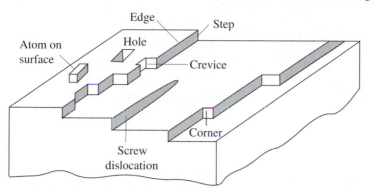

Edge

Step

Hole

Atom on
surface

Crevice

Screw
dislocation

Corner

Figure 1.53   Typically, a crystal surface has many types of imperfections, such as steps, ledges, kinks, crevices, holes, and dislocations.

Figure 1.53. The surface has ledges, kinks, and various imperfections such as holes, dislocations, etc., as well as impurities which can diffuse to and from the surface. The dimensions of the various imperfections (e.g., the step size) depend on the process that generated the surface.

VACANCY CONCENTRATION IN ALUMINUM   Calculate the fractional concentration of vacancies in the aluminum crystal, at room temperature and near the melting temperature, given that the energy of formation of a vacancy in aluminum is about 0.75 eV/atom. | **Example 1.9**

Solution

Equation 1.23 gives the fractional concentration of vacancies.
At 300 K,

$$\frac{n_v}{N} = \exp\left(-\frac{E_v}{kT}\right) = \exp\left[\frac{-(0.75 \text{ eV})(1.6 \times 10^{-19} \text{ C})}{(1.38 \times 10^{-23} \text{ J K}^{-1})(300 \text{ K})}\right]$$

$$= 2.5 \times 10^{-13}$$

which is a very small fraction.
At 660 °C,

$$\frac{n_v}{N} = \exp\left(-\frac{E_v}{kT}\right) = \exp\left[\frac{-0.75 \times 1.6 \times 10^{-19}}{(1.38 \times 10^{-23})(933)}\right] = 9 \times 10^{-5}$$

That is, almost one in 10,000 atomic sites is a vacancy.

VACANCY CONCENTRATION IN A SEMICONDUCTOR   The energy E_v required to create a vacancy in Si is 2.4 eV/atom. What is the concentration of vacancies in an Si crystal at room temperature (300 K) and at 1000 °C? (In device fabrication, Si is frequently doped by the diffusion of impurities at this temperature.) The atomic mass M_{at} and density ρ of Si are 28.09 (g mole^{-1}) and 2.33 g cm^{-3}, respectively. | **Example 1.10**

Solution

The number of atoms per unit volume is

$$N = \frac{(N_A \rho)}{M_{at}} = \frac{(6.022 \times 10^{23})(2.33)}{28.09} = 5.0 \times 10^{22} \text{ cm}^{-3}$$

Equation 1.23 gives the concentration of vacancies. For Si at 300 K (27 °C),

$$n_v = N \exp\left(\frac{-E_v}{kT}\right) = (5 \times 10^{22} \text{ cm}^{-3}) \exp\left[\frac{-2.4 \times 1.6 \times 10^{-19}}{(1.38 \times 10^{-23})(300)}\right]$$

$$= 2.6 \times 10^{-18} \text{ cm}^{-3}$$

For Si at 1273 K (1000 °C),

$$n_v = N \exp\left(\frac{-E_v}{kT}\right) = (5 \times 10^{22}) \exp\left[\frac{-2.4 \times 1.6 \times 10^{-19}}{(1.38 \times 10^{-23})(1273)}\right]$$

$$= 1.6 \times 10^{13} \text{ cm}^{-3}$$

Clearly, there is a dramatic increase, by a factor of 10^{31}, in the vacancy concentration from 27 °C to 1000 °C.

1.7.5   Stoichiometry, Nonstoichiometry, and Defect Structures

Stoichiometric compounds are those that have an integer ratio of atoms, for example, as in CaF_2 where two F atoms bond with one Ca atom. Similarly, in the compound ZnO, if there is one O atom for every Zn atom, the compound is stoichiometric, as schematically illustrated in Figure 1.54a. Since there are equal numbers of O^{2-} anions and Zn^{2+} cations, the crystal overall is neutral. It is also possible to have a nonstoichiometric ZnO in which there is excess zinc. This may result if, for example, there is insufficient oxygen during the preparation of the compound. The Zn^{2+} ion has a radius of 0.074 nm, which is about 1.9 times smaller than the O^{2-} anion (radius of 0.14 nm), so it is much easier for a Zn^{2+} ion to enter an interstitial site than the O^{2-} ion or the Zn atom itself, which has a radius of

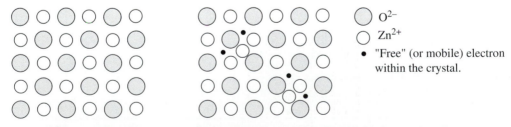

(a) Stoichiometric ZnO crystal with equal number of anions and cations and no free electrons.

(b) Nonstoichiometric ZnO crystal with excess Zn in interstitial sites as Zn^{2+} cations.

◯ O^{2-}

◯ Zn^{2+}

● "Free" (or mobile) electron within the crystal.

Figure 1.54   Stoichiometry and nonstoichiometry and the resulting defect structure.

0.133 nm. Excess Zn atoms therefore occupy interstitial sites as Zn^{2+} cations. Even though the excess zinc atoms are still ionized within the crystal, their lost electrons cannot be taken by oxygen atoms, which are all O^{2-} anions, as indicated in Figure 1.54b. Thus, the nonstoichiometric ZnO with excess Zn has Zn^{2+} cations in interstitial sites and mobile electrons within the crystal, which can contribute to the conduction of electricity. Overall, the crystal is neutral, as the number of Zn^{2+} ions is equal to the number of O^{2-} ions plus two electrons from each excess Zn. The structure shown in Figure 1.54b is a defect structure, since it deviates from the stoichiometry.

1.8   SINGLE-CRYSTAL CZOCHRALSKI GROWTH

The fabrication of discrete and integrated-circuit (IC) solid-state devices requires semiconductor crystals with impurity concentrations as low as possible and crystals that contain very few imperfections. A number of laboratory techniques are available for growing high-purity semiconductor crystals. Generally, they involve either solidification from the melt or condensation of atoms from the vapor phase. The initial process in IC fabrication requires large single-crystal wafers that are typically 15 cm in diameter and 0.6 mm thick. These wafers are cut from a long, cylindrical single Si crystal (typically, 2 meters in length).

Large, single, Si crystals for IC fabrication are often grown by the **Czochralski method,** which involves growing a single crystal ingot from the melt, using solidification on a seed crystal, as schematically illustrated in Figure 1.55. Molten Si

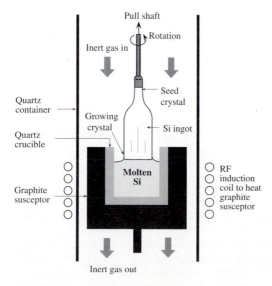

Figure 1.55   Schematic illustration of the growth of a single crystal Si ingot by the Czochralski technique.

is held in a quartz (crystalline SiO_2) crucible in a graphite susceptor, which is heated by a radio frequency induction coil (a process called **RF heating**).[9] A small dislocation-free crystal, called a **seed,** is lowered to touch the melt and then slowly pulled out of the melt; a crystal grows by solidifying on the seed crystal. The seed is rotated during the pulling stage, to obtain a cylindrical ingot. To suppress evaporation from the melt and prevent oxidation, argon gas is passed through the system.

Initially, as the crystal is withdrawn, its cross-sectional area increases; it then reaches a constant value determined by the temperature gradients, heat losses, and the rate of pull. As the melt solidifies on the crystal, heat of fusion is released and must be conducted away; otherwise, it will raise the temperature of the crystal and remelt it. The area of the melt–crystal interface determines the rate at which this heat can be conducted away through the crystal, whereas the rate of pull determines the rate at which latent heat is released. Although the analysis is not a simple one, it is clear that to obtain an ingot with a large cross-sectional area, the pull speed must be slow. Typical growth rates are a few millimeters per minute.

The sizes and diameters of crystals grown by the Czochralski method are obviously limited by the equipment, though crystals 15–20 cm in diameter and 1–2 meters in length are routinely grown for the IC fabrication industry. Also, the crystal orientation of the seed and its flatness with melt surface are important engineering requirements. For example, for very large scale integration (VLSI), the seed is placed with its (100) plane flat to the melt, so that the axis of the cylindrical ingot is along the [100] direction.

Following growth, the Si ingot is usually ground to a specified diameter. Using x-ray diffraction, the crystal orientation is identified and either a flat or an edge is ground along the ingot, as shown in Figure 1.56. Subsequently, the ingot is cut into

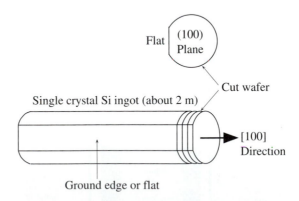

Figure 1.56   The crystallographic orientation of the silicon ingot is marked by grounding a flat.

The ingot can be as long as 2 m. Wafers are cut using a rotating annular diamond saw. Typical wafer thickness is 0.6–0.7mm.

| [9] The induced eddy currents in the graphite give rise to I^2R heating of the graphite susceptor.

thin wafers by a rotating annular diamond saw. To remove any damage to the wafer surfaces caused by sawing and obtain flat, parallel surfaces, the wafers are lapped (ground flat with alumina powder and glycerine), chemically etched, and then polished. The wafers are then used in IC fabrication, usually as a substrate for the growth of a thin layer of crystal from the vapor phase.

The Czochralski technique is also used for growing Ge, GaAs, and InP single crystals, through each case has its own particular requirements. The main drawback of the Czochralski technique is that the final Si crystal inevitably contains oxygen impurities dissolved from the quartz crucible.

1.9   GLASSES AND AMORPHOUS SEMICONDUCTORS

1.9.1   Glasses and Amorphous Solids

A characteristic property of the crystal structure is its periodicity and degree of symmetry. For each atom, the number of neighbors and their exact orientations are well defined; otherwise, the periodicity would be lost. There is therefore a **long-range order** resulting from strict adherence to a well-defined bond length and **relative bond angle** (or exact orientation of neighbors). Figure 1.57a schematically illustrates the presence of a clear, long-range order in a hypothetical two-dimensional crystal. Taking an arbitrary origin, we can predict the position of each atom anywhere in the crystal. We can perhaps use this to represent crystalline SiO_2 (silicon dioxide), for example, in two dimensions. In reality, an Si atom bonds with

● Silicon (or Arsenic) atom   ○ Oxygen (or Selenium) atom

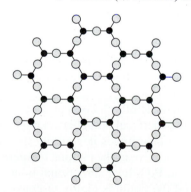

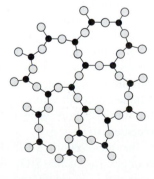

(a) A crystalline solid reminiscent of crystalline SiO_2. (Density = 2.6 g cm^{-3})

(b) An amorphous solid reminiscent of vitreous silica (SiO_2) cooled from the melt. (Density = 2.2 gcm^{-3})

Figure 1.57   Crystalline and amorphous structures illustrated schematically in two dimensions.

four oxygen atoms to form a tetrahedron, and the tetrahedra are linked at the corners to create a three-dimensional crystal structure.

Not all solids exhibit crystallinity. Many substances exist in a noncrystalline or amorphous form, due to their method of formation. For example, SiO_2 can have an amorphous structure, as illustrated schematically in two dimensions in Figure 1.57b. In the amorphous phase, SiO_2 is called **vitreous silica,** a form of glass, which has wide engineering applications, including optical fibers. The structure shown in the figure for vitreous silica is essentially that of a frozen liquid, or a **supercooled liquid.** Vitreous silica is indeed readily obtained by cooling the melt.

Many amorphous solids are formed by rapidly cooling or quenching the liquid to temperatures where the atomic motions are so sluggish that crystallization is virtually halted. (The cooling rate is measured relative to the crystallization rate, which depends on atomic diffusion.) We refer to these solids as **glasses.** In the liquid state, the atoms have sufficient kinetic energy to break and make bonds frequently and to bend and twist their bonds. There are bond angle variations, as well as rotations of various atoms around bonds (**bond twisting).** Thus, the bonding geometry around each atom is not necessarily identical to that of other atoms, which leads to the loss of long-range order and the formation of an amorphous structure, as illustrated in Figure 1.57b for the same material in Figure 1.57a. We may view Figure 1.57b as a snapshot of the structure of a liquid. As we move away from a reference atom, after the first and perhaps the second neighbors, random bending and twisting of the bonds is sufficient to destroy long-range order. The amorphous structure therefore lacks the long-range order of the crystalline state.

To reach the glassy state, the temperature is rapidly dropped well below the melting temperature where the atomic diffusion processes needed for arranging the atoms into a crystalline structure are infinitely slow on the time scale of the observation. The liquid structure thus becomes frozen. Figure 1.57b shows that for an amorphous structure, the coordination of each atom is well defined, because each atom must satisfy its chemical bonding requirement, but the whole structure lacks long-range order. Therefore, there is only a **short-range order** in an amorphous solid. The structure is a continuous random network of atoms (often called a CRN model of an amorphous solid). As a consequence of the lack of long-range order, amorphous materials do not possess such crystalline imperfections as grain boundaries and dislocations, which is a distinct advantage in certain engineering applications.

Whether a liquid forms a glass or a crystal structure on cooling depends on a combination of factors, such as the nature of the chemical bond between the atoms or molecules, the viscosity of the liquid (which determines how easily the atoms move), the rate of cooling, and the temperature relative to the melting temperature. For example, the oxides SiO_2, B_2O_3, GeO_2, and P_2O_5 have directional bonds that are a mixture of covalent and ionic bonds and the liquid is highly viscous. These oxides readily form glasses on cooling from the melt. On the other hand, it is virtually impossible to quench a pure metal, such as copper, from the melt, bypass crystallization, and form a glass. The metallic bonding is due to an electron gas permeating the space between the copper ions, and that bonding is nondirectional, which means that on cooling, copper ions are readily (and hence, quickly) shifted with respect to each other to form the crystal. There are, however, a number of metal–metal ($Cu_{66}Zr_{33}$) and metal–metalloid alloys ($Fe_{80}B_{20}$, $Pd_{80}Si_{20}$) that form glasses if quenched at ultrahigh cooling rates of 10^6–10^8 °C per second. In practice, such cooling rates are achieved by squirting a thin jet of the molten metal against a fast-rotating, cooled copper cylinder. On impact, the melt is frozen within a few milliseconds, producing a long ribbon of metallic glass. The process is known as **melt spinning** and is depicted in Figure 1.58.

Many solids used in various applications have an amorphous structure. The ordinary window glass, $(SiO_2)_{0.8}(Na_2O)_{0.2}$, and the majority of glassware are common examples. Vitreous silica (SiO_2) mixed with germania (GeO_2) is used extensively in optical fibers. The insulating oxide layer grown on the Si wafer during IC fabrication is the amorphous form of SiO_2. Some intermetallic alloys, such as $Fe_{0.8}B_{0.2}$, can be rapidly quenched from the liquid (as shown in Figure 1.58) to obtain a glassy metal used in low-loss transformer cores. Arsenic triselenide, As_2Se_3, has a crystal structure that resembles the two-dimensional sketch in Figure 1.57a, where an As atom (valency III) bonds with three Se atoms, and an Se atom (valency VI) bonds with two As atoms. In the amorphous phase, this crystal structure looks like the sketch in Figure 1.57b, in which the bonding requirements are only locally satisfied. The crystal can be prepared by condensation from the vapor phase, or by cooling the melt. The vapor-grown films of amorphous As_2Se_3 are used in some photoconductor drums in the photocopying industry.

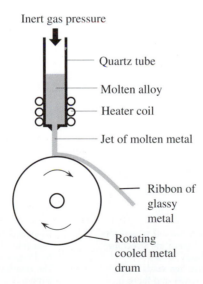

Inert gas pressure

Quartz tube

Molten alloy

Heater coil

Jet of molten metal

Ribbon of glassy metal

Rotating cooled metal drum

Figure 1.58   It is possible to rapidly quench a molten metallic alloy, thereby bypassing crystallization, and forming a glassy metal commonly called a metallic glass. The process is called *melt spinning*.

1.9.2   Crystalline and Amorphous Silicon

A silicon atom in the silicon crystal forms four tetrahedrally oriented, covalent bonds with four neighbors, and the repetition of this exact bonding geometry with a well-defined bond length and angle leads to the diamond structure shown in Figure 1.5. A simplified two-dimensional sketch of the Si crystal is shown in Figure 1.59. The crystal has a clear long-range order. Single crystals of Si are commercially grown by the Czochralski crystal pulling technique.

It is also possible to grow amorphous silicon, denoted a-Si, by the condensation of Si vapor onto a solid surface, called a substrate. For example, an electron beam is used to vaporize a silicon target in vacuum; the Si vapor then condenses on a metallic substrate to form a thin layer of solid noncrystalline silicon. The technique, which is schematically depicted in Figure 1.60, is referred to as **electron beam deposition.** The structure of amorphous Si (a-Si) lacks the long-range order of crystalline Si (c-Si), even though each Si atom in a-Si, on average, prefers to bond with four neighbors. The difference is that the relative angles between the Si−Si bonds in a-Si deviate considerably from those in the crystal, which obey a strict geometry. Therefore, as we move away from a reference atom in a-Si, eventually the periodicity for generating the crystalline structure is totally lost, as illustrated schematically in Figure 1.59. Furthermore, because the Si−Si bonds do not follow the equilibrium geometry, the bonds are strained and some are even missing, simply because the formation of a bond causes substantial bond bending. Consequently, the a-Si structure has many voids and incomplete bonds, or **dangling bonds,** as schematically depicted in Figure 1.59.

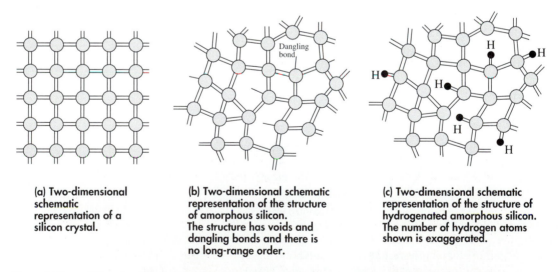

(a) Two-dimensional schematic representation of a silicon crystal.

(b) Two-dimensional schematic representation of the structure of amorphous silicon. The structure has voids and dangling bonds and there is no long-range order.

(c) Two-dimensional schematic representation of the structure of hydrogenated amorphous silicon. The number of hydrogen atoms shown is exaggerated.

Figure 1.59   Silicon can be grown as a semiconductor crystal or as an amorphous semiconductor film. Each line represents an electron in a bond. A full covalent bond has two lines and a broken bond has one line.

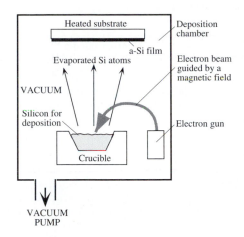

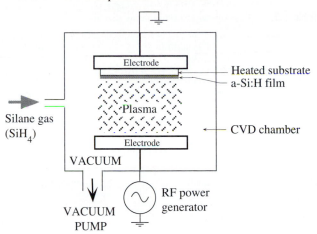

Figure 1.60   Amorphous silicon, a-Si, can be prepared by an electron beam evaporation of silicon.

Silicion has a high melting temperature so that an energetic electron beam is used to melt the crystal in the crucible locally and thereby vaporize Si atoms. Si atoms condense on a substrate placed above the crucible, to form a film of a-Si.

Figure 1.61   Hydrogenated amorphous silicon, a-Si:H, is generally prepared by the decomposition of silane molecules in a radio frequency (RF) plasma discharge.

Si and H atoms condense on a substrate to form a film of a-Si:H.

One way to reduce the density of dangling bonds is simply to terminate a dangling bond using hydrogen. Since hydrogen only has one electron, it can attach itself to a dangling bond, that is, passivate the dangling bond. The structure resulting from hydrogen in amorphous silicon is called **hydrogenated amorphous Si (a-Si:H).**

Many electronic devices, such as a-Si:H solar cells, are based on a-Si being deposited with H to obtain a-Si:H, in which the hydrogen concentration is typically 10 at.% (atomic %). The process involves the decomposition of silane gas, SiH_4, in an electrical plasma in a vacuum chamber. Called **plasma-enhanced chemical vapor deposition (PECVD),** the process is illustrated schematically in Figure 1.61. The silane gas molecules are dissociated in the plasma, and the Si and H atoms then condense onto a substrate to form a film of a-Si:H. If the substrate temperature is too hot, the atoms on the substrate surface will have sufficient kinetic energy, and hence the atomic mobility, to orient themselves to form a polycrystalline structure. Typically, the substrate temperature is ~250 °C. The advantage of a-Si:H is that it can be grown on large areas, for such applications as photovoltaic cells and the photoconductor drums used in some photocopying machines. Table 1.5 summarizes the properties of crystalline and amorphous silicon, in terms of structure and applications.

Table 1.5   Crystalline and amorphous silicon

	Crystalline Si (c-Si)	Amorphous Si (a-Si)	Hydrogenated a-Si (a-Si:H)
Structure	Diamond cubic.	Short-range order only. On average, each Si covalently bonds with four Si atoms. Has microvoids and dangling bonds.	Short-range order only. Structure typically contains 10% H. Hydrogen atoms passivate dangling bonds and relieve strain from bonds.
Typical preparation	Czochralski technique.	Electron beam evaporation of Si.	Chemical vapor deposition of silane gas by RF plasma.
Density g cm^{-3}	2.33	About 3–10% less dense.	About 1–3% less dense.
Electronic applications	Discrete and integrated electronic devices.	None	Large-area electronic devices such as solar cells and the photoconduc drums used in photocopying.

1.10   SOLID SOLUTIONS AND TWO-PHASE SOLIDS

1.10.1   Isomorphous Solid Solutions: Isomorphous Alloys

A phase of a material has the same composition, structure, and properties everywhere, so it is a homogeneous portion of the chemical system under consideration. In a given chemical system, one phase may be in contact with another phase. For example, at 0 °C, iced water will have solid and liquid phases in contact. Each phase, ice and water, has a distinct structure.

A bartender knows that alcohol and water are totally miscible; he can dilute whisky with as much water as he likes. When the two liquids are mixed, the molecules are randomly mixed with each other and the whole liquid is a homogenous mixture of the molecules. The liquid therefore has one phase; the properties of the liquid are the same everywhere. The same is not true when we try to mix water and oil. The mixture consists of two distinctly separate phases, oil and water, in contact. Each phase has a different composition, even though both are liquids.

Many solids are a homogeneous mixture of two types of separate atoms. For example, when nickel atoms are added to copper, Ni atoms substitute directly for the Cu atoms, and the resulting solid is a **solid solution,** as depicted in Figure 1.62a. The structure remains an FCC crystal whatever the amount of Ni we add, from 100% Cu to 100% Ni. The solid is a homogenous mixture of Cu and Ni atoms, with the same structure everywhere in the solid solution, which is called an **isomorphous solid solution.** The atoms in the majority make up the solvent, whereas the atoms in the minority are the solute, which is dissolved in the solvent. For a Cu–Ni alloy with an Ni content of less than 50 at.%, copper is the solvent and nickel is the solute.

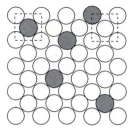

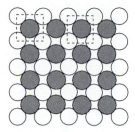

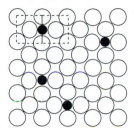

(a) Disordered substitutional solid solution. Example: Cu–Ni alloys ({100} planes)

(b) Ordered substitutional solid solution. Example: Cu–Zn alloy of composition 50% Cu–50% Zn. ({110} planes)

(c) Interstitial solid solution. Example: Small number of C atoms in FCC Fe (austenite). ({100} planes)

Figure 1.62   Solid solutions can be disordered substitutional, ordered substitutional, and interstitial substitutional. Only one phase within the alloy has the same composition, structure, and properties everywhere.

The substitution of solute atoms for solvent atoms at various lattice sites of the solvent can be either random (disordered) or ordered. The two cases are schematically illustrated in Figure 1.62a and b, respectively. In many solid solutions, the substitution is random, but for certain compositions, the substitution becomes ordered. There is a distinct ordering of atoms around each solute atom such that the crystal structure resembles that of a compound. For example, β' brass has the composition 50 at.% Cu–50 at.% Zn. Each Zn atom is surrounded by eight Cu atoms and vice versa, as depicted in two dimensions in Figure 1.62b. The structure is that of a metallic compound between Cu and Zn.

Another type of solid solution is the **interstitial solid solution,** in which solute atoms occupy interstitial sites, or voids between atoms, in the crystal. Figure 1.62c shows an example in which a small number of carbon atoms have been dissolved in a γ-iron crystal (FCC) at high temperatures.

1.10.2   Phase Diagrams: Cu–Ni and Other Isomorphous Alloys

The Cu–Ni alloy is isomorphous. Unlike pure copper or pure nickel, when a Cu–Ni alloy melts, its melting temperature is not well defined. The alloy melts over a range of temperatures in which both the liquid and the solid coexist as a heterogeneous mixture. It is therefore instructive to know the phases that exist in a chemical system at various temperatures as a function of composition, and this need leads to the use of phase diagrams.

Suppose we take a crucible of molten copper and allow it to cool. Above its melting temperature (1083 °C), there is only the liquid phase. The temperature drops with time, as shown in Figure 1.63a, until at the melting or fusion temperature at point L_0 when copper crystals begin to **nucleate** (solidify) in the crucible.

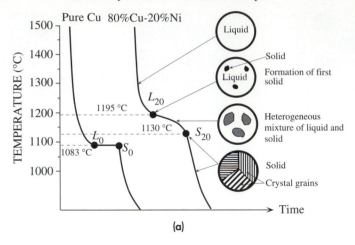

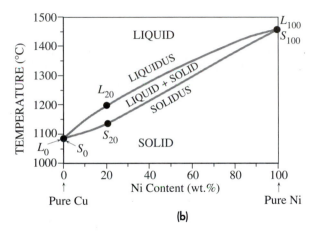

Figure 1.63   Solidification of an isomorphous alloy such as Cu–Ni.
(a) Typical cooling curves.
(b) The phase diagram marking the regions of existence for the phases.

During solidification, the temperature remains constant. As long as we have both the liquid and solid phases coexisting, the temperature remains constant at 1083 °C. During this time, heat is given off as the Cu atoms in the melt attach themselves to the Cu crystals. This heat is called the **heat of fusion.** Once all the liquid has solidified, (point S_0), the temperature begins to drop as the solid cools. There is therefore a sharp melting temperature for copper, at 1083 °C.

If we were to cool pure nickel from its melt, we would observe a behavior similar to that of pure copper, with a well-defined melting temperature at 1453 °C.

Now suppose we cool the melt of a Cu–Ni alloy with a composition[10] of 80 wt.% Cu and 20 wt.% Ni. In the melt, the two species of atoms are totally

---

| [10] In Materials Science, we generally prefer to give alloy compositions in wt.%, which henceforth will simply be %.

miscible, and there is only a single liquid phase. As the cooling proceeds, we reach the temperature 1195 °C, identified as point L_{20} in Figure 1.63a, where the first crystals of Cu–Ni alloy begin to appear. In this case, however, the temperature does not remain constant until the liquid is solidified, but continues to drop. Thus, there is no single melting temperature, but a range of temperatures over which both the liquid and the solid phases coexist in a heterogeneous mixture. We find that when the temperature reaches 1130 °C, corresponding to point S_{20}, all the liquid has solidified. Below 1130 °C, we have a single-phase solid that is an isomorphous solid solution of Cu and Ni. If we repeat these experiments for other compositions, we find a similar behavior, that is, freezing occurs over a transition temperature range. The beginning and end of solidification, at points L and S, respectively, depend on the specific composition of the alloy.

To characterize the freezing or melting behavior of other compositions of Cu–Ni alloys, we can plot the temperatures for the beginning and end of so-lidification versus the composition and identify those temperature regions where various phases exist, as shown in Figure 1.63b. When we join all the points corresponding to the beginning of freezing, that is, all the L points, we obtain what is called the **liquidus curve.** For any given composition, only the liquid phase can exist above the liquidus curve. If we join all the points where the liquid has totally solidified, that is, all the S points, we have a curve called the **solidus curve.** At any temperature and composition below the solidus curve, we can only have the solid phase. The region between the liquidus and solidus curves marks where a heteroge-neous mixture of liquid and solid phases exists.

Let's follow the cooling behavior of the 80% Cu–20% Ni alloy from the melt at 1300 °C down to the solid state at 1000 °C, as shown in Figure 1.64. The vertical

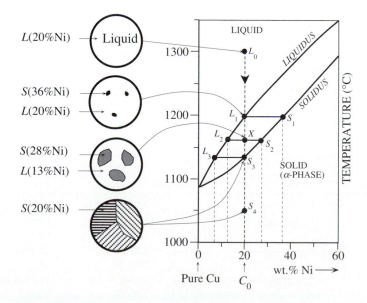

Figure 1.64   Cooling of a 80% Cu–20% Ni alloy from the melt to the solid state.

dashed line at 20% Ni represents the overall composition of the alloy (the whole chemical system) and the cooling process corresponds to movement down this dashed line, starting from the liquid phase at L_0.

When the Cu–Ni alloy begins to solidify at 1195 °C, at point L_1, the first solid that forms is richer in Ni content. The only solid that can exist at this temperature has a composition S_1, which has a greater Ni content than the liquid, as shown in Figure 1.64. Intuitively, we can see this by noting that Cu, the component with the lower melting temperature, prefers to remain in the liquid, whereas Ni, which has a higher melting temperature, prefers to remain in the solid. When the temperature drops further, say to 1160 °C (indicated by X in the figure), the alloy is a heterogeneous mixture of liquid and solid. At this temperature, the only solid that can coexist with the liquid has a composition S_2. The liquid has the composition L_2. Since the liquid has lost some of its Ni atoms, the liquid composition is less than that at L_1. The liquidus and solidus curves therefore give the compositions of the liquid and solid phases coexisting in the heterogeneous mixture during melting.

At 1160 °C, the overall composition of the alloy (the whole chemical system) is still 20% Ni and is represented by point X in the phase diagram. When the temperature reaches 1130 °C, nearly all the liquid has been solidified. The solid has the composition S_3, which is 20% Ni, as we expect since the whole alloy is almost all solid. The last drops of the liquid in the alloy have the composition L_3, since at this temperature, only the liquid with this composition can coexist with the solid at S_3. Table 1.6 summarizes the phases and their compositions, as observed during the cooling process depicted in Figure 1.64. By convention, all solid phases that can exist are labelled by different Greek letters. Since we can only have one solid phase, this is labelled the α-phase.

During the solidification process depicted in Figure 1.64, the solid composition changes from S_1 to S_2 to S_3. We tacitly assume that the cooling is sufficiently slow to allow time for atomic diffusion to change the composition of the whole solid. Therefore, the phase diagram in Figure 1.63, which assumes near equilibrium conditions during cooling, is termed an **equilibrium phase diagram.** If the cooling is fast, there will be limited time for atomic diffusion in the solid phase, and the resulting solid will have a composition variation. The inner core will correspond to the solidification at S_1 and the outer region will correspond to that at S_3. The solid structure will be **cored,** as depicted in Figure 1.65. The cooling process is then said

Table 1.6   Phase in the 80% Cu–20% Ni isomorphous alloy

Temperature	Phases	Composition (wt.%)	Amount (wt.%)
1300 °C	Liquid only	$L_0 = 20\%$ Ni	100%
1195 °C	Liquid and solid	$L_1 = 20\%$ Ni	100%
		$S_1 = 36\%$ Ni	First solid appears
1160 °C	Liquid and solid	$L_2 = 13\%$ Ni	53.3%
		$S_2 = 28\%$ Ni	46.7%
1130 °C	Liquid and solid	$L_3 = 7\%$ Ni	The last liquid drop
		$S_3 = 20\%$ Ni	100%
1050 °C	Solid only	$S_4 = 20\%$ Ni	100%

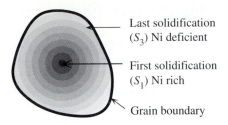

Figure 1.65   Segregation in a grain due to rapid cooling (nonequilibrium cooling).

to have occurred under nonequilibrium conditions, which leads to a segregation of the elements in the grains. Under nonequilibrium cooling conditions we cannot quantitatively use the equilibrium phase diagram in Figure 1.63. The diagram can only serve as a qualitative guide.

The amounts of liquid and solid in the mixture can be determined from the phase diagram using the **lever rule,** which is based on the fact that the total mass of the alloy remains the same throughout the entire cooling process. Let W_L and W_S be the **weight (or mass) fraction** of the liquid and solid phases in the alloy mixture. The compositions of the liquid and solid are denoted as C_L and C_S, respectively. The overall composition of the alloy is denoted C_O, which is the overall weight fraction of Ni in the alloy.

If we take the alloy to have a weight of unity, then the conservation of mass means that

$$W_L + W_S = 1$$

Further, the weight fraction of Ni in both the liquid and solid must add up to the composition C_O of Ni in the whole alloy, or

$$C_L W_L + C_S W_S = C_O$$

We can substitute for W_S in the above equation to find the weight fraction of the liquid and then that of the solid phase, as follows:

$$W_L = \frac{C_S - C_O}{C_S - C_L} \quad \text{and} \quad W_S = \frac{C_O - C_L}{C_S - C_L} \qquad \text{[1.24]} \quad \textit{Lever rules}$$

To apply Equation 1.24, we first draw a line (called a **tie line**) from L_2 to S_2 corresponding to C_L and C_S, as shown in Figure 1.64. The line represents a "horizontal lever" and point X at C_O at this temperature is the lever's fulcrum. The lengths of the lever arms from the fulcrum to the liquidus and solidus curves are $(C_O - C_L)$ and $(C_S - C_O)$, respectively. The lever must be balanced by the weights W_L and W_S attached to the ends. The total length of lever is $(C_S - C_L)$. At 1160 °C, $C_L = 0.13$ (13% Ni)   and $C_S = 0.28$ (28% Ni), so the weight fraction of the liquid phase is

$$W_L = \frac{C_S - C_O}{C_S - C_L} = \frac{0.28 - 0.20}{0.28 - 0.13} = 0.533 \quad \text{or} \quad 53.3\%$$

Similarly, the weight fraction of the solid phase is $1 - 0.533$ or 0.467.

1.10.3   Zone Refining and Pure Silicon Crystals

Zone refining is used for the production of high-purity crystals. Silicon, for example, has a high melting temperature, so any impurities present in the crystal decrease the melting temperature. This is similar to the depression of the melting temperature of pure Ni by the addition of Cu, as shown by the right-hand side of Figure 1.63. We can represent the phase diagram of Si with small impurities as shown in Figure 1.66. Consider what happens if we have a rod of the solid and we melt only the left end by applying heat locally (using RF heating, for example). At the same time, we move the melted zone toward the right by moving the heater. We therefore melt the solid at A and refreeze it at B, as shown in Figure 1.67a.

The solid has an impurity concentration of C_O; when it melts at A, the melt initially also has the same concentration $C_L = C_O$. However, at temperature T_B, the melt begins to solidify. At the start of solidification the solid that freezes has a composition C_B, which is considerably less than C_O, as is apparent in Figure 1.66. The cooling at B occurs rapidly, so the concentration C_B cannot adjust to the equilibrium value at the end of freezing. Thus, the solid that freezes at B has a lower concentration of impurities. The impurities have been pushed out of the solid at B and into the melt, whose impurity concentration increases from C_L to $C_{L'}$.

Next, refreezing at B', shown in Figure 1.67b, occurs at a lower temperature $T_{B'}$, because the melt concentration $C_{L'}$ is now greater than C_O. The solid that freezes at B' has the concentration $C_{B'}$, shown in Figure 1.66, which is greater than C_B but less than C_O. As the melted zone is floated toward the right, the melt that is solidified at B, B', etc., has a higher and higher impurity concentration, until its impurity content reaches that of the impure solid, at which point the concentration remains at C_O. When the melted zone approaches the far right where the freezing is halted, the impurities in the final melt appear in the last frozen region at the far right. The resulting impurity concentration profile is schematically depicted in Figure 1.67c. The region of impurity concentration below C_O is the **zone refined** section of the rod. The zone refining procedure can be repeated again, starting

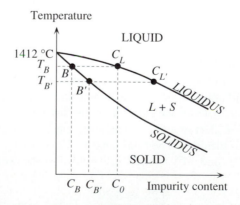

Figure 1.66   The phase diagram of Si with impurities near the low-concentration region.

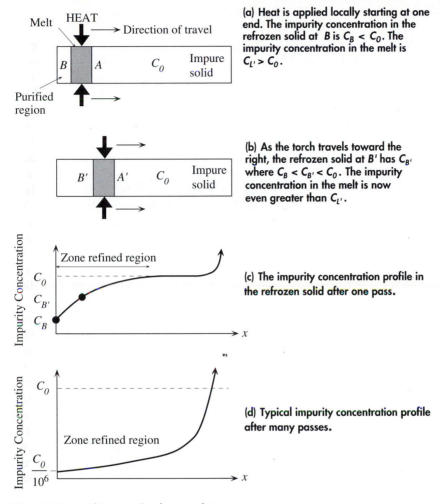

(a) Heat is applied locally starting at one end. The impurity concentration in the refrozen solid at B is $C_B < C_0$. The impurity concentration in the melt is $C_{L'} > C_0$.

(b) As the torch travels toward the right, the refrozen solid at B' has $C_{B'}$ where $C_B < C_{B'} < C_0$. The impurity concentration in the melt is now even greater than $C_{L'}$.

(c) The impurity concentration profile in the refrozen solid after one pass.

(d) Typical impurity concentration profile after many passes.

Figure 1.67 The principle of zone refining.

from the left toward the right, to reduce the impurity concentration even lower. The impurity concentration profile after many passes is sketched in Figure 1.67d. Although the profile is nonuniform, due to the segregation effect, the impurity concentrations in the zone refined section may be as low as a factor of 10^{-6}.

Zone refining, specifically the floating zone technique, is used fruitfully to grow ultrahigh-purity Si crystals, as shown in Figure 1.68. A vertical polycrystalline Si rod is held in contact with a small single crystal (a seed) of Si at the bottom. The rod is melted locally by a narrow coil heater, starting from the seed end and moving upward. The melted zone resolidifies on the seed and therefore grows a new single crystal on this seed. As the molten zone is slowly moved up, more of the polycrystalline rod is melted and then resolidified. The single crystal grows as the zone is floated up. At the end of the crystal growth process, the heater can be run from bottom to top several times to achieve zone refining. In the

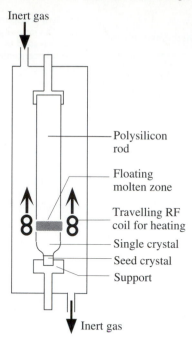

Inert gas

Polysilicon rod

Floating molten zone

Travelling RF coil for heating

Single crystal

Seed crystal

Support

Inert gas

Figure 1.68   The principle of the floating zone crystal growth technique.

Czochralski technique, the molten Si is held in a quartz crucible, which inevitably allows oxygen from the quartz to be dissolved as impurities into the melt, whereas crystallization in the floating zone process does not suffer from this drawback. Further, the floating zone technique achieves even lower impurity concentrations in the grown crystal.

1.10.4   Binary Eutectic Phase Diagrams and Pb–Sn Solders

When we dissolve salt in water, we obtain a brine solution. If we continue to add more salt, we eventually reach the solubility limit of salt in the solution, and the excess salt remains as a solid at the bottom of the container. We then have two coexisting phases: brine (liquid solution) and salt (solid), as shown in Figure 1.69. The solubility limit of the one component in another in a mixture is represented by a **solvus curve** shown schematically in Figure 1.69 for salt in brine. In the solid state, there are many elements that can only be dissolved in small amounts in another solid.

Lead in the solid phase has an FCC crystal structure, and tin has a BCT[11] structure. Although the two elements are totally miscible in any proportion when melted, this is not so in the solid state. We can only dissolve so much Sn in solid

[11] Body-centered tetragonal.

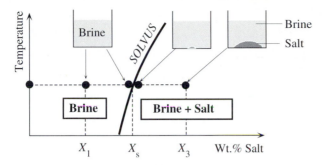

Figure 1.69   We can only dissolve so much salt in brine (solution of salt in water).

Eventually, we reach the solubility limit at X_S, which depends on the temperature. If we add more salt, the excess salt does not dissolve; it coexists with the brine. Past X_S we have two phases, brine (solution) and salt (solid).

Pb, and vice versa. We quickly reach the solubility limit, and the resulting solid is a mixture of two distinctly different solid phases. One solid phase, labeled α, is Pb rich and has the FCC structure with some Sn atoms dissolved in the crystal. The amount of Sn dissolved in α is given by the solvus curve of Sn in α at that temperature. The other phase, labeled β, is Sn rich and has the BCT structure with some Pb atoms dissolved in it. The amount of Sn dissolved in β is given by the solvus curve of Pb in β at that temperature.

The existence of various phases and their compositions as a function of temperature are given by the equilibrium phase diagram for the Pb–Sn alloy, shown in Figure 1.70. This is called an equilibrium **eutectic phase diagram.** The liquidus and solidus curves, as usual, mark the borders for the liquid and solid phases. Between the liquidus and solidus curves, we have a heterogeneous mixture of melt and solid. Unlike the Cu–Ni case, the melting temperature of both elements here is depressed with alloying. The liquidus and solidus curves thus decrease from both ends, starting at A and B. They meet at a point E, called the **eutectic point,** at 61.9% Sn and 183 °C. This point has a special significance: No liquid can exist below this temperature, so 183 °C is the lowest melting temperature of the alloy.

In addition, we must insert the solvus curves at both the Pb and Sn ends to mark the extent of solid state solubility and hence identify the two-phase solid region. The solvus curve for the solubility limit of Sn in Pb meets the solidus curve at point C, 19% Sn. Similarly, the solubility limit of Pb in Sn meets the solvus curve at D. A characteristic feature of this phase diagram is that CD is a straight line through E at 183 °C. Below 183 °C, between the two solvus curves, we have a solid with two phases, α and β. This is identified as $(\alpha + \beta)$ in the diagram.

The usefulness of such a phase diagram is best understood by examining the phase transformations and microstructures during the cooling of a melt of a given composition alloy. Consider a 90% Pb–10% Sn alloy being cooled from the melt at 350 °C (point L) where there is only one phase, the liquid phase. At point M, 315 °C, few nuclei of the α-phase appear in the liquid. The composition of the α-phase is given by the solidus curve at 350 °C and is about 5% Sn. At point N,

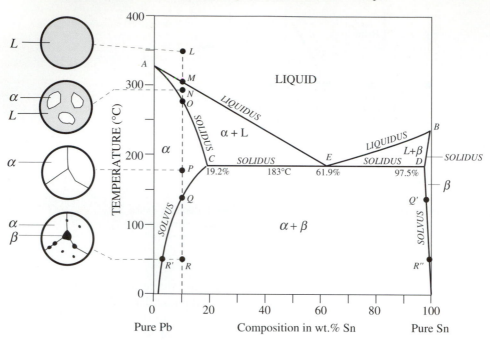

Figure 1.70   The equilibrium phase diagram of the Pb–Sn alloy.

The microstructures on the left show the observations at various points during the cooling of a 90% Pb–10% Sn from the melt along the dashed line (the *overall alloy composition* remains constant at 10% Sn).

290 °C, there is more α-phase in the mixture. The compositions of the liquid and α-phases are given respectively by the liquidus and solidus curves at 290 °C. At point O, 275 °C, all liquid has been solidified into the α-phase, which then has the composition 20% Sn.

Between M and O, the alloy is a coexistent mixture of the liquid phase (melt) and the solid α-phase. At point P, 175 °C, we still have only the α-phase. When we reach the solvus curve at point Q, 140 °C, we can no longer keep all the Sn dissolved in the α-phase, as we have reached the solubility limit of Sn in α. Some of the Sn atoms must diffuse out from the α-phase; they do so by forming a second solid phase, which is the β-phase. The β-phase nucleates within the α-phase (usually at the grain boundaries, where atomic diffusion occurs readily). The β-phase will contain as much dissolved Pb as is allowed by the solubility of Pb in the β-phase, which is given by the solvus curve on the Sn side and marked as point Q', about 98% Sn. Thus, the microstructure is now a mixture of the α and β phases.

As cooling proceeds, the two phases continue to coexist, but their relative proportions change. At R, 50 °C, the alloy is a mixture of the α-phase given by R'(4% Sn) and the β-phase given by R''(99% Sn). The relative amounts of α and β phases are given by the lever rule. Figure 1.70 illustrates the microstructure of the 90% Pb–10% Sn alloy as it is cooled.

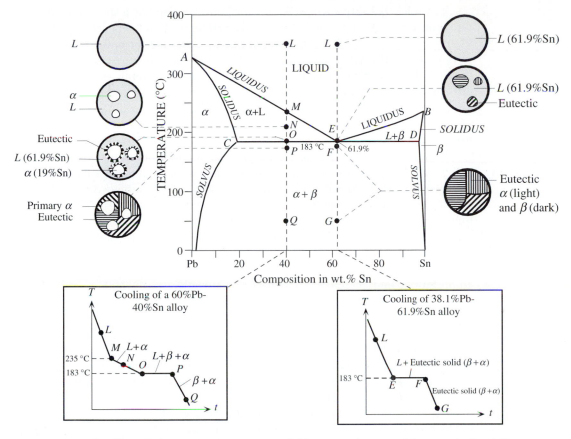

Figure 1.71   The alloy with the eutectic composition cools like a pure element, exhibiting a single solidification temperature at 183 °C. The solid has the special eutectic structure. The alloy with the composition 60% Pb–40% Sn when solidified is a mixture of primary α and eutectic solid.

An interesting phenomenon can be observed when we cool an alloy of the eutectic composition 38.1% Pb–61.9% Sn from the melt. The cooling process and the observed microstructures are illustrated in Figure 1.71; the microstructures are on the right. The temperature–time profile is also depicted in Figure 1.71. At point L, 350 °C, the alloy is all liquid; as it cools, its temperature drops until point E at 183 °C. At E, the temperature remains constant and a solid phase nucleates within the melt. With time, the amount of solid grows until all the liquid is solidified and the temperature begins to drop again. This behavior is much like that of a pure element, for which melting occurs at a well-defined temperature. This behavior only occurs for the eutectic composition (61.9% Sn), because this is the composition at which the liquidus and solidus curves meet at one temperature. Generally, the liquid with the eutectic composition will solidify through the eutectic transformation at the eutectic temperature, or

$$L_{61.9\% \, Sn} \rightarrow \alpha_{19.2\% \, Sn} + \beta_{97.5\% \, Sn} \quad (183 \, °C) \qquad \textbf{[1.25]}$$

The solid that forms from the eutectic solidification has a special microstructure, consisting of alternating plates, or **lamellae,** of α and β phases, as shown in Figure 1.71. This is called the **eutectic microstructure** (or **eutectic solid**). The formation of a Pb-rich α-phase and an Sn-rich β-phase from the 61.9% Sn liquid requires the redistribution of the two types of atoms by atomic diffusion. Atomic diffusions are easier in the liquid than in the solid. The formation of a solid with alternating α and β layers allows the Pb and Sn atoms to diffuse in the liquid without having to move over long distances. The eutectic structure is not a phase itself, but a mixture of the two phases, α and β.

When cooled from the melt, an alloy with a composition between 19.2% Sn and 61.9% Sn solidifies into a mixture of α-phase and a eutectic solid (a mixture of α and β phases). Consider the cooling of an alloy with a composition of 40% Sn, starting from the liquid phase L at 350 °C, as shown in Figure 1.71. At point M, the first solid, the α-phase, nucleates. Its composition is about 15% Sn. At N, 210 °C, the alloy is a mixture of liquid, composition 50% Sn, and α-phase, composition 18% Sn. The composition of the liquid thus moves along the liquidus line from M toward E. At 183 °C, the liquid has the composition 61.9% Sn, or the eutectic composition, and therefore undergoes the eutectic transformation indicated in Equation 1.25. There is still α-phase in the alloy, but its composition is now 19.2% Sn; it does not take part in the eutectic transformation of the liquid. During the eutectic transformation, the temperature remains constant. When all the liquid has been solidified, we have a mixture of the preexisting α-phase, called **primary α** (or **proeutectic α**), and the newly formed eutectic solid. The final microstructure is shown in Figure 1.71 and consists of a primary α and a eutectic solid; therefore, two solid phases, α and β, coexist.

During cooling between points M and O, the alloy 60% Pb–40% Sn is a mixture of melt and α-phase, and it exhibits plastic-like characteristics while solidifying. Further, the temperature range for the solidification is about 183 °C to 235 °C, or about 50 °C. Such an alloy is preferable for such uses as soldering wiped joints to join pipes together, giving the plumber sufficient play for adjusting and wiping the joint. On the other hand, a solder with the eutectic composition (commercially, this is 40% Pb–60% Sn solder, which is close to the eutectic) has the lowest melting temperature and solidifies quickly. The liquid also has good wetting properties. Therefore, 40% Pb–60% Sn is widely used for soldering semiconductor devices, where good wetting and minimal exposure to high temperature are required.

Example 1.11 | **THE 60% Pb–40% Sn ALLOY**   Consider the solidification of the 60% Pb–40% Sn alloy. What are the phases, compositions, and weight fractions of various phases existing in the alloy at 250 °C, 210 °C, 183.5 °C (just above 183 °C), and 182.5 °C (just below 183 °C)?

Solution

We again refer to the phase diagram in Figure 1.71 to identify which phases exist at what temperatures. At 250 °C, we only have the liquid phase. At 210 °C, point N, the liquid and the α-phase are in equilibrium. The composition of the α-phase is given by the solidus line; at 210 °C, $C_\alpha = 18\%$ Sn. The composition of the liquid is given by the liquidus line; at

210 °C, C_L = 50% Sn. To find the weight fraction of α in the alloy, we use the lever rule,

$$W_\alpha = \frac{C_L - C_O}{C_L - C_\alpha} = \frac{50 - 40}{50 - 18} = 0.313$$

From $W_\alpha + W_L = 1$, we obtain the weight fraction of the liquid phase, $W_L = 1 - 0.313$ or 0.687.

At 183.5 °C, point O, the composition of the α-phase is 19.2% Sn corresponding to C and that of the liquid is 61.9% Sn corresponding to E. The liquid therefore has the eutectic composition. The weight fractions are

$$W_\alpha = \frac{C_L - C_O}{C_L - C_\alpha} = \frac{61.9 - 40}{61.9 - 19.2} = 0.513$$

$$W_L = 1 - 0.513 = 0.487$$

As expected, the amount of α-phase increases during solidification; at the same time, its composition changes along the solidus curve. Just above 183 °C, about half the alloy is the solid α-phase and the other half is liquid with the eutectic composition. Thus, on solidification, the liquid undergoes the eutectic transformation and forms the eutectic solid. Just below 183 °C, therefore, the microstructure is the primary α-phase and the eutectic solid. Stated differently, below 183 °C, the α and β phases coexist, and β is in the eutectic structure. The weight fraction of the eutectic phase is the same as that of the liquid just above 183 °C, from which it was formed. The weight fractions of α and β in the whole alloy are given by the lever rule applied at point P, or

$$W_\alpha = \frac{C_\beta - C_O}{C_\beta - C_\alpha} = \frac{97.5 - 40}{97.5 - 19.2} = 0.734$$

$$W_\beta = \frac{C_O - C_\alpha}{C_\beta - C_\alpha} = \frac{40 - 19.2}{97.5 - 19.2} = 0.266$$

The microstructure at room temperature will be much like that just below 183 °C, at which the alloy is a two-phase solid because atomic diffusions in the solid will not be sufficiently fast to allow the compositions to change. Table 1.7 summarizes the phases that exist in this alloy at various temperatures.

Table 1.7   The 60%, Pb–40% Sn alloy.

Temperature (°C)	Phases	Composition	Mass (g)	Microstructure and Comment
250	L	40% Sn	100	
235	L	40% Sn	100	The first solid (α phase) nucleates in the
	α	15% Sn	0	liquid.
210	L	50% Sn	68.7	Mixture of liquid and α phases. More solid
	α	18% Sn	31.3	forms. Compositions change.
183.5	L	61.9% Sn	48.7	Liquid has the eutectic composition.
	α	19.2% Sn	51.3	
182.5	α	19.2% Sn	73.4	Eutectic (α and β phases) and primary α
	β	97.5% Sn	26.6	phase.

| Assume mass of the alloy is 100 grams.

IMPORTANT TERMS

Activated state is the state that occurs temporarily during a transformation or reaction when the reactant atoms or molecules come together to form a particular arrangement (intermediate between reactants and products) that has a higher potential energy than the reactants. The potential energy barrier between the activated state and the reactants is the activation energy.

Activation energy is the potential energy barrier against the formation of a product. In other words, it is the minimum energy that the reactant atom or molecule must have to be able to reach the activated state and hence form a product.

Amorphous solid is a solid that exhibits no crystalline structure or long-range order. It only possesses a short-range order in the sense that the nearest neighbors of an atom are well defined by virtue of chemical bonding requirements.

Anion is an atom that has gained negative charge by virtue of accepting one or more electrons. Usually, atoms of nonmetallic elements can gain electrons easily to become anions. Anions become attracted to the anode (positive terminal) in ionic conduction. Typical anions are the halogen ions F^-, Cl^-, Br^-, and I^-.

Atomic mass (or **relative atomic mass** or **atomic weight**) M_{at} of an element is the average atomic mass, in atomic mass units (amu), of all the naturally occurring isotopes of the element. Atomic masses are listed in the Periodic Table. The amount of an element that has 6.022×10^{23} atoms (the Avogadro number of atoms) has a mass in grams equal to the atomic mass.

Atomic mass unit (amu) is a convenient mass measurement equal to one-twelfth of the mass of a neutral carbon atom that has a mass number of $A = 12$ (6 protons and 6 neutrons). It has been found that amu $= 1.66054 \times 10^{27}$ kg, which is equivalent to $10^{-3}/N_A$, where N_A is Avogadro's number.

Atomic packing factor (APF) is the fraction of volume actually occupied by atoms in a crystal.

Avogadro's number (N_A) is the number of atoms in exactly 12 grams of carbon-12. It is 6.022×10^{23}. Since atomic mass is defined as one-twelfth of the mass of the carbon-12 atom, the N_A number of

atoms of any substance has a mass equal to the atomic mass M_{at}, in grams.

Bond energy or **binding energy** is the work (or energy) needed to separate two atoms infinitely from their equilibrium separation in the molecule or solid.

Cation is an atom that has gained positive charge by virtue of loosing one or more electrons. Usually, metal atoms can lose electrons easily to becomes cations. Cations become attracted to the cathode (negative terminal) in ionic conduction, as in gaseous discharge. The alkali metals, Li, Na, K, . . . , easily lose their valence electron to become cations, Li^+, Na^+, K^+, . . .

Crystal is a three-dimensional periodic arrangement of atoms, molecules, or ions. A characteristic property of the crystal structure is its periodicity and a degree of symmetry. For each atom, the number of neighbors and their exact orientations are well defined; otherwise the periodicity will be lost. Therefore, a long-range order results from strict adherence to a well-defined bond length and relative bond angle (that is, exact orientation of neighbors).

Crystallization is a process by which crystals of a substance are formed from another phase of that substance. Examples are solidification just below the fusion temperature from the melt, or condensation of the molecules from the vapor phase onto a substrate. The crystallization process initially requires the formation of small crystal nuclei, which contain a limited number (perhaps 10^3–10^4) of atoms or molecules of the substance. Following nucleation, the nuclei grow by atomic diffusion from the melt or vapor.

Coordination number is the number of nearest neighbors around a given atom in the crystal.

Covalent bond is the sharing of a pair of valence electrons between two atoms. For example, in H_2, the two hydrogen atoms share their electrons, so that each has a closed shell.

Diffusion is the migration of atoms by virtue of their random thermal motions.

Diffusion coefficient is a measure of the rate at which atoms diffuse. The rate depends on the nature of the diffusion process and is typically temperature dependent. The diffusion coefficient is defined as the magnitude of diffusion flux per unit concentration gradient.

Dislocation is a line imperfection within a crystal that extends over many atomic distances.

Edge dislocation is a line imperfection within a crystal that occurs when an additional, short plane of atoms does not extend as far as its neighbors. The edge of this short plane constitutes a line of atoms where the bonding is irregular, that is, a line of imperfection called an edge dislocation.

Elastic modulus or **Young's modulus** (Y) is a measure of the ease with which a solid can be elastically deformed. The greater Y is, the more difficult it is to deform the solid elastically. When a solid of length ℓ is subjected to a tensile stress σ (force per unit area), the solid will extend elastically by an amount $\delta\ell$, where $\delta\ell/\ell$ is the strain ϵ. Stress and strain are related by $\sigma = Y\epsilon$, so Y is the stress needed per unit elastic strain.

Electric dipole moment is a vector that exists when a positive charge $+Q$ is separated from a negative charge $-Q$. Even though the net charge is zero, there is an electric dipole moment $\mathbf{p}$, given by $\mathbf{p} = Q\mathbf{x}$, where $\mathbf{x}$ is the distance vector from $-Q$ to $+Q$. Just as two charges exert a coulombic force on each other, two dipoles exert a force on each other, which depends on the magnitudes of the dipoles, their separation, and their orientation.

Electronegativity is a relative measure of the ability of an atom to attract electrons to form an anion. Fluorine is the most highly electronegative atom, as it most easily accepts an electron to become an anion (F^-). Typically, halogens (F, Cl, Br, I) are electronegative elements. In HCl, for example, the Cl atom is more electronegative than the H atom and it therefore attracts the electrons more readily than the H-proton. The Cl atom thus acquires a net negative charge and the hydrogen proton becomes exposed. The molecule therefore has a permanent dipole moment.

Equilibrium state of a system is the state in which the pressure and temperature in the system are uniform throughout. We say that the system possesses mechanical and thermal equilibrium.

Equilibrium between two systems requires mechanical, thermal, and chemical equilibrium. Mechanical equilibrium means that the pressure should be the same in the two systems, so that one does not expand at the expense of the other. Thermal equilibrium implies that both have the same temperature. Equilibrium within a single-phase substance (e.g., steam only or hydrogen gas only) implies uniform pressure and temperature within the system.

Eutectic composition is an alloy composition of two elements that results in the lowest melting temperature compared to any other composition. A eutectic solid has a structure that is a mixture of two phases. The eutectic structure is usually special, such as alternating lamellae.

Face centered cubic (FCC) lattice is a cubic lattice that has one lattice point at each corner of a cube and one at the center of each face. If there is a chemical species (atom or a molecule) at each lattice point, then the structure is an FCC crystal structure.

Flux is the number of particles crossing a unit area per unit time. If ΔN particles cross an area A in time Δt, then the particle flux Γ is defined as $\Gamma = \Delta N/(A\ \Delta t)$.

Frenkel defect is an ionic crystal imperfection that occurs when an ion moves into an interstitial site, thereby creating a vacancy in its original site. The imperfection is therefore a pair of point defects.

Grain is an individual crystal within a polycrystalline material. Within a grain, the crystal structure and orientation are the same everywhere and the crystal is oriented in one direction only.

Grain boundary is a surface region between differently oriented, adjacent, grain crystals. The grain boundary contains a lattice mismatch between adjacent grains.

Heat is the amount of energy transferred from one system to another (or between the system and its surroundings) as a result of a temperature difference. Heat is not a new form of energy, but rather the transfer of energy from one body to another by virtue of the random motions of their molecules. When a hot body is in contact with a cold body, heat is transferred from the hot body to the cold one. What is actually transferred is the excess mean kinetic energy of the molecules in the hot body. Molecules in the hot body have a higher mean kinetic energy and vibrate more violently. As a result of the collisions between the molecules, there is a net transfer of energy from the hot body to the cold one, until the molecules in both bodies have the same mean kinetic energy; that is, until their temperatures become equal.

Heterogeneous mixture is a mixture in which the individual components remain physically separate and possess different chemical and physical properties; that is, a mixture of different phases.

Homogeneous mixture is a mixture of two or more chemical species in which the chemical properties (e.g., composition) and physical properties (e.g., density, heat capacity) are uniform throughout. A homogeneous mixture is a solution.

Interstitial site (interstice) is an unoccupied space between the atoms (or ions, or molecules) in a crystal.

Isomorphous is a structure that is the same everywhere (from *iso,* uniform, and *morphology,* structure).

Isotropic substance is a material that has the same property in all directions.

Lattice is a regular array of points in space with a discernible periodicity. There are 14 distinct lattices possible in three-dimensional space. When an atom or molecule is placed at each lattice point, the resulting regular structure is a crystal structure.

Lattice parameters are: (a) the lengths of the sides of the unit cell, and (b) the angles between the sides.

Metallic bonding is the binding of metal atoms in a crystal through the attraction between the positive metal ions and the mobile valence electrons in the crystal. The valence electrons permeate the space between the ions.

Miller indices *(hkl)* are indices that conveniently identify parallel planes in a crystal. Consider a plane with the intercepts, x_1, y_1, and z_1, in terms of lattice parameters a, b, and c. (For a plane passing through the origin, we shift the origin or use a parallel plane.) Then, *(hkl)* are obtained by taking the reciprocals of x_1, y_1, and z_1 and clearing all fractions.

Miscibility of two substances is a measure of the mutual solubility of those two substances when they are in the same phase, such as liquid.

Mole of a substance is that amount of the substance that contains N_A number of atoms (or molecules), where N_A is Avogadro's number (6.023×10^{-23}).

One mole of a substance has a mass equal to its atomic (molecular) mass, in grams. For example, 1 mole of copper contains 6.023×10^{23} atoms and has a mass of 63.55 grams.

Phase of a system is a homogeneous portion of the chemical system that has the same composition, structure, and properties everywhere. In a given chemical system, one phase may be in contact with another phase of the system. For example, iced water at 0 °C will have solid and liquid phases in contact. Each phase, solid ice and liquid water, has a distinct structure.

Phase diagram is a temperature versus composition diagram in which the existence and coexistence of various phases are identified by regions and lines. Between the liquidus and solidus lines, for example, the material is a heterogeneous mixture of the liquid and solid phases.

Polarization is the separation of positive and negative charges in a system, which means there is a net electric dipole moment per unit volume.

Polymorphism or **allotropy** is a material attribute that allows the material to possess more than one crystal structure. Each possible crystal structure is called a polymorph. Generally, the structure of the polymorph depends on the temperature and pressure, as well as on the method of preparation of the solid. (For example, diamond can be prepared from graphite by the application of very high pressures.)

Primary bond is a strong interatomic bond, typically greater then 1 eV/atom, that involves ionic, covalent, or metallic bonding.

Property is a system characteristic or an attribute that we can measure. Pressure, volume, temperature, mass, energy, electrical resistivity, magnetization, polarization, and color are all properties of matter. Properties such as pressure, volume, and temperature can only be attributed to a system of many particles (which we treat as a continuum). Note that heat and work are not properties of a substance; instead, they represent energy transfers involved in producing changes in the properties.

Saturated solution is a solution that has the maximum possible amount of solute dissolved in a given amount of solvent at a specified temperature and pressure.

Saturated solution is a solution that has the maximum possible amount of solute dissolved in a given amount of solvent at a specified temperature and pressure.

Schottky defect is an ionic crystal imperfection that occurs when a pair of ions is missing; that is, a cation and anion pair vacancy.

Screw dislocation is a crystal defect that occurs when one portion of a perfect crystal is twisted or skewed with respect to another portion on only one side of a line.

Secondary bond is a weak bond, typically less than 0.1 eV/atom, which is due to dipole–dipole interactions between the atoms or molecules.

Solid solution is a homogeneous crystalline phase that contains two or more chemical components.

Solute is the minor chemical component of a solution; the component that is usually added in small amounts to a solvent to form a solution.

Solvent is the major chemical component of a solution.

Stoichiometric compounds are compounds with an integer ratio of atoms, as in CaF_2, in which two fluorine atoms bond with one calcium atom.

Unit cell is the most convenient small cell in a crystal structure that carries the characteristics of the crystal. The repetition of the unit cell in three dimensions generates the whole crystal structure.

Vacancy is a point defect in a crystal, where a normally occupied lattice site is missing an atom.

Valence electrons are the electrons in the outer shell of an atom. Since they are the farthest away from the nucleus, they are the first electrons involved in atom-to-atom interactions.

QUESTIONS AND PROBLEMS

1.1 Virial Theorem and H-atom   In a given system, the potential energy PE of interaction between the various charges is electrostatic only. If we neglect the magnetic interactions between moving charges within the atom, then the PE of interaction between the electrons themselves and between electrons and the nucleus also is electrostatic only. The **Virial Theorem** relates the average values of the PE, and kinetic energy KE, and the overall (total) energy E when the interactive forces follow the inverse square law, as in electrostatics:

$$\overline{E} = \overline{PE} + \overline{KE} \qquad [1]$$

and

$$\overline{KE} = -\frac{1}{2}\,\overline{PE} \qquad [2]$$

a. It takes 13.6 eV to ionize the hydrogen atom, that is, to remove the electron to infinity. If the latter condition defines the zero reference energy, then the total energy of the electron within the H atom is -13.6 eV. Calculate the average PE and average KE of the electron.

b. Assume the electron is in a stable orbit of radius r_o around the positive nucleus. What is the coulombic PE of the electron? Hence what is the average radius r_o of the electron orbit?

c. What is the velocity of the electron?

d. What is the frequency of oscillation?

1.2 The covalent bond   In this problem, we consider the H_2 molecule as two touching H atoms, as depicted in Figure 1.72. Does this arrangement have a lower energy than two separated H atoms? Suppose the electrons totally correlate their motions; that is, they move in the same direction and go around both nuclei such that at one instant, the snapshot will be as shown in Figure 1.72. Use the Virial Theorem stated in Question 1.1 to do the following.

a. Calculate the *PE* (in eV) of the electron when it is in the center.

b. Calculate the *PE* (in eV) of the electron when it is at the far left (or right).

c. From (a) and (b), calculate the average *PE* (in eV) of an electron in the H–H molecule.

d. Calculate the change in the average *PE* of an electron with respect to the *PE* when the electron was in the isolated H atom [In the isolated H atom, the *PE* of the electron is 2×-13.6 eV.]

e. What is the change in the *PE* per electron?

f. What is the change in the total energy per electron? What is the covalent bond energy? How does this compare with 4.51 eV?

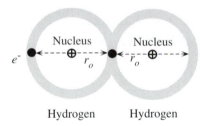

Figure 1.72   A simplified view of the covalent bond in H_2: a snapshot at one instant.

The electrons correlate their motions (move in the same direction and circle both nuclei).

1.3 Ionic bonding and NaCl   The interaction energy between Na^+ and Cl^- ions in the NaCl crystal can be written as

$$E(r) = -\frac{4.03 \times 10^{-28}}{r} + \frac{6.97 \times 10^{-96}}{r^8}$$

where the energy is given in joules per ion pair, and the interionic separation r is in meters. Calculate the binding energy and the equilibrium separation between the Na^+ and Cl^- ions. Also estimate the elastic modulus Y of NaCl given that

$$Y \approx \frac{1}{6r_o}\left[\frac{d^2E}{dr^2}\right]_{r=r_o}$$

*1.4 **van der Waals bonding**   Below 24.5 K, Ne is a crystalline solid with an FCC structure. The interatomic interaction energy per atom can be written as

$$E(r) = -2\varepsilon\left[14.45\left(\frac{\sigma}{r}\right)^6 - 12.13\left(\frac{\sigma}{r}\right)^{12}\right] \quad \text{(eV/atom)}$$

where ε and σ are constants that depend on the polarizability, the mean dipole moment, and the extent of overlap of core electrons. For crystalline Ne, $\varepsilon = 3.121 \times 10^{-3}$ eV and $\sigma = 0.274$ nm.

a. Show that the equilibrium separation between the atoms in an inert gas crystal is given by $r_o = (1.090)\sigma$. What is the equilibrium interatomic separation in the Ne crystal?

b. Find the bonding energy per atom in solid Ne.

c. Calculate the density of solid Ne (atomic mass = 20.18).

1.5 **Kinetic Molecular Theory** Calculate the effective (rms) speeds of the He and Ne atoms in the He–Ne gas laser tube at room temperature (300 K).

*1.6 **Vacuum deposition** Consider air as composed of nitrogen molecules N_2.

a. What is the concentration n (number of molecules per unit volume) of N_2 molecules at 1 atm. and 27 °C?

b. Estimate the mean separation between the N_2 molecules.

c. Assume each molecule has a finite size that can be represented by a sphere of radius r. Also assume that ℓ is the **mean free path,** defined as the mean distance a molecule travels before colliding with another molecule, as illustrated in Figure 1.73. If we consider the motion of one N_2 molecule, with all the others stationary, it is aparent that if the path of the travelling

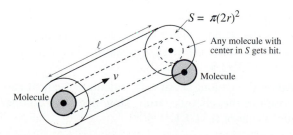

Figure 1.73   A molecule moving with a velocity v travels a mean distance ℓ between collisions. Since the collision cross-sectional area is S, in the volume $S\ell$, there must be at least one molecule. Consequently, $n(S\ell) = 1$.

molecule crosses the cross-sectional area $S = \pi(2r)^2$, there will be a collision. Since ℓ is the mean distance between collisions, there must be at least one stationary molecule within the volume $S\ell$, as shown in Figure 1.73. Since n is the concentration, we must have $n(S\ell) = 1$ or $\ell = 1/(\pi 4r^2 n)$. However, this must be corrected for the fact that all the molecules are in motion, which only introduces a numerical factor, so that

$$\ell = \frac{1}{2^{1/2} 4\pi r^2 n}$$

Assuming a radius r of 0.1 nm, calculate the mean free path of N_2 molecules between collisions at 27 °C and 1 atm.

d. Assume that an Au film is to be deposited onto the surface of an Si chip to form metallic interconnections between various devices. The deposition process is generally carried out in a vacuum chamber and involves the condensation of Au atoms from the vapor phase onto the chip surface. In one procedure, a gold wire is wrapped around a tungsten filament, which is heated by passing a large current through the filament (analogous to the heating of the filament in a light bulb). The Au wire melts and wets the filament, but as the temperature of the filament increases, the gold evaporates to form a vapor. Au atoms from this vapor then condense onto the chip surface, to solidify and form the metallic connections. Suppose that the source (filament)-to-substrate (chip) distance L is 10 cm. Unless the mean free path of air molecules is much longer than L, collisions between the metal atoms and air molecules will prevent the deposition of the Au onto the chip surface. Taking the mean free path ℓ to be $100L$, what should be the pressure inside the vacuum system? (Assume the same r for Au atoms).

1.7 Heat capacity

a. Calculate the heat capacity per mole and per gram of N_2 gas. How does this compare with the experimental value of 0.743 J g^{-1} K^{-1}?

b. Calculate the heat capacity per mole and per gram of CO_2 gas, neglecting the vibrations of the molecule. How does this compare with the experimental value of 0.648 J K^{-1} g^{-1}?

c. Based on the Dulong–Petit rule, calculate the heat capacity per mole and per gram of solid silver. How does this compare with the experimental value of 0.235 J K^{-1} g^{-1}?

d. Based on the Dulong–Petit rule, calculate the heat capacity per mole and per gram of the silicon crystal. How does this compare with the experimental value of 0.71 J K^{-1} g^{-1}?

*1.8 **Thermal fluctuations** The cross section of a typical moving-coil ammeter is shown in Figure 1.74. The rectangular moving coil (extended into the paper) is placed in the air gap between the poles of a permanent magnet and a fixed, cylindrical iron core. The iron core ensures a strong magnetic field through

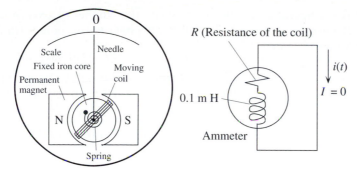

Figure 1.74   The cross section of a moving coil ammeter and its circuit representation in a closed circuit.

the coil. The coil is hinged to rotate in the air gap and around the iron core. When a current I is passed through the coil, it experiences a torque and tries to rotate, but the rotation is restricted by the spring. A needle attached to coil indicates the amount of rotation. The deflection θ of the needle is proportional to the torque τ, by virtue of Hooke's Law

$$\theta = K\tau$$

where, from electromagnetic theory, the torque is proportional to the product of the current I and the magnetic field B. Thus, θ measures the current I.

A given ammeter has a coil inductance of 0.1 mH and a coil resistance of 100 Ω. The spring constant K is 10^{11} rad N^{-1} m^{-1} and the length of the needle is 10 cm.

a. What is the rms fluctuation in the position of the ammeter needle?

b. Assume the ammeter is placed in a closed circuit where the dc current is zero ($I = 0$), as shown in Figure 1.74. Sketch the instantaneous current $i(t)$ versus time. What is the minimum current this ammeter can measure (noise in the current)? How can this be improved?

1.9 **Electrical noise**   Consider an amplifier with a bandwidth B of 5 kHz, corresponding to a typical speech bandwidth. Assume the input resistance of the amplifier is 1 MΩ. What is the rms noise voltage at the input? What will happen if the bandwidth is doubled to 10 kHz? What is your conclusion?

1.10 **Electrical noise**   The Y-channel input of a typical oscilloscope has an input resistance of 1 MΩ and an input capacitance of 10 pF. What is the minimum voltage that can be displayed (provided amplification is available) without corruption by noise?

1.11 **Thermal activation**   A certain chemical oxidation process (e.g., SiO_2) has an activation energy of 2 eV atom^{-1}.

a. Consider the material exposed to pure oxygen gas at a pressure of 1 atm. at 27 °C. Estimate how many oxygen molecules per unit volume will have energies in excess of 2 eV?

b. If the temperature is 900 °C, estimate the number of oxygen molecules with energies more than 2 eV. What happens to this concentration if the pressure is doubled?

1.12 Diffusion in Si   The diffusion coefficient of boron (B) atoms in a single crystal of Si has been measured to be 1.5×10^{-18} m^2 s^{-1} at 1000 °C and 1.1×10^{-16} m^2 s^{-1} at 1200 °C.

a. What is the activation energy for the diffusion of B, in eV/atom?

b. What is the preexponential constant D_o?

c. What is the rms distance (in micrometers) diffused in 1 hour by the B atom in the Si crystal at 1200 °C and 1000 °C?

d. The diffusion coefficient of B in polycrystalline Si has an activation energy of 2.4–2.5 eV atom^{-1} and $D_o = (1.5$–$6) \times 10^{-7}$ m^2 s^{-1}. What constitutes the diffusion difference between the single crystal sample and the polycrystalline sample?

1.13 Diffusion in SiO$_2$ The diffusion coefficient of P atoms in SiO$_2$ has an activation energy of 2.30 eV atom^{-1} and $D_o = 5.73 \times 10^{-9}$ m^2 s^{-1}. What is the rms distance diffused in one hour by P atoms in SiO$_2$ at 1200 °C?

*1.14 BCC and FCC crystals

a. Consider iron below 912 °C, where its structure is BCC. Given the density of iron as 7.86 g cm^{-3} and its atomic mass as 55.85, calculate the lattice parameter of the unit cell and the radius of the Fe atom.

b. At 912 °C, iron changes from the BCC (α-Fe) to the FCC (γ-Fe) structure. The radius of the Fe atom correspondingly changes from 0.1258 nm to 0.1291 nm. Calculate the density of γ-Fe and explain whether there is a volume expansion or contraction during this phase change.

c. Identify the most densely packed crystal planes in the BCC and FCC crystal structures.

1.15 Diamond and zinc blende   Si has the diamond and GaAs has the zinc blende crystal structure. Given the lattice parameters of Si and GaAs, $a = 0.543$ nm and $a = 0.565$ nm, respectively, and the atomic masses of Si, Ga, and As as 28.08, 69.73, and 74.92, respectively, calculate the density of Si and GaAs. What is the atomic concentration (atoms per unit volume) in each crystal?

1.16 Crystallographic directions and planes   Consider the cubic crystal system.

a. Show that the line $[hk\ell]$ is perpendicular to the $(hk\ell)$ plane.

b. Show that the spacing between adjacent $(hk\ell)$ planes is given by

$$d = \frac{a}{\sqrt{h^2 + k^2 + \ell^2}}$$

1.17 Si and SiO$_2$

a. Given the Si lattice parameter $a = 0.543$ nm, calculate the number of Si atoms per unit volume, in nm^{-3}.

b. Calculate the number of atoms per m^2 and per nm^2 on the (100), (110), and (111) planes in the Si crystal. Which plane has the most number of atoms per unit area?

c. The density of SiO$_2$ is 2.27 g cm^{-3}. Given that its structure is amorphous, calculate the number of molecules per unit volume, in nm^{-3}. Compare your result with (a) and comment on what happens when the surface of an Si crystal oxidizes. The atomic masses of Si and O are 28.09 and 16, respectively.

1.18 Vacancies   Estimate the atomic vacancy concentration in the Si crystal at room temperature and at 1200 °C, given that the energy of vacancy formation for Si is 2.4 eV/atom.

*1.19 Dislocations   The presence of a dislocation causes a strain field around the dislocation line, as shown in Figure 1.46. The atoms are in compression above the line and in tension below the line. Consider the screw dislocation shown in Figure 1.47, where the upper portion of the crystal has been distorted (sheared) with respect to the lower portion by a displacement b (called the **Burgers vector**). The largest displacement, b resulting from the shearing distortion is typically the nearest-neighbor separation. In general, the elastic energy per unit volume is given by

$$\text{Elastic energy per unit volume} = \frac{1}{2} (\text{Shear modulus})(\text{Shear strain})^2$$

However, the dislocation-induced shear strain depends on the distance from the dislocation line. Therefore, the expression given here cannot be applied directly to the dislocation region. We must integrate the expression around the dislocation line of length L to find the total energy. We then obtain the dislocation energy per unit length of dislocation, or $E_L \approx Gb^2$ (in Joules per meter), where G is the shear modulus.

a. Consider a typical metal such as copper, for which $G \approx 50$ GPa and $b = 0.25$ nm. Calculate the energy in eV per nanometer of dislocation line length. Taking the number of atoms along [100] to be $1/a$, where $a = 0.3615$ nm, calculate the energy per atom. Will the thermal fluctuations produce a dislocation? How will they be produced?

b. Calculate the dislocation energy per unit length (per nm) in the Si crystal for which $G = 75$ GPa   and $b = 0.271$ nm ($= a/2$).

*1.20 A two-phase alloy and an analogy with Canada   In this problem, we use an analogy to treat a two-phase alloy, such as that occurring during the melting of a Cu–Ni alloy (liquid and solid phases coexisting), or Pb–Sn solder (α and β phases coexisting) at room temperature.

Table 1.8   Number of atoms in the phases

	α-phase (English Canada)	β-phase (Quebec)
Number of A atoms (Anglophones)	19 million	1 million
Number of B atoms (Francophones)	1 million	4 million

Consider Canada as an alloy with two "phases," English Canada (the α-phase) and French Canada or Quebec (the β-phase), coexisting. Atoms A correspond to Anglophones and atoms B correspond to Francophones. The α-phase is A rich, that is, has more A atoms; the β-phase is B-rich, which means it has more B atoms.

If we measure the composition in terms of B atoms, and round up the numbers (with exaggeration), we can represent the populations as shown in Table 1.8.

Let us also assume that the A atom is 10% heavier than the B atom (the author chose this at random, with no reference to the weight ratio of an Anglophone to a Francophone).

a. What is the overall composition C_O of the alloy (Canada)?

b. What is the composition of the β-phase (Quebec)?

c. What is the composition of the α-phase (English Canada)?

d. What is the weight fraction of the α-phase (English Canada) in the alloy (Canada)?

e. What is the weight fraction of the β-phase (Quebec) in the alloy (Canada)?

1.21 Pb–Sn solder   Consider the soldering of two copper components. When the solder melts, it wets both metal surfaces. If the surfaces are not clean or have an oxide layer, the molten solder cannot wet the surfaces and the soldering fails. Assume that soldering takes place at 250 °C, and consider the diffusion of Sn atoms into the copper (the Sn atom is smaller than the Pb atom and hence diffuses more easily).

a. The diffusion coefficient of Sn in Cu at two temperatures is $D = 1.69 \times 10^{-9}$ cm^2 hr^{-1} at 400 °C   and $D = 2.48 \times 10^{-7}$ cm^2 hr^{-1} at 650 °C. Calculate the rms distance diffused by an Sn atom into the copper, assuming the cooling process takes 10 seconds.

b. What should be the composition of the solder if it is to begin freezing at 250 °C?

c. What are the components (phases) in this alloy at 200 °C? What are the compositions of the phases and their relative weights in the alloy?

d. What is the microstructure of this alloy at 25 °C? What are weight fractions of the α and β phases assuming near equilibrium cooling?

1.22 Pb–Sn solder   Consider 50% Pb–50% Sn solder.

a. Sketch the temperature-time profile and the microstructure of the alloy at various stages as it is cooled from the melt.

b. At what temperature does the solid melt?

c. What is the temperature range over which the alloy is a mixture of melt and solid? What is the structure of the solid?

d. Consider the solder at room temperature following cooling from 182 °C. Assume that the rate of cooling from 182 °C to room temperature is faster than the atomic diffusion rates needed to change the compositions of the α and β phases in the solid. Assuming the alloy is 1 kg, calculate the masses of the following components in the solid:
1. The primary α.
2. α in the whole alloy.
3. α in the eutectic solid.
4. β in the alloy. (Where is the β-phase?)

e. Calculate the specific heat of the solder given the atomic masses of Pb (207.2) and Sn (118.71).

2

Electrical and Thermal Conduction in Solids

Electrical conduction involves the motion of charges in a material under the influence of an applied electric field. A material can generally be classified as a conductor if it contains a large number of "free" or mobile charge carriers. In metals, due to the nature of metallic bonding, the valence electrons from the atoms form a sea of electrons that are free to move within the metal and are therefore called conduction electrons. In this chapter, we will treat the conduction electrons in metal as "free charges" that can be accelerated by an applied electric field. In the presence of an electric field, the conduction electrons attain an average velocity, called the drift velocity, that depends on the field. By applying Newton's Second Law to electron motion and using such concepts as mean free time between electron collisions with lattice vibrations, crystal defects, impurities, etc., we will derive the fundamental equations that govern electrical conduction in solids. A key concept will be the drift mobility, which is a measure of the ease with which charge carriers in the solid drift under the influence of an external electric field.

Good electrical conductors, such as metals, are also known to be good thermal conductors. The conduction of thermal energy from higher to lower temperature regions in a metal involves the conduction electrons carrying the energy. Consequently, there is an innate relationship between the electrical and thermal conductivities, which is supported by theory and experiments.

2.1 CLASSICAL THEORY: THE DRUDE MODEL

2.1.1 Metals and Conduction by Electrons

The electric current density J is defined as the net amount of charge flowing across a unit area per unit time, that is,

$$J = \frac{\Delta q}{A\,\Delta t}$$

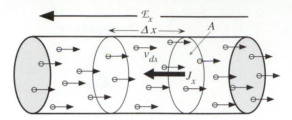

Figure 2.1   Drift of electrons in a conductor in the presence of an applied electric field.

Electrons drift with an average velocity v_{dx} in the x direction.

where Δq is the net quantity of charge flowing through an area A in time Δt. Figure 2.1 shows the net flow of electrons in a conductor section of cross-sectional area A in the presence of an applied field $\mathcal{E}_x$. Notice that the direction of electron motion is opposite to that of the electric field $\mathcal{E}_x$ and of conventional current, because the electrons experience a Coulombic force $e\mathcal{E}_x$ in the x direction, due to their negative charge.

We know that the conduction electrons are actually moving around randomly[1] in the metal, but we will assume that as a result of the application of the electric field $\mathcal{E}_x$, they all acquire a net velocity in the x direction. Otherwise, there would be no net flow of charge through area A.

The average velocity of the electrons in the x direction at time t is denoted $v_{dx}(t)$. This is called the **drift velocity,** which is the instantaneous velocity v_x in the x direction averaged over many electrons, (perhaps, $\sim 10^{28}$ m^{-3}); that is

Definition of drift velocity

$$v_{dx} = \frac{1}{N}[v_{x1} + v_{x2} + v_{x3} + \cdots + v_{xN}] \qquad \text{[2.1]}$$

where v_{xi} is the x direction velocity of ith electron, and N is the number of conduction electrons in the metal. Suppose that n is the number of electrons per unit volume in the conductor ($n = N/V$). In time Δt, electrons move a distance $\Delta x = v_{dx}\,\Delta t$, so the total charge Δq crossing the area A is $enA\,\Delta x$. This is valid because all the electrons within distance Δx pass through A; thus, $n(A\,\Delta x)$ is the total number of electrons crossing A in time Δt.

The current density in the x direction is

$$J_x = \frac{\Delta q}{A\,\Delta t} = \frac{enAv_{dx}\,\Delta t}{A\,\Delta t} = env_{dx}$$

This general equation relates J_x to the average velocity v_{dx} of the electrons. It must be appreciated that the average velocity at one time may not be the same as at another time, because the applied field, for example, may be changing: $\mathcal{E}_x = \mathcal{E}_x(t)$. We therefore allow for a time-dependent current by writing

Current density and drift velocity

$$J_x(t) = env_{dx}(t) \qquad \text{[2.2]}$$

[1] All the conduction electrons are "free" within the metal and move around randomly, being scattered from vibrating metal ions, as we discuss in this chapter.

To relate the current density J_x to the electric field $\mathcal{E}_x$, we must examine the effect of the electric field on the motion of the electrons in the conductor. To do so, we will consider the copper crystal.

The copper atom has a single valence electron in its $4s$ subshell, and this electron is loosely bound. The solid metal consists of positive ion cores, Cu^+, at regular sites, in the face-centered cubic (FCC) crystal structure. The valence electrons detach themselves from their parents and wander around freely in the solid, forming a kind of electron cloud or gas. These mobile electrons are free to respond to an applied field, creating a current density J_x. The valence electrons in the electron gas are therefore conduction electrons.

The attractive forces between the negative electron cloud and the Cu^+ ions are responsible for metallic bonding and the existence of the solid metal. (This simplistic view of metal was depicted in Figure 1.6 for copper.) The electrostatic attraction between the conduction electrons and the positive metal ions, like the electrostatic attraction between the electron and the proton in the hydrogen atom, results in the conduction electron having both potential energy PE and kinetic energy KE. The conduction electrons move about the crystal lattice in the same way that gas atoms move randomly in a cylinder. Although the average KE for gas atoms is $(3/2)\,kT$, this is not the case for electrons in metal, because these electrons strongly interact with the metal ions, as a result of electrostatic attraction.

The metallic bond energy per atom is typically $1-10$ eV. The magnitude of the attractive electrostatic PE between the conduction electrons and the positive metal ions must be greater than this bond energy, because this attractive PE must be stronger (i.e., greater in magnitude) that the repulsive PE between the positive metal ions, the KE of the electrons, and the mutual repulsion between the electrons themselves. So the net PE per atom is negative (a requirement of bond formation) and $1-10$ eV per atom. As an electron moves about the crystal, its PE changes, depending on its location with respect to various positive ions. When an electron moves toward a positive ion, its PE decreases and its KE consequently increases. The mean KE of a conduction electron depends on the electrostatic interactions with the metal ions, and the value of this mean KE is smaller than the mean PE of attraction. However, its order of magnitude is comparable to the magnitude of the mean PE. In fact, from classical mechanics[2] we can guess that the value of the mean KE is about half the magnitude of the mean PE. If we take $(1/2)\,m_e u^2$ to be the mean KE and typically to be a few electron volts, where u is the mean speed of the electron, u turns out typically to be $\sim 10^6$ m s^{-1}. Since the mean KE is primarily determined by the electrostatic interaction of the conduction electrons with the metal ions, its temperature dependence is weak compared with other factors that control the behavior of the conduction electrons in the metal crystal, and can thus be neglected. This purely classical and intuitive reasoning is not sufficient, however, to show that the mean speed u is relatively temperature insensitive and much

[2] There is a theorem in classical mechanics called the Virial theorem, which states that for a collection of particles, the mean KE has half the magnitude of the mean PE if the only forces acting on the particles are such that they follow an inverse square law dependence on the particle-particle separation (as in Coulombic and gravitation forces).

greater than that expected from kinetic molecular theory. The true reasons are quantum mechanical and are discussed in Chapter 4. (They arise from what is called the **Pauli exclusion principle.**)

In general, the copper crystal will not be perfect and the atoms will not be stationary. There will be crystal defects, vacancies, dislocations, impurities, etc., which will scatter the conduction electrons. More importantly, due to their thermal energy, the atoms will vibrate about their lattice sites (equilibrium positions), as depicted in Figure 2.2a. An electron will not be able to avoid collisions with vibrating atoms; consequently, it will be "scattered" from one atom to another. In the absence of an applied field, the path of an electron may be visualized as illustrated in Figure 2.2, where scattering from lattice vibrations causes the electron to move randomly in the lattice. On those occasions when the electron reaches a crystal surface, it becomes "deflected" (or "bounced") back into the crystal. Therefore, in the absence of a field, after some duration of time, the electron crosses its initial x plane position again. Over a long time, the electrons therefore show no net displacement in any one direction.

When the conductor is connected to a battery and an electric field is applied to the crystal, as shown in Figure 2.2b, the electron experiences an acceleration in the x direction in addition to its random motion, so that after some time, it will drift a finite distance in the x direction. The electron accelerates along the x direction under the action of the force $e\mathcal{E}_x$, and then it suddenly collides with a vibrating atom and loses the gained velocity. Therefore, there is an average velocity in the x direction, which, if calculated, determines the current via Equation 2.2. Note that since the electron experiences an acceleration in the x direction, its trajectory

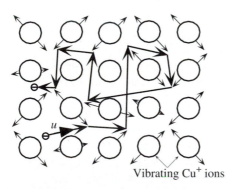

Vibrating Cu$^+$ ions

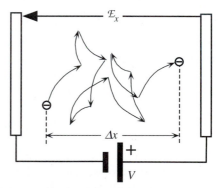

(a) A conduction electron in the electron gas moves about randomly in a metal (with a mean speed u) being frequently and randomly scattered by thermal vibrations of the atoms. In the absence of an applied field there is no net drift in any direction.

(b) In the presence of an applied field, $\mathcal{E}_x$, there is a net drift along the x direction. This net drift along the force of the field is superimposed on the random motion of the electron. After many scattering events the electron has been displaced by a net distance, Δx, from its initial position toward the positive terminal

Figure 2.2   Motion of a conduction electron in a metal

between collisions is a parabola, like the trajectory of a golf ball experiencing acceleration due to gravity.

To calculate the drift velocity v_{dx} of the electrons due to applied field $\mathcal{E}_x$, we first consider the velocity v_{xi} of the ith electron in the x direction at time t. Suppose its last collision was at time t_i; therefore, for time $(t - t_i)$, it accelerated *free of collisions,* as indicated in Figure 2.3. Let u_{xi} be the velocity of electron i in the x direction just after the collision. We will call this the initial velocity. Since $e\mathcal{E}_x/m_e$ is the acceleration of the electron, the velocity v_{xi} in the x direction at time t will be

$$v_{xi} = u_{xi} + \frac{e\mathcal{E}_x}{m_e}(t - t_i)$$

However, this is only for the ith electron. We need the average velocity v_{dx} for all such electrons along x. We average the expression for $i = 1$ to N electrons, as in Equation 2.1. We assume that immediately after a collision with a vibrating ion, the electron may move in any random direction, that is, it can just as likely move along the negative or positive x, so that u_{xi} averaged over many electrons is zero. Thus,

$$v_{dx} = \frac{1}{N}[v_{x1} + v_{x2} + \cdots + v_{xN}] = \frac{e\mathcal{E}_x}{m_e}\overline{(t - t_i)}$$

where $\overline{(t - t_i)}$ is the **average free time** for N electrons between collisions.

Suppose that τ is the mean free time, or the **mean time between collisions** (also known as the **mean scattering time**). For some electrons, $(t - t_i)$ will be greater than τ, and for others, it will be shorter, as shown in Figure 2.3. Averaging $(t - t_i)$ for N electrons will be the same as τ. Thus, we can substitute τ for $\overline{(t - t_i)}$ in the previous expression to obtain

$$v_{dx} = \frac{e\tau}{m_e}\mathcal{E}_x \tag{2.3}$$

Equation 2.3 shows that the drift velocity increases linearly with the applied field. The constant of proportionality, $e\tau/m_e$, has been given a special name and

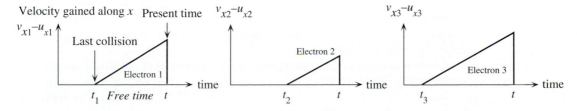

Figure 2.3   Velocity gained in the x direction at time t from the electric field ($\mathcal{E}_x$) for three electrons. There will be N electrons to consider in the metal.

symbol. It is called the **drift mobility** μ_d, which is defined as

*Definition of
drift mobility*

$$v_{dx} = \mu_d \mathcal{E}_x \qquad [2.4]$$

where

*Drift mobility
and mean free
time*

$$\mu_{dx} = \frac{e\tau}{m_e} \qquad [2.5]$$

Equation 2.5 relates the drift mobility of the electrons to their mean scattering time τ. To reiterate, τ which is also called the **relaxation time,** is directly related to the microscopic processes that cause the scattering of the electrons in the metal; that is, lattice vibrations, crystal imperfections, and impurities, to name a few.

From the expression for the drift velocity v_{dx}, the current density J_x follows immediately by substituting Equation 2.4 into 2.2, that is,

$$J_x = en\mu_d\mathcal{E}_x \qquad [2.6]$$

Therefore, the current density is proportional to the electric field and the conductivity σ is the term multiplying $\mathcal{E}_x$, that is,

*Unipolar
conductivity*

$$\sigma = en\mu_d \qquad [2.7]$$

It is gratifying that by treating the electron as a particle and applying classical mechanics ($F = ma$), we are able to derive Ohm's Law. We should note, however, that we assumed τ to be independent of the field.

Drift mobility is important because it is a widely used electronic parameter in semiconductor device physics. The drift mobility gauges how fast electrons will drift when driven by an applied field. If the electron is not highly scattered, then the mean free time between collisions will be long, τ will be large, and by Equation 2.5, the drift mobility will also be large; the electrons are therefore highly mobile. A small field results in a large drift velocity (Equation 2.4). However, a large drift mobility does not necessarily imply high conductivity, because σ also depends on the concentration of conduction electrons, n.

The mean time between collisions τ has further significance. Its reciprocal $1/\tau$ represents the **mean frequency of collisions** or **scattering events;** that is, $1/\tau$ is the mean probability per unit time that the electron will be scattered (see Example 2.1). Therefore, during a small time interval δt, the probability of scattering will be $\delta t/\tau$. The probability of scattering per unit time $1/\tau$ is time independent and depends only on the nature of the scattering mechanism, regardless of whether the scattering involves lattice vibrations, impurities, dislocations, etc.

There is one important assumption in the derivation of the drift velocity v_{dx} in Equation 2.3. We obtained v_{dx} by averaging the velocities v_{xi} of N electrons along x at one instant, as defined in Equation 2.1. The drift velocity therefore represents the average velocity of *all* the electrons along x at one instant; that is, v_{dx} is a number average at one instant. Figure 2.2b shows that after many collisions, after a time interval $\Delta t \gg \tau$, an electron would have been displaced by a net distance Δx

along x. The term $\Delta x/\Delta t$ represents the effective velocity with which the electron drifts along x. It is an average velocity for one electron over many collisions, that is, over a long time (hence, $\Delta t \gg \tau$), so $\Delta x/\Delta t$ is a time average. Provided that Δt contains many collisions, it is reasonable to expect that the drift velocity $\Delta x/\Delta t$ from the time average for one electron is the same as the drift velocity v_{dx} per electron from averaging for all electrons at one instant, as in Equation 2.1, or

$$\frac{\Delta x}{\Delta t} = v_{dx} \qquad\qquad [2.8]$$

The two velocities are the same only under steady-state conditions ($\Delta t \gg \tau$). The proof may be found in more advanced texts.

PROBABILITY OF SCATTERING PER UNIT TIME AND THE MEAN FREE TIME   If $1/\tau$ is defined as the mean probability per unit time that an electron is scattered, show that the mean time between collisions is τ. | **Example 2.1**

Solution

Consider an infinitesimally small time interval dt at time t. Let N be the number of unscattered electrons at time t. The probability of scattering during dt is $(1/\tau)\,dt$, and the number of scattered electrons during dt is $N(1/\tau)\,dt$. The change dN in N is thus

$$dN = -N\left(\frac{1}{\tau}\right)dt$$

The negative sign indicates a reduction in N because, as electrons become scattered, N decreases. Integrating this equation, we can find N at any time t, given that at time $t = 0$, N_0 is the total number of unscattered electrons. Therefore,

$$N = N_0 \exp\left(\frac{-t}{\tau}\right)$$

This equation represents the number of unscattered electrons at time t. It reflects an exponential decay law for the number of unscattered electrons. The mean free time $\bar{t}$ can be calculated from the mathematical definition of $\bar{t}$,

$$\bar{t} = \frac{\int_0^\infty tN\,dt}{\int_0^\infty N\,dt} = \tau$$

where we have used $N = N_0 \exp(-t/\tau)$. Clearly, $1/\tau$ is the mean probability of scattering per unit time.

ELECTRON DRIFT MOBILITY IN METALS   Calculate the drift mobility of electrons in copper at room temperature, given that the conductivity of copper is $5.9 \times 10^5\ \Omega^{-1}\ \mathrm{cm}^{-1}$. The density of copper is $8.93\ \mathrm{gm\ cm^{-3}}$ and its atomic mass is $63.5\ \mathrm{g\ mol^{-1}}$. | **Example 2.2**

Solution

We can calculate μ_d from $\sigma = en\mu_d$ because we already know the conductivity σ. The number of free electrons n per unit volume can be taken as equal to the number of Cu atoms per unit volume, if we assume that each Cu atom donates one electron to the conduction electron gas in the metal. One mole of copper has N_A (6.02×10^{23}) atoms and a mass of 63.5 grams. Therefore, the number of copper atoms per unit volume is

$$n = \frac{dN_A}{M_{at}}$$

where d = density = 8.93 g cm^{-3}, and M_{at} = atomic mass = 63.5 (g mol^{-1}). Substituting for d, N_A, and M_{at}, and we find $n = 8.5 \times 10^{22}$ electrons cm^{-3}.

The electron drift mobility is therefore

$$\mu_d = \frac{\sigma}{en} = \frac{5.9 \times 10^5 \ \Omega^{-1} \ \text{cm}^{-1}}{[(1.6 \times 10^{-19} \ \text{C})(8.5 \times 10^{22} \ \text{cm}^{-3})]}$$

$$= 43.4 \ \text{cm}^2 \ \text{V}^{-1} \ \text{s}^{-1}$$

Experimentally (from Hall effect measurement), $\mu_d = 32$ cm^2 V^{-1} s^{-1}. The difference is due to the fact that, on average, more than one conduction electron is donated per Cu atom.

Example 2.3 | **DRIFT VELOCITY AND THE MEAN SPEED**   What is the applied electric field that will impose a drift velocity equal to 10% of the mean speed u ($\sim 1.2 \times 10^6$ m s^{-1}) of the conduction electrons in copper?

Solution

The drift velocity of the conduction electrons is $v_{dx} = \mu_d \mathcal{E}_x$, where μ_d is the drift mobility, which for copper is 32 cm^2 V^{-1} s^{-1} (see Example 2.1). With $v_{dx} = 0.1u = 1.2 \times 10^5$ m s^{-1}, we have

$$\mathcal{E} = \frac{v_{dx}}{\mu_d} = \frac{1.2 \times 10^5 \ \text{m s}^{-1}}{3.2 \times 10^{-3} \ \text{m}^2 \ \text{V}^{-1} \ \text{s}^{-1}} = 3.8 \times 10^7 \ \text{V m}^{-1} = 38 \ \text{MV m}^{-1}$$

It is clear from this example that for all practical purposes, even under the highest working voltages, the drift velocity is much smaller than the mean speed of the electrons. Consequently, when an electric field is applied to a conductor, for all practical purposes the mean speed is unaffected.

2.2   TEMPERATURE DEPENDENCE OF RESISTIVITY: IDEAL PURE METALS

When the conduction electrons are only scattered by thermal vibrations of the metal ions, then τ in the mobility expression $\mu_d = e\tau/m_e$ refers to the mean time between scattering events by this process. The resulting conductivity and resistivity are denoted by σ_T and ρ_T, where the subscript T represents "thermal vibration scattering."

To find the temperature dependence of σ, we first consider the temperature dependence of the mean free time τ, since this determines the drift mobility. An electron moving with a mean speed u is scattered when its path crosses the cross-sectional area S of a scattering center, as depicted in Figure 2.4. The scattering center may be a vibrating atom, impurity, vacancy, or other crystal defect. Since τ is the mean time taken for one scattering process, the **mean free path** ℓ of the electron between scattering processes is $u\tau$. If N_s is the concentration of scattering centers, then in the volume $S\ell$, there is one scattering center, that is, $(Su\tau)N_s = 1$. Thus, the mean free time is given by

$$\tau = \frac{1}{SuN_s}$$
[2.9]   *Mean free time between collisions*

The mean speed u of conduction electrons in a metal can be shown to be only slightly temperature dependent.[3] In fact, electrons wander randomly around in the metal crystal with an almost constant mean speed that depends largely on their concentration and hence on the crystal material. Taking the number of scattering centers per unit volume to be the atomic concentration, the temperature dependence of τ then arises essentially from that of the cross-sectional area S. Consider what a free electron "sees" as it approaches a vibrating crystal atom as in Figure 2.4. Because the atomic vibrations are random, the atom covers a cross-sectional area πa^2, where a is the amplitude of the vibrations. If the electron's path crosses πa^2, it gets scattered. Therefore, the mean time between scattering events τ is inversely proportional to the area πa^2 that scatters the electron, that is, $\tau \propto 1/\pi a^2$.

The thermal vibrations of the atom can be considered to be simple harmonic vibration, much the same way as that of a mass M attached to a spring. The average

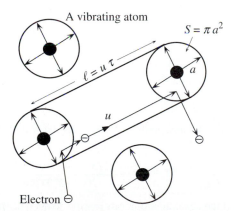

A vibrating atom

$S = \pi a^2$

$\ell = u\tau$

a

u

Electron $\ominus$

Figure 2.4   Scattering of an electron from the thermal vibrations of the atoms.

The electron travels a mean distance $\ell = u\tau$ between collisions. Since the scattering cross-sectional area is S, in the volume $S\ell$, there must be at least one scatterer, $N_s(Su\tau) = 1$.

[3] The fact that the mean speed of electrons in a metal is only weakly temperature dependent can be proved from what is called the Fermi–Dirac statistics for the collection of electrons in a metal (see Chapter 4). This result contrasts sharply with the kinetic molecular theory of gases (Chapter 1), which predicts that the mean speed of molecules is proportional to $\sqrt{T}$. For the time being, we simply use a constant mean speed u.

kinetic energy of the oscillations is $(1/4)\,Ma^2\omega^2$, where ω is the oscillation frequency. From the kinetic theory of matter, this average kinetic energy must be on the order of $(1/2)\,kT$. Therefore,

$$Ma^2\omega^2 = 2kT$$

so $a^2 \propto T$. Intuitively, this is correct because raising the temperature increases the amplitude of the atomic vibrations. Thus,

$$\tau \propto \frac{1}{\pi a^2} \propto \frac{1}{T} \quad \text{or} \quad \tau = \frac{C}{T}$$

where C is a temperature-independent constant. Substituting for τ in $\mu_d = e\tau/m_e$, we obtain

$$\mu_d = \frac{eC}{m_e T}$$

So, the resistivity of a metal is

$$\rho_T = \frac{1}{\sigma_T} = \frac{1}{en\mu_d} = \frac{m_e T}{e^2 n C}$$

that is,

$$\rho_T = AT \qquad\qquad [2.10]$$

where A is a temperature-independent constant. This shows that the resistivity of a pure metal wire increases linearly with the temperature, and that the resistivity is due simply to the scattering of conduction electrons by the thermal vibrations of the atoms. We term this conductivity **lattice-scattering-limited conductivity.**

Example 2.4 | **TEMPERATURE DEPENDENCE OF RESITIVITY**   What is the percentage change in the resistance of a pure metal wire from Saskatchewan's summer to winter, neglecting the changes in the dimensions of the wire?

Solution

Assuming $20\,°C$ for the summer and perhaps $-30\,°C$ for the winter, from $R \propto \rho = AT$, we have

$$\frac{R_{summer} - R_{winter}}{R_{summer}} = \frac{T_{summer} - T_{winter}}{T_{summer}} = \frac{(20 + 273) - (-30 + 273)}{(20 + 273)} = 0.171 \quad \text{or} \quad 17\%$$

Notice that we have used the absolute temperature for T. How will the outdoor cable power losses be affected?

Example 2.5 | **DRIFT MOBILITY AND RESISTIVITY DUE TO LATTICE VIBRATIONS**   Given that the mean speed of conduction electrons in copper is 1.25×10^6 m s^{-1} and the frequency of vibration of the copper atoms at room temperature is about 4×10^{12} s^{-1}, estimate the drift mobility of electrons and the conductivity of copper. The density d of copper is 8.96 g cm^{-3} and the atomic mass M_{at} is 63.56 g mol^{-1}.

Solution

The method for calculating the drift mobility and hence the conductivity is based on evaluating the mean free time τ via Equation 2.9, that is, $\tau = 1/SuN_s$. Since τ is due to scattering from atomic vibrations, N_s is the atomic concentration,

$$N_s = \frac{dN_A}{M_{at}} = \frac{(8.96 \times 10^3 \text{ kg m}^{-3})(6.02 \times 10^{23} \text{ mol}^{-1})}{63.56 \times 10^{-3} \text{ kg mol}^{-1}}$$

$$= 8.5 \times 10^{28} \text{ m}^{-3}$$

The cross-sectional area $S = \pi a^2$ depends on the amplitude a of the thermal vibrations Figure 2.4. The average kinetic energy KE_{av} associated with a vibrating mass M attached to a spring is given by $KE_{av} = (1/4)Ma^2\omega^2$, where ω is the angular frequency of the vibration ($\omega = 2\pi 4 \times 10^{12}$ rads s^{-1}). Applying this equation to the vibrating atom and equating the average kinetic energy KE_{av} to $(1/2)kT$, by virtue of equipartition of energy theorem, we have $a^2 = 2kT/M\omega^2$ and thus

$$S = \pi a^2 = \frac{2\pi kT}{M\omega^2} = \frac{2\pi(1.38 \times 10^{-23} \text{ J K}^{-1})(300 \text{ K})}{\left(\dfrac{63.56 \times 10^{-3} \text{ kg mol}^{-1}}{6.022 \times 10^{23} \text{ mol}^{-1}}\right)(2\pi \times 4 \times 10^{12} \text{ rad s}^{-1})^2}$$

$$= 4 \times 10^{-22} \text{ m}^2$$

Therefore,

$$\tau = \frac{1}{(SuN_s)} = \frac{1}{[(4 \times 10^{-22} \text{ m}^2)(1.25 \times 10^6 \text{ m s}^{-1})(8.5 \times 10^{28} \text{ m}^{-3})]} = 2.35 \times 10^{-14} \text{ s}$$

The drift mobility is

$$\mu_d = \frac{e\tau}{m_e} = \frac{(1.6 \times 10^{-19} \text{ C})(2.35 \times 10^{-14} \text{ s})}{(9.1 \times 10^{-31} \text{ kg})}$$

$$= 4.13 \times 10^{-3} \text{ m}^2 \text{ V}^{-1} \text{ s}^{-1} = 41.3 \text{ cm}^2 \text{ V}^{-1} \text{ s}^{-1}$$

The conductivity is then

$$\sigma = en\mu_d = (1.6 \times 10^{-19} \text{ C})(8.5 \times 10^{22} \text{ cm}^{-3})(41.3 \text{ cm}^2 \text{ V}^{-1} \text{ s}^{-1})$$

$$= 5.6 \times 10^5 \text{ }\Omega^{-1} \text{ cm}^{-1}$$

The experimentally measured values for the drift mobility and conductivity are 32 cm^2 V^{-1} s^{-1} and 5.9×10^5 Ω^{-1} cm^{-1}, respectively, which are in surprisingly close agreement with the crude calculation based on Equation 2.9. (As we might have surmised, the agreement is brought about by using reasonable values for the mean speed u and the atomic vibrational frequency, which were taken from quantum mechanical calculations. So our evaluation for τ was not truly based on classical concepts.)

2.3   MATTHIESSEN'S RULE

2.3.1   Matthiessen's Rule and the Temperature Coefficient of Resistivity (α)

The theory of conduction that considers scattering from lattice vibrations only works well with pure metals; unfortunately, it fails for metallic alloys. Their resistivities are only weakly temperature dependent. We must therefore search for a different type of scattering mechanism.

Consider a metal alloy that has randomly distributed impurity atoms. An electron can now be scattered by the impurity atoms because they are not identical to the host atoms, as illustrated in Figure 2.5. The impurity atom need not be larger than the host atom; it can be smaller. As long as the impurity atom results in a local distortion of the crystal lattice, it will be effective in scattering. One way of looking at the scattering process from an impurity is to consider the scattering cross section. What actually scatters the electron is a local, unexpected change in the potential energy *PE* of the electron as it approaches the impurity, because the force experienced by the electron is given by

$$F = \frac{-d(PE)}{dx}$$

For example, when an impurity atom of a different size compared to the host atom is placed into the crystal lattice, the impurity atom distorts the region around it, either by pushing the host atoms farther away, or by pulling them in, as depicted in Figure 2.5. The cross section that scatters the electron is the lattice region that has been elastically distorted by the impurity (the impurity atom itself and its neigh-

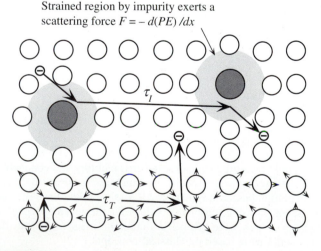

Figure 2.5   Two different types of scattering processes involving scattering from impurities alone and from thermal vibrations alone.

boring host atoms), so that in this zone, the electron suddenly experiences a force $F = -d(PE)/dx$ due to a sudden change in the *PE*. This region has a large scattering cross section, since the distortion induced by the impurity may extend a number of atomic distances. These impurity atoms will therefore hinder the motion of the electrons, thereby increasing the resistance.

We now effectively have two types of mean free times between collisions: one, τ_T, for scattering from thermal vibrations only, and the other, τ_I, for scattering from impurities only. We define τ_T as the mean time between scattering events arising from thermal vibrations alone and τ_I as the mean time between scattering events arising from collisions with impurities alone. Both are illustrated in Figure 2.5.

In general, an electron may be scattered by both processes, so the effective mean free time τ between any two scattering events will be less than the individual scattering times τ_T and τ_I. The electron will therefore be scattered when it collides with either an atomic vibration or an impurity atom. Since in unit time, $1/\tau$ is the net probability of scattering, $1/\tau_T$ is the probability of scattering from lattice vibrations alone, and $1/\tau_I$ is the probability of scattering from impurities alone, then within the realm of elementary probability theory for independent events, we have

$$\frac{1}{\tau} = \frac{1}{\tau_T} + \frac{1}{\tau_I}$$ [2.11]

In writing Equation 2.11 for the various probabilities, we make the reasonable assumption that, to a greater extent, the two scattering mechanisms are essentially independent. Here, the effective mean scattering time τ is clearly smaller than both τ_T and τ_I. We can also interpret Equation 2.11 as follows: In unit time, the overall number of collisions $(1/\tau)$ is the sum of the number of collisions with thermal vibrations alone $(1/\tau_T)$ and the number of collisions with impurities alone $(1/\tau_I)$.

The drift mobility μ_d depends on the effective scattering time τ via $\mu_d = e\tau/m_e$, so Equation 2.11 can also be written in terms of the drift mobilities determined by the various scattering mechanisms. In other words,

$$\frac{1}{\mu_d} = \frac{1}{\mu_L} + \frac{1}{\mu_I}$$ [2.12]

where μ_L is the lattice-scattering-limited drift mobility, and μ_I is the impurity-scattering-limited drift mobility. By definition, $\mu_L = e\tau_T/m_e$ and $\mu_I = e\tau_I/m_e$. The effective (or overall) resistivity ρ of the material is simply $1/en\mu_d$, or

$$\rho = \frac{1}{en\mu_d} = \frac{1}{en\mu_L} + \frac{1}{en\mu_T}$$

which can be written

$$\rho = \rho_T + \rho_I$$ [2.13]   *Matthiessen's Rule*

where $1/en\mu_L$ is defined as the resistivity due to scattering from thermal vibrations, and $1/en\mu_I$ is the resistivity due to scattering from impurities, or

$$\rho_T = \frac{1}{en\mu_L} \quad \text{and} \quad \rho_I = \frac{1}{en\mu_I}$$

The final result in Equation 2.13 simply states that the effective resistivity ρ is the sum of two contributions. First, $\rho_T = 1/en\mu_L$ is the resistivity due to scattering by thermal vibrations of the host atoms. For those near-perfect pure metal crystals, this is the dominating contribution. As soon as we add impurities, however, there is an additional resistivity, $\rho_I = 1/en\mu_I$, which arises from the scattering of the electrons from the impurities. The first term is temperature dependent because $\tau_T \propto T^{-1}$ (see Section 2.2), but the second term is not.

The mean time τ_I between scattering events involving electron collisions with impurity atoms depends on the separation between the impurity atoms and there-fore on the concentration of those atoms (see Figure 2.5). If ℓ_I is the mean separation between the impurities, then the mean free time between collisions with impurities alone will be ℓ_I/u, which is temperature independent because ℓ_I is determined by the impurity concentration N_I (i.e., $\ell_I = N_I^{1/3}$), and the mean speed of the electrons u is nearly constant in a metal. In the absence of impurities, τ_I is infinitely long, and thus $\rho_I = 0$. The summation rule of resistivities from different scattering mechanisms, as shown by Equation 2.13 is called **Matthiessen's Rule.**

There may also be electrons scattering from dislocations and other crystal defects, as well as from gain boundaries. All of these scattering processes add to the resistivity of a metal, just as the scattering process from impurities. We can therefore write the effective resistivity of a metal as

Matthiessen's Rule

$$\rho = \rho_T + \rho_R \qquad [2.14]$$

where ρ_R is called the **residual resistivity** and is due to the scattering of electrons by impurities, dislocations, interstitial atoms, vacancies, grain boundaries, etc. (which means that ρ_R also includes ρ_I). The residual resistivity shows very little temperature dependence, whereas $\rho_T = AT$, so the effective resistivity ρ is given by

$$\rho \approx AT + B \qquad [2.15]$$

where A and B are temperature-independent constants.

Equation 2.15 indicates that the resistivity of a metal varies almost linearly with the temperature, with A and B depending on the material. Instead of listing A and B in resistivity tables, we prefer to use a temperature coefficient that refers to small, normalized changes around a reference temperature. The **temperature coefficient of resistivity (TCR)** α_0 is defined as the fractional change in the resistivity per unit temperature increase at the reference temperature T_0, that is,

Definition of temperature coefficient of resistivity

$$\alpha_0 = \frac{1}{\rho_0}\left[\frac{\delta\rho}{\delta T}\right]_{T=T_0} \qquad [2.16]$$

where ρ_0 is the resistivity at the reference temperature T_0, usually 273 K (0 °C) or 293 K (20 °C), and $\delta\rho = \rho - \rho_0$ is the change in the resistivity due to a small increase in temperature, $\delta T = T - T_0$.

When the resistivity follows the behavior $\rho \approx AT + B$ in Equation 2.15, then according to Equation 2.16, α_0 is constant over a temperature range T_0 to T, and Equation 2.16 leads to the well-known equation,

Temperature dependence of resistivity

$$\rho = \rho_0[1 + \alpha_0(T - T_0)] \qquad [2.17]$$

Equation 2.17 is actually only valid when α_0 is constant over the temperature range of interest, which requires Equation 2.15 to hold. Over a limited temperature range, this will usually be the case. Although it is not obvious from Equation 2.17, we should note that α_0 depends on the reference temperature T_0, by virtue of ρ_0 depending on T_0.

The equation $\rho = AT$, which we used for perfect, pure-metal crystals to find the change in the resistance with temperature, is only approximate; nonetheless, for pure metals, it is useful to recall in the absence of tabulated data. To determine how good the formula $\rho = AT$ is, put it in Equation 2.17, which leads to $\alpha_0 = T_0^{-1}$. If we take the reference temperature T_0 as 273 K (0 °C), then α_0 is simply 1/273 K; stated differently, Equation 2.17 is then equivalent to $\rho = AT$.

Table 2.1 shows that $\rho \propto T$ is not a bad approximation for some of the familiar pure metals used as conductors (Cu, Al, Au, etc.), but it fails badly for others, such as indium, antimony and, in particular, the magnetic metals, iron and nickel.

Table 2.1   Resistivity, thermal coefficient of resistivity α_0 at 273 K (0 °C) for various metals. The resistivity index n in $\rho \propto T^n$ for some of the metals is also shown.

Metal	$\rho_0(n\Omega\ m)$	$\alpha_0\left(\dfrac{1}{K}\right)$	n	Comment
Aluminum, Al	25.0	$\dfrac{1}{233}$		
Antimony, Sb	38	$\dfrac{1}{196}$		
Copper, Cu	15.7	$\dfrac{1}{232}$	1.15	
Gold, Au	22.8	$\dfrac{1}{251}$		
Indium, In	78.0	$\dfrac{1}{196}$		
Platinum, Pt	98	$\dfrac{1}{255}$	0.94	
Silver, Ag	14.6	$\dfrac{1}{244}$	1.11	
Tantalum, Ta	117	$\dfrac{1}{294}$	0.93	
Tin, Sn	110	$\dfrac{1}{247}$	1.11	
Tungsten, W	50	$\dfrac{1}{202}$	1.20	
Iron, Fe	84.0	$\dfrac{1}{152}$	1.80	Magnetic metal; $273 < T < 1043$ K
Nickel, Ni	59.0	$\dfrac{1}{125}$	1.72	Magnetic metal; $273 < T < 627$ K

ǀ NOTE: Data was extracted and combined from several sources.

The temperature dependence of the resistivity of various metals is shown in Figure 2.6, where it is apparent that except for the magnetic materials, such as iron and nickel, the linear relationship, $\rho \propto T$ seems to be approximately obeyed almost all the way to the melting temperature for many pure metals. It should also be noted that for the alloys, such as nichrome (Ni–Cr), the resistivity is essentially dominated by the residual resistivity, so the resistivity is relatively temperature insensitive, with a very small TCR.

Frequently, the resistivity versus temperature behavior of pure metals can be empirically represented by a power law of the form

Resistivity of pure metal

$$\rho = \rho_0 \left[\frac{T}{T_0} \right]^n$$

[2.18]

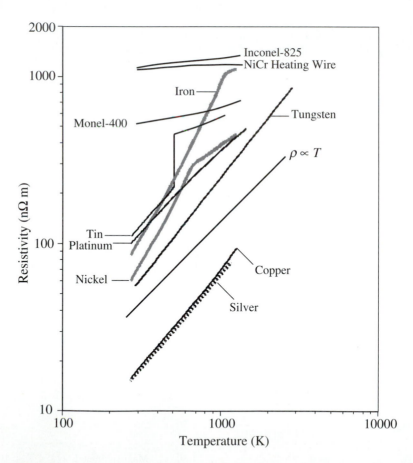

Figure 2.6   The resistivity of various metals as a function of temperature above 0 °C.

Tin melts at 505 K, whereas nickel and iron go through a magnetic-to-non-magnetic (Curie) transformation at about 627 K and 1043 K, respectively. The theoretical behavior ($\rho \sim T$) is shown for reference.

Data selectively extracted from various sources, including sections in *Metals Handbook*, 10th ed., 2 and 3. Metals Park, Ohio: ASM, 1991.

where ρ_0 is the resistivity at the reference temperature T_0, and n is a characteristic index that best fits the data. Table 2.1 lists some typical n values for various pure metals above 0 °C. It is apparent that for the nonmagnetic metals, n is close to unity, whereas it is closer to 2 than 1 for the magnetic metals Fe and Ni. In iron, for example, the conduction electron is not scattered simply by atomic vibrations, as in copper, but is affected by its magnetic interaction with the Fe ions in the lattice. This leads to a complicated temperature dependence.

Although our oversimplified theoretical analysis predicts a linear $\rho = AT + B$ behavior for the resistivity down to the lowest temperatures, this is not true in reality, as depicted for copper in Figure 2.7. As the temperature decreases, typically below ~100 K for many metals, our simple and gross assumption that all the atoms are vibrating with a constant frequency fails. Indeed, the number of atoms that are vibrating with sufficient energy to scatter the conduction electrons starts to decrease rapidly with decreasing temperature, so the resistivity due to scattering from thermal vibrations becomes more strongly temperature dependent. The mean free time $\tau = 1/SuN_s$ becomes longer and strongly temperature dependent, leading to a smaller resistivity than the $\rho \propto T$ behavior. A full theoretical analysis, which is beyond the scope of this book, shows that $\rho \propto T^5$. Thus, at the lowest tempera-

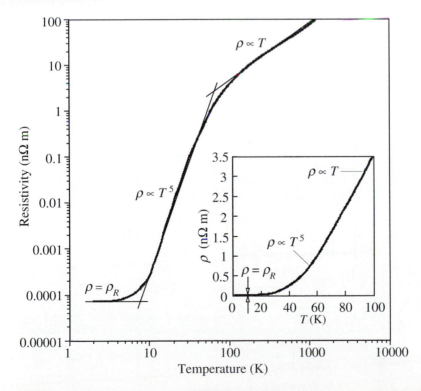

Figure 2.7   The resistivity of copper from lowest to highest temperatures (near melting temperature, 1358 K) on a log-log plot.

Above about 100 K, $\rho \propto T$, whereas at low temperatures, $\rho \propto T^5$, and at the lowest temperatures ρ approaches the residual resistivity ρ_R. The inset shows the ρ vs. T behavior below 100 K on a linear plot. (ρ_R is too small on this scale.)

tures, from Matthiessen's rule, the resistivity becomes $\rho = DT^5 + \rho_R$, where D is a constant. Since the slope of ρ vs. T is $d\rho/dT = 5DT^4$, which tends to zero as T becomes small, we have ρ curving toward ρ_R as T decreases toward 0 K. This is borne out by experiments, as shown in Figure 2.7 for copper. Therefore, at the lowest temperatures of interest, the resistivity is limited by scattering from impurities and crystal defects.[4]

| **Example 2.6** | **MATTHIESSEN'S RULE**   Explain the typical resistivity versus temperature behavior of annealed and cold-worked (deformed) copper containing various amounts of Ni as shown in Figure 2.8. |

Solution

When small amounts of nickel are added to copper, the resistivity increases by virtue of Matthiessen's rule, $\rho = \rho_T + \rho_R + \rho_I$, where ρ_T is the resistivity due to scattering from thermal vibrations, ρ_R is the residual resistivity of the copper crystal due to scattering from crystal defects, dislocations, trace impurities, etc., and ρ_I is the resistivity arising from Ni addition alone (scattering from Ni impurity regions). Since ρ_I is temperature independent, for small amounts of Ni addition, ρ_I will simply shift up the ρ versus T curve for copper, by an amount proportional to the Ni content, $\rho_I \propto N_{Ni}$, where N_{Ni} is the Ni impurity concentration. This is apparent in Figure 2.8, where the resistivity of Cu–2. 16% Ni is almost twice that of Cu–1.12% Ni. Cold working (CW) or deforming a metal results in a higher concentration

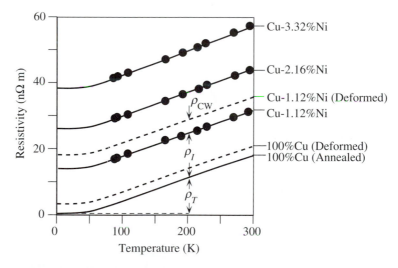

Figure 2.8   Typical temperature dependence of the resistivity of annealed and cold-worked (deformed) copper containing various amounts of Ni in atomic percentage.

| SOURCE: Data adapted from J.O. Linde, *Ann Pkysik*, 5, 219 (Germany, 1932).

---

[4] At sufficiently low temperatures (typically, below 10–20 K for many metals and below ~135 K for certain ceramics) certain materials exhibit superconductivity in which the resistivity vanishes ($\rho = 0$), even in the presence of impurities and crystal defects. Superconductivity and its quantum mechanical origin will be explained in Chapter 9.

of dislocations and therefore increases the residual resistivity ρ_R by ρ_{CW}. Thus, cold-worked samples have a resistivity curve that is shifted up by an additional amount ρ_{CW} that depends on the extent of cold working.

TEMPERATURE COEFFICIENT OF RESISTIVITY α AND RESISTIVITY INDEX n   If α_0 is the tempera- | **Example 2.7**
ture coefficient of resistivity (TCR) at temperature T_0 and the resistivity obeys the equation

$$\rho = \rho_0 \left[\frac{T}{T_0} \right]^n$$

show that

$$\alpha_0 = \frac{n}{T_0} \left[\frac{T}{T_0} \right]^{n-1}$$

What is your conclusion?

Experiments indicate that $n = 1.2$ for W. What is its α_0 at 20 °C? Given that, experimentally, $\alpha_0 = 0.00393$ for Cu at 20 °C, what is n?

Solution

Since the resistivity obeys $\rho = \rho_0(T/T_0)^n$, we substitute this equation into the definition of TCR,

$$\alpha_0 = \frac{1}{\rho_0} \left[\frac{d\rho}{dT} \right] = \frac{n}{T_0} \left[\frac{T}{T_0} \right]^{n-1}$$

It is clear that, in general, α_0 depends on the temperature T, as well as on the reference temperature T_0. The TCR is only independent of T when $n = 1$.

At $T = T_0$, we have

$$\frac{\alpha_0 T_0}{n} = 1 \qquad \text{or} \qquad n = \alpha_0 T_0$$

For W, $n = 1.2$, so at $T = T_0 = 293$ K, we have $\alpha_{293 \text{ K}} = 0.0041$, which agrees reasonably well with $\alpha_{293 \text{ K}} = 0.0045$, frequently found in data books.

For Cu, $\alpha_{293 \text{ K}} = 0.00393$, so that $n = 1.15$, which agrees with the experimental value of n.

TCR AT DIFFERENT REFERENCE TEMPERATURES   If α_1 is the temperature coefficient of resistiv- | **Example 2.8**
ity (TCR) at temperature T_1 and α_0 is the TCR at T_0, show that

$$\alpha_1 = \frac{\alpha_0}{1 + \alpha_0(T_1 - T_0)}$$

Solution

Consider the resistivity at temperature T in terms of α_0 and α_1:

$$\rho = \rho_0[1 + \alpha_0(T - T_0)] \qquad \text{and} \qquad \rho = \rho_1[1 + \alpha_1(T - T_1)]$$

These equations are expected to hold at any temperature, T, so the first and second equations at T_1 and T_0 respectively, give

$$\rho_1 = \rho_0[1 + \alpha_0(T_1 - T_0)] \qquad \text{and} \qquad \rho_0 = \rho_1[1 + \alpha_1(T_0 - T_1)]$$

These two equations can be readily solved to eliminate ρ_0 and ρ_1 to obtain

$$\alpha_1 = \frac{\alpha_0}{1 + \alpha(T_1 - T_0)}$$

Example 2.9 | **TEMPERATURE OF THE FILAMENT OF A LIGHT BULB**

a. Consider a 40 W, 120 V incandescent light bulb. The tungsten filament is 0.381 m long and has a diameter of 33 μm. Its resistivity at room temperature is 5.51×10^{-8} Ω m. Given that the resistivity of the tungsten filament varies at $T^{1.2}$, estimate the temperature of the bulb when it is operated at the rated voltage, that is, it is lit directly from a power outlet, as shown schematically in Figure 2.9. Note that the bulb dissipates 40 W at 120 V.

b. Assume that the electrical power dissipated in the tungsten wire is radiated from the surface of the filament. The radiated electromagnetic power at the absolute temperature T can be described by **Stefan's Law,** as follows:

$$P_{\text{radiated}} = \epsilon \sigma_S A(T^4 - T_0^4)$$

where σ_S is Stefan's constant (5.6×10^{-8} W m^{-2} K^{-4}), ϵ is the emissivity of the surface (0.35 for tungsten), A is the surface area of the tungsten filament, and T_0 is the room temperature (293 K), For $T > T_0$, the equation becomes

$$P_{\text{radiated}} = \epsilon \sigma_S A T^4$$

Assuming that all of the electrical power is radiated as electromagnetic waves from the surface, estimate the temperature of the filament and compare it with your answer in part (a).

Solution

a. When the bulb is operating at 120 V, it is dissipating 40 W, which means that the current is

$$I = \frac{P}{V} = \frac{40 \text{ W}}{120 \text{ V}} = 0.333 \text{ A}$$

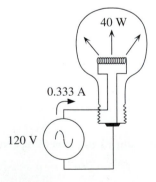

Figure 2.9   Power radiated from a light bulb at 2408 °C is equal to the electrical power dissipated in the filament.

The resistance of the filament at the operating temperature T must be

$$R = \frac{V}{I} = \frac{120}{0.333} = 360 \ \Omega$$

Since $R = \rho L/A$, the resistivity of tungsten at the operating temperature T must be

$$\rho(T) = \frac{R(\pi D^2/4)}{L} = \frac{360 \ \Omega \ \pi(33 \times 10^{-6} \ \text{m})^2}{4(0.381 \ \text{m})} = 8.08 \times 10^{-7} \ \Omega \ \text{m}$$

But, $\rho(T) = \rho_0(T/T_0)^{1.2}$, so that

$$T = T_0 \left(\frac{80.8 \times 10^{-8}}{5.51 \times 10^{-8}}\right)^{1/1.2}$$

$$T = 2746 \ \text{K} \quad \text{or} \quad 2473 \ °\text{C} \qquad \text{(melting temperature of W is about 3680 K)}$$

b. To calculate T from the radiation law, we note that $T = [P_{\text{radiated}}/\epsilon \sigma_S A]^{1/4}$. The surface area is

$$A = L(\pi D) = (0.381)(\pi \, 33 \times 10^{-6}) = 3.95 \times 10^{-5} \ \text{m}^2$$

Then,

$$T = \left[\frac{P_{\text{radiated}}}{\epsilon \sigma_S A}\right]^{1/4} = \left[\frac{40 \ \text{W}}{(0.35)(5.6 \times 10^{-8} \ \text{W m}^{-2} \ \text{K}^{-4})(3.95 \times 10^{-5} \ \text{m}^2)}\right]^{1/4}$$

$$= [5.167 \times 10^{13}]^{1/4} = 2681 \ \text{K} \quad \text{or} \quad 2408 \ °\text{C}$$

2.3.2   Solid Solutions and Nordheim's Rule

In an isomorphous alloy of two metals, that is, a binary alloy that forms a solid solution, we would expect Equation 2.11 to apply, with the temperature-independent impurity contribution ρ_I increasing with the concentration of solute atoms. This means that as the alloy concentration increases, the resistivity ρ increases and becomes less temperature dependent as ρ_I overwhelms ρ_T, leading to $\alpha \ll 1/273$. This is the advantage of alloys in resistive components. Table 2.2 shows that when 80% nickel is alloyed with 20% chromium, the resistivity of Ni increases almost sixteen times. In fact, the alloy is called **nichrome** and is widely used as a heater wire in household appliances and industrial furnaces.

Table 2.2   The effect of alloying on the resistivity

Material	Resistivity at 20 °C $n\Omega$ m	α at 20 °C $\frac{1}{K}$
Nickel	69	0.006
Chrome	129	0.003
Nichrome	1120	0.0003

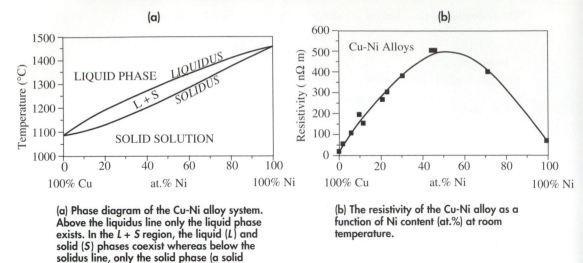

(a) Phase diagram of the Cu-Ni alloy system. Above the liquidus line only the liquid phase exists. In the L + S region, the liquid (L) and solid (S) phases coexist whereas below the solidus line, only the solid phase (a solid solution) exists.

(b) The resistivity of the Cu-Ni alloy as a function of Ni content (at.%) at room temperature.

Figure 2.10   The Cu-Ni alloy system.

SOURCE: Data extracted from *Metals Handbook*, 10th ed., 2 and 3, Metals Park, Ohio: ASM, 1991, and *Constitution of Binary Alloys*, M. Hansen, and K. Anderko, New York: McGraw-Hill, 1958.

As a further example of the resistivity of a solid solution consider the copper–nickel alloy. The phase diagram for this alloy system is shown in Figure 2.10a. It is clear that the alloy forms a one-phase solid solution for all compositions. Both Cu and Ni have the same FCC crystal structure, and since the Cu atom is only slightly larger than the Ni atom by about ~3% (easily checked on the Periodic Table), the Cu–Ni alloy will therefore still be FCC, but with Cu and Ni atoms randomly mixed, resulting in a solid solution. When Ni is added to copper, the impurity resistivity ρ_I in Equation 2.11 will increase with the Ni concentration. Experimental results for this alloy system are shown in Figure 2.10b. It should be apparent that when we reach 100% Ni, we again have a pure metal whose resistivity must be small. Therefore, ρ versus Ni concentration must pass through a maximum, which for the Cu–Ni alloy seems to be at around ~50% Ni.

There are other binary solid solutions that reflect very similar behavior to that depicted in Figure 2.10, such as Cu–Au, Ag–Au, Pt–Pd, Cu–Pd, to name a few. Quite often, the use of an alloy for a particular application is necessitated by the mechanical properties, rather than the desired electrical resistivity alone. For example, brass, which is 70% Cu–30% Zn in solid solution has a higher strength compared to pure copper; as such, it is a suitable metal for the prongs of an electrical plug.

An important semiempirical equation that can be used to predict the resistivity of an alloy is the **Nordheim rule** which relates the impurity resistivity ρ_I to the atomic fraction X of solute atoms in a solid solution, as follows:

The Nordheim rule for solid solutions

$$\rho_I = CX(1 - X) \qquad \text{[2.19]}$$

where C is the constant termed the **Nordheim coefficient,** which represents the effectiveness of the solute atom in increasing the resistivity. Obviously, for small

Table 2.3   Nordheim coefficient (at 20 °C) for dilute alloys

Solute in Solvent (element in matrix)	Nordheim Coefficient (nΩ m)	Maximum Solubility at 25 °C (at.%)
Au in Cu matrix	5500	100
Mn in Cu matrix	2900	24
Ni in Cu matrix	1250	100
Sn in Cu matrix	2900	0.6
Zn in Cu matrix	300	30
Cu in Au matrix	450	100
Mn in Au matrix	2410	25
Ni in Au matrix	790	100
Sn in Au matrix	3360	5
Zn in Au matrix	950	15

SOURCES: *Electronics Engineers' Handbook,* 2nd ed., ed. D.G. Fink, and D. Christiansen. New York: McGraw-Hill, 1982. Section 6, J.K. Stanley, *Electrical and Magnetic Properties of Metals,* Metals Park, Ohio: American Society for Metals, 1963. Solubility data from M. Hansen, and K. Anderko, *Constitution of Binary Alloys,* 2nd ed., New York: McGraw-Hill, 1985.

amounts of impurity, $X \ll 1$ and $\rho_I = CX$, which explains the initial approximately equal increments of rise in the resistivity of copper with 1.11% and 2.16% Ni additions as shown in Figure 2.8.

Table 2.3 lists some typical Nordheim coefficients for various additions to copper and gold. The value of the Nordheim coeffcient depends on the type of solute and the solvent. A solute atom that is drastically different in size to the solvent atom will result in a bigger increase in ρ_I and will therefore lead to a larger C. An important assumption in the Nordheim rule in Equation 2.19 is that the alloying does not significantly vary the number of conduction electrons per atom in the alloy. Although this will be true for alloys with the same valency, that is, from the same column in the Periodic Table (e.g., Cu–Au, Pd–Pt), it will not be true for alloys of different valency, such Cu and Zn. In pure copper, there is just one conduction electron per atom, whereas each Zn atom can donate two conduction electrons. As the Zn content in brass is increased, more conduction electrons become available per atom. Consequently, the resistivity predicted by Equation 2.17 at high Zn contents is greater than the actual value. Nonetheless, the Nordheim rule is still useful, for predicting the resistivities of dilute alloys, particularly in the low-concentration region.

With the Nordheim rule in Equation 2.19, the resistivity of an alloy of composition X is

$$\rho = \rho_{\text{matrix}} + CX(1 - X) \qquad \text{[2.20]}$$

Combined Matthiessen and Nordheim rules

where $\rho_{\text{matrix}} = \rho_T + \rho_R$ is the resistivity of the matrix due to scattering from thermal vibrations and from other defects, in the absence of alloying elements. To reiterate, the value of C depends on the alloying element and the matrix. For example, C for gold in copper would be different than C for copper in gold, as shown in Table 2.3.

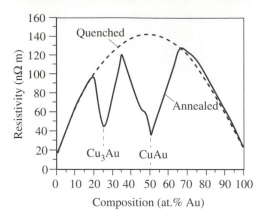

Figure 2.11   Electrical resistivity vs. composition at room temperature in Cu-Au alloys.

The quenched sample (dashed curve) is obtained by quenching the liquid, and the Cu and Au atoms are randomly mixed. The resistivity obeys the Nordheim rule. When the quenched sample is annealed or the liquid is slowly cooled (solid curve), certain compositions (Cu_3Au and CuAu) result in an ordered crystalline structure in which the Cu and Au atoms are positioned in an ordered fashion in the crystal and the scattering effect is reduced.

In solid solutions, at some concentrations of certain binary alloys, such as 75% Cu–25% Au   and   50% Cu–50% Au, the annealed solid has an orderly structure; that is, the Cu and Au atoms are not randomly mixed, but occupy regular sites. In fact, these compositions can be viewed as pure compound-like the solids Cu_3Au and CuAu. The resistivities of Cu_3Au and CuAu will therefore be less than the same composition random alloy that has been quenched from the melt. As a consequence, the resistivity ρ versus composition X curve does not follow the dashed parabolic curve throughout; rather, it exhibits sharp falls at these special compositions, as illustrated Figure 2.11.

Example 2.10   **NORDHEIM'S RULE**   The alloy 90 wt.% Au–10 wt.% Cu is sometimes used in low-voltage dc electrical contacts, because pure gold is mechanically soft and the addition of copper increases the hardness of the metal without sacrificing the corrosion resistance. Predict the resistivity of the alloy and compare it with the experimental value of 108 nΩ m.

Solution

We apply Equation 2.20, $\rho(X) = \rho_{Au} + CX(1 - X)$ but with 10 wt.% Cu converted to the atomic fraction for X. If w is the weight fraction of solute (Cu), then the atomic fraction X is given by

$$X = \frac{M_{Au}w}{(1 - w)M_{Cu} + wM_{Au}}$$

where M_{Au} and M_{Cu} are the atomic masses of Au and Cu. Substituting $w = 0.1$, $M_{Au} = 197$, and $M_{Cu} = 63.55$, we find that $X = 0.256$ (25.6 at.%).
   Given that $\rho_{Au} = 22.8$ nΩ m   and $C = 450$ nΩ m,

$$\rho = \rho_{Au} + CX(1 - X) = (22.8 \text{ n}\Omega \text{ m}) + (450 \text{ n}\Omega \text{ m})(0.256)(1 - 0.256)$$

$$= 108.5 \text{ n}\Omega \text{ m}$$

This value is only 0.5% different from the experimental value.

RESISTIVITY DUE TO IMPURITIES   Estimate the worst-case resistivity when the concentration of impurities in copper is 1 at.%. From quantum mechanics, the speed of conduction electrons in Cu (those that contribute to conductivity) is 1.58×10^6 m s^{-1}. The atomic concentration in copper is 8.5×10^{28} m^{-3}. Assume that each Cu atom donates one conduction electron. | **Example 2.11**

Solution

We assume that each impurity atom is effective in the scattering process. This will be the case when the impurity atom differs appreciably in size or valency from the host atom. The separation of the impurity atoms is given by $\ell_I = N_I^{-1/3}$, where $N_I = 0.01(8.5 \times 10^{28} \text{ m}^{-3})$. Thus, $\ell_I = 1.06 \times 10^{-9}$ m. The mean speed u of the conduction electrons is 1.58×10^6 m s^{-1}. Therefore, the mean free time between scattering events involving impurities is

$$\tau_I = \frac{\ell_I}{u} = \frac{1.06 \times 10^{-9} \text{ m}}{1.58 \times 10^6 \text{ m s}^{-1}} = 6.7 \times 10^{-16} \text{ s}$$

Compared with the mean free time between scattering events from lattice vibrations, which is about 10^{-14} s, τ_I is ~15 times smaller and hence dominates the resistivity of the metal. The reistivity is thus

$$\rho = \rho_T + \rho_I \approx \rho_I = \frac{1}{en\mu_I} = \frac{m_e}{e^2 n\tau_I}$$

$$= \frac{9.1 \times 10^{-31} \text{ kg}}{(1.6 \times 10^{-19} \text{ C})^2 (8.5 \times 10^{28} \text{ m}^{-3})(6.7 \times 10^{-16} \text{ s})} = 6.24 \times 10^{-7} \text{ } \Omega \text{ m}$$

Compare this with the addition of gold to copper. Gold atoms are effective scatterers in the copper lattice because gold not only has higher electronegativity but also a larger radius, so we would expect an appreciable increase in the resistivity. Using the Nordheim rule $\rho_I = CX(1 - X)$ with $C = 5500$ nΩ m, or 5.50×10^{-6} Ω m, we get

$$\rho_I = 5.5 \times 10^{-6}(0.1)(0.9) = 4.95 \times 10^{-7} \text{ } \Omega \text{ m}$$

This result shows that our estimate for the worst case compares reasonably well for an impurity that is quite effective as a scatterer in the copper matrix.

2.4   MIXTURE RULES AND ELECTRICAL SWITCHES

2.4.1   Heterogeneous Mixtures

Nordheim's rule only applies to solid solutions that are single-phase solids. In other words, it is valid for homogeneous mixtures in which the atoms are mixed at the atomic level throughout the solid, as in the Cu–Ni alloy. The classic problem of determining the effective resistivity of a multiphase solid is closely related to the evaluation of the effective dielectric constant, effective thermal conductivity, effective elastic modulus, effective Poisson's ratio, etc., for a variety of mixtures, including such composite materials as fiberglass. Indeed, many of the mixture rules are identical.

Consider a material with two distinct phases α and β, which are stacked in layer fashion as illustrated in Figure 2.12a. Let us evaluate the effective resistivity for current flow in the x direction. Since the layers are in series, the effective resistance R_{eff} for the whole material is

$$R_{eff} = \frac{L_\alpha \rho_\alpha}{A} + \frac{L_\beta \rho_\beta}{A} \qquad [2.21]$$

where L_α is the total length (thickness) of the α-phase layers, and L_β is the total length of the β-phase layers, $L_\alpha + L_\beta = L$ is the length of the sample, and A is the cross-sectional area. If χ_α and χ_β are the volume fractions of the α and β phases, then the effective resistance is

$$R_{eff} = \frac{L \rho_{eff}}{A}$$

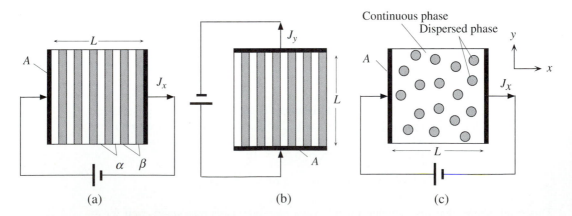

Figure 2.12   The effective resistivity of a material with a layered structure.
(a) Along a direction perpendicular to the layers.
(b) Along a direction parallel to the plane of the layers.
(c) Materials with a dispersed phase in a continuous matrix.

where ρ_{eff} is the **effective resistivity** in Equation 2.21. Using $\chi_\alpha = L_\alpha/L$ and $\chi_\beta = L_\beta/L$, we then get

$$\rho_{eff} = \chi_\alpha \rho_\alpha + \chi_\beta \rho_\beta \qquad \text{[2.22]}$$

Resistivity–mixture rule

which is called the **resistivity–mixture rule** (or the **series rule of mixtures**).

If we are interested in the effective resistivity in the y direction, as shown in Figure 2.12b, obviously the α and β layers are in parallel, so an effective conductivity could be calculated in the same way as we did for the series case to find the **parallel rule of mixtures,** that is,

$$\sigma_{eff} = \chi_\alpha \sigma_\alpha + \chi_\beta \sigma_\beta \qquad \text{[2.23]}$$

Conductivity–mixture rule

where σ is the electrical conductivity of those phases identified by the subscript. Notice that the parallel rule uses the conductivity, and the series rule uses the resistivity. Equation 2.23 is often referred to as the **conductivity–mixture rule.**

Although these two rules refer to special cases, in general, for a random mixture of phase α and phase β, we would not expect either equation to apply rigorously. When the resistivities of two randomly mixed phases are not markedly different, the series mixture rule can be applied at least approximately, as we will show in Example 2.12.

However, if the resistivity of one phase is appreciably different than the other, there are two semiempirical rules that are quite useful in materials engineering.[5] Consider a heterogeneous material that has a dispersed phase (labeled d), in the form of particles, in a continuous phase (labeled c) that acts as a matrix, as depicted in Figure 2.12c. Assume that ρ_c and ρ_d are the resistivities of the continuous and dispersed phases, and χ_c and χ_d are their volume fractions. If the dispersed phase is much more resistive with respect to the matrix, that is, $\rho_d > 10\rho_c$, then

$$\rho_{eff} = \rho_c \frac{(1 + \frac{1}{2}\chi_d)}{(1 - \chi_d)} \qquad (\rho_d > 10\rho_c) \qquad \text{[2.24]}$$

On the other hand, if $\rho_d < (\rho_c/10)$, then

$$\rho_{eff} = \rho_c \frac{(1 - \chi_d)}{(1 + 2\chi_d)} \qquad (\rho_d < 0.1\rho_c) \qquad \text{[2.25]}$$

We therefore have at least four mixture rules at our disposal, the uses of which depend on the mixture geometry and the resistivities of the various phases. The problem is identifying which one to use for a given material, which in turn requires a knowledge of the microstructure and properties of the constituents. It should be emphasized that, at best, Equations 2.22 to 2.25 provide only a reasonable estimate of the effective resistivity of the mixture.[6]

---

[5] Over the years, the task of predicting the resistivity of a mixture has challenged many theorists and experimentalists, including Lord Rayleigh who, in 1892, published an excellent exposition on the subject in the *Philosophical Magazine*. An extensive treatment of mixtures can be found in a paper by J. A. Reynolds and J. M. Hough published in 1957 (*Proceedings of the Physical Society*, 70, no. 769, London), which contains nearly all the mixture rules for the resistivity.

[6] More accurate mixture rules have been established for various types of mixtures with components possessing widely different properties, which the keen reader can find in P. L. Rossiter, *The Electrical Resistivity of Metals and Alloys* (Cambridge University Press, Cambridge, 1987).

Example 2.12   **THE RESISTIVITY-MIXTURE RULE**   Consider a two-phase alloy consisting of phase α and phase β randomly mixed as shown in Figure 2.13a. The solid consists of a random mixture of two types of resistivities, ρ_α of α and ρ_β of β. We can divide the solid into a bundle of N parallel fibers of length L and cross-sectional area A/N, as shown in Figure 2.13b. In this fiber (infinitesimally thin), the α and β phases are in series, so that if $\chi_\alpha = V_\alpha/V$ is the volume fraction of phase α and χ_β is that of β, then the total length of all α regions present in the fiber is $\chi_\alpha L$, and the total length of β regions is $\chi_\beta L$. The two resistances are in series, so the fiber resistance is

$$R_{\text{fiber}} = \frac{\rho_\alpha(\chi_\alpha L)}{(A/N)} + \frac{\rho_\beta(\chi_\beta L)}{(A/N)}$$

But the resistance of the solid is made up of N such fibers in parallel, that is,

$$R_{\text{solid}} = \frac{R_{\text{fiber}}}{N} = \frac{\rho_\alpha \chi_\alpha L}{A} + \frac{\rho_\beta \chi_\beta L}{A}$$

By definition, $R_{\text{solid}} = \rho_{eff} L/A$, where ρ_{eff} is the effective resistivity of the material, so

$$\frac{\rho_{eff} L}{A} = \frac{\rho_\alpha \chi_\alpha L}{A} + \frac{\rho_\beta \chi_\beta L}{A}$$

Thus, for a two-phase solid, the effective resistivity will be

$$\rho_{eff} = \chi_\alpha \rho_\alpha + \chi_\beta \rho_\beta$$

If the densities of the two phases are not too different, we can use weight fractions instead of volume fractions. The series rule fails when the resistivities of the phases are vastly different. A major (and critical) tacit assumption here is that the current flow lines are all parallel, so that no current crosses from one fiber to another. Only then we can say that the effective resistance is R_{fiber}/N.

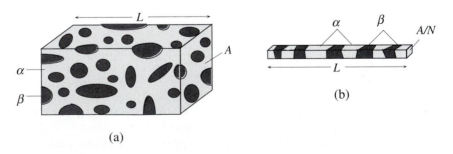

Figure 2.13   **(a)** A two-phase solid.   **(b)** A thin fiber cutout from the solid.

Example 2.13   **A COMPONENT WITH DISPERSED AIR PORES**   What is the effective resistivity of 95/5 (95% Cu–5% Sn) bronze, which is made from powdered metal containing dispersed pores at 15^{v}/o (volume percent, vol.%). The resistivity of 95/5 bronze is 1×10^{-7} Ω m.

Solution

Pores are infinitely more resistive ($\rho_d = \infty$) than the bronze matrix, so we use Equation 2.24,

$$\rho_{eff} = \rho_c \frac{1 + \frac{1}{2}\chi_d}{1 - \chi_d} = (1 \times 10^{-7}\ \Omega\ \text{m}) \frac{1 + \frac{1}{2}0.15}{1 - 0.15} = 1.27 \times 10^{-7}\ \ \Omega\ \text{m}$$

COMBINED NORDHEIM AND MIXTURE RULES   Suppose that a 90% Cu–10% Zn brass component is made from powdered metal containing dispersed pores at 15v/o (vol. %). Given that the Nordheim coefficient for Zn in the copper matrix is about 300 nΩ m, predict the effective resistivity of this brass component.

Example 2.14

Solution

We first calculate the resistivity of the alloy without the pores, which forms the continuous phase in the powdered material. From Nordheim's rule,

$$\rho_{\text{brass}} = \rho_{\text{copper}} + CX(1 - X)$$

where $\rho_{\text{copper}} = 1.7 \times 10^{-8}\ \Omega$ m, $C = 300 \times 10^{-9}\ \Omega$ m, and $X = 0.1$. Substituting, we obtain $\rho_{\text{brass}} = 4.4 \times 10^{-8}\ \Omega$ m. The experimental value, $3.9 \times 10^{-8}\ \Omega$ m, is actually less, because Zn has valency of 2 and when a Zn atom replaces a host Cu atom, it donates two electrons instead of one. We can roughly adjust the calculated resistivity by noting that a 10% Zn addition increases the conduction electron concentration by 10% and hence reduces the resistivity by 10% to $4 \times 10^{-8}\ \Omega$ m.

The powdered metal has $\chi_d = 0.15$, which is the volume fraction of the dispersed phase, that is, the air pores, and $\rho_c = \rho_{\text{brass}} = 4 \times 10^{-8}\ \Omega$ m is the resistivity of the continuous matrix. The effective resistivity of the powdered metal is given by

$$\rho_{eff} = \rho_c \frac{1 + \frac{1}{2}\chi_d}{1 - \chi_d} = (4 \times 10^{-8}\ \Omega\ \text{m}) \frac{1 + \frac{1}{2}0.15}{1 - 0.15} = 5.06 \times 10^{-8}\ \Omega\ \text{m}$$

On the other hand, if we use the simple conductivity-mixture rule, the effective resistivity is $4.70 \times 10^{-8}\ \Omega$ m and is underestimated.

2.4.2   Two-Phase Alloy (Ag–Ni) Resistivity and Electrical Contacts

Certain binary alloys, such as Pb–Sn and Cu–Ag, only exhibit a single-phase alloy structure over very small composition ranges. For most compositions, these alloys form a two-phase heterogenous mixture of phases α and β. A typical phase diagram for such a eutectic binary alloy system is shown in Figure 2.14a, which could be a schematic scheme for the Cu–Ag system or the Pb–Sn system. The phase diagram identifies the phases existing in the alloy at a given temperature and composition. If the overall composition X is less than X_1, then at T_1, the alloy will consist of phase α only. This phase is Cu rich. When the composition X is between X_1 and X_2, then the alloy will consist of the two phases α and β randomly mixed. The phase α is Cu rich (that is, it has composition X_1) and the phase β is Ag rich (composition X_2). The relative amounts of each phase are determined by the well known **lever rule,** which means that we can determine the volume fractions of α and β, χ_α and χ_β, as the alloy composition is changed from X_1 to X_2.

that our cut gives $I_1 = I_2 = (1/2)I$. Obviously, I_1 flowing in the inner conductor is threaded (or linked) by both B_1 and B_2. (Remember that B_1 is just inside the conductor in Figure 2.15b, so it threads at least 99% of I_1). On the other hand, the outer conductor is only threaded by B_2, simply because I_2 flows in the hollow cylinder and there is no current in the hollow, which means that B_1 is not threaded by I_2. Clearly, I_1 threads more magnetic field than I_2 and thus conductor (b) has a higher inductance than (c). Recall that inductance is defined as the total magnetic flux threaded per unit current. Consequently, an ac current will prefer paths near the surface where the inductive impedance is smaller. As the frequency increases, the current is confined more and more to the surface region.

For a given conductor, we can assume that most of the current flows in the surface region of depth δ, as indicated in Figure 2.16. In the central region, the current will be negligibly small. The "skin depth" will obviously depend on the frequency ω. To find δ, we must solve Maxwell's equations in a conductive medium, a tedious task that, fortunately, has been done by others. We can therefore simply take the result that the skin depth δ is given by

Skin depth for conduction

$$\delta = \frac{1}{\sqrt{\frac{1}{2}\omega\sigma\mu}}$$

[2.26]

where ω is the angular frequency of the current (or applied field), the σ is the conductivity (σ is constant from dc up to $\sim 10^{14}$ Hz in metals), and μ is the magnetic permeability of the medium, which is the product of the absolute (free space) permeability μ_o and the relative permeability μ_r.

Intuitively, those factors that enhance inductive impedance will also tend to emphasize the skin effect and will hence tend to decrease the skin depth. For example, the greater the permeability of the conducting medium, the stronger the magnetic field inside the conductor, and hence the larger the inductance of the central region. The higher the frequency of the current, the greater the inductive impedance. If we imagine the central conductor as a resistance and an inductance in series, the greater the conductivity, the more important the inductive impedance; the skin effect is an ac effect.

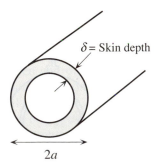

$\delta =$ Skin depth

$2a$

Figure 2.16   At high frequencies, the current flows in the surface region of a conductor approximately defined by the skin depth δ.

With the skin depth known, the effective cross-sectional area is given approximately by

$$A = \pi a^2 - \pi(a - \delta)^2 \approx 2\pi a \delta$$

where δ^2 is neglected ($\delta \ll a$). The ac resistance r_{ac} of the conductor per unit length is therefore

$$r_{ac} = \frac{\rho}{A} \approx \frac{1}{2\pi a \delta} \qquad [2.27]$$

HF resistance per unit length due to skin effect

where ρ is the ac resistivity at the frequency of interest, which for all practical purposes is equal to the dc resistivity of the metal. Equation 2.26 clearly shows that as ω increases, δ decreases, by virtue of $\delta \propto \omega^{-1/2}$ and, as a result, r_{ac} increases.

From this discussion, it is obvious that the skin effect arises because the magnetic field of the ac current in the conductor restricts the current flow to the surface region within a depth of $\delta < a$. Since the current can only flow in the surface region, there is an effective increase in the resistance due to a decrease in the cross-sectional area for current flow. Taking this effective area for current flow as $2\pi a \delta$ leads to Equation 2.27.

The skin effect plays an important role in electronic engineering because it limits the use of solid-core conductors in high-frequency applications. As the signal frequencies reach and surpass the gigahertz (10^9 Hz) range, the transmission of the signal over a long distance becomes almost impossible through an ordinary, solid-metal conductor. We must then resort to pipes (or waveguides).

SKIN EFFECT FROM DIMENSIONAL ANALYSIS   Using dimensional analysis, obtain the general form of the equation for the skin depth δ in terms of the angular frequency of the current ω, conductivity σ, and permeability μ.

Example 2.15

Solution

The skin effect depends on the angular frequency ω of the current, the conductivity σ, and the magnetic permeability μ of the conducting medium. In the most general way, we can group these effects as

$$[\delta] = [\omega]^x [\sigma]^y [\mu]^z$$

where the indices x, y, and z are to be determined. We then substitute the dimensions of each quantity in this expression. The dimensions of each, in terms of the fundamental units, are as follows:

Quantity	Units	Fundamental Units	Comment
δ	m	m	
ω	s^{-1}	s^{-1}	
σ	$\Omega^{-1}.m^{-1}$	$C^2.s.kg^{-1}.m^{-3}$	$\Omega = V.A^{-1} = (J.C^{-1})(C.s^{-1})^{-1}$ $= N.m.s.C^{-2} = (kg.m.s^{-2})(m.s.C^{-2})$
μ	$Wb.A^{-1}.m^{-1}$	$kg.m.C^{-2}$	$Wb = T.m^2 = (N.A^{-1}m^{-1})(m^2)$ $= (kg.m.s^{-2})(C^{-1}.s)(m)$

Therefore,

$$[m] = [s^{-1}]^x [C^2.s.kg^{-1}.m^{-3}]^y [kg.m.C^{-2}]^z$$

Matching the dimensions of both sides, we see that $y = z$; otherwise C and kg do not cancel.

For m $1 = -3y + z$

For s $0 = -x + y$

For C or kg $0 = 2y - 2z$ or $0 = -y + z$

Clearly, $x = y = z = -1/2$ is the only possibility. Then, $\delta \propto [\omega\sigma\mu]^{-1/2}$. It should be reemphasized that the dimensional analysis is not a proof of the skin depth expression, but a consistency check that assures confidence in the equation.

Example 2.16

SKIN EFFECT IN AN INDUCTOR   What is the change in the dc resistance of a copper wire of radius 1 mm for an ac signal at 10 MHz? What is the change in the dc resistance at 1 GHz? Copper has $\rho_{dc} = 1.70 \times 10^{-8}\ \Omega$ m or $\sigma_{dc} = 5.9 \times 10^7\ \Omega^{-1}\ m^{-1}$ and a relative permeability near unity.

Solution

Per unit length, $r_{dc} = \rho_{dc}/\pi a^2$ and at high frequencies, from Equation 2.27, $r_{ac} = \rho_{dc}/2\pi a\delta$. Therefore, $r_{ac}/r_{dc} = a/2\delta$.

We need to find δ. From Equation 2.26, at 10 MHz we have

$$\delta = [\tfrac{1}{2}\omega\sigma_{dc}\mu]^{-1/2} = [\tfrac{1}{2} \times 2\pi \times 10 \times 10^6 \times 5.9 \times 10^7 \times 1.257 \times 10^{-6}]^{-1/2}$$

$$= 2.07 \times 10^{-5}\ m = 20.7\ \mu m$$

Thus

$$\frac{r_{ac}}{r_{dc}} = \frac{a}{2\delta} = \frac{(10^{-3}\ m)}{(2 \times 2.07 \times 10^{-5}\ m)} = 24.13$$

The resistance has increased by 24 times. At 1 GHz, the increase is 240 times. Furthermore, the current is confined to a surface region of about $\sim 2 \times 10^{-5}$ (20 μm) at 10 MHz and $\sim 2 \times 10^{-6}$ m (2 μm) at 1 GHz, so most of the material is wasted. This is exactly the reason why solid conductors would not be used for high-frequency work. As very high frequencies, in the gigahertz range and above, are reached, the best bet would be to use pipes (wave guides).

One final comment is appropriate. An inductor wound from a copper wire would have a certain Q (quality factor) value[7] that depends inversely on its resistance. At high frequencies, Q would drop, because the current would be limited to the surface of the wire. One way to overcome this problem is to use a thick conductor that has a surface coating of higher-conductivity metal, such as silver. This is what the early radio engineers practiced. In fact, tank circuits of high-power radio transmitters often have coils made from copper tubes with a coolant flowing inside.

---

[7] The Q value refers to the quality factor of an inductor, which is defined by $Q = \omega_0 L/R$, where ω_0 is the resonant frequency, L is the inductance, and R is the resistance due to the losses in the inductor.

2.6   THE HALL EFFECT AND HALL DEVICES

An important phenomenon that we can comfortably explain using the "electron as a particle" concept is the Hall effect, which is illustrated in Figure 2.17. When we apply a magnetic field in a perpendicular direction to the applied field (which is driving the current), we find there is a transverse field in the sample that is perpendicular to the direction of both the applied field $\mathcal{E}_x$ and the magnetic field B_z, that is, in the y direction. Putting a voltmeter across the sample, as in Figure 2.17, gives a voltage reading V_H. The applied field $\mathcal{E}_x$ drives a current J_x in the sample. The electrons move in the $-x$ direction, with a drift velocity v_{dx}. Because of the magnetic field, there is a force (called the **Lorentz force**) acting on each electron and given by $F_y = -ev_{dx}B_z$. The direction of this Lorentz force is the $-y$ direction, which we can show by applying the corkscrew rule, because, in vector notation, the force F acting on a charge q moving with a velocity $\mathbf{v}$ in a magnetic field $\mathbf{B}$ is given through the vector product

$$\mathbf{F} = q\mathbf{v} \times \mathbf{B} \qquad \text{[2.28]}$$

All moving charges experience the Lorentz force in Equation 2.28 as shown schematically in Figure 2.18. In our example of a metal in Figure 2.17, this Lorentz force is the $-y$ direction, so it pushes the electrons downward, as a result of which there is a negative charge accumulation near the bottom of the sample and a positive charge near the top of the sample, due to exposed metal ions (e.g., Cu^+).

The accumulation of electrons near the bottom results in an internal electric field $\mathcal{E}_H$ in the $-y$ direction. This is called the **Hall field** and gives rise to a Hall voltage V_H between the top and bottom of the sample. Electron accumulation

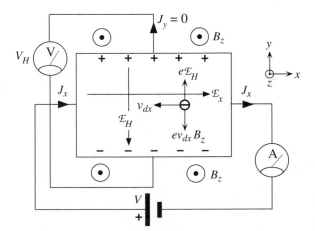

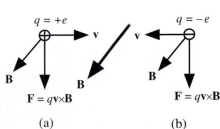

Figure 2.17   Illustration of the Hall effect.
The z direction is out of the plane of the paper. The externally applied magnetic field is along the z direction.

Figure 2.18   A moving charge experiences a Lorentz force in a magnetic field.
(a) A positive charge moving in the x direction experiences a force downward. **(b)** A negative charge moving in the $-x$ direction also experiences a force downward.

continues until the increase in $\mathcal{E}_H$ is sufficient to stop the further accumulation of electrons. When this happens, the magnetic-field force $ev_{dx}B_z$ that pushes the electrons down just balances the force $e\mathcal{E}_H$ that prevents further accumulation. Therefore, in the steady state,

$$e\mathcal{E}_H = ev_{dx}B_z$$

However, $J_x = env_{dx}$. Therefore, we can substitute for v_{dx} to obtain $e\mathcal{E}_H = J_xB_z/n$ or

$$\mathcal{E}_H = \left(\frac{1}{en}\right)J_xB_z \tag{2.29}$$

A useful parameter called the **Hall coefficient** R_H is defined as

Definition of
Hall coefficient

$$R_H = \frac{\mathcal{E}_y}{J_xB_z} \tag{2.30}$$

The quantity R_H measures the resulting Hall field, along y, per unit transverse applied current and magnetic field. The larger R_H, the greater $\mathcal{E}_y$ for a given J_x and B_z. Therefore, R_H is a gauge of the magnitude of the Hall effect. A comparison of Equations 2.29 and 2.30 shows that for metals,

Hall coefficient
for electron
conduction

$$R_H = -\frac{1}{en} \tag{2.31}$$

The reason for the negative sign is that $\mathcal{E}_H = -\mathcal{E}_y$, which means that $\mathcal{E}_H$ is in the $-y$ direction.

Inasmuch as R_H depends inversely of the free electron concentration, its value in metals in much less than that in semiconductors. In fact, Hall-effect devices (such as magnetometers) always employ a semiconductor material, simply because the R_H is larger. Table 2.4 lists the Hall coefficients of various metals. Note that this is negative for most metals, although a few metals exhibit a positive Hall coefficient

Table 2.4   Hall coefficient and Hall mobility ($\mu_H = |\sigma R_H|$) of selected metals

Metal	n [m^{-3}] ($\times 10^{28}$)	R_H (Experimental) [m^3 A^{-1} s^{-1}] ($\times 10^{-11}$)	$\mu_H = |\sigma R_H|$ [m^2 V^{-1} s^{-1}] ($\times 10^{-4}$)
Ag	5.85	−9.0	57
Al	18.06	−3.5	13
Au	5.90	−7.2	31
Be	24.2	+3.4	?
Cu	8.45	−5.5	32
Ga	15.3	−6.3	3.6
In	11.49	−2.4	2.9
Mg	8.60	−9.4	22
Na	2.56	−25	53

Data from various sources, including C. Nording and J. Osterman, *Physics Handbook*, Bromley, England: Chartwell-Bratt Ltd., 1982.

(see Be in the table). The reasons for the latter involve the band theory of solids, which we will discuss in Chapter 4.

Since the Hall voltage depends on the product of two quantities, the current density J_x and the transverse applied magnetic field B_z, we see that the effect naturally multiplies two independently-variable quantities. Therefore, it provides a means of carrying out a multiplication process. One obvious application is measuring the power dissipated in a load, where the load current and voltage are multiplied. There are many instances when it is necessary to measure magnetic fields, and the Hall effect is ideally suited to such applications. Commercial Hall-effect magnetometers can measure magnetic fields in the range of 10 μT to 1 T full-scale, which should be compared to the earth's magnetic field of ~50 μT and that of a strong magnet, typically ~1 T. Depending on the application, manufacturers use different semiconductors to obtain the desired sensitivity.

Hall-effect semiconductor devices are generally inexpensive, small, and reliable. Typical commercial, linear Hall-effect sensor devices are the TL170–173 from Texas Instruments and the UG3500–3600 series from Sprague or Allegro MicroSystems. For example, the TL173 is capable of providing a Hall voltage of 15 mV per mT of applied magnetic field.

The Hall effect is also widely used in magnetically actuated electronic switches. The application of a magnetic field, say from a magnet, results in a Hall voltage that is amplified to trigger an electronic switch. The switches invariably use Si and are readily available from Texas Instruments, Micro Switch of Honeywell, Allegra MicroSystems, and other companies. Hall-effect electronic switches are used as noncontacting keyboard and panel switches that last almost forever, as they have no mechanical contact assembly. Another advantage is that the electrical contact is "bounce" free. There are a variety of interesting applications for Hall-effect switches, ranging from ignition systems, to speed controls, position detectors, alignment controls, brushless dc motor commutators, etc.

HALL EFFECT WATTMETER   The Hall effect can be used to implement a wattmeter to measure | **Example 2.17**
electrical power dissipated in a load. The schematic sketch of the Hall-effect wattmeter is shown in Figure 2.19, where the Hall-effect sample is typically a semicondutor material

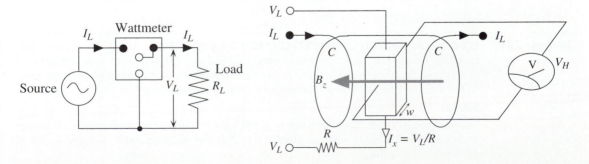

Figure 2.19   Wattmeter based on the Hall effect.

Load voltage and load current have *L* as subscript; *C* denotes the current coils for setting up a magnetic field through the Hall-effect sample (semiconductor).

(usually Si). The load current I_L passes through two coils, which are called current coils and are shown as C in the figure. These coils sets up a magnetic field B_z such that $B_z \propto I_L$. The Hall-effect sample is positioned in this field between the coils. The voltage V_L across the load drives a current $I_x = V_L/R$ through the sample, where R is a series resistance that is much larger than the resistance of the sample and that of the load. Normally, the current I_x is very small and negligible compared to the load current. If w is the width of the sample, then the measured Hall voltage is

$$V_H = w\mathcal{E}_H = wR_H J_x B_z \propto I_x B_z \propto V_L I_L$$

which is the electric power dissipated in the load. The voltmeter that measures V_H can now be calibrated to read directly the power dissipated in the load.

Example 2.18 | **HALL MOBILITY**   Show that if R_H is the Hall coefficient and σ is the conductivity of a metal, then the drift mobility of the conduction electrons is given by

$$\mu_d = |\sigma R_H| \qquad\qquad [2.32]$$

The Hall coefficient and conductivity of copper at 300 K have been measured to be $-0.55 \times 10^{-10}\,\text{m}^3\,\text{A}^{-1}\,\text{s}^{-1}$ and $5.9 \times 10^7\,\Omega^{-1}\,\text{m}^{-1}$ respectively. Calculate the drift mobility of electrons in copper.

Solution

Consider the expression for

$$R_H = \frac{-1}{en}$$

Since the conductivity is given by $\sigma = en\mu_d$, we can substitute for en to obtain

$$R_H = \frac{-\mu_d}{\sigma} \qquad \text{or} \qquad \mu_d = -R_H\sigma$$

which is Equation 2.32. The drift mobility can thus be determined from R_H and σ.

The product of σ and R_H is called the **Hall mobility** μ_H. Some values for the Hall mobility of electrons in various metals are listed in Table 2.4. From the expression in Equation 2.32, we get

$$\mu_d = -(-0.55 \times 10^{-10}\,\text{m}^3\,\text{A}^{-1}\,\text{s}^{-1})(5.9 \times 10^7\,\Omega^{-1}\,\text{m}^{-1}) = 3.2 \times 10^{-3}\,\text{m}^2\,\text{V}^{-1}\,\text{s}^{-1}$$

It should be mentioned that Equation 2.32 is an oversimplification. The actual relationship involves a numerical factor that multiplies the right term in Equation 2.32. The factor depends on the charge carrier scattering mechanism that controls the drift mobility.

Example 2.19 | **CONDUCTION ELECTRON CONCENTRATION FROM THE HALL EFFECT**   Using the electron drift mobility from Hall effect measurements (Table 2.4), calculate the concentration of conduction electrons in copper, and then determine the average number of electrons contributed to the free electron gas per copper atom in the solid.

Solution

The number of conduction electrons is given by $n = \sigma/e\mu$. The conductivity of copper is $5.9 \times 10^7\,\Omega^{-1}\,\text{m}^{-1}$, whereas from Table 2.4, the electron drift mobility is $3.2 \times 10^{-3}\,\text{m}^2\,\text{V}^{-1}\,\text{s}^{-1}$. So,

$$n = \frac{(5.9 \times 10^7\,\Omega\,\text{m})}{[(1.6 \times 10^{-19}\,\text{C})(3.2 \times 10^{-3}\,\text{m}^2\,\text{V}^{-1}\,\text{s}^{-1})]} = 1.15 \times 10^{29}\,\text{m}^{-3}$$

Since the concentration of copper atoms is 8.5×10^{28} m^{-3}, the average number of electrons contributed per atom is $(1.15 \times 10^{29}$ m$^{-3})/(8.5 \times 10^{28}$ m$^{-3}) \approx 1.36$.

2.7   THERMAL CONDUCTIVITY

Experience tells us that metals are both good electrical and good thermal conductors. We may therefore surmise that the free conduction electrons in a metal must also play a role in heat conduction. Our conjecture is correct for metals, but not for other materials. At room temperature and above, the transport of heat in a metal is mainly accomplished by the electron gas, whereas at low temperatures, the conduction is due to lattice vibrations.

When a metal piece is heated at one end, the amplitude of the atomic vibrations, and thus the average kinetic energy of the electrons, in this region increases, as depicted in Figure 2.20. Electrons gain energy from energetic atomic vibrations when the two collide. By virtue of their increased random motion, these energetic electrons then transfer the extra energy to the colder regions by colliding with the atomic vibrations there. Thus, electrons act as "energy carriers."

The thermal conductivity of a material, as its name implies, measures the ease with which heat, that is, thermal energy, can be transported through the medium. Consider the metal rod shown in Figure 2.21, which is heated at one end. Heat will

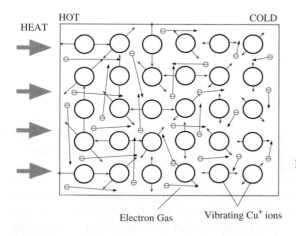

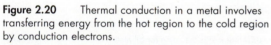

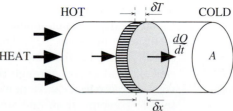

Figure 2.20   Thermal conduction in a metal involves transferring energy from the hot region to the cold region by conduction electrons.

More energetic electrons (shown with longer velocity vectors) from the hotter regions arrive at cooler regions, collide with lattice vibrations, and transfer their energy. Lengths of arrowed lines on atoms represent the magnitudes of atomic vibrations.

Figure 2.21   Heat flow in a metal rod heated at one end.

Consider the rate of heat flow, dQ/dt, across a thin section δx of the rod. The rate of heat flow is proportional to the temperature gradient $\delta T/\delta x$ and the cross-sectional area A.

flow from the hot end to the cold end. Experiments show that the rate of heat flow, $Q' = dQ/dt$, through a thin section of thickness δx is proportional to the temperature gradient $\delta T/\delta x$ and the cross-sectional area A, so that

Fourier's law of thermal conduction

$$Q' = -A\kappa \frac{\delta T}{\delta x}$$
[2.33]

where κ is a material-dependent **constant of proportionality** that we call the **thermal conductivity.** The negative sign indicates that the heat flow direction is that of decreasing temperature. Equation 2.33 is often referred to as **Fourier's Law** of heat conduction and is a defining equation for κ. The driving force for the heat flow is the temperature gradient $\delta T/\delta x$. If we compare Equation 2.33 with Ohm's Law for the electric current I, we see that

Ohm's law of electrical conduction

$$I = -A\sigma \frac{\delta V}{\delta x}$$
[2.34]

which shows that in this case, the driving force is the potential gradient, that is, the electric field.[8] In metals, electrons participate in the processes of charge and heat transport, which are characterized by σ and κ, respectively. Therefore, it is not surprising to find that the two coefficients are related by the **Wiedemann–Franz–Lorenz Law,**[9] which is

Wiedemann–Franz–Lorenz Law

$$\frac{\kappa}{\sigma T} = C_{\text{WFL}}$$
[2.35]

where $C_{\text{WFL}} = \pi^2 k^2/3e^2 = 2.45 \times 10^{-8} \text{ W } \Omega \text{ K}^{-2}$ is a constant called the **Lorenz number** (or the Wiedemann–Franz–Lorenz coefficient).

Experiments on a wide variety of metals, ranging from pure metals to various alloys, show that Equation 2.35 is reasonably well obeyed at close to room temperature and above, as illustrated in Figure 2.22. Since the electrical conductivity of pure metals is inversely proportional to the temperature, we can immediately conclude that the thermal conductivity of these metals must be relatively temperature independent at room temperature and above.

Figure 2.23 shows the temperature dependence of κ for copper and aluminum down to the lowest temperatures. It can be seen that for these two metals, above $\sim$100 K, the thermal conductivity becomes temperature independent, in agreement with Equation 2.35. The unusual behavior at the lowest temperatures is directly related to the nature of lattice vibrations and their involvement in transferring thermal energy through the crystal. Qualitatively, above $\sim$100K, κ is constant, because heat conduction depends essentially on the rate at which the electron transfers energy from one atomic vibration to another as it collides with them (Figure 2.20). This rate of energy transfer depends on the mean speed of the

---

[8] Recall that $J = \sigma E$ which is equivalent to Equation 2.34.

[9] Historically, Wiedemann and Franz noted in 1853 that κ/σ is the same for all metals at the same temperature. Lorenz in 1881 showed that κ/σ is proportional to the temperature with a proportionality constant that is nearly the same for many metals. The law stated in Equation 2.35 reflects both observations. By the way, Lorenz, who was a Dane, should not be confused with Lorentz, who was Dutch.

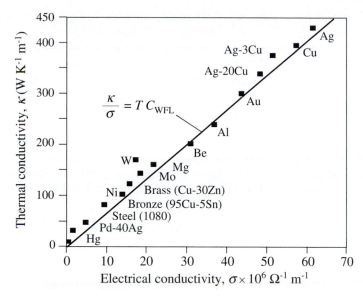

Figure 2.22   Thermal conductivity κ vs. electrical conductivity, σ, for various metals (elements and alloys) at 20 °C.

The solid line represents the WFL law, with an average $C_{WFL} \approx 2.3 \times 10^8$ W Ω K^{-2}.

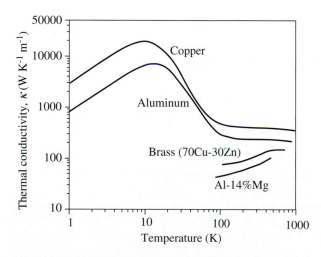

Figure 2.23   Thermal conductivity versus temperature for two pure metals (Cu and Al) and two alloys (brass and Al-14% Mg).

SOURCE: Data extracted from Y. S. Touloukian et al., *Thermophysical Properties of Matter*, vol. 1: "Thermal Conductivity, Metallic Elements and Alloys," New York: Plenum, 1970.

electron u, which increases only fractionally with the temperature. In fact, the fractionally small increase in u is more than sufficient to carry the energy from one collision to another an thereby excite more energetic lattice vibrations in the colder regions.

In an analogy with an electrical resistance, we may define a thermal resistance θ by

Thermal resistance

$$\theta = \frac{L}{A\kappa} \qquad \text{[2.36]}$$

where L is the length of the specimen and A is the cross-sectional area. Taking the heat flow Equation 2.33, at point x and integrating it across the specimen length, assuming no surface heat loss (that is, that the surface has been insulated), we obtain

$$Q' = A\kappa\frac{\Delta T}{L} = \frac{\Delta T}{\theta} \qquad \text{[2.37]}$$

where ΔT is the temperature difference between the ends of the sample.

Compare this with Ohm's Law for electrical circuits, that is,

$$I = \frac{\Delta V}{R} \qquad \text{[2.38]}$$

The rate of heat flow Q' and the temperature difference ΔT correspond to the current I and potential difference ΔV, respectively. Hence, thermal resistance is the thermal analog of electrical resistance.

Example 2.20 | **THERMAL CONDUCTIVITY**   A 95/5 (95% Cu–5% Sn) bronze bearing made of powdered metal contains 15v/o (vol.%) porosity. Calculate its thermal conductivity, given that the electrical conductivity of 95/5 bronze is $10^7\,\Omega^{-1}\,m^{-1}$.

Solution

Recall that in Example 2.13, we found the electrical resistivity of the same bronze by using the mixture rule in Equation 2.24 in section 2.4. We can use the same mixture rule again here, but we need the thermal conductivity of 95/5 bronze. From $\kappa/\sigma T = C_{WFL}$, we have

$$\kappa = \sigma T C_{WFL} = (1 \times 10^7)(300)(2.45 \times 10^{-8}) = 73.5\ \text{W m}^{-1}\ \text{K}^{-1}$$

Thus, the effective thermal conductivity is

$$\frac{1}{\kappa_{eff}} = \frac{1}{\kappa_c}\left[\frac{1 + \frac{1}{2}\chi_d}{1 - \chi_d}\right] = \frac{1}{(73.5\ \text{W m}^{-1}\ \text{K}^{-1})}\left[\frac{1 + \frac{1}{2}(0.15)}{1 - 0.15}\right]$$

so that

$$\kappa_{eff} = 58.1\ \text{W m}^{-1}\ \text{K}^{-1}$$

Example 2.21 | **THERMAL RESISTANCE**   A brass disk of electrical resistivity 50 nΩ m conducts heat from a heat source to a heat sink at a rate of 10 W. If its diameter is 20 mm and its thickness is 30 mm, what is the temperature drop across the disk, neglecting the heat losses from the surface?

Solution

We first determine the thermal conductivity:

$$\kappa = \sigma T C_{\text{WFL}} = (5 \times 10^{-8} \ \Omega \ \text{m})^{-1}(300 \ \text{K})(2.45 \times 10^{-8} \ \text{W} \ \Omega \ \text{K}^{-2}) = 147 \ \text{W} \ \text{m}^{-1} \ \text{K}^{-1}$$

The thermal resistance is

$$\theta = \frac{L}{\kappa A} = \frac{(30 \times 10^{-3} \ \text{m})}{\pi(10 \times 10^{-3} \ \text{m})^2(147 \ \text{W} \ \text{m} \ \text{K}^{-1})} = 0.650 \ \text{K} \ \text{W}^{-1}$$

Therefore, the temperature drop is

$$\Delta T = \theta Q' = (0.650 \ \text{K} \ \text{W}^{-1})(10 \ \text{W}) = 6.5 \ ^\circ\text{C}$$

ADDITIONAL TOPICS

2.8 THIN METAL FILMS AND INTEGRATED CIRCUIT INTERCONNECTIONS

The resistivity of a material, as listed in materials tables and in our analysis of conduction, refers to the resistivity of the material in bulk form; that is, any dimension of the specimen is much larger than the mean free path ℓ for electron scattering. In certain applications, notably in microelectronics, metal films are widely used to provide electrical conduction paths to and from semiconductor devices. The metal film could be a gate of a **metal-oxide-semiconductor (MOS)** device, or interconnections on an integrated circuit. In practice, the metal film is deposited onto a high-resistivity material, such as a semiconductor or an insulator (e.g., SiO_2), by evaporating the metal from a heated source in a vacuum. As the metal atoms, evaporated from the source, impinge on and adhere to the semiconductor surface, they form a metal film, which is often highly polycrystalline. We are therefore interested in the resistivity of a metal film in which the thickness of the film, or the average size of the grain boundaries, is either comparable to or smaller than the mean distance between scattering events ℓ (i.e., the mean free path).

The resistivity of many thin metal films is greater than the bulk value. Furthermore, that resistivity generally increases with decreasing thickness, for two main reasons. First, due to the deposition process, the film structure is normally more disordered, or grainy, with smaller grains that have a higher dislocation and defect density. In the grain boundary region, there is a discontinuity in the crystal orientation; therefore, there is a sharp variation in the potential energy PE of the electron, as depicted in Figure 2.24a. The electron in the grain boundary region thus experiences a force $F = d(PE)/dx$ and becomes scattered. In a highly polycrystalline metal film, the electrons are more frequently scattered at the grain

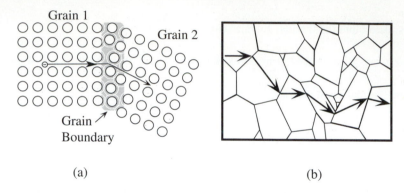

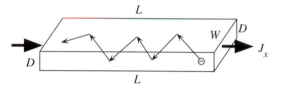

(a)   (b)

Figure 2.24   Grain boundaries cause scattering of the electron and therefore add to the resistivity, by Matthissen's rule.

For a very grainy solid, the electron is scattered from grain boundary to grain boundary, and the mean free path is approximately equal to the mean grain diameter.

Figure 2.25   Conduction in thin films may be controlled by scattering from the surfaces.

boundaries than by the thermal vibrations of the lattice, as shown in Figure 2.24b. Consequently, these grain boundaries and other structural defects increase the residual resistivity ρ_R. By virtue of Mathiessen's rule, $\rho = \rho_T + \rho_R$, we therefore expect the overall resistivity of the thin film to be greater than that of a bulk sample.

The second reason for the increased resistivity is that in very thin films, the electron can also become scattered from the surfaces, as illustrated in Figure 2.25. This process is particularly prominent when the film thickness D is comparable to the mean free path ℓ_B in the bulk sample (due to scattering from thermal vibrations). Indeed, in thin films for which $D/\ell_B < 0.1$, such that D is typically 10 nm or less, the residual resistivity is appreciably larger, by a factor of perhaps 2 or 3. We note in passing that very thin films, typically less than 1 nm, can be discontinuous and can therefore lose all electrical conduction.

Table 2.5 lists typical ranges of resistivities for several types of thin metal films used in various micoelectronic applications. The film thickness is normally greater than the mean free path, so the resistivity of the film is primarily determined by the morphology of the film (e.g., polycrystallinity, grain size, etc.) and hence the fabrication process.

Table 2.5   Typical resistivity values for thin metal films (polycrystalline) in microelectronics

Metal	Bulk ρ (Ω m) ($\times 10^{-8}$)	Film ρ (Ω m) ($\times 10^{-8}$)	Comment
Aluminum	2.67	2.8–3.3	Vacuum evaporated
Gold	2.2	2.4	Vacuum evaporated or sputtered
Nickel	6.9	12	Vacuum evaporated or sputtered
Molybdenum	5.7	10	Sputtered and annealed

Data from C.R.M. Grovenor, *Microelectronic Materials*, Bristol, U.K.: Adam Hilger, 1989, p. 197; and K.L. Chopra, and I. Kaur, *Thin Film Device Applications*, Chapter 4, New York: Plenum Press, 1983, p. 130.

IMPORTANT TERMS

Alloy is a metal that contains more than one element. Brass is an alloy of Cu and Zn.

Brass is a copper-rich Cu–Zn alloy.

Bronze is a copper-rich Cu–Sn alloy.

Drift mobility is the drift velocity per unit applied field. If μ_d is the drift mobility, then the defining equation is $v_d = \mu_d \mathcal{E}$, where v_d is the drift velocity and $\mathcal{E}$ is the field.

Drift velocity is the average electron velocity, over all the conduction electrons in the conductor, in the direction of an applied electrical force ($F = -e\mathcal{E}$ for electrons). In the absence of an applied field, all the electrons move around randomly, and the average velocity over all the electrons in any direction is zero. With an applied field $\mathcal{E}_x$, there is a net velocity per electron v_{dx}, in the direction opposite to the field, where v_{dx} depends on $\mathcal{E}_x$ by virtue of $v_{dx} = \mu_d \mathcal{E}_x$, where μ_d is the drift mobility.

Electrical conductivity (σ) is a property of a material that quantifies the ease with which charges flow inside the material along an applied electric field or a voltage gradient. The conductivity is the inverse of electrical resistivity ρ. Since charge flow is caused by a voltage gradient, σ is the rate of charge flow across a unit area per unit voltage gradient, $J = \sigma\mathcal{E}$.

Fourier's law states that the rate of heat flow Q' through a sample, due to thermal conduction, is proportional to the temperature gradient dT/dx and the cross-sectional area A, that is, $Q' = -\kappa A(dT/dx)$, where κ is the thermal conductivity.

Hall coefficient (R_H) is a parameter that gauges the magnitude of the Hall effect. If $\mathcal{E}_y$ is the electric field in the y direction, due to a current density J_x along x and a magnetic field B_z along z, then $R_H = \mathcal{E}_y/J_x B_z$.

Hall effect is a phenomenon that occurs in a conductor carrying a current when the conductor is placed in a magnetic field perpendicular to the current. The charge carriers in the conductor are deflected by the magnetic field, giving rise to an electric field (Hall field) that is perpendicular to both the current and the magnetic field. If the current density J_x is along x and the magnetic field, B_z, is along z, then the Hall field is along either $+y$ or $-y$, depending on the polarity of the charge carriers in the material.

Isomorphous phase diagram is a phase diagram for an alloy that has unlimited solid solubility.

Joule's Law relates the power dissipated per unit volume P_{vol} by a current-carrying conductor to the applied field $\mathcal{E}$ and the current density J, such that $P_{vol} = J\mathcal{E} = \sigma\mathcal{E}^2$.

Lorentz force is the force experienced by a moving charge in a magnetic field. When a charge q is moving with a velocity **v** in a magnetic field **B**, the charge experiences a force **F** that is proportional to the magnitude of its charge q, its velocity $\mathbf{v}$, and the field $\mathbf{B}$, such that $\mathbf{F} = q\mathbf{v} \times \mathbf{B}$.

Magnetic field, magnetic flux density, or magnetic induction (B) is a vector field quantity that describes the magnitude and direction of the *magnetic force* exerted on a moving charge or a current-carrying conductor. The magnetic force is essentially the Lorentz force and excludes the electrostatic force $q\mathcal{E}$.

Magnetic permeability (μ) or simply permeability is a property of the medium that characterizes the effectiveness of a medium in generating as much magnetic field as possible for given external currents. It is the product of the permeability of free space (vacuum) or absolute permeability (μ_o) and relative permeability of the medium (μ_r), i.e. $\mu = \mu_o \mu_r$.

Magnetometer is an instrument for measuring the magnitude of a magnetic field. Some instruments can also measure the direction of the magnetic field, as well.

Matthiessen's rule gives the overall resistivity of a metal as the sum of individual resistivities due to scattering from thermal vibrations, impurities, and crystal defects. If the resistivity due to scattering from thermal vibrations is denoted ρ_T and the resistivities due to scattering from crystal defects and impurities can be lumped into a single resistivity term called the residual resistivity ρ_R, then $\rho = \rho_T + \rho_R$.

Mean free time is the average time it takes to scatter a conduction electron. If t_i is the free time between collisions (between scattering events) for an electron labeled i, then $\tau = \overline{t_i}$ averaged over all the electrons. The drift mobility is related to the mean free time by $\mu_d = e\tau/m_e$. The reciprocal of the mean free time is the mean probability per unit time that a conduction electron will be scattered; in other words, the mean frequency of scattering events.

Mean free path is the mean distance traversed by an electron between scattering events. If τ is the mean free time between scattering events and u is the mean speed of the electron, then the mean free path is $\ell = u\tau$.

Nordheim's rule states that the resistivity of a solid solution (an isomorphous alloy) due to impurities ρ_I is proportional to the concentrations of the solute X and the solvent $(1 - X)$.

Phase (in materials science) is a physically homogeneous portion of a materials system that has uniform physical and chemical characteristics.

Relaxation time is an equivalent term for the mean free time between scattering events.

Residual resistivity (ρ_R) is the contribution to the resistivity arising from scattering processes other than thermal vibrations of the lattice, for example, impurities, grain boundaries, point defects, etc.

Skin effect is an electromagnetic phenomenon that, at high frequencies, restricts ac current flow to near the surface of a conductor to reduce the energy stored in the magnetic field.

Solid solution is a crystalline material that is a homogeneous mixture of two or more chemical species. The mixing occurs at the atomic scale, as in mixing alcohol and water. Solid solutions can be substitutional (as in Cu–Ni) or interstitial (for example, C in Fe).

Stefan's Law is a phenomenological description of the energy radiated (as electromagnetic waves) from a surface per second. When a surface is heated to a temperature T, it radiates net energy at a rate given by $P_{\text{radiated}} = \epsilon\sigma_S A(T^4 - T_0^4)$, where σ_s is Stefan's constant (5.67×10^{-8} W m^{-2} K^{-4}), ϵ is the emissivity of the surface, A is the surface area, and T_0 is the ambient temperature.

Temperature coefficient of resistivity (TCR) (α_0) is defined as the fractional change in the electrical resistivity of a material per unit increase in the temperature with respect to some reference temperature T_0.

Thermal conductivity (κ) is a property of a material that quantifies the ease with which heat flows along the material from higher to lower temperature regions. Since heat flow is due to a temperature gradient, κ is the rate of heat flow across a unit area per unit temperature gradient.

Thermal resistance (θ) is a measure of the difficulty with which heat conduction takes place along a material sample. The thermal resistance is defined as the temperature drop per unit heat flow, $\theta = \Delta T/Q'$. It depends on both the material and its geometry. If the heat losses from the surfaces are negligible, then $\theta = L/\kappa A$, where L is the length of the sample (along heat flow) and A is the cross-sectional area.

QUESTIONS AND PROBLEMS

2.1 **Electrical conduction** Na is a monovalent metal with a density of 0.9712 g cm^{-3}. Its atomic mass is 22.99 g mol^{-1}. The drift mobility of electrons in Na is 53 cm^2 V^{-1}

   a. Consider the collection of conduction electrons in the solid. If each Na atom donates one electron to the electron sea, what is the mean separation between the electrons?

   b. What is the approximate mean separation between an electron (e^-) and a metal ion (Na$^+$), assuming that this is comparable to the equilibrium atomic separation? What is the approximate Coulombic interaction energy (in eV) between an electron and an Na$^+$ ion?

   c. How does this electron/metal-ion interaction energy compare with the average thermal energy per particle, according to the kinetic molecular theory of matter? Do you expect the theory to be applicable to the conduction electrons in Na? If the mean electron/metal-ion interaction energy is of the same order of magnitude as the mean KE of the electrons, what is the mean speed of electrons in Na? Why should the mean kinetic energy be comparable to the mean electron/metal-ion interaction energy?

   d. Calculate the electrical conductivity of Na and compare this with the experimental value of 2.1×10^7 Ω^{-1} m^{-1} and comment on the difference.

2.2 **Electrical conduction** The resistivity of aluminum at 25 °C has been measured to be 2.65×10^{-8} Ω m. The thermal coefficient of resistivity of aluminum at 0 °C is 4.29×10^{-3} K^{-1}. Aluminum has a valency of 3, a density of 2.70 g cm^{-3}, and an atomic mass of 27.

   a. Calculate the resistivity of aluminum at -40 °C.

   b. What is the thermal coefficient of resistivity at -40 °C?

   c. Estimate the mean free time between collisions for the conduction electrons in aluminum at 25 °C.

   d. If the mean speed of the conduction electrons is $\sim 1.5 \times 10^6$ m s^{-1}, calculate the mean free path and compare this with the interatomic separation in Al. What should be the thickness of an Al film that has deposited on an IC chip such that its resistivity is the same as that of bulk Al?

   e. What is the percentage change is the power loss due to Joule heating of the aluminum wire when the temperature drops from 25 °C to -40 °C?

2.3 **TCR and Matthiessen's rule** Determine the temperature coefficient of resistivity of pure iron and of electrotechnical steel (Fe with 4% C), which are used in various electrical machinery, at two temperatures: 0 °C and 500 °C. Comment on the similarities and differences in the resistivity versus temperature behavior shown in Figure 2.26 for the two materials.

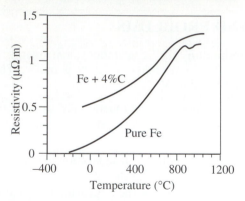

Figure 2.26   Resistivity versus temperature for pure iron and 4% C steel.

*2.4   TCR of isomorphous alloys

a.   For an isomorphous alloy $A\%-B\%$ ($B\%$ solute in $A\%$ solvent), the temperature coefficient of resistivity α_{AB} is given by

$$\alpha_{AB} = \frac{\alpha_A \rho_A}{\rho_{AB}}$$

where ρ_{AB} is the resistivity of the alloy (AB) and ρ_A and α_A are the resistivity and TCR of pure A. What are the assumptions behind this equation?

b.   Estimate the composition of the Cu–Ni alloy that will have a TCR of 4×10^{-4} K^{-1}, that is, a TCR that is an order of magnitude less than that of Cu.

2.5   **Constantan** Constantan has the composition 45% Ni–55% Cu. Cu–Ni alloys show complete solid solubility. As an alloy, constantan is widely used in resistor applications (up to 500 °C), in strain gauges, and as one of the thermocouple metal pairs. Given that the resistivity and TCR of copper at 20 °C are 17 nΩ m and 0.004 K^{-1}, respectively, and the Nordheim coefficient of Ni dissolved in Cu is 1250 nΩ m, calculate the resistivity ρ, TCR (α), and thermal conductivity κ of constantan and compare the values with the experimental measurements: $\rho(20\ °C) = 5 \times 10^{-7}$ Ω m, $\alpha(20\ °C) = 2 \times 10^{-5}$ K^{-1}, $\kappa = 21$ W m^{-1} K^{-1}. What are the reasons for the differences between the calculated and experimental values?

2.6   **Experimental Nordheim coefficient for Pd in Ag** Silver and palladium form a complete solid solution and, as alloys, they are widely used in various electrical switches. Pd improves the wear resistance of Ag and the alloy has good fabricability. Table 2.6 shows the resistivity of Ag–Pd alloys for various

Table 2.6

wt.% Pd in Ag	0	1	3	10	30
Resistivity, nΩ m	16.1	21.8	38.3	63.9	149.3

compositions. Using a suitable plot, obtain the Nordheim coefficient for Pd in Ag. [Note: The atomic masses of Pd and Ag are very close, 106.42 and 107.87, respectively. You can therefore assume that the atomic % = weight %.]

2.7 **Electrical and thermal conductivity of In** Electron drift mobility in indium has been measured to be 6 cm^2 V^{-1} s^{-1}. The room temperature (27 °C) resistivity of In is 8.37×10^{-8} Ω m, and its atomic mass and density are 114.82 amu or g mol^{-1} and 7.31 g cm^{-3}, respectively.

a. Based on the resistivity value, determine how many free electrons are donated by each In atom in the crystal. How does this compare with the position of In the Periodic Table (Group IIIB)?

b. If the mean speed of conduction electrons in In is 1.74×10^8 cm s^{-1}, what is the mean free path?

c. Calculate the thermal conductivity if In. How does this compare with the experimental value of 81.6 W m^{-1} K^{-1}?

2.8 **Electrical and thermal conductivity of Ag** The electron drift mobility in silver has been measured to be 56 cm^2 V^{-1} s^{-1} at 27 °C. The atomic mass and density of Ag are given as 107.87 amu or g mol^{-1} and 10.50 g cm^{-3}, respectively.

a. Assuming that each Ag atom contributes one conduction electron, calculate the resistivity of Ag at 27 °C. Compare this value with the measured value of 1.6×10^{-8} Ω m at the same temperature and suggest reasons for the difference.

b. Calculate the thermal conductivity of silver at 27 °C and at 0 °C.

2.9 **Mixture rules** A 70% Cu–30% Zn brass electrical component has been made of powdered metal and contains 15 vol.% porosity. Assume that the pores are dispersed randomly. Estimate the effective electrical resistivity of the brass component.

2.10 **Mixture rules**

a. Consider a mixture that consists of a continuous conducting phase, with a conductivity σ_c, and dispersed spheres of another phase, of conductivity σ_d and volume fraction χ. The effective conductivity of the mixture (according to Reynolds and Hough, 1957) is given by

$$\frac{\sigma - \sigma_c}{\sigma + 2\sigma_c} = \chi \frac{\sigma_d - \sigma_c}{\sigma_d + 2\sigma_c} \qquad \text{[2.39]} \qquad \textit{A mixture rule for dispersed spheres}$$

Assume that the spheres are randomly dispersed in the material. Show that if $\sigma_d \ll \sigma_c$ (dispersed phase is very resistive), then

$$\sigma = \sigma_c \frac{1 - \chi}{1 + \frac{1}{2}\chi} \qquad \text{[2.40]} \qquad \textit{High resistivity dispersed spheres}$$

What is your conclusion?

b. A certain carbon electrode used in electrical arcing applications is 47% porous. Given that the resistivity of graphite (in polycrystalline form) at room temperature is about 91 nΩ m, estimate the effective resistivity of the Carbon electrode, from Equation 2.40 and from the simple conductivity-mixture rule. Compare your estimates with the measured value of 180 nΩ m and comment on the difference.

c. Graphite at room temperature has a resistivity of about 91 nΩ m, whereas silver has a resistivity of 14.6 nΩ m. Silver particles are dispersed in a graphite paste to increase the effective conductivity of the paste. If the volume fraction of dispersed silver is 30%, what is the effective conductivity of this paste? What is your conclusion?

2.11 Thermal conduction

a. What is the defining equation for thermal conductivity κ?

b. An 80%–20%(% means at %) brass disk of 40 mm diameter and 5 mm thickness is used to conduct heat from a heat source to a heat sink.
  (1) Calculate the thermal resistance of the brass disk.
  (2) If the disk is conducting heat at a rate of 100 W, calculate the temperature drop along the disk.

c. What should be the composition of brass if the temperature drop across the disk is to be halved?

2.12 Q factor of an inductor and the skin effect

Consider the skin effect on the resistance of a copper wire and the quality factor Q of an RF inductor. The inductance $\mathcal{L}$ of a coil (inductor) with an air core is given by the radio engineer's inductor equation:

$$\mathcal{L}(\mu H) = \frac{D^2 N^2}{18D + 40L}$$

where $\mathcal{L}$ is the inductance in microhenry (μH), D is the diameter of the coil in inches, and L is the length of the coil in inches, as depicted in Figure 2.27. The expression assumes that the core is air and that there is a single layer of

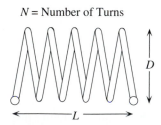

N = Number of Turns

Figure 2.27 An RF inductor with an air core.

The coil has a length L and diameter D. The number of turns is N.

windings. In addition, $L > 0.4D$. Consider a coil with 48 turns, wound at 32 turns per inch. The diameter of the coil is 3/4 in. The coil is tuned in with a 25 pF capacitor. Enameled wire of American Wire Gauge (AWG) size 21 (i.e., wire size AWG.21) corresponds to a diameter of 28.5 mil, or 28.5×10^{-3} in.) Calculate the Q factor of the tank circuit with and without the skin effect and comment on the difference. How can you double the Q factor of the coil?

Note: The quality factor, Q of a coil is defined by

$$Q = \frac{\omega_o \mathscr{L}}{R}$$

where ω_o is the resonance frequency, $\mathscr{L}$ is the inductance, and R is the resistance of the coil, which represents all the losses in the coil. In the present case, R is simply the resistance of the coil wire.

2.13 Skin effect at 60 Hz

a. What is the skin depth for a copper wire carrying a current at 60 Hz? The resistivity of copper at 27 °C is 17 nΩ m. Its relative permeability is $\mu_r \approx 1$. Is there any sense in using a conductor with a diameter of more than 2 cm for power transmission?

b. What is the skin depth for an iron wire carrying a current at 60 Hz? The resistivity of iron at 27 °C is 97 nΩ m. Assume that its relative permeability is $\mu_r \approx 700$. How does this compare with the copper wire? Considering that iron is 100 times cheaper than copper, is there any economic advantage in using iron?

2.14 The Hall effect Consider a rectangular sample, a metal or an n-type semiconductor, with a length L, width W, and thickness D. A current I is passed along L, perpendicular to the cross-sectional area WD. The face $W \times L$ is exposed to a magnetic field density B. A voltmeter is connected across the width, as shown in Figure 2.28, to read the Hall voltage V_H.

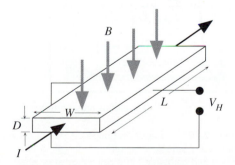

Figure 2.28   Hall effect in a rectangular material with length L, width W, and thickness D.

The voltmeter is across the width W.

a.   Show that the Hall voltage recorded by the voltmeter is

$$V_H = \frac{IB}{Den}$$

Is it important to have a rectangular sample?

b.   Consider a 1-micron-thick strip of gold layer on an insulating substrate that is a candidate for a Hall probe sensor. If the current through the film is maintained at constant 100 mA, what is the magnetic field that can be recorded per μV of Hall voltage?

2.15 The strain gauge A **strain gauge** is a transducer attached to a body to measure its fractional elongation $\Delta L / L$ under an applied load (force) F. The gauge is a grid of many folded runs of a thin, resistive wire glued to a flexible backing, as depicted in Figure 2.29. The gauge is attached to the body under test such that the resistive wire length is parallel to the strain.

a.   Assume that the elongation does not change the resistivity and show that the change in the resistance ΔR is related to the strain $\varepsilon = \Delta L / L$ by

$$\Delta R \approx R(1 + 2v)\varepsilon \qquad \text{[2.41]}$$

where v is the Poisson ratio, which is defined by

$$v = -\frac{\text{Transverse strain}}{\text{Longitudinal strain}} = -\frac{\varepsilon_t}{\varepsilon_l} \qquad \text{[2.42]}$$

where ε_l is the strain along the applied load, that is, $\varepsilon_l = \Delta L / L = \varepsilon$, and ε_t is the strain in the transverse direction, that is, $\varepsilon_t = \Delta D / D$, where D is the diameter (thickness) of the wire.

b.   Explain why a nichrome wire would be a better choice than copper for the strain gauge (consider the TCR).

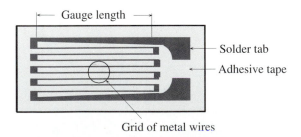

Gauge length

Solder tab

Adhesive tape

Grid of metal wires

Figure 2.29   The strain gauge consists of a long, thin wire folded several times along its length to form a grid as shown and embedded in a self-adhesive tape.

The ends of the wire are attached to terminals (solder pads) for external connections. The tape is stuck on the component for which the strain is to be measured.

c. How do temperature changes affect the response of the gauge? Consider the effect of temperature on ρ. Also consider the differential expansion of the specimen with respect to the gauge wire such that even if there is no applied load, there is still strain, which is determined by the differential expansion coefficient, $\lambda_{specimen} - \lambda_{gauge}$, where λ is the thermal coefficient of linear expansion: $L = L_0[1 + \lambda(T - T_0)]$, where T_0 is the reference temperature.

d. The gauge factor for a transducer is defined as the fractional change in the measured property $\Delta R/R$ per unit input signal (ε). What is the gauge factor for a metal-wire strain gauge, given that for most metals, $v \approx 1/3$?

e. Consider a strain gauge that consists of a nichrome wire of resistivity $1 \ \mu\Omega$ m, a total length of 1 m, and a diameter of 25 μm. What is ΔR for a strain of 10^{-3}? Assume that $v \approx 1/3$.

f. What will ΔR be if constantan wire is used (see Question 2.5)?

2.16 Thermal coefficients of expansion and resistivity

a. Consider a thin metal wire of length L and diameter D. Its resistance is $R = \rho L/A$, where $A = \pi D^2/4$. By considering the temperature dependence of L, A, and ρ individually, show that

$$\frac{1}{R}\frac{dR}{dT} = \alpha_0 - \lambda_0$$

where α_0 is the temperature coefficient of resistivity (TCR), and λ_0 is the temperature coefficient of linear expansion (thermal expansion coefficient or expansivity), that is,

$$L = L_0[1 + \lambda_0(T - T_0)] \quad \text{and} \quad D = D_0[1 + \lambda_0(T - T_0)]$$

Note that we can define λ_0 by

$$\lambda_0 = L_0^{-1}\left(\frac{dL}{dT}\right)$$

or

$$\lambda_0 = D_0^{-1}\left(\frac{dD}{dT}\right)$$

Given that typically, for most pure metals, $\alpha_0 \approx 1/273 \ \text{K}^{-1}$ and $\lambda_0 \approx 2 \times 10^{-5} \ \text{K}^{-1}$, confirm that the temperature dependence of ρ controls R, rather than the temperature dependence of the geometry. Is it necessary to modify the given equation for a wire with a noncircular cross section?

b. Is it possible to design a resistor from a suitable alloy such that its temperature dependence is almost nil? Consider the TCR of an alloy of two metals A and B, for which $\alpha_{AB} \approx \alpha_A \rho_A/\rho_{AB}$.

2.17 Temperature of a light bulb filament

a. Consider a 100 W, 120 V incandescent bulb (lamp). The tungsten filament has a length of 0.579 m and a diameter of 63.5 μm. Its resistivity at room temperature is 56 nΩ m. Given that the resistivity of the filament can be represented as

Resistivity of W

$$\rho = \rho_0 \left[\frac{T}{T_0} \right]^n \tag{2.43}$$

where T is the temperature in K, ρ_0 is the resistance of the filament at T_0 K, and $n = 1.2$, estimate the temperature of the bulb when it is operated at the rated voltage, that is, directly from the mains outlet. Note that the bulb dissipates 100 W at 120 V.

b. Suppose that the electrical power dissipated in the tungsten wire is totally radiated from the surface of the filament. The radiated power at the absolute temperature T can be described by Stefan's Law

Radiated power

$$P_{\text{radiated}} = \epsilon \sigma_s A (T^4 - T_0^4) \tag{2.44}$$

where σ_s is Stefan's constant (5.67×10^{-8} W m^{-2} K^{-4}), ϵ is the emissivity of the surface (0.35 for tungsten), A is the surface area of the tungsten filament, and T_0 is room temperature (293 K). Obviously, for $T > T_0$, $P_{\text{radiated}} = \epsilon \sigma_s A T^4$.

Assuming that all of the electrical power is radiated from the surface, estimate the temperature of the filament and compare it with your answer in part (a).

c. If the melting temperature of W is 3407 °C, what is the voltage that guarantees that the light bulb will blow?

*2.18 **Radiation theory of the electrical fuse** Consider the principle of operation of an electrical fuse, such as that schematically depicted in Figure 2.30. As current is passed through the fuse, electrical energy is released in the wire by Joule heating. Part of this electrical energy escapes from the surface by radiation (Stefan's Law), and the remaining energy, the net energy released,

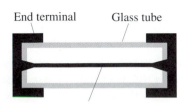

End terminal   Glass tube

Fuse wire (e.g., Sn, In, or Ag)

Figure 2.30   Schematic diagram of a typical fuse for an electronic appliance.

The fuse wire is contained in a glass tube and attached to rigid end terminals.

increases the heat content (internal energy) of the wire, that is, it increases the temperature by an amount determined by the mass m and the specific heat capacity (heat capacity per unit mass) c of the specimen. The final (eventual) temperature T_f of the wire is determined by the steady-state condition:

$$\text{Electrical power input} = \text{Power loss by radiation}$$

or

$$I^2 R = \epsilon \sigma_s A(T_f^4 - T_0^4) \qquad \text{[2.45]}$$

Steady state; final temperature

where σ_s is Stefan's constant (5.67×10^{-8} W m^{-2} K^{-4}), ϵ is the emissivity of the surface, A is the surface area of the wire, and T_0 is the room temperature (293 K). If T_f is greater than the melting temperature T_m, the fuse metal will melt and break the circuit.

During a small time interval dt, the current I through the fuse wire will release an energy $I^2 R dt$ as heat. Some of this energy will escape from the surface by radiation, and the remainder will raise the temperature of the wire by an amount dT that depends on the mass m and specific heat capacity c of the metal. Thus, energy balance during time dt requires

$$mc\,dT = (I^2 R)dt - [\epsilon \sigma_s A(T^4 - T_0^4)]dt \qquad \text{[2.46]}$$

Energy balance

In Equation 2.46, the resistance of the wire depends on T as

$$R = \frac{L\rho_0[1 + \alpha_0(T - T_0)]}{\pi r^2} \qquad \text{[2.47]}$$

where L and r are the length and radius of the fuse wire, ρ_0 is the resistivity at room temperature, and α_0 is the temperature coefficient of resistivity. We can eliminate m and L in Equations 2.46 and 2.47 by noting that the density D is

$$D = \frac{Mass}{Volume} = \frac{m}{\pi r^2 L} \qquad \text{[2.48]}$$

a.   Show that the minimum current that will blow the fuse is given by

$$I_{min} = \sqrt{\frac{2\pi^2 r^3 \epsilon \sigma_s(T_m^4 - T_0^4)}{\rho_0[1 + \alpha_0(T_m - T_0)]}} \qquad \text{[2.49]}$$

Minimum fusing current

where T_m is the melting temperature of the fuse metal.

b.   If the current $I(> I_{min})$ is constant, and t_{fuse} is the time taken to reach the melting temperature T_m of the fuse material, show that

$$t_{fuse} = Dc\pi^2 r^4 \int_{T_0}^{T_m} \frac{dT}{I^2 \rho_0[1 + \alpha_0(T - T_0)] - 2\pi^2 r^3 \epsilon \sigma_s[T^4 - T_0^4]} \qquad \text{[2.50]}$$

Radiation theory of fuse

c.   What are the limitations of the given fuse theory? When would you expect the theory to fail?

Table 2.7   Fuse wires (diameter = 1 mm)

Property	Sn	In	Ag
Resistivity, ρ, nΩ m	126	88	16
TCR, α_0, K^{-1}	0.0047	0.0052	0.0041
Density, D, g cm^{-3}	7.30	7.31	10.5
Specific heat capacity, c, J K^{-1} kg^{-1}	230	233	235
Melting temperature, K	505	430	1234
Emissivity, ϵ	0.05	0.05	0.03
I_{min}	?	?	?
Fusing time for $5I_{min}$	?	?	?

d. Consider three candidate fuse materials, Sn, In, and Ag, whose properties are listed in Table 2.7. Suppose that each wire is 1 mm thick. What is the minimum fusing current I_{min} and t_{fuse} when the current I is $5I_{min}$? Equation 2.50 can only be integrated numerically.

e. Based on Equation 2.50, sketch schematically the expected dependence of t_{fuse} on the current I through the fuse wire.

f. What are major assumptions and limitations of the radiation theory of the electrical fuse? Consider the heat of fusion (or the latent heat of melting) which is 7.2, 3.27 and 11.3 kJ/mole for Sn, In and Ag respectively.

chapter
3

Elementary Quantum Physics

The triumph of modern physics is the triumph of quantum mechanics. Even the simplest experimental observation that the resistivity of a metal depends linearly on the temperature can only be explained by quantum physics, simply because we must take the mean speed of the conduction electrons to be nearly independent of temperature. The modern definitions of voltage and ohm, adopted in January 1990 and now part of the IEEE standards, are based on Josephson and quantum Hall effects, both of which are quantum mechanical phenomena.

One of the most important discoveries in physics has been the wave–particle duality of nature. The electron, which we have so far considered to be a particle and hence to be obeying Newton's second law ($F = ma$), can also exhibit wave-like properties quite contrary to our intuition. An electron beam can give rise to diffraction patterns and interference fringes, just like a light wave. Interference and diffraction phenomena displayed by light can only be explained by treating light as an electromagnetic wave. But light can also exhibit particle-like properties in which it behaves as if it were a stream of discrete entities ("photons"), each carrying a linear momentum and each interacting discretely with electrons in matter (just like a particle colliding with another particle).

3.1 PHOTONS

In introductory physics courses, light is considered to be a wave. Indeed, such phenomena as interference, diffraction, refraction, and reflection can all be explained by the theory of waves. In all these phenomena, a ray of light is considered to be an electromagnetic (EM) wave with a given frequency, as depicted in Figure 3.1. The electric and magnetic fields, $\mathcal{E}_y$ and B_z, of this wave are perpendicular

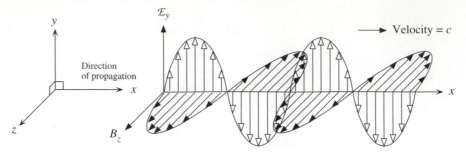

Figure 3.1   The classical view of light as an electromagnetic wave.

An electromagnetic wave is a traveling wave with time-varying electric and magnetic fields that are perpendicular to each other and to the direction of propagation.

to each other and to the direction of propagation x. The electric field $\mathcal{E}_y$ at position x at time t may be described by

$$\mathcal{E}_y(x, t) = \mathcal{E}_{yo} \sin(kx - \omega t) \qquad [3.1]$$

where k is the wave number (propagation constant) related to the wavelength λ by $k = 2\pi/\lambda$, and ω is the angular frequency of the wave (or $2\pi v$, where v is the frequency). A similar equation describes the variation of the magnetic field B_z (directed along z) with x at any time t. Equation 3.1 represents a traveling wave in the x direction, which, in the present example, is a sinusoidally varying function (Figure 3.1). The velocity of the wave (strictly the phase velocity) is

$$c = \frac{\omega}{k} = v\lambda$$

where v is the frequency. The instantaneous intensity, that is, the instantaneous energy flowing per unit area per second, of the wave represented by Equation 3.1 is given by

$$I = c\varepsilon_o \mathcal{E}^2$$

which has the average value

$$I_{av} = \frac{1}{2} c\varepsilon_o \mathcal{E}_o^2$$

Understanding the wave nature of light is fundamental to understanding interference and diffraction, two phenomena that we experience with sound waves almost on a daily basis. Figure 3.2 illustrates how the interference of secondary waves from the two slits S_1 and S_2 gives rise to the dark and bright fringes (called **Young's fringes**) on a screen placed at some distance from the slits. At point P on the screen, the waves emanating from S_1 and S_2 interfere constructively, if they are in phase. This is the case if the path difference between the two rays is an integer multiple of the wavelength λ, or

$$S_1 P - S_2 P = n\lambda$$

where n is an integer. If the two waves are out of phase by a path difference of $\lambda/2$, or

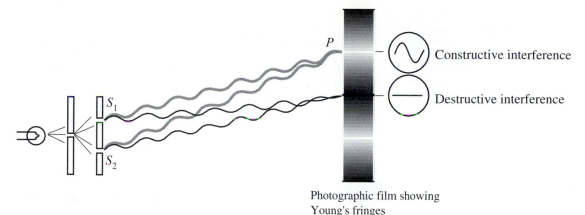

Photographic film showing
Young's fringes

Figure 3.2   Schematic illustration of Young's double-slit experiment.

$$S_1 P - S_2 P = \left(n + \frac{1}{2} \right) \lambda$$

then the waves interfere destructively and the intensity at point P vanishes. Thus, in the y direction, the observer sees a pattern of bright and dark fringes.

When X-rays are incident on a crystalline material, they give rise to typical diffraction patterns on a photographic plate, as shown in Figure 3.3a and b, which can only be explained by using wave concepts. For simplicity, consider two waves, *1* and *2*, in an X-ray beam. The waves are initially in phase, as shown in Figure 3.3c. Suppose that wave *1* is "reflected" from the first plane of atoms in the crystal, whereas wave *2* is "reflected" from the second plane.[1] After reflection, wave *2* has traveled an additional distance equivalent to $2d \sin \theta$ before reaching wave *1*. The path difference between the two waves is $2d \sin \theta$, where d is the separation of the atomic planes. For constructive interference, this must be $n\lambda$, where n is an integer. Otherwise, waves *1* and *2* will interfere destructively and will cancel each other. Waves reflected from adjacent atomic planes interfere constructively to constitute a diffracted beam *only* when the path difference between the waves is an integer multiple of the wavelength, and this will only be the case for certain directions. Therefore the *condition* for the existence of a diffracted beam is

$$2d \sin \theta = n\lambda \qquad n = 1, 2, 3, \ldots \qquad \text{[3.2]}$$

*Bragg
diffraction
condition*

The condition expressed in Equation 3.2, for observing a diffracted beam, forms the whole basis for identifying and studying various crystal structures (the science of crystallography). The equation is referred to as **Bragg's Law,** and arises from the constructive interference of waves.

Aside from exhibiting wave-like properties, light can behave like a stream of "particles" of zero rest-mass. As it turns out, the only way to explain a vast number of experiments is to view light as a stream of discrete entities or energy packets

---

[1] Strictly, one must consider the scattering of waves from the electrons in individual atoms (e.g., atoms A and B in Figure 3.3c) and examine the constructive interference of these scattered waves, which leads to the same condition as that derived in Equation 3.2.

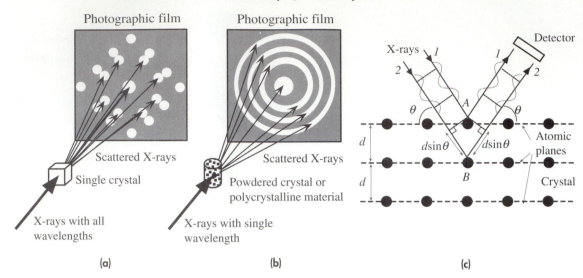

Figure 3.3   Diffraction patterns obtained by passing X-rays through crystals can only be explained by using ideas based on the interference of waves

(a) Diffraction of X-rays from a single crystal gives a diffraction pattern of bright spots on a photographic film.
(b) Diffraction of X-rays from a powdered crystalline material or a polycrystalline material gives a diffraction pattern of bright rings on a photographic film.
(c) X-ray diffraction involves the constructive interference of waves being "reflected" by various atomic planes in the crystal.

called **photons,** each carrying a quantum of energy $h\upsilon$, and momentum, h/λ, where h is a universal constant that can be determined experimentally, and υ is the frequency of light. This photonic view of light is drastically different than the simple wave picture and must be examined closely to understand its origin.

3.1.1   The Photoelectric Effect

Consider a quartz glass vacuum tube with two metal electrodes, a photocathode and an anode, which are connected externally to a voltage supply V (variable and reversible) via an ammeter, as schematically illustrated in Figure 3.4. When the cathode is illuminated with light, if the frequency υ of the light is greater than a certain critical value υ_0, the ammeter registers a current I, even when the anode voltage is zero (i.e., the supply is bypassed). When light strikes the cathode, electrons are emitted with sufficient kinetic energy to reach the opposite electrode. Applying a positive voltage to the anode helps to collect more of the electrons and thus increases the current, until it saturates because all the photoemitted electrons have been collected. The current, then, is limited by the rate of supply of photoemitted electrons. If, on the other hand, we apply a negative voltage to the anode, we can "push" back the photoemitted electrons and hence reduce the current I. Figure 3.5 shows the dependence of the photocurrent on the anode voltage, for one particular frequency of light.

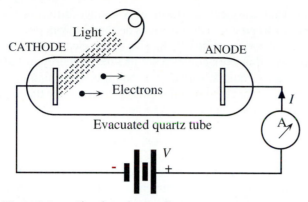

Figure 3.4   The photoelectric effect.

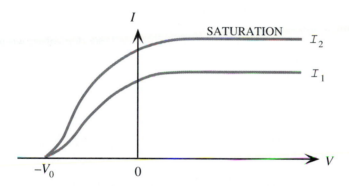

(a) Photoelectric current versus voltage when the cathode is illuminated with light of identical wavelength but different intensities (I). The saturation current is proportional to the light intensity.

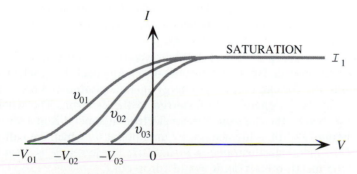

(b) The stopping voltages and therefore the maximum kinetic energies of the emitted electrons depend on the frequency of light, v.

Figure 3.5   Results from the photoelectric experiment.

Recall that when an electron traverses a voltage difference V, its potential energy changes by eV (potential difference is defined as work done per unit charge). When a negative voltage is applied to the anode, the electron has to do work to get to this electrode, and this work comes from its kinetic energy just after photoemission. When the negative anode voltage V is equal to V_0, which just "extinguishes" the current I, we know that the potential energy "gained" by the electron is just the kinetic energy lost by the electron, or

$$eV_0 = \frac{1}{2}m_e v^2 = KE_m$$

where v is the velocity and KE_m is the kinetic energy of the electron just after photoemission. Therefore, we can conveniently measure the maximum kinetic energy KE_m of the emitted electrons.

For a given frequency of light, increasing the intensity of light $\mathcal{I}$ requires the *same* voltage V_0 to extinguish the current; that is, the KE_m of emitted electrons is independent of the light intensity $\mathcal{I}$. This is quite surprising. However, increasing the intensity does increase the saturation current. Both of these effects are noted in the $I–V$ results shown in Figure 3.5.

Since the magnitude of the saturation photocurrent depends on the light intensity $\mathcal{I}$, whereas the KE of the emitted electron is independent of $\mathcal{I}$, we are forced to conclude that only the *number* of electrons ejected depends on the light intensity. Furthermore, if we plot KE_m (from the V_0 value) against the light frequency v for different electrode metals for the cathode, we find the typical behavior shown in Figure 3.6. This shows that the KE of the emitted electron depends on the frequency of light. The experimental results shown in Figure 3.6 can be summarized by a statement that relates the KE_m of the electron to the frequency of light and the electrode metal, as follows:

$$KE_m = hv - hv_0 \qquad \text{[3.3]}$$

where h is the slope of the straight line and is independent of the type of metal, whereas v_0 depends on the electrode material for the photocathode (e.g., v_{01}, v_{02}, etc). Equation 3.3 is essentially a succinct statement of the experimental observations of the photoelectric effect as exhibited in Figure 3.6. The constant h is called **Planck's constant,** which, from the slope of the straight lines in Figure 3.6, can be shown to be about 6.6×10^{-34} J s. This was beautifully demonstrated by Millikan in 1915, in an excellent series of photolectric experiments using different photocathode materials.

The successful interpretation of the photoelectric effect was first given in 1905 by Einstein, who proposed that light consists of "energy packets," each of which has the magnitude hv. We can call these energy quanta **photons.** When one photon strikes an electron, its energy is transferred to the electron. The whole photon becomes absorbed by the electron. Yet, an electron in a metal is in a lower state of potential energy (PE) than in vacuum, by an amount Φ, which we call the **work function** of the metal, as illustrated in Figure 3.7. The lower PE is what keeps the electron in the metal; otherwise, it would "drop out."

This lower PE is a result of the Coulombic attraction interaction between the electron and the positive metal ions. Some of the photon energy hv therefore goes toward overcoming this PE barrier. The energy that is left ($hv - \Phi$) gives the

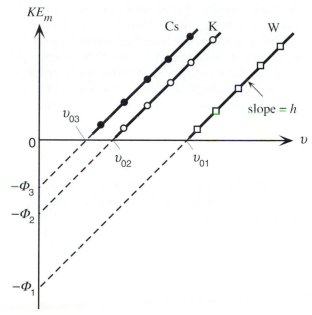

Figure 3.6   The effect of varying the frequency of light and the cathode material in the photoelectric experiment.

The lines for the different materials have the same slope h but different intercepts.

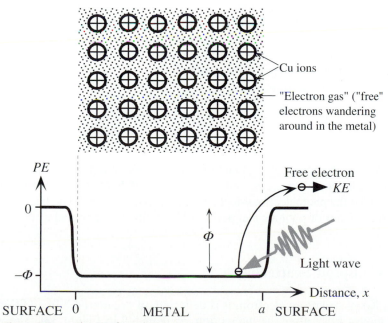

Figure 3.7   The PE of an electron inside the metal is lower than outside by an energy called the workfunction of the metal.

Work must be done to remove the electron from the metal.

electron its *KE*. The work function Φ changes from one metal to another. Photo-emission only occurs when $h\upsilon$ is greater than Φ. This is clearly borne out by experiment, since a critical frequency υ_o is needed to register a photocurrent. When υ is less than υ_0, even if we use an extremely intense light, no current exists because no phtoemission occurs, as demonstrated by the experimental results in Figure 3.6. Inasmuch as Φ depends on the metal, so does υ_o. Therefore, in Einstein's interpretation $h\upsilon_0 = \Phi$. In fact, the measurement of υ_o constitutes one method of determining the work function of the metal.

This explanation for the photoelectric effect is further supported by the fact that the work function Φ from $h\upsilon_o$ is in good agreement with that from thermionic emission experiments. There is an apparent similarity between the *I–V* characteristics of the phototube and that of the vacuum tube used in early radios. The only difference is that in the vacuum tube, the emission of electrons from the cathode is achieved by heating the cathode. Thermal energy ejects some of electrons over the *PE* barrier Φ. The measurement of Φ by this thermionic emission process agrees with that from photoemission experiments.

In the photonic interpretation of light, we still have to resolve the meaning of the intensity of light, because the classical average intensity expression

$$I_{av} = \frac{1}{2} c \varepsilon_o \mathcal{E}_o^2$$

is obviously not acceptable. Increasing the intensity of illumination in the photoelectric experiment increases the saturation current, which means that more electrons are emitted per unit time. We therefore infer that the cathode must be receiving more photons per unit time at higher intensities. By definition, "intensity" refers to the amount of energy flowing through unit area per unit time. If the number of photons crossing a unit area per unit time is the photon flux, denoted by Γ_{ph}, then the flow of energy through a unit area per unit time is the product of this photon flux and the energy per photon, that is,

Light intensity

$$I = \Gamma_{ph} h\upsilon \tag{3.4}$$

where

Photon flux

$$\Gamma_{ph} = \frac{\Delta N_{ph}}{A \, \Delta t} \tag{3.5}$$

in which ΔN_{ph} is the net number of photons crossing an area A in time Δt. With the energy of a photon given as $h\upsilon$ and the intensity of light defined as $\Gamma_{ph} h\upsilon$, the explanation for the photoelectric effect becomes self-consistent. The interpretation of light as a stream of photons can perhaps be intuitively imagined as depicted in Figure 3.8.

3.1.2   Compton Scattering

When an X-ray strikes an electron, it is deflected, or "scattered." In addition, the electron moves away after the interaction, as depicted in Figure 3.9. The wavelength of the incoming and scattered X-rays can readily be measured. The frequency υ' of the scattered X-ray is less than the frequency υ of the incoming X-ray.

When the *KE* of the electron is determined, we find that

$$KE = h\upsilon - h\upsilon'$$

Since the electron now also has a momentum p_e, then from the conservation of linear momentum law, we are forced to accept that the X-ray also has a momentum. The Compton effect experiments showed that the momentum of the photon is related to its wavelength by

$$p = \frac{h}{\lambda}$$

[3.6] *Momentum of a photon*

We see that a photon not only has an energy $h\upsilon$, but also a momentum p, and it interacts as if it were a discrete entity like a particle. Therefore, when discussing the properties of a photon, we must consider its energy and momentum as if it were a particle.

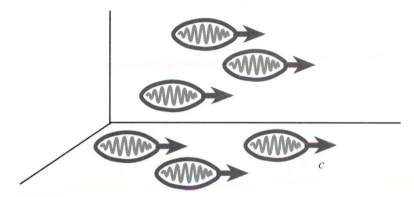

Figure 3.8   Intuitive visualization of light consisting of a stream of photons (not to be taken too literally).

SOURCE: R. Serway, C. J. Moses, and C. A. Moyer, *Modern Physics*, Saunders College Publishing, 1989, p. 56, Figure 2.16(b).

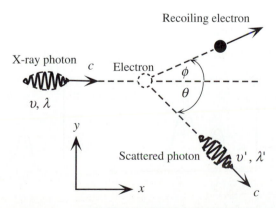

Figure 3.9   Scattering of an X-ray photon by a "free" electron in a conductor.

We should mention that the description of the Compton effect shown in Figure 3.9 is, in fact, the inference from a more practical experiment involving the scattering of X-rays from a metal target. A collimated monochromatic beam of X-rays of wavelength λ_0 strikes a conducting target, such as graphite, as illustrated in Figure 3.10a. A conducting target contains a large number of nearly "free" electrons (conduction electrons), which can scatter the X-rays. The scattered X-rays are detected at various angles θ with respect to the original direction, and their wavelength λ' is measured. The result of the experiment is therefore the scattered wavelength λ' measured at various scattering angles θ, as shown in Figure 3.10b. It turns out that the λ' versus θ results agree with the conservation of linear momentum law applied to an X-ray photon colliding with an electron with the momentum of the photon given precisely by Equation 3.6.

The photoelectric experiment and the Compton effect are just two convincing experiments in modern physics that force us to accept that light can have particle-like properties. We already know that it can also exhibit wave-like properties, in such experiments as Young's interference fringes. We are then faced with what is known as the wave–particle dilemma. How do we know whether light is going to

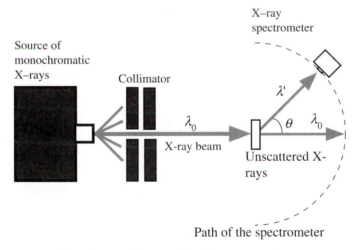

(a) A schematic diagram of the Compton experiment.

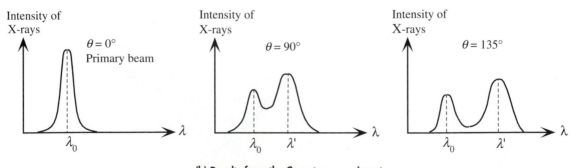

(b) Results from the Compton experiment.

Figure 3.10   The Compton experiment and its results.

behave like a wave or a particle? The properties exhibited by light depend very much on the nature of the experiment. Some experiments will require the wave model, whereas others may use the particulate interpretation of light. There have been no experiments in which both wave and particle properties have been required simultaneously. We should perhaps view the two interpretations as two complementary ways of modelling the behavior of light when it interacts with matter, accepting the fact that light has a dual nature. Both models are needed for a full description of the behavior of light.

The expressions for the energy and momentum of the photon, $E = h\upsilon$ and $p = h/\lambda$, can also be written in terms of the angular frequency ω and the wavenumber k, as follows:

$$E = \hbar\omega \quad \text{and} \quad p = \hbar k$$

where $\hbar$ is defined as $h/2\pi$.

3.1.3   Black Body Radiation

Experiments indicate that all objects emit and absorb energy in the form of radiation, and the intensity of this radiation depends on the radiation wavelength and temperature of the object. This radiation is frequently termed **thermal radiation.** When the object is in thermal equilibrium with its surroundings, that is, at the same temperature, the object absorbs as much radiation energy as its emits. On the other hand, when the temperature of the object is above the temperature of its surroundings, there is a net emission of radiation energy. The maximum amount of radiation energy that can be emitted by an object is called the **black body radiation.** Although, in general, the intensity of the radiated energy depends on the material's surface, the radiation emitted from a cavity with a small aperture is independent of the material of the cavity and corresponds very closely to black body radiation.

The intensity of the emitted radiation has the spectrum (i.e., intensity vs. wavelength characteristic), and the temperature dependence illustrated in Figure 3.11. It is useful to define a **spectral irradiance** I_λ as the emitted radiation intensity (power per unit area) per unit wavelength, so that $I_\lambda\,\delta\lambda$ is the intensity in a small range of wavelengths $\delta\lambda$. Figure 3.11 shows the typical I_λ versus λ behavior of black body radiation at two temperatures. We assume that the characteristics of the radiation emerging from the aperture represent those of the radiation within the cavity.

Classical physics predicts that the acceleration and deceleration of the charges due to various thermal vibrations, oscillations, or motions of the atoms in the surface region of the cavity material result in electromagnetic waves of the emissions. These waves then interfere with each other, giving rise to many types of standing electromagnetic waves with different wavelengths in the cavity. Each wave contributes an energy kT to the emitted intensity. If we calculate the number of standing waves within a small range of wavelength, the classical prediction leads to the **Rayleigh–Jeans Law** in which $I_\lambda \propto 1/\lambda^4$ and $I_\lambda \propto T$, which are not in agreement with the experiment, especially in the short-wavelength range (see Figure 3.11).

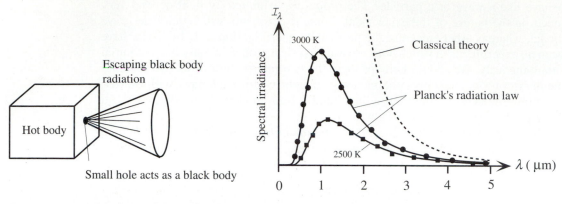

Figure 3.11   Schematic illustration of black body radiation and its characteristics.

Spectral irradiance versus wavelength at two temperatures (3000 K is about the temperature of the incandescent tungsten filament in a light bulb).

Max Planck (1900) was able to show that the experimental results can be explained if we assume that the radiation within the cavity involves the emission and absoprtion of discrete amounts of light energy by the oscillation of the molecules of the cavity material. He assumed that oscillating molecules emit and absorb a quantity of energy that is an integer multiple of a discrete energy quantum that is determined by the frequency v of the radiation and given by hv. This is what we now call a photon. He then considered the energy distribution (the statistics) in the molecular oscillations and took the probability of an oscillator possessing an energy nhv (where n is an integer) to be proportional to the Boltzmann factor, $\exp(-nhv/kT)$. He eventually derived the mathematical form of the black body radiation characteristics in Figure 3.11. Planck's black body radiation formula for I_λ is generally expressed as

*Planck's
radiation law*

$$I_\lambda = \frac{2\pi hc^2}{\lambda^5\left[\exp\left(\dfrac{hc}{\lambda kT}\right) - 1\right]}$$ [3.7]

where k is the Boltzman constant. Planck's radiation law based on the emission and absorption of photons is in excellent agreement with all observed black body radiation characteristics as depicted in Figure 3.11.

Example 3.1 | **ENERGY OF A BLUE PHOTON**   What is the energy of a blue photon which has a wavelength of 450 nm.

Solution

The energy of the photon is given by

$$E_{ph} = hv = \frac{hc}{\lambda} = \frac{(6.6 \times 10^{-34} \text{ J s})(3 \times 10^8 \text{ m s}^{-1})}{450 \times 10^{-9} \text{ m}} = 4.4 \times 10^{-19} \text{ J}$$

Generally, with such small energy values, we prefer electron–volts (eV), so the energy of the photon is

$$\frac{4.4 \times 10^{-19} \text{ J}}{1.6 \times 10^{-19} \text{ C}} = 2.75 \text{ eV}$$

THE PHOTOELECTRIC EXPERIMENT   In the photoelectric experiment, green light, with a wavelength 522 nm, is the longest-wavelength radiation that can cause the photoemission of electrons from a clean sodium surface.   **Example 3.2**

a.   What is the work function of sodium, in electron–volts?

b.   If UV (ultraviolet) radiation of wavelength 250 nm is incident to the sodium surface, what will be the kinetic energy of the photoemitted electrons, in electron–volts?

c.   Suppose that the UV light of wavelength 250 nm has an intensity of 20 mW cm^{-2}. If the emitted electrons are collected by applying a positive bias to the opposite electrode, what will be the photoelectric current density?

Solution

a.   At threshold, the photon energy just causes photoemissions; that is, the electron just overcomes the potential barrier Φ. Thus, $hc/\lambda_0 = e\Phi$, where Φ is the work function in eV, and λ_0 is the longest wavelength.

$$\Phi = \frac{hc}{e\lambda_0} = \frac{(6.626 \times 10^{-34} \text{ J s}^{-1})(3 \times 10^8 \text{ m s}^{-1})}{[(1.6 \times 10^{-19} \text{ C})(522 \times 10^{-9} \text{ m})]} = 2.38 \text{ eV}$$

b.   The energy of the incoming photon E_{ph} is (hc/λ), so the excess energy over $e\Phi$ goes to the kinetic energy of the electron. Thus,

$$KE = \frac{hc}{e\lambda} - \Phi = 4.96 \text{ eV} - 2.38 \text{ eV} = 2.58 \text{ eV}$$

c.   The light intensity (defined as energy flux) is given by $\mathcal{I} = \Gamma_{ph}(hc/\lambda)$, where Γ_{ph} is the number of photons arriving per unit area per unit time; that is, photon flux and (hc/λ) is the energy per photon. Thus, if each photon releases one electron, the electron flux will be equal to the photon flux, and the current density, which is the charge flux, will be

$$J = e\Gamma_{ph} = \frac{e\mathcal{I}\lambda}{hc}$$

$$= \frac{(1.6 \times 10^{-19} \text{ C})(20 \times 10^{-3} \times 10^4 \text{ J s}^{-1} \text{ m}^{-2})(250 \times 10^{-9} \text{ m})}{(6.626 \times 10^{-34} \text{ J s})(3 \times 10^8 \text{ m s}^{-1})}$$

$$= 40.3 \text{ A m}^{-2} \quad \text{or} \quad 4.0 \text{ mA cm}^{-2}$$

3.2   THE ELECTRON AS A WAVE

3.2.1   De Broglie Relationship

It is apparent from the photoelectric and Compton effects that light, which we thought was a wave, can behave as if it were a stream of particulate-like entities called photons. Can electrons exhibit wave-like properties? Again, this depends on the experiment and on the energy of the electrons.

When the interference and diffraction experiments in Figures 3.2 and 3.3 are repeated with an electron beam, very similar results are found to those obtainable with light and X-rays. When we use an electron beam in Young's double-slit experiment, we observe high and low intensity regions (i.e., Young's fringes), as illustrated in Figure 3.12. The interference pattern is viewed on a fluorescent TV screen. When an energetic electron beam hits an Al polycrystalline sample, it produces diffraction rings on a fluorescent screen (Figure 3.13), just like X-rays do on a photographic plate. The diffraction pattern obtained with an electron beam (Figure 3.13) means that the electrons are obeying the Bragg diffraction condition $2d \sin \theta = n\lambda$ just as much as the X-ray waves.

Since we know the interatomic spacing d and we can measure the angle of diffraction 2θ, we can readily evaluate the wavelength λ associated with the wave-like behavior of the electrons. Furthermore, from the accelerating voltage V in the electron tube, we can also determine the momentum of the electrons, because the kinetic energy gained by the electrons, $(p^2/2m_e)$, is equal to eV. Simply by adjusting the accelerating voltage V, we can therefore study how the wavelength of the electron depends on the momentum.

As a result of such studies and other similar experiments, it has been found that an electron traveling with a momentum p behaves like a wave of wavelength λ given by

Wavelength of the electron

$$\lambda = \frac{h}{p} \qquad\qquad\qquad [3.8]$$

This is just the reverse of the equation for the momentum of a photon given its wavelength. The same equation therefore relates wave-like and particle-like properties to and from each other. Thus,

$$\lambda = \frac{h}{p} \qquad \text{or} \qquad p = \frac{h}{\lambda}$$

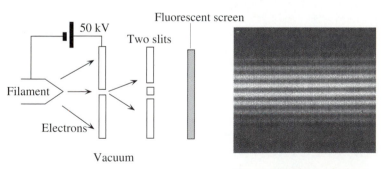

Electron diffraction fringes on the screen

Figure 3.12   Young's double-slit experiment with electrons involves an electron gun and two slits in a cathode ray tube (CRT) (hence, in vacuum).

Electrons from the filament are accelerated by a 50 kV anode voltage to produce a beam that is made to pass through the slits. The electrons then produce a visible pattern when they strike a fluoresecent screen (e.g., a TV screen), and the resulting visual pattern is photographed.

Pattern from C. Jönsson, D. Brandt, and S. Hirschi, *Am. J. Physics*, 42. 1974, p. 9, Figure 8. Used with permission.

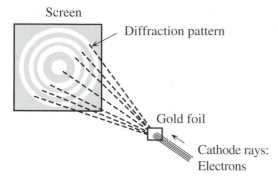

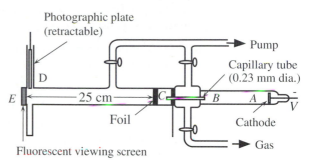

(a) Thomson diffracted electrons by using a thin gold foil and produced a diffraction pattern on the screen of his apparatus in (b). The foil was polycrystalline, so the diffraction pattern was circular rings.

(b) In Thomson's electron diffraction apparatus a beam of electrons is generated in tube *A*, passed through collimating tube *B*, and made to impinge on a thin gold foil *C*. The transmitted electrons impinge on the fluorescent screen *E*, or a photographic plate *D*, which could be lowered into the path. The entire apparatus was evacuated during the experiment.

(c) Electron diffraction pattern obtained by G. P. Thomson using a gold foil target.

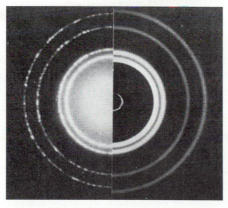

(d) Composite photograph showing diffraction patterns produced with an aluminum foil by X-rays and electrons of similar wavelength. Left: X-rays of $\lambda = 0.071$ nm. Right: Electrons of energy 600 eV.

(e) Diffraction pattern produced by 40 keV electrons passing through zinc oxide powder. The distortion of the pattern was produced by a small magnet placed between the sample and the photographic plate. An X-ray diffraction pattern would not be affected by a magnetic field.

Figure 3.13   The diffraction of electrons by crystals gives typical diffraction patterns that would be expected if waves were being diffracted, as in X-ray diffraction with crystals. [(b) from G.P. Thomson, *Proceedings of the Royal Society*, A117, no. 600, 1928. (c) and (d) from A.P. French and F. Taylor, *An Introduction to Quantum Mechanics* (Norton, New York, 1978) p 75; (e) from R.B. Leighton, *Principles of Modern Physics* (McGraw-Hill, 1959) p.84]

is an equation that exposes the wave–particle duality of nature. It was first hypothesized by De Broglie in 1924. As an example, we can calculate the wavelengths of a number of particle-like objects:

a.   A 50-gram golf ball traveling at a velocity of 20 m s^{-1}.
     The wavelength is

$$\lambda = \frac{h}{mv} = \frac{6.63 \times 10^{-34} \text{ J s}}{(50 \times 10^{-3} \text{ kg})(20 \text{ m s}^{-1})} = 6.63 \times 10^{-34} \text{ m}$$

The wavelength is so small that this golf ball will not exhibit any wave effects. Firing a stream of golf balls at a wall will not result in "diffraction rings" of golf balls.

b.   A proton traveling at 2200 m s^{-1}.

Using $m_p = 1.67 \times 10^{-27}$ kg, we have $\lambda = (h/mv) \approx 0.18$ nm. This is only slightly smaller than the interatomic distance in crystals, so firing protons at a crystal can result in diffraction. (Recall that to get a diffraction peak, we must satisfy the Bragg condition, $2d \sin \theta = n\lambda$.) Protons, however, are charged, so they can penetrate only a small distance into the crystal. Hence, they are not used in crystal diffraction studies.

c.   Electron accelerated by 100 volts.

This voltage accelerates the electron to a *KE* equal to *eV*. From $KE = p^2/2m_e = eV$, we can calculate p and hence $\lambda = h/p$. The result is $\lambda = 0.123$ nm. Since this is comparable to typical interatomic distances in solids, we would see a diffraction pattern when an electron beam strikes a crystal. The actual pattern is determined by the Bragg diffraction condition.

3.2.2   Time-Independent Schrödinger Equation

The experiments in which electrons exhibit interference and diffraction phenomena show quite clearly that, under certain conditions, the electron can behave as a wave; in other words, it can exhibit wave-like properties. There is a general equation that describes this wave-like behavior and, with the appropriate potential energy and boundary conditions, will predict the results of the experiments. The equation is called the **Schrödinger equation** and it forms the foundations of quantum theory. Its fundamental nature is analogous to the classical physics assertion of Newton's second law, $F = ma$, which of course cannot be proved. As a fundamental equation, Schrödinger's has been found to successfully predict every observable physical phenomenon at the atomic scale. Without this equation, we will not be able to understand the principles of operation of many semiconductor devices. We introduce the equation through an analogy.

A traveling electromagnetic wave resulting from sinusoidal current oscillations, or the traveling voltage wave on a long transmission line, can generally be described by a traveling-wave equation of the form

$$\mathcal{E}(x, t) = \mathcal{E}_o \exp j(kx - \omega t) = \mathcal{E}(x) \exp(-j\omega t) \qquad \textbf{[3.9]}$$

where $\mathcal{E}(x) = \mathcal{E}_0 \exp(jkx)$ represents the spatial dependence, which is separate from the time variation. We assume that no transients exist to upset this perfect sinusoidal propagation. We note that the time dependence is harmonic and therefore predictable. For this reason, in ac circuits we put aside the $\exp(-j\omega t)$ term until we need the instantaneous magnitude of the voltage.

The average intensity $\mathcal{I}_{av} = (1/2)c\varepsilon_o\mathcal{E}_o^2$ depends on the square of the amplitude. In Young's double-slit experiment, the intensity varies along the y direction, which means that $\mathcal{E}_o^2$ for the resultant wave depends on y. In the electron version of this experiment in Figure 3.12, what changes in the y direction is the probability of observing electrons; that is, there are peaks and troughs in the probability of finding electrons along y, just like $\mathcal{E}_o^2$ variation along y. We should therefore attach some probability interpretation to the wave description of the electron.

In 1926, Max Born suggested a probability wave interpretation for the wave-like behavior of the electron.

$$\mathcal{E}(x, t) = \mathcal{E}_o \sin(kx - \omega t)$$

is a plane traveling **wavefunction** for an electric field, experimentally, we measure and interpret the *intensity* of a wave, namely $|\mathcal{E}(x, t)|^2$. There may be a similar wave function for the electron, which we can represent by a function $\Psi(x, t)$. According to Born, the significance of $\Psi(x, t)$ is that its amplitude squared represents the probability of finding the electron per unit distance. Thus, in three dimensions, if $\Psi(x, y, z, t)$ represents the wave property of the electron, it must have the following interpretation:

$|\Psi(x, y, z, t)|^2$ is the probability of finding the electron per unit volume at x, y, z at time t, or

$|\Psi(x, y, z, t)|^2 \, dx \, dy \, dz$ is the probability of finding the electron in a small elemental volume $dx \, dy \, dz$ at x, y, z at time t.

If we are just considering one dimension, then the wavefunction is $\Psi(x, t)$, and $|\Psi(x, t)|^2 \, dx$ is the probability of finding the electron between x and $(x + dx)$ at time t.

We should note that since only $|\Psi|^2$ has meaning, not Ψ, the latter function need not be real; it can be a complex function with real and imaginary parts. For this reason, we tend to use $\Psi^* \Psi$, where Ψ^* is the complex conjugate of Ψ, instead of $|\Psi|^2$, to represent the probability per unit volume.

To obtain the wavefunction $\Psi(x, t)$ for the electron, we need to know how the electron interacts with its environment. This is embodied in its potential energy function $V = V(x, t)$, because the net force the electron experiences is given by

$$F = -dV/dx.$$

For example, if the electron is attracted by a positive charge (e.g., the proton in a hydrogen atom), then it clearly has an electrostatic potential energy given by

$$V(r) = \frac{-e^2}{4\pi\varepsilon_o r}$$

where $r = \sqrt{(x^2 + y^2 + z^2)}$ is the distance between the electron and the proton.

If the *PE* of the electron is time independent, which means that $V = V(x)$ in one dimension, then the spatial and time dependences of $\Psi(x, t)$ can be separated, just as in Equation 3.9, and the **total wavefunction** $\Psi(x, t)$ of the electron can be written as

$$\Psi(x, t) = \psi(x) \exp\left(-\frac{jEt}{\hbar}\right) \tag{3.10}$$

where $\psi(x)$ is the electron wavefunction that describes only the spatial behavior, and E is the energy of the electron. The temporal behavior is simply harmonic, by virtue of $\exp(-jEt/\hbar)$, which corresponds to $\exp(-j\omega)$ with an angular frequency $\omega = E/\hbar$. The fundamental equation that describes the electron's behavior by determining $\psi(x)$ is called the **time-independent Schrödinger equation.** It is given by the famous equation

Schrödinger's equation for one dimension

$$\frac{d^2\psi}{dx^2} + \frac{2m}{\hbar^2}(E - V)\psi = 0 \tag{3.11}$$

where m is the mass of the electron.

This is a second-order differential equation. It should be re-emphasized that the potential energy V in Equation 3.11 depends only on x. If the potential energy of the electron depends on time as well, that is, if $V = V(x, t)$, then in general $\Psi(x, t)$ cannot be written as $\psi(x) \exp(-jEt/\hbar)$. Instead, we must use the full version of the Schrödinger equation, which is discussed in Section 3.10.

In three dimensions, there will be derivatives of ψ with respect to x, y, and z. We use the calculus notation $(\partial\psi/\partial x)$, diffrentiating $\psi(x, y, z)$ with respect to x but keeping y and z constant. Similar notations $\partial\psi/\partial y$ and $\partial\psi/\partial z$ are used for derivatives with respect to y alone and with respect to z alone, respectively. In three dimensions, Equation 3.11 becomes

Schrödinger's equation for three dimensions

$$\frac{\partial^2\psi}{\partial x^2} + \frac{\partial^2\psi}{\partial y^2} + \frac{\partial^2\psi}{\partial z^2} + \frac{2m}{\hbar^2}(E - V)\psi = 0 \tag{3.11a}$$

where $V = V(x, y, z)$ and $\psi = \psi(x, y, z)$.

Equation 3.11a is a fundamental equation, called the time-independent Schrödinger equation, the solution of which gives the steady-state behavior of the electron in a time-independent potential energy environment described by $V = V(x, y, z)$. By solving Equation 3.11a, we will know the probability distribution and the energy of the electron. Once $\psi(x, y, z)$ has been determined, the total wavefunction for the electron is given by Equation 3.10 so that

$$|\Psi(x, y, z, t)|^2 = |\psi(x, y, z)|^2$$

which means that the steady-state probability distribution of the electron is simply $|\psi(x, y, z)|^2$.

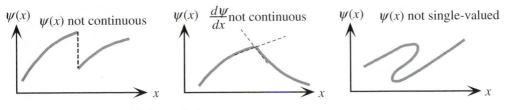

Figure 3.14   Unacceptable forms of $\psi(x)$

The time-independent Schrödinger equation can be viewed as a "mathematical crank". We input the potential energy of the electron and the boundary conditions, turn the crank, and get the probability distribution and the energy of the electron under steady-state conditions.

Two important boundary conditions are often used to solve the Schrödinger equation. First, as an analogy, when we stretch a string between to fixed points and put it into a steady-state vibration, there are no discontinuities or kinks along the string. We can therefore intelligently guess that because $\psi(x)$ represents wave-like behavior, it must be a smooth function without any discontinuities.

The first boundary condition is that Ψ must be continuous, and the second is that $d\Psi/dx$ must be continuous. In the steady state, these two conditions translate directly to ψ and $d\psi/dx$ being continuous. Since the probability of finding the electron is represented by $|\psi|^2$, this function must be single-valued and smooth, without any discontinuities, as illustrated in Figure 3.14. The enforcement of these boundary conditions results in strict requirements on the wavefunction $\psi(x)$, as a result of which only certain wavefunctions are acceptable. These wavefunctions are called the **eigenfunctions** (characteristic functions) of the system, and they deter-mine the behavior and energy of the electron under steady-state conditions. The eigenfunctions $\psi(x)$ are also called **stationary states,** inasmuch as we are only considering steady-state behavior.

It is important to note that the Schrödinger equation is generally applicable to all matter, not just the electron. For example, the equation can also be used to describe the behavior of a proton, if the appropriate potential energy $V(x, y, z)$ and mass (m_{proton}) are used. Wavefunctions associated with particles are frequently called **matter waves.**

THE FREE ELECTRON   Solve the Schrödinger equation for a free electron whose energy is E. What is the uncertainty in the position of the electron and the uncertainty in the momentum of the electron? | **Example 3.3**

Solution

Since the electron is free, its potential energy is zero $V = 0$. In the Schrödinger equation, this leads to

$$\frac{d^2\psi}{dx^2} + \frac{2m}{\hbar^2}E\psi = 0$$

We can write this as

$$\frac{d^2\psi}{dx^2} + k^2\psi = 0$$

where we defined $k^2 = (2m/\hbar^2)E$. Solving the differential equation, we get

$$\psi(x) = A \exp(jkx) \qquad \text{or} \qquad B \exp(-jkx)$$

The total wavefunction is obtained by multiplying $\psi(x)$ by $\exp(-jEt/\hbar)$. We can define a fictitious frequency for the electron by $\omega = E/\hbar$ and multiply $\psi(x)$ by $\exp(-j\omega t)$:

$$\Psi(x, t) = A \exp j(kx - \omega t) \qquad \text{or} \qquad B \exp j(-kx - \omega t)$$

Each of these is a traveling wave. The first solution is a traveling wave in the $+x$ direction, and the second one is in the $-x$ direction. Thus, the free electron has a traveling wave solution with a wavenumber $k = 2\pi/\lambda$, that can have any value. The energy E of the electron is simply KE, so that

$$KE = E = \frac{(\hbar k)^2}{2m}$$

When we compare this with the classical physics expression $KE = (p^2/2m)$, we see that the momentum is given by

$$p = \hbar k \qquad \text{or} \qquad p = \frac{h}{\lambda}$$

This is the de Broglie relationship. The latter therefore results naturally from the Schrödinger equation for a free electron.

The probability distribution for the electron is

$$|\psi(x)|^2 = |A \exp j(kx)|^2 = A^2$$

which is constant over the entire space. Thus, the electron can be anywhere between $x = -\infty$ and $x = +\infty$. The uncertainty Δx in its position is infinite. Since the electron has a well-defined wavenumber k, its momentum p is also well-defined by virtue of $p = \hbar k$. The uncertainty Δp in its momentum is thus zero.

Example 3.4

WAVELENGTH OF AN ELECTRON BEAM   Electrons are accelerated through a 100 V potential difference to strike a polycrystalline aluminum sample. The diffraction pattern obtained indicates that the highest intensity and smallest angle diffraction, corresponding to diffraction from the (111) planes, has a diffraction angle of 30.4 degrees. From X-ray studies, the separation of the (111) planes is 0.234 nm. What is the wavelength of the electron and how does it compare with that from the de Broglie relationship?

Solution

Since we know the angle of diffraction 2θ ($= 30.4°$) and the interplanar separation d ($= 0.234$ nm), we can readily calculate the wavelength of the electron from the Bragg condition for diffraction, $2d \sin \theta = n\lambda$. With $n = 1$,

$$\lambda = 2d \sin \theta = 2(0.234 \text{ nm}) \sin(15.2°) = 0.1227 \text{ nm}$$

This is the wavelength of the electron.

When an electron is accelerated through a voltage V, it gains KE equal to eV, so that $p^2/2m = eV$ and $p = [2meV]^{1/2}$. This is the momentum imparted by the potential difference V. From the de Broglie relationship, the wavelength should be

$$\lambda = \frac{h}{p} = \frac{h}{[2meV]^{1/2}}$$

or

$$\lambda = \left[\frac{h^2}{2meV}\right]^{1/2}$$

Substituting for e, h, and m, we obtain

$$\lambda = \frac{1.226 \text{ nm}}{V^{1/2}}$$

The experiment uses 100 V, so the de Broglie wavelength is

$$\lambda = \frac{1.226 \text{ nm}}{V^{1/2}} = \frac{1.226 \text{ nm}}{100^{1/2}} = 0.1226 \text{ nm}$$

which is in excellent agreement with that determined from the Bragg condition.

3.3   INFINITE POTENTIAL WELL: A CONFINED ELECTRON

Consider the behavior of the electron when it is confined to a certain region, $0 < x < a$. Its PE is zero inside that region and infinite outside, as shown in Figure 3.15. The electron cannot escape, because it would need an infinite PE. Clearly the probability $|\psi|^2$ of finding the electron per unit volume is zero outside $0 < x < a$. Thus, $\psi = 0$ when $x \le 0$ and $x \ge a$, and ψ is determined by the Schrödinger equation in $0 < x < a$ with $V = 0$. Therefore, in the region $0 < x < a$

$$\frac{d^2\psi}{dx^2} + \frac{2m}{\hbar^2}E\psi = 0 \qquad \text{[3.12]}$$

This is a second-order linear differential equation. As a general solution, we can take

$$\psi(x) = A \exp(jkx) + B \exp(-jkx)$$

where k is some constant (to be determined) and substitute this in Equation 3.12 to find k. We first note that $\psi(0) = 0$; therefore, $B = -A$, so that

$$\psi(x) = A[\exp(jkx) - \exp(-jkx)] = 2Aj \sin kx \qquad \text{[3.13]}$$

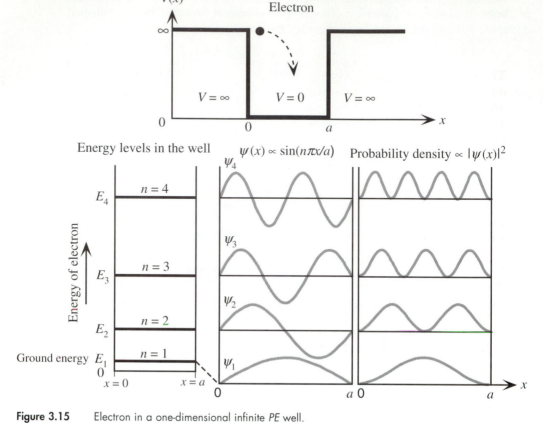

Figure 3.15   Electron in a one-dimensional infinite *PE* well.
The energy of the electron is quantized. Possible wavefunctions and the probability distributions for the electron are shown.

We now substitute this into the Schrödinger Equation 3.12 to relate the energy E to k. Thus, Equation 3.12 becomes

$$-2Ajk^2(\sin kx) + \left(\frac{2m}{\hbar^2}\right)E(2Aj \sin kx) = 0$$

which can be rearranged to obtain the energy of the electron:

$$E = \frac{\hbar^2 k^2}{2m} \qquad\qquad \text{[3.14]}$$

Since the electron has no *PE* within the well, its total energy E is kinetic energy *KE*, and we can write

$$E = KE = \frac{p_x^2}{2m}$$

where p_x is its momentum. Comparing this with Equation 3.14, we see that the momentum of the electron must be

$$p_x = \pm \hbar k \qquad \text{[3.15]}$$

The momentum p_x may be in the $+x$ direction or the $-x$ direction (which is the reason for $\pm$), so the **average momentum** is actually zero, $p_{av} = 0$.

We have already seen this relationship, when we defined k as $2\pi/\lambda$ (wavenumber) for a free traveling wave. So the constant k here is a wavenumber-type quantity even though there is no distinct traveling wave. Its value is determined by the boundary condition at $x = a$ where $\psi = 0$, or

$$\psi(a) = 2Aj \sin ka = 0$$

The solution to $\sin ka = 0$ is simply $ka = n\pi$, where $n = 1, 2, 3, \ldots$ is an integer. We exclude $n = 0$ because it will result in $\psi = 0$ everywhere (no electron at all).

We notice immediately that k, and therefore the energy of the electron, can only have certain values; they are **quantized** by virtue of n being an integer. Here, n is called a **quantum number.** For each n, there is a special wavefunction

$$\psi_n(x) = 2Aj \sin\left(\frac{n\pi x}{a}\right) \qquad \text{[3.16]}$$

which is called an eigenfunction.[2] All ψ_n for $n = 1, 2, 3 \ldots$ constitute the eigenfunctions of the system. Each eigenfunction identifies a possible state for the electron. For each n, there is one special k value, $k_n = n\pi/a$, and hence a special energy value E_n, since

$$E_n = \frac{\hbar^2 k_n^2}{2m}$$

that is,

$$E_n = \frac{\hbar^2(\pi n)^2}{2ma^2} = \frac{h^2 n^2}{8ma^2} \qquad \text{[3.17]}$$

The energies E_n defined by Equation 3.17 with $n = 1, 2, 3 \ldots$ are called **eigenenergies** of the system.

We still have not completely solved the problem, because A has yet to be determined. To find A, we use what is called the **normalization condition.** The total probability of finding the electron in the whole region $0 < x < a$ is unity,

| [2]From the German meaning "characteristic function."

because we know the electron is somewhere in this region. Thus, $|\psi|^2 \, dx$ summed between $x = 0$ to $x = a$ must be unity, or

$$\int_{x=0}^{x=a} |\psi(x)|^2 \, dx = \int_{x=0}^{x=a} \left| 2Aj \sin\left(\frac{n\pi x}{a}\right) \right|^2 \, dx = 1$$

Carrying out the simple integration, we find

$$A = \left(\frac{1}{2a}\right)^{1/2}$$

The resulting wavefunction for the electron is thus

$$\psi_n(x) = j\left(\frac{2}{a}\right)^{1/2} \sin\left(\frac{n\pi x}{a}\right) \qquad \text{[3.18]}$$

We can now summarize the behavior of an electron in a one-dimensional *PE* well. Its wavefunction and energy, shown in Figure 3.15, are given by Equations 3.18 and 3.17, respectively. Both depend on the quantum number n. The energy of the electron increases as n^2, so the minimum energy of the electron corresponds to $n = 1$. This is called the **ground state,** and the energy of the ground state is the lowest energy the electron can possess. Note also that the energy of the electron in this potential well cannot be zero, even though the *PE* is zero. Thus, the electron always has *KE*, even when it is in the ground state.

It may seem surprising that the energy of the electron is quantized; that is, that it can only have finite values, given by Equation 3.17. The electron cannot be made to take on any value of energy, as in the classical case. If the electron behaved like a particle, then an applied force F could impart any value of energy to it, because $F = dp/dt$ (Newton's second law), or $p = \int F \, dt$. By applying a force F for a time t, we can give the electron a *KE* of

$$E = \frac{p^2}{2m} = \left(\frac{1}{2}m\right)\left[\int F \, dt\right]^2$$

However, Equation 3.17 tells us that, in the microscopic world, the energy can only have quantized values. The two conflicting views can be reconciled if we consider the energy difference between two consecutive energy levels, as follows:

$$\Delta E = E_{n+1} - E_n = \frac{h^2(2n + 1)}{8ma^2}$$

As a increases to macroscopic dimensions, $a \to \infty$, the electron is completely free and $\Delta E \to 0$. Since $\Delta E = 0$, the energy of a completely free electron ($a = \infty$) is continuous. The energy of a confined electron, however, is quantized, and ΔE depends on the dimension (or size) of the potential well confining the electron.

In general, an electron will be "contained" in a spatial region of three dimensions, within which the *PE* will be lower (hence the confinement). We must then solve the Schrödinger equation in three dimensions. The result is three quantum numbers that characterize the behavior of the electron.

ELECTRON CONFINED WITHIN ATOMIC DIMENSIONS   Consider an electron in an infinite | **Example 3.5**
potential well of size of 0.1 nm (typical size of an atom). What is the ground energy of the
electron? What is the energy required to put the electron at the third energy level? How can
this energy be provided?

Solution

The electron is confined in an infinite potential well, so its energy is given by

$$E_n = \frac{h^2 n^2}{8ma^2}$$

We use $n = 1$ for the ground level and $a = 0.1$ nm. Therefore,

$$E_1 = \frac{(6.6 \times 10^{-34} \text{ J s})^2(1)}{8(9.1 \times 10^{-31} \text{ kg})(0.1 \times 10^{-9} \text{ m})^2} = 6.025 \times 10^{-18} \text{ J} \quad \text{or} \quad 37.6 \text{ eV}$$

The frequency of the electron associated with this energy is

$$\omega = \frac{E}{\hbar} = \frac{6.025 \times 10^{-18} \text{ J}}{1.055 \times 10^{-34} \text{ J s}} = 5.71 \times 10^{16} \text{ rad s}^{-1} \quad \text{or} \quad v = 9.092 \times 10^{15} \text{ s}^{-1}$$

The third energy level E_3 is

$$E_3 = E_1 n^2 = (37.6 \text{ eV})(3)^2 = 338.4 \text{ eV}$$

The energy required to take the electron from 37.6 eV to 338.4 eV is 300.8 eV. This can be
provided by a photon of exactly that energy; no less, and no more. Since the photon energy
is $E = hv = hc/\lambda$, or

$$\lambda = \frac{hc}{E} = \frac{(6.6 \times 10^{-34} \text{ J s})(3 \times 10^8 \text{ m s}^{-1})}{300.8 \text{ eV} \times 1.6 \times 10^{-19} \text{ C}}$$

$$= 4.12 \text{ nm}$$

which is an X-ray photon.

ENERGY OF AN APPLE IN A CRATE   Consider a macroscopic object of mass 100 g (say, an | **Example 3.6**
apple) confined to move between two rigid walls separated by 1 m (say, a typical size of a
large apple crate). What is the minimum speed of the object? What should the quantum
number n be if the object is moving with a speed 1 m s^{-1}? What is the separation of the
energy levels of the object moving with that speed?

Solution

Since the object is within rigid walls, we take the *PE* outside the walls as infinite and use

$$E_n = \frac{h^2 n^2}{8ma^2}$$

to find the ground level energy. With $n = 1$, $a = 1$ m, $m = 0.1$ kg, we have

$$E_1 = \frac{(6.6 \times 10^{-34} \text{ J s})^2(1)^2}{8(0.1 \text{ kg})(1 \text{ m})^2} = 5.45 \times 10^{-67} \text{ J} = 3.4 \times 10^{-48} \text{ eV}$$

Since this is kinetic energy, $\frac{1}{2}mv_1^2 = E_1$, so the minimum speed is

$$v_1 = \sqrt{\frac{2E_1}{m}} = \sqrt{\frac{2(5.45 \times 10^{-67} \text{ J})}{0.1 \text{ kg}}} = 3.3 \times 10^{-33} \text{ m s}^{-1}$$

This speed cannot be measured by any instrument; therefore, for all practical purposes, the apple is at rest in the crate (a relief for the fruit grocer). The time required for the object to move a distance of 1 mm is 3×10^{29} s or 10^{21} years, which is more than the present age of the universe!

When the object is moving with a speed 1 m s^{-1},

$$KE = \frac{1}{2}mv^2 = \frac{1}{2}(0.1 \text{ kg})(1 \text{ m s}^{-1})^2 = 0.05 \text{ J}$$

This must be equal to $E_n = h^2 n^2 / 8ma^2$ for some value of n

$$n = \left[\frac{8ma^2 E_n}{h^2}\right]^{1/2} = \left[\frac{8(0.1 \text{ kg})(1 \text{ m})^2(0.05 \text{ J})}{(6.6 \times 10^{-34} \text{ J s})^2}\right]^{1/2} = 3.01 \times 10^{32}$$

which is an enormous number. The separation between two energy levels corresponds to a change in n from 3.01×10^{32} to $3.01 \times 10^{32} + 1$. This is such a negligibly small change in n that for all practical purposes, the energy levels form a continuum. Thus,

$$\Delta E = E_{n+1} - E_n = \frac{h^2(2n + 1)}{8ma^2}$$

$$= \frac{[(6.626 \times 10^{-34} \text{ J s})^2(2 \times 3.01 \times 10^{32} + 1)]}{[8(0.1 \text{ kg})(1 \text{ m})^2]}$$

$$= 3.31 \times 10^{-34} \text{ J} \quad \text{or} \quad 2.07 \times 10^{-15} \text{ eV}$$

This energy separation is not detectable by any instrument. So for all practical purposes, the energy of the object changes continuously.

We see from this example that in the limit of large quantum numbers, quantum predictions agree with the classical results. This is the essence of **Bohr's correspondence principle.**

3.4   HEISENBERG'S UNCERTAINTY PRINCIPLE

The wavefunction of a free electron corresponds to a traveling wave with a single wavelength λ, as shown in Example 3.3. The traveling wave extends over all space, along all x, with the same amplitude, so the probability distribution function is uniform throughout the whole of space. The uncertainty Δx in the position of the electron is therefore infinite. Yet, the uncertainty Δp_x in the momentum of the electron is zero, because λ is well-defined, which means that we know p_x exactly from the de Broglie relationship, $p_x = h/\lambda$.

For an electron trapped in a one-dimensional infinite PE well, the wavefunction extends from $x = 0$ to $x = a$, so the uncertainty in the position of the electron is a. We know that the electron is within the well, but we cannot pinpoint with certainty exactly where it is. The momentum of the electron is either $p_x = \hbar k$ in the $+x$ direction or $-\hbar k$ in the $-x$ direction. The uncertainty Δp_x in the momentum is therefore $2\hbar k$; that is, $\Delta p_x = 2\hbar k$. For the ground-state wavefunction, which corresponds to $n = 1$, we have $ka = \pi$. Thus, $\Delta p_x = 2\hbar\pi/a$. Taking the product of the uncertainties in x and p, we get

$$(\Delta x)(\Delta p_x) = (a)\left(\frac{2\hbar\pi}{a}\right) = h$$

In other words, the product of the position and momentum uncertainties is simply h. This relationship is fundamental; and it constitutes a limit to our knowledge of the behavior of a system. *We cannot exactly and simultaneously know both the position and momentum of a particle along a given coordinate.* In general, if Δx and Δp_x are the respective uncertainties in the simultaneous measurement of the position and momentum of a particle along a particular coordinate (such as x), the **Heisenberg uncertainty principle** states that

$$\Delta x\, \Delta p_x \gtrsim \hbar \qquad \text{[3.19]}$$

Heisenberg uncertainty principle for position and momentum

We are therefore forced to conclude that as previously stated, because of the wave nature of quantum mechanics, we are unable to determine exactly and simultaneously the position and momentum of a particle along a given coordinate. There will be an uncertainty Δx in the position and an uncertainty Δp_x in the momentum of the particle and these uncertainties will be related by Heisenberg's uncertainty relationship in Equation 3.19.

These uncertainties are not in any way a consequence of the accuracy of a measurement or the precision of an instrument. Rather, they are the theoretical limits to what we can determine about a system. They are part of the quantum nature of the universe. In other words, even if we build the most perfectly engineered instrument to measure the position and momentum of a particle at one instant, we will still be faced with position and momentum uncertainties Δx and Δp_x such that $\Delta x\, \Delta p_x > \hbar$.

There is a similar uncertainty relationship between the uncertainty ΔE in the energy E (or angular frequency ω) of the particle and the time duration Δt during which it possesses the energy (or during which its energy is measured). We know that the kx part of the wave leads to the uncertainty relation $\Delta x\, \Delta p_x \geq \hbar$ or $\Delta x\, \Delta k \geq 1$. By analogy we should expect a similar relationship for the ωt part, or $\Delta\omega\, \Delta t \geq 1$. This hypothesis is true, and since $E = \hbar\omega$, we have the uncertainty relation for the particle energy and time:

$$\Delta E\, \Delta t \gtrsim \hbar \qquad \text{[3.20]}$$

Heisenberg uncertainty principle for energy and time

Note that the uncertainty relationships in Equations 3.19 and 3.20 have been written in terms of $\hbar$, rather than h, as implied by the electron in an infinite potential energy well ($\Delta x\, \Delta p_x \geq h$). In general there is also a numerical factor of $1/2$ multiplying $\hbar$ in Equations [3.19] and [3.20] which comes about when we consider a Gaussian spread for all possible position and momentum values. The proof is not presented here, but can be found in advanced quantum mechanics books.

It is important to note that the uncertainty relationship applies only when the position and momentum are measured in the same direction (such as the x direction). On the other hand, the exact momentum, along, say, the y direction and the exact position, along, say, the x direction can be determined exactly, since $\Delta x \, \Delta p_y$ need not satisfy the Heisenberg uncertainty relationship (in other words, $\Delta x \, \Delta p_y$ can be zero).

Example 3.7

THE MEASUREMENT TIME AND THE FREQUENCY OF WAVES: AN ANALOGY WITH $\Delta E \, \Delta t \geq \hbar$
Consider the measurement of the frequency of a sinusoidal wave of frequency 1000 Hz (or cycles/s). Suppose we can only measure the number of cycles to an accuracy of 1 cycle, because we need to receive a whole cycle to record it as one complete cycle. Then, in a time interval of $\Delta t = 1$ s, we will register 1000 ± 1 cycles. The uncertainty Δf in the frequency is 1 cycle/1 s or 1 Hz. If Δt is 2 s, we will measure 2000 ± 1 cycles, and the uncertainty Δf will be 1 cycle/2 s or $(1/2)$ cycle/s or $(1/2)$ Hz. Thus, Δf decreases with Δt.

Suppose that in a time interval Δt, we measure $N \pm 1$ cycles. Since the uncertainty is 1 cycle in a time interval Δt, the uncertainty in f will be

$$\Delta f = \frac{(1 \text{ cycle})}{\Delta t} = \frac{1}{\Delta t} \text{ Hz}$$

Since $\omega = 2\pi f$, we have

$$\Delta \omega \, \Delta t = 2\pi$$

In quantum mechanics, under steady-state conditions, an object has a time-oscillating wavefunction with a frequency ω which is related to its energy E by $\omega = E/\hbar$ (see Equation 3.10). Substituting this into the previous relationship gives

$$\Delta E \, \Delta t = h$$

The uncertainty in the energy of a quantum object is therefore related, in a fundamental way, to the time duration during which the energy is observed. Notice that we again have h, as for $\Delta x \, \Delta p_x = h$, though the quantum mechanical uncertainty relationship in Equation 3.20 has $\hbar$.

Example 3.8

THE UNCERTAINTY PRINCIPLE ON THE ATOMIC SCALE   Consider an electron confined to a region of size 0.1 nm, which is the typical dimension of an atom. What will be the uncertainty in its momentum and hence its kinetic energy?

Solution
We apply the Heisenberg uncertainty relationship, $\Delta x \, \Delta p_x \sim \hbar$, or

$$\Delta p_x \sim \frac{\hbar}{\Delta x} = \frac{1.054 \times 10^{-34} \text{ J s}}{0.1 \times 10^{-9} \text{ m}} = 1.054 \times 10^{-24} \text{ kg m s}^{-1}$$

The uncertainty in the velocity is therefore

$$\Delta v = \frac{\Delta p_x}{m_e} = \frac{1.054 \times 10^{-24} \text{ kg m s}^{-1}}{9.1 \times 10^{-31} \text{ kg}} = 1.16 \times 10^6 \text{ m s}^{-1}$$

We can take this uncertainty to represent the order of magnitude of the actual speed. The kinetic energy associated with this momentum is

$$KE = \frac{\Delta p_x^{\;2}}{2m_e} = \frac{(1.054 \times 10^{-24} \text{ kg m s}^{-1})^2}{2(9.1 \times 10^{-31} \text{ kg})}$$

$$= 6.10 \times 10^{-19} \text{ J} \quad \text{or} \quad 3.81 \text{ eV}$$

THE UNCERTAINTY PRINCIPLE WITH MACROSCOPIC OBJECTS   Estimate the minimum velocity | **Example 3.9**
of an apple of mass 100 g confined to a crate of size 1 m.

Solution

Taking the uncertainty in the position of the apple as 1 m, then somewhere in the crate, we have

$$\Delta p_x \sim \frac{\hbar}{\Delta x} = \frac{1.05 \times 10^{-34} \text{ J s}}{1 \text{ m}} = 1.05 \times 10^{-34} \text{ kg m s}^{-1}$$

So the minimum uncertainty in the velocity is

$$\Delta v_x = \frac{\Delta p_x}{m} = \frac{1.05 \times 10^{-34} \text{ m s}^{-1}}{0.1 \text{ kg}} = 1.05 \times 10^{-33} \text{ m s}^{-1}$$

The quantum nature of the universe implies that the apple in the crate is moving with a velocity on the order of 10^{-33} m s^{-1}. This cannot be measured by any instrument; indeed, it would take the apple $\sim 10^{19}$ years to move an atomic distance of 0.1 nm.

3.5   TUNNELING PHENOMENON: QUANTUM LEAK

To understand the tunneling phenomenon, let us examine the thrilling events experienced by the roller coaster shown in Figure 3.16a. Consider what the roller coaster can do when released from rest at a height A. The conservation of energy means that the carriage can reach B and at most C, but certainly not beyond C and definitely not D and E. Classically, there is no possible way the carriage will reach E at the other side of the potential barrier D. An extra energy corresponding to the height difference, $D - A$, is needed. Anyone standing at E will be quite safe. Ignoring frictional losses, the roller coaster will go back and forth between A and C.

Now, consider an analogous event on an atomic scale. An electron moves with an energy E in a region $x < 0$ where the potential energy PE is zero; therefore, E is solely kinetic energy. The electron then encounters a potential barrier of "height" V_o, which is greater than E at $x = 0$. The extent (width) of the potential barrier is a. On the other side of the potential barrier, $x > a$, the PE is again zero. What will the electron do? Classically, just like the roller coaster, the electron should bounce back and thus be confined to the region $x < 0$, because its total energy E is less

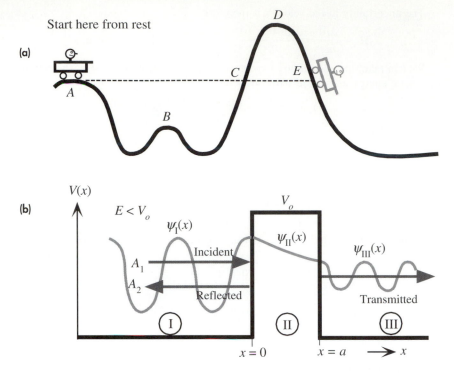

Figure 3.16

(a) The roller coaster released from *A* can at most make it to *C*, but not to *E*. Its *PE* at *A* is less than the *PE* at *D*. When the car is at the bottom, its energy is totally *KE*. *CD* is the energy barrier that prevents the car from making it to *E*. In quantum theory, on the other hand, there is a chance that the car could tunnel (leak) through the potential energy barrier between *C* and *E* and emerge on the other side of the hill at *E*.

(b) The wavefunction for the electron incident on a potential energy barrier (V_o). The incident and reflected waves interfere to give $\psi_I(x)$. There is no reflected wave in region III. In region II, the wavefunction decays with *x* because $E < V_o$.

than V_o. In the quantum world, however, there is a distinct possibility that the electron will "tunnel" through the potential barrier and appear on the other side; it will leak through.

To show this, we need to solve the Schrödinger equation for the present choice of $V(x)$. Remember that the only way the Schrödinger equation will have the solution $\psi(x) = 0$ is if the *PE* is infinite, that is, $V = \infty$. Therefore, within any zero or finite *PE* region, there will always be a solution $\psi(x)$ and there always will be some probability of finding the electron.

We can divide the electron's space into three regions, I, II, and III, as indicated in Figure 3.16b. We can then solve the Schrödinger equation for each region, to obtain three wavefunctions $\psi_I(x)$, $\psi_{II}(x)$, and $\psi_{III}(x)$. In regions I and III, $\psi(x)$ must be travelling waves, as there is no *PE* (the electron is free and moving with a kinetic energy E). In zone II, however, $E - V_o$ is negative, so the general solution of the Schrödinger equation is the sum of an exponentially decaying function and an exponentially increasing function. In other words,

$$\psi_I(x) = A_1 \exp(jkx) + A_2 \exp(-jkx)$$ [3.21a]

$$\psi_{II}(x) = B_1 \exp(\alpha x) + B_2 \exp(-\alpha x)$$ [3.21b]

$$\psi_{III}(x) = C_1 \exp(jkx) + C_2 \exp(-jkx)$$ [3.21c]

are the wavefunctions in which

$$k^2 = \frac{2mE}{\hbar^2}$$ [3.22]

and

$$\alpha^2 = \frac{2m(V_o - E)}{\hbar^2}$$ [3.23]

Both k^2 and α^2, and hence k and α, in Equations 3.21a to c are positive numbers. This means that $\exp(jkx)$ and $\exp(-jkx)$ represent traveling waves in opposite directions, and $\exp(-\alpha x)$ and $\exp(\alpha x)$ represent an exponential decay and rise, respectively. We see that in region I, $\psi_I(x)$ consists of the incident wave $A_1 \exp(jkx)$ in the $+x$ direction, and a reflected wave $A_2 \exp(-jkx)$, in the $-x$ direction. Furthermore, because the electron is travelling toward the right in region III, there is no reflected wave, so $C_2 = 0$.

We must now apply the boundary conditions and the normalization condition to determine the various constants A_1, A_2, B_1, B_2, and C_1. In other words, we must match the three waveforms in Equations 3.21a to c at their boundaries ($x = 0$ and $x = a$) so that they form a continuous single-valued wavefunction. With the boundary conditions enforced onto the wavefunctions $\psi_I(x)$, $\psi_{II}(x)$, and $\psi_{III}(x)$, all the constants can be determined in terms of the amplitude A_1 of the incoming wave. The relative probability that the electron will tunnel from region I through to III is defined as the **transmission coefficient T**, and this depends very strongly on both the relative *PE* barrier height ($V_o - E$) and the width a of the barrier. The final result that comes out from a tedious application of the boundary conditions is

$$T = \frac{|\psi_{III}(x)|^2}{|\psi_I(incident)|^2} = \frac{C_1^2}{A_1^2} = \frac{1}{1 + D \sinh^2(\alpha a)}$$ [3.24]

where

$$D = \frac{V_o^2}{4E(V_o - E)}$$ [3.25]

and α is the rate of decay of $\psi_{II}(x)$ as expressed in Equation 3.23. For a wide or high barrier, using $\alpha a \gg 1$ in Equation 3.24 and $\sinh(\alpha a) \approx (1/2) \exp(\alpha a)$, we can deduce

$$T = T_o \exp(-2\alpha a)$$ [3.26]

Probability of tunneling through

where

$$T_o = \frac{16E(V_o - E)}{V_o^2}$$ [3.27]

By contrast, the relative probability of reflection is determined by the ratio of the square of the amplitude of the reflected wave to that of the incident wave. This quantity is the **reflection coefficient R,** which is given by

$$R = \frac{A_2^2}{A_1^2} = 1 - T$$ [3.28]

We can now summarize the entire tunneling affair as follows. When an electron encounters a potential energy barrier of height V_o greater than its energy E, there is a finite probability that it will leak through that barrier. This probability depends sensitively on the energy and width of the barrier. For a wide potential barrier, the probability of tunneling is proportional to $\exp(-2\alpha a)$, as in Equation 3.26. The wider or higher the potential barrier, the smaller the chance of the electron tunneling.

One of the most remarkable technological uses of the tunneling effect is in the scanning tunneling microscope (STM), which elegantly maps out the surfaces of solids. A conducting probe is brought so close to the surface of a solid that electrons can tunnel from the surface of the solid to the probe, as illustrated in Figure 3.17. When the probe is far removed, the wavefunction of an electron decays exponentially outside the material, by virute of the potential energy barrier being finite (the work function is ~ 10 eV). When the probe is brought very close to the surface, the wavefunction penetrates into the probe and, as a result, the electron can tunnel from the material into the probe. Without an applied voltage, there will be as many electrons tunneling from the material to the probe as there are going in the opposite direction from the probe to the material, so the net current will be zero.

On the other hand, if a positive bias is applied to the probe with respect to the material, as shown in Figure 3.17, an electron tunneling from the material to the probe will see a lower potential barrier than one tunneling from the probe to the material. Consequently, there will be a net current from the probe to the material and this current will depend very sensitively on the separation a of the probe from the surface, by virtue of Equation 3.26.

Because the tunneling current is extremely sensitive to the width of the potential barrier, the tunneling current is essentially dominated by electrons tunneling to the probe atom nearest to the surface. Thus, the probe tip has an atomic dimension. By scanning the surface of the material with the probe and recording the tunneling current the user can map out the surface topology of the material with a resolution comparable to the atomic dimension. The probe motion along the surface, and also perpendicular to the surface, is controlled by piezoelectric transducers to provide sufficiently small and smooth displacements. Figure 3.18 shows an STM image of

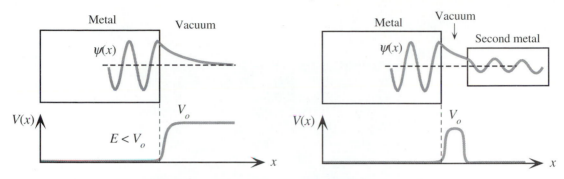

(a) The wavefunction decays exponentially as we move away from the surface because the *PE* outside the metal is V_o and the energy of the electron, $E < V_o$.

(b) If we bring a second metal close to the first metal, then the wavefunction can penetrate into the second metal. The electron can tunnel from the first metal to the second.

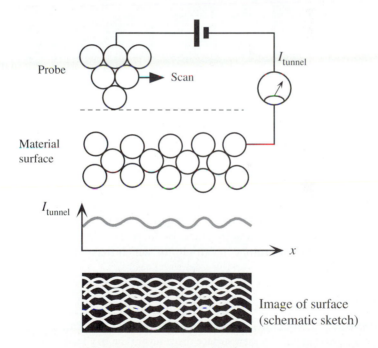

Image of surface
(schematic sketch)

(c) The principle of the scanning tunneling microscope. The tunneling current depends on $\exp(-2\alpha a)$ where a is the distance of the probe from the surface of the specimen and α is a constant.

Figure 3.17

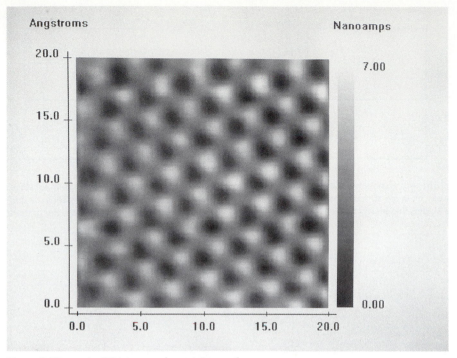

Figure 3.18   An STM image of a graphite surface.

We can clearly see the hexagonal symmetry of atomic arrangements on the surface. The x and y axes represent distances on the graphite surface and both are in angstroms (1Å $= 10^{-10}$m).

| Courtesy of Burleigh Instruments, Inc.

a graphite surface, on which the hexagonal symmery of atomic arrangements can be clearly seen. STM was invented by Gerd Binning and Heinrich Rohrer at the IBM Research Laboratory in Zurich, for which they were awarded the 1986 Nobel prize.[3]

Example 3.10 | **TUNNELING CONDUCTION THROUGH METAL-TO-METAL CONTACTS**   Consider two copper wires separated only by their surface oxide layer (CuO). Classically, since the oxide layer is an insulator, no current should be possible through the two copper wires. Suppose that for the conduction ("free") electrons in copper, the surface oxide layer looks like a square potential energy barrier of height 10 eV. Consider an oxide layer thickness of 5 nm and evaluate the transmission coefficient for conduction electrons in copper, which have a kinetic energy of about 7 eV. What will be the transmission coefficient if the oxide barrier is 1 nm?

---

[3] The IBM Research Laboratory in Zurich, Switzerland, received both the 1986 and the 1987 Nobel prizes. The first was for the scanning tunneling microscope by Gerd Binning and Heinrich Rohrer. The second was awarded to Georg Bednorz and Alex Müller for the discovery of high temperature superconductors which we will examine in Chapter 8.

Solution

We can calculate α from

$$\alpha = \left[\frac{2m(V_o - E)}{\hbar^2} \right]^{1/2}$$

$$= \left[\frac{2(9.1 \times 10^{-31} \text{ kg})(10 \text{ eV} - 7 \text{ eV})(1.6 \times 10^{-19} \text{ J/eV})}{(1.05 \times 10^{-34} \text{ J s})^2} \right]^{1/2}$$

$$= 8.9 \times 10^9 \text{ m}^{-1}$$

so that

$$\alpha a = (8.9 \times 10^9 \text{ m}^{-1})(5 \times 10^{-9} \text{ m}) = 44.50$$

Since this is greater than unity, we use the wide-barrier transmission coefficient in Equation 3.26

Now,

$$T_o = \frac{16E(V_o - E)}{V_o^2} = \frac{16(7 \text{ eV})(10 \text{ eV} - 7 \text{ eV})}{(10 \text{ eV})^2} = 3.36$$

Thus,

$$T = T_o \exp(-2\alpha a)$$

$$= 3.36 \exp[-2(8.9 \times 10^9 \text{ m}^{-1})(5 \times 10^{-9} \text{ m})] = 3.36 \exp(-89)$$

$$\approx 7.4 \times 10^{-39}$$

an incredibly small number.

With $a = 1$ nm,

$$T = 3.36 \exp[-2(8.9 \times 10^9 \text{ m}^{-1})(1 \times 10^{-9} \text{ m})]$$

$$= 3.36 \exp(-17.8) \approx 6.2 \times 10^{-8}$$

Notice that reducing the layer thickness by five times increases the transmission probability by 10^{31}! Small changes in the barrier width lead to enormous changes in the transmission probability. We should note that when a voltage is applied across the two wires, the potential energy height is altered ($PE = \text{charge} \times \text{voltage}$), which results in a large increase in the transmission probability and hence results in a current.

SIGNIFICANCE OF A SMALL $\hbar$ Estimate the probability that a roller coaster carriage that weighs 100 kg released from point A in Figure 3.16a from a height at 10 m can reach point E over a hump that is 15 m high and 10 m wide. What will this probability be in a universe where $\hbar \sim 10$ kJ s? | **Example 3.11**

Solution

The total energy of the carriage at height A is

$$E = PE = mg(height) = (100 \text{ kg})(10 \text{ m s}^{-2})(10 \text{ m}) = 10^4 \text{ J}$$

Suppose that as a first approximation, we can approximate the hump as a square hill of height 15 m and length 10 m. The PE required to reach the peak would be

$$V_o = mg(height) = (100 \text{ kg})(10 \text{ m s}^{-2})(15 \text{ m}) = 1.5 \times 10^4 \text{ J}$$

Applying this, we have

$$\alpha^2 = \frac{2m(V_o - E)}{\hbar^2} = \frac{2(100 \text{ kg})(1.5 \times 10^4 \text{J} - 10^4 \text{J})}{(1.05 \times 10^{-34} \text{ J s})^2} = 9.07 \times 10^{73}$$

and so

$$\alpha = 9.52 \times 10^{36} \text{ m}^{-1}$$

With $a = 10$ m, we have $\alpha a \gg 1$, so we can use the wide-barrier tunneling equation,

$$T = T_o \exp(-2\alpha a)$$

where

$$T_o = \frac{16[E(V_o - E)]}{V_o^2} = 3.56$$

Thus,

$$T = 3.56 \exp[-2(9.52 \times 10^{36} \text{ m}^{-1})(10 \text{ m})] = 3.56 \exp(-1.9 \times 10^{38})$$

which is a fantastically small number, indicating that it is impossible for the carriage to tunnel through the hump.

Suppose that $\hbar \sim 10$ kJ s. Then

$$\alpha^2 = \frac{2m(V_o - E)}{\hbar^2} = \frac{2(100 \text{ kg})(1.5 \times 10^4 \text{ J} - 10^4 \text{ J})}{(10^4 \text{ J s})^2} = 0.01$$

so that $\alpha = 0.1$. Clearly, $\alpha a = 1$, so we must use

$$T = [1 + D \sinh^2(\alpha a)]^{-1}$$

where

$$D = \frac{V_o^2}{[4E(V_o - E)]} = 1.125$$

Thus,

$$T = [1 + 1.125 \sinh^2(1)]^{-1} = 0.39$$

Thus, after three goes, the carriage would tunnel to the other side (giving the person standing at E the shock of his life).

3.6   POTENTIAL BOX: THREE QUANTUM NUMBERS

To examine the properties of a particle confined to a region of space, we take a three-dimensional space with a volume marked by a, b, c along the x, y, z axes. The *PE* is zero ($V = 0$) inside the space and is infinite on the outside, as illustrated in Figure 3.19. This is a three-dimensional potential energy well. The electron essentially lives in the "box." What will the behavior of the electron be in this box? In

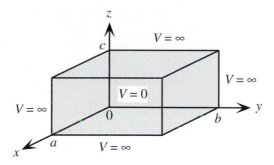

Figure 3.19   Electron confined in three dimensions by a three-dimensional infinite *PE* box.

Everywhere inside the box, *V* = 0, but outside, *V* = ∞. The electron cannot escape from the box.

this case we need to solve the three-dimensional version of the Schrödinger equation[4], which is

$$\frac{\partial^2 \psi}{\partial x^2} + \frac{\partial^2 \psi}{\partial y^2} + \frac{\partial^2 \psi}{\partial z^2} + \frac{2m}{\hbar^2}(E - V)\psi = 0 \qquad \textbf{[3.29]}$$

with $V = 0$ in $0 < x < a$, $0 < y < b$, and $0 < z < c$, and V infinite outside. We can try to solve this by separating the variables via $\psi(x, y, z) = \psi_x(x)\, \psi_y(y)\, \psi_z(z)$. Substituting this back into Equation 3.29, we can obtain three ordinary differential equations, each just like the one for the one-dimensional potential well. Having found $\psi_x(x)$, $\psi_y(y)$, and $\psi_z(z)$ we know that the total wavefunction is simply the product,

$$\psi(x, y, z) = A \sin(k_x x)\, \sin(k_y y)\, \sin(k_z z) \qquad \textbf{[3.30]}$$

where k_x, k_y, k_z, and A are constants to be determined. We can then apply the boundary conditions at $x = a$, $y = b$, and $z = c$ to determine the constants k_x, k_y, and k_z in the same way we found k for the one-dimensional potential well. If $\psi(x, y, z) = 0$ at $x = a$, then k_x will be quantized via

$$k_x a = n_1 \pi$$

where n_1 is a quantum number, $n_1 = 1, 2, 3, \ldots$ Similarly, if $\psi(x, y, z) = 0$ at $y = b$ and $z = c$, then k_y and k_z will be quantized, so that, overall, we will have

$$k_x = \frac{n_1 \pi}{a} \qquad k_y = \frac{n_2 \pi}{b} \qquad k_z = \frac{n_3 \pi}{c} \qquad \textbf{[3.31]}$$

where n_1, n_2, and n_3 are quantum numbers, each of which can be any integer except zero.

---

[4] The term $\partial \psi / \partial x$ simply means differentiating $\psi(x, y, z)$ with respect to x while keeping y and z constant, just like $d\psi/dx$ in one dimension.

We notice immediately that in three dimensions, we have three quantum numbers, n_1, n_2, and n_3, associated with $\psi_x(x)$, $\psi_y(y)$, and $\psi_z(z)$. The eigenfunctions of the electron, denoted by the quantum numbers n_1, n_2, and n_3, are now given by

$$\psi_{n_1 n_2 n_3}(x, y, z) = A \sin\left(\frac{n_1 \pi x}{a}\right) \sin\left(\frac{n_2 \pi y}{b}\right) \sin\left(\frac{n_3 \pi z}{c}\right)$$ [3.32]

Notice that these consist of the products of infinite one-dimensional *PE* well-type wavefunctions, one for each dimension, and each has its own quantum number *n*. Each possible eigenfunction can be labeled a **state** for the electron. Thus, ψ_{111} and ψ_{121} are two possible states.

To find the constant *A* in Equation 3.32, we need to use the normalization condition that $|\psi_{n_1 n_2 n_3}(x, y, z)|^2$ integrated over the volume of the box must be unity, since the electron is somewhere in the box. The result for a square box is $A = (2/a)^{3/2}$.

We can find the energy of the electron by substituting the wavefunction in Equation 3.30 into the Schrödinger Equation 3.29. The energy as a function of k_x, k_y, k_z is then found to be

$$E = E(k_x, k_y, k_z) = \frac{\hbar^2}{2m}[k_x^2 + k_y^2 + k_z^2]$$

which is quantized by virtue of k_x, k_y, and k_z being quantized. We can write this energy in terms of n_1^2, n_2^2, and n_3^2 by using Equation 3.31, as follows:

$$E_{n_1 n_2 n_3} = \frac{h^2}{8m}\left[\frac{n_1^2}{a^2} + \frac{n_2^2}{b^2} + \frac{n_3^2}{c^2}\right]$$

For a square box for which $a = b = c$, the energy is

$$E_{n_1 n_2 n_3} = \frac{h^2[n_1^2 + n_2^2 + n_3^2]}{8ma^2} = \frac{h^2 N^2}{8ma^2}$$ [3.33]

where $N^2 = [n_1^2 + n_2^2 + n_3^2]$, which can only have certain integer values. It is apparent that the energy now depends on three quantum numbers. Our conclusion is that in three dimensions, we have three quantum numbers, each one arising from boundary conditions along one of the coordinates. They quantize the energy of the electron via Equation 3.33 and its momentum in a particular direction, such as $p_x = \pm \hbar k_x = \pm (hn_1/2a)$, though the average momentum is zero.

The lowest energy for the electron is obviously equal to E_{111}, not zero. The next energy level corresponds to E_{211}, which is the same as E_{121} and E_{112}, so there are three states (i.e., ψ_{211}, ψ_{121}, ψ_{112}) for this energy. The number of states that have the same energy is termed the **degeneracy** of that energy level. The second energy level E_{211} is thus **three-fold degenerate.**

Example 3.12

NUMBER OF STATES WITH THE SAME ENERGY   How many states (eigenfunctions) are there at energy level E_{443} for a square potential energy box?

Solution

This energy level corresponds to $n_1 = 4$, $n_2 = 4$, and $n_3 = 3$, but the energy depends on

$$N^2 = n_1^2 + n_2^2 + n_3^2 = 4^2 + 4^2 + 3^2 = 41$$

via Equation 3.33. As long as $N^2 = 41$ for any choice of (n_1, n_2, n_3), not just $(4, 4, 3)$, the energy will be the same.

The value $N^2 = 41$ can be obtained from $(4, 4, 3)$, $(4, 3, 4)$, and $(3, 4, 4)$ as well as $(6, 2, 1)$, $(6, 1, 2)$, $(2, 6, 1)$, $(2, 1, 6)$, $(1, 6, 2)$, and $(1, 2, 6)$. There are thus three states from $(4, 4, 3)$ combinations and six from $(6, 2, 1)$ combinations, giving nine possible states, each with a distinct wavefunction, $\psi_{n_1 n_2 n_3}$. However, all these $\psi_{n_1 n_2 n_3}$ for the electron have the same energy E_{443}.

3.7   HYDROGENIC ATOM

3.7.1   Electron Wavefunctions

Consider the behavior of the electron in a hydrogenic atom, which has a nuclear charge of $+Ze$, as depicted in Figure 3.20. For the hydrogen atom, $Z = 1$, whereas

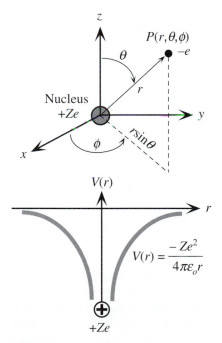

Figure 3.20   The electron in the hydrogenic atom is attracted by a central force that is always directed toward the positive nucleus.

Spherical coordinates centered at the nucleus are used to describe the position of the electron. The *PE* of the electron depends only on r.

for an ionized helium atom He^+, $Z = 2$. For a doubly ionized lithium atom Li^{++}, $Z = 3$, and so on. The electron is attracted by a positive nuclear charge and therefore has a Coulombic *PE*,

$$V(r) = \frac{-Ze^2}{4\pi \varepsilon_o r}$$ [3.34]

Since force $F = -dV/dr$, Equation 3.34 is simply a statement of Coulomb's force between the positive charge $+Ze$ of the nucleus and the negative charge $-e$ of the electron. The task of finding $\psi(x, y, z)$ and the energy E of the electron now involves putting $V(r)$ from Equation 3.34 into the Schrödinger equation with $r = \sqrt{(x^2 + y^2 + z^2)}$ and solving it.

Fortunately, the problem has a spherical symmetry, and we can solve the Schrödinger equation by transforming it into the r, θ, ϕ coordinates shown in Figure 3.20. Even then, obtaining a solution is not easy. We must then ensure that the solution for $\psi(r, \theta, \phi)$ satisfies all the boundary conditions, as well being single-valued and continuous with a continuous derivative. For example, when we go 2π around the ϕ coordinate, $\psi(r, \theta, \phi)$ should come back to its original value, or $\psi(r, \theta, \phi) = \psi(r, \theta, \phi + 2\pi)$, as is apparent from an examination of Figure 3.20. Along the radial coordinate, we need $\psi(r, \theta, \phi) \rightarrow 0$ as $r \rightarrow \infty$; otherwise, the total probability will diverge when $|\psi(r, \theta, \phi)|^2$ is integrated over all space. In an analogy with the three-dimensional potential well, there should be three quantum numbers to characterize the wavefunction, energy, and momentum of the electron. The three quantum numbers are called the **principal, orbital angular momentum,** and **magnetic quantum numbers** and are respectively denoted by n, ℓ, and m_ℓ. Unlike the three-dimensional potential well, however, not all the quantum numbers run as independent positive integers.

The solution to the Schrödinger equation $\psi(r, \theta, \phi)$ depends on three variables, r, θ, ϕ. The wavefunction $\psi(r, \theta, \phi)$ can be written as the product of two functions

$$\psi(r, \theta, \phi) = R(r) \, Y(\theta, \phi)$$

where $R(r)$ is a radial function depending only on r, and $Y(\theta, \phi)$ is called the **spherical harmonic,** which expresses the angular dependence of the wavefunction. These functions are characterized by the quantum numbers n, ℓ, m_ℓ. The radial part, $R(r)$, depends on n and ℓ, whereas the spherical harmonic depends on ℓ and m_ℓ, so that

$$\psi(r, \theta, \phi) = \psi_{n, \ell, m_\ell}(r, \theta, \phi) = R_{n, \ell}(r) \, Y_{\ell, m_\ell}(\theta, \phi)$$ [3.35]

By solving the Schrödinger equation, these functions have already been evaluated. It turns out that we can only assign certain values to the quantum number n, ℓ, and m_ℓ to obtain acceptable solutions, that is, $\psi_{n, \ell, m_\ell}(r, \theta, \phi)$ that are well behaved: single-valued, and ψ and gradient of ψ continuous. We can summarize the allowed values of n, ℓ, m_ℓ as follows:

Principal quantum number	$n = 1, 2, 3, \ldots$
Obital angular momentum quantum number	$\ell = 0, 1, 2, \ldots, (n - 1) < n$
Magnetic quantum number	$m_\ell = -\ell, -(\ell - 1), \ldots, 0, \ldots, (\ell - 1), \ell$
	or $|m_\ell| \leq \ell$

The ℓ values carry a special notation inherited from spectroscopic terms. The first four ℓ values are designated by the first letters of the terms *sharp, principal, diffuse,* and *fundamental,* whereas the higher ℓ values follow from *f* onwards, as *g, h, i,* etc. For example, any state ψ_{n,ℓ,m_ℓ} that has $\ell = 0$ is called an *s*-state, whereas that which has $\ell = 1$ is termed a *p*-state. We can also use n as a prefix to ℓ to identify n. Thus ψ_{n,ℓ,m_ℓ} with $n = 2$ and $\ell = 0$ corresponds to the 2*s* state. The notation for identifying the ℓ value and labeling a state is summarized in Table 3.1.

Table 3.2 summarizes the functional forms of $R_{n,\ell}(r)$ and $Y_{\ell,m_\ell}(\theta, \phi)$. For $\ell = 0$ (the *s*-states), the angular dependence of $Y_{0,0}(\theta, \phi)$ is constant, which means that $\psi(r, \theta, \phi)$ is spherically symmetrical about the nucleus. For the $\ell = 1$ and higher states, there is a strong directionality to the wavefunctions with respect to each other. The radial part, $R_{n,\ell}(r)$, is sketched in Figure 3.21a for two choices of n and ℓ. Notice that $R_{n,\ell}(r)$ is largest at $r = 0$, when $\ell = 0$. However, this does not mean that the electron will be mainly at $r = 0$, because the probability that the electron will be within a certain spherical volume at a distance r actually depends on $r^2|R_{n,\ell}(r)|^2$, which vanishes as $r \to 0$.

Let us examine the probability of finding the electron at a distance r within a thin spherical shell of radius r and thickness δr (assumed to be very small). The directional dependence of the probability will be determined by the function $Y_{\ell,m_\ell}(\theta, \phi)$. We can average this over all directions (all angles θ and ϕ) to obtain

Table 3.1   Labeling of various $n\ell$ possibilities

ℓ	0	1	2	3	4
n					
1	1*s*				
2	2*s*	2*p*			
3	3*s*	3*p*	3*d*		
4	4*s*	4*p*	4*d*	4*f*	
5	5*s*	5*p*	5*d*	5*f*	5*g*

Table 3.2   The radial and spherical harmonic parts of the wavefunction in the hydrogen atom ($a_o = 0.0529$ nm)

n	ℓ	$R(r)$	m_ℓ	$Y(\theta, \phi)$
1	0	$\left(\dfrac{1}{a_o}\right)^{3/2} 2\exp\left(-\dfrac{r}{a_o}\right)$	0	$\dfrac{1}{2\sqrt{\pi}}$
2	0	$\left(\dfrac{1}{2a_o}\right)^{3/2}\left(2 - \dfrac{r}{a_o}\right)\exp\left(-\dfrac{r}{2a_o}\right)$	0	$\dfrac{1}{2\sqrt{\pi}}$
2	1	$\left(\dfrac{1}{2a_o}\right)^{3/2}\left(\dfrac{r}{\sqrt{3}\,a_o}\right)\exp\left(-\dfrac{r}{2a_o}\right)$	0	$\dfrac{1}{2}\sqrt{\dfrac{3}{\pi}}\cos\theta$
			1	$-\dfrac{1}{2}\sqrt{\dfrac{3}{2\pi}}\sin\theta\, e^{j\phi}$
			-1	$\dfrac{1}{2}\sqrt{\dfrac{3}{2\pi}}\sin\theta\, e^{-j\phi}$

$\propto \sin\theta\cos\phi$ and $\propto \sin\theta\sin\phi$ Correspond to $m_\ell = -1$ and $+1$.

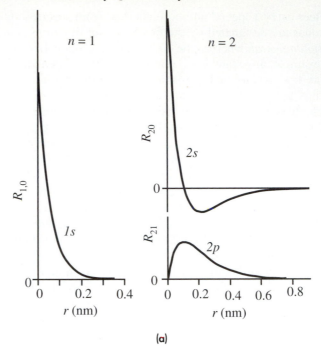

(a)

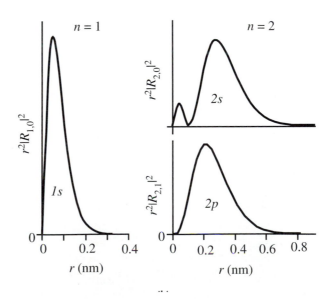

Figure 3.21
(a) Radical wavefunctions of the electron in a hydrogenic atom for various n and ℓ values.
(b) $r^2|R_{n,\ell}|^2$ gives the radial probability density. (Vertical axis scales are linear, in arbitrary units.)

$\overline{Y_{\ell,m_\ell}(\theta, \phi)}$, which turns out to be simply $1/4\pi$. The volume of the spherical shell is $\delta V = 4\pi^2 \delta r$. The probability of finding the electron in this shell is then

$$| (\overline{Y_{\ell,m_\ell}(\theta, \phi)})(R_{n,\ell}(r)) |^2 \times (4\pi r^2 \delta r)$$

If $\delta P(r)$ represents the probability that the electron is in this spherical shell of thickness δr, then

$$\delta P(r) = | R_{n,\ell}(r) |^2 r^2 \delta r \qquad [3.36]$$

The radial probability density $P_{n,\ell}(r)$ is defined as probability per unit radial distance, that is, dP/dr which from Equation 3.36 is $| R_{n,\ell}(r) |^2 r^2$. The latter vanishes at the nucleus and peaks at certain locations, as shown in Figure 3.21b. This behavior implies that the probability of finding the electron within a thin spherical shell close to the nucleus also disappears. For $n = 1$, and $\ell = 0$, for example, the maximum probability is at $r = a_o (0.0529 \text{ nm})$, which is called the **Bohr radius.** Therefore, if the electron is in the $1s$ state, it spends most of its time at a distance a_o. Notice that the probability distribution does not depend on m_ℓ, but only on n and ℓ.

Table 3.2 summarizes the nature of the functions $R_{n,\ell}(r)$ and $Y_{\ell,m_\ell}(\theta, \phi)$ for various n, ℓ, m_ℓ values. Each possible wavefunction $\psi_{n,\ell,m_\ell}(r, \theta, \phi)$ with a particular choice of n, ℓ, m_ℓ constitutes a **quantum state** for the electron. The function $\psi_{n,\ell,m_\ell}(r, \theta, \phi)$ basically describes the behavior of the electron in the atom in probabilistic terms, as distinct from a well-defined line orbit for the electron, as one might expect from classical mechanics. For this reason, $\psi_{n,\ell,m_\ell}(r, \theta, \phi)$ is often referred to as an **orbital,** in contrast to the classical theory, which assigns an orbit to the electron.

Figure 3.22a shows the polar plots of $Y_{\ell,m_\ell}(\theta, \phi)$ for s and p orbitals. The radial distance from the origin in the polar plot represents the magnitude of $Y_{\ell,m_\ell}(\theta, \phi)$, which depends on the angles θ and ϕ. The polar plots of the probability distribution $| Y_{\ell,m_\ell}(\theta, \phi) |^2$ are shown in Figure 3.22b. Although for the s states, $Y_{1,0}(\theta, \phi)$ is spherically symmetric, resulting in a spherically symmetrical probability distribution around the nucleus, this is not so for $\ell = 1$ and higher states.

For example, each of the p states has a distinctly directional character, as illustrated in the polar plots in Figure 3.22. The angular dependence of $| \psi_{2,1,0}(r, \theta, \phi) |$, for which $m_\ell = 0$, is such that most of the probability is oriented along the z axis. This wavefunction is referred to as the $2p_z$ orbital. The two wavefunctions for $m_\ell = \pm 1$ are often represented by $\psi_{2p_x}(r, \theta, \phi)$ and $\psi_{2p_y}(r, \theta, \phi)$, or more simply, $2p_x$ and $2p_y$ orbitals, which do not possess a specific m_ℓ individually, but together represent the two $m_\ell = \pm 1$ wavefunctions. The angular dependence of $2p_x$ and $2p_y$ are essentially along the x and y directions. Thus, the three orbitals for $m_\ell = 0, \pm 1$ are all oriented perpendicular to each other, as depicted in Figure 3.22.

It should be noted that the probability distributions in Figures 3.21b and 3.22b do not depend on time. As previously mentioned, under steady-state conditions, the magnitude of the total wavefunction is

$$| \Psi(r, \theta, \phi, t) | = \left| \psi(r, \theta, \phi) \exp\left(-\frac{jEt}{\hbar} \right) \right| = | \psi(r, \theta, \phi) |$$

which is independent of time.

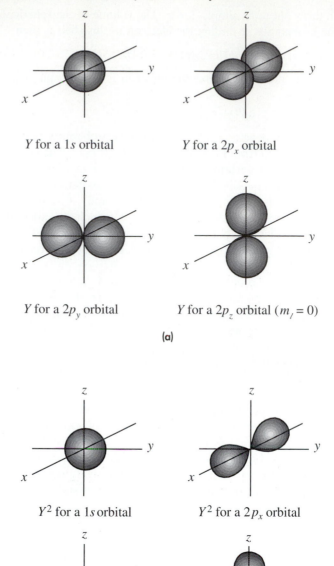

Y for a 1*s* orbital

Y for a 2p_x orbital

Y for a 2p_y orbital

Y for a 2p_z orbital ($m_l = 0$)

(a)

Y^2 for a 1*s* orbital

Y^2 for a 2p_x orbital

Y^2 for a 2p_y orbital

Y for a 2p_z orbital ($m_l = 0$)

(b)

Figure 3.22
(a) The polar plots of $Y_{n,\ell}(\theta, \phi)$ for 1*s* and 2*p* states.
(b) The angular dependence of the probability distribution, which is proportional to $|Y_{n,\ell}(\theta, \phi)|^2$.

3.7.2 Quantized Electron Energy

Once the wavefunctions, $\psi_{n,\ell,m_\ell}(r, \theta, \phi)$, have been found, they can be substituted into the Schrödinger equation to find the possible energies of the electron. These turn out to depend only on the principal quantum number n. The energy is given by

$$E_n = -\frac{me^4 Z^2}{8\varepsilon_o^2 h^2 n^2} \qquad \text{[3.37a]}$$

or

$$E_n = -\frac{Z^2 E_I}{n^2} = -\frac{Z^2 (13.6 \text{ eV})}{n^2} \qquad \text{[3.37b]}$$

where

$$E_I = \frac{me^4}{8\varepsilon_o^2 h^2} = 2.18 \times 10^{-18} \text{ J} \quad \text{or} \quad 13.6 \text{ eV}$$

This corresponds to the energy required to remove the electron in the hydrogen atom ($Z = 1$) from the lowest energy level E_1 (at $n = 1$) to infinity; hence, it represents the **ionization energy.** The energy E_n in Equation (3.37b) is negative with respect to that for the electron completely isolated from the nucleus (at $r = \infty$, therefore $V = 0$). Thus, when the electron is in the vicinity of the nucleus, $+Ze$, it has a lower energy, which is a favorable situation (hence, formation of the hydrogenic atom is energetically favorable).

Since the energy is quantized, the lowest energy of the electron corresponds to $n = 1$, which is -13.6 eV. The next higher energy value it can have is $E_2 = -3.40$ eV when $n = 2$, and so on, as sketched in Figure 3.23. Normally, the electron will take up a state corresponding to $n = 1$, because this has the lowest energy, called the **ground energy.** Its wavefunction corresponds to $\psi_{100}(r, \theta, \phi)$, which has a probability peak at $r = a_o$ and no angular dependence, as indicated in Figures 3.21 and 3.22.

The electron can only become excited to the next energy level if it is supplied by the right amount of energy, $E_2 - E_1$. A photon of energy $hv = E_2 - E_1$ can readily supply this energy when it strikes the electron. The electron then gets excited to the state with $n = 2$ by absorbing the photon, and its wavefunction changes to $\psi_{210}(r, \theta, \phi)$, which has the maximum probability at $r = 4a_o$. The electron thus spends most of its time in this excited state, at $r = 4a_o$. It can return from the excited state at E_2 to the ground state at E_1 by emitting a photon of energy $hv = E_2 - E_1$.

By virtue of the quantization of energy, we see that the emission of light from excited atoms can only have certain wavelengths: those corresponding to transitions from higher quantum-number states to lower ones. In fact, in spectroscopic analysis, these wavelengths can be used to identify the elements, since each element has its unique set of emission and absorption wavelengths arising from a unique set of energy levels. Figure 3.24 illustrates the origin of the emission and absorption spectra of atoms, which are a direct consequence of the quantization of the energy.

The electrons in atoms can also be excited by other means, for example, by collisions with other atoms as a result of heating a gas. Figure 3.25 depicts how

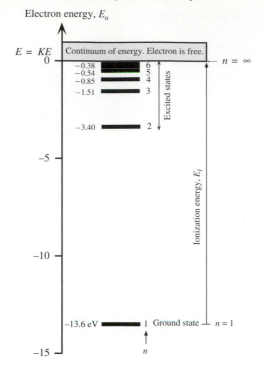

Figure 3.23   The energy of the electron in the hydrogen atom $(Z = 1)$.

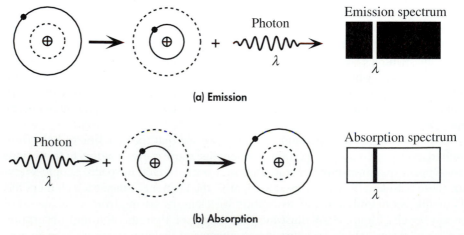

(a) Emission

(b) Absorption

Figure 3.24   The physical origin of spectra.

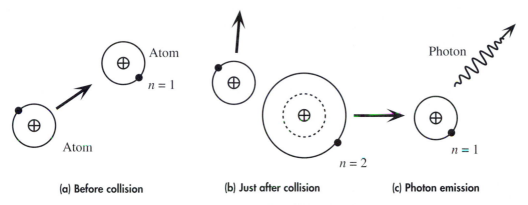

(a) Before collision   (b) Just after collision   (c) Photon emission

Figure 3.25   An atom can become excited by a collision with another atom. When it returns to its ground energy state, the atom emits a photon.

collisions with other atoms can excite an electron to higher energies. If an imping-ing atom has sufficient kinetic energy, it can impart just the right energy to excite the electron to a higher energy level. Since the total energy must be conserved, the incoming atom will lose some of its kinetic energy in the process. The excited electron can later return to its ground state by emitting a photon. Excitation by atomic collisions is the process by which we obtain light from an electrical dis-charge in gases, a quantum phenomenon we experience every day as we read a neon sign. Indeed, this is exactly how the Ne atoms in the common laboratory HeNe laser are excited, via atomic collisions between Ne and He atoms.

Since the principal quantum number determines the energy of the electron and also the position of maximum probability, as we noticed in Figure 3.21, various n values define electron **shells,** within which we can most likely find the electron. These shells are customarily labelled $K, L, M, N, \ldots$, corresponding to $n = 1, 2, 3, \ldots$. For each n value, there are a number of ℓ values that determine the spatial distribution of the electron. For a given n, each ℓ value constitutes a **subshell.** For example, we often talk about $3s, 3p, 3d$ subshells within the M shell. From the radial dependence of the electron's wavefunction $\psi_{n,\ell,m_\ell}(r, \theta, \phi)$, shown in Figure 3.21, we see that for higher values of n, which correspond to more energetic states, the mean distance of the electron from the nucleus increases. In fact, we observe from Figure 3.21 that an orbital with $\ell = n - 1$ (e.g., $1s, 2p$) exhibits a single maximum in its radial probability distribution, and this maximum rapidly moves farther away from the nucleus as n increases. By examining the electron wavefunctions, we can show that the location of the maxima for these $\ell = n - 1$ states are at

$$r_{max} = \frac{n^2 a_o}{Z} \qquad \ell = n - 1 \qquad \text{[3.38]}$$

where a_o is the radius of the ground state (0.0529 nm). The maximum probability radius r_{max} in Equation 3.38 is the Bohr radius. Note that r_{max} in Equation 3.38 is for $\ell = n - 1$ states only. For other ℓ values, there are multiple maxima, and we must think in terms of the average position of the electron from the nucleus. When we evaluate the average position from $\psi_{n,\ell,m_\ell}(r, \theta, \phi)$, we see that it depends on both n and ℓ; strongly on n and weakly on ℓ.

3.7.3   Orbital Angular Momentum and Space Quantization

The electron in the atom has an orbital angular momentum L. The electron is attracted to the nucleus by a central force, just like the earth is attracted by the central gravitational force of the sun and thus possesses an orbital angular momentum. It is well known that in classical mechanics, under the action of a central force, both the total energy ($KE + PE$) and the orbital angular momentum (L) of an orbiting object are conserved. In quantum mechanics, the orbital angular momentum of the electron, like its energy, is also quantized, but by the quantum number ℓ. The magnitude of L is given by

Orbital angular momentum

$$L = \hbar[\ell(\ell + 1)]^{1/2} \qquad [3.39]$$

where $\ell = 0, 1, 2, \ldots < n$. Thus, for an electron in the ground state, $n = 1$ and $\ell = 0$, the angular momentum is zero, which is surprising since we always think of the electron as orbiting the nucleus. In the ground state, the spherical harmonic is a constant, independent of the angles θ and ϕ, so the electron has a spherically symmetrical probability distribution that depends only on r.

The quantum numbers n and ℓ quantize the energy and the magnitude of the orbital angular momentum. What is the significance of m_ℓ? In the presence of an external magnetic field B_z, taken arbitrarily in the z direction, the component of the angular momentum along the z axis, L_z, is also quantized and is given by

Orbital angular momentum along B_z

$$L_z = m_\ell \hbar \qquad [3.40]$$

Therefore, the quantum number m_ℓ quantizes the component of the angular momentum along the direction of an external magnetic field B_z, which for reference purposes is taken along z, as illustrated in Figure 3.26. Therefore, m_ℓ is appropriately called the **magnetic quantum number.** For any given ℓ, quantum mechanics requires that m_ℓ must have values in the range $-\ell, -(\ell - 1), \ldots, -1, 0, 1, \ldots, (\ell - 1), \ell$. We see that $|m_\ell| \le \ell$. Moreover, m_ℓ can be negative, since L_z can be negative or positive, depending on the orientation of the angular momentum vector **L**. Since $|m_\ell| \le \ell$, **L** can never align with the magnetic field along z; instead, it makes an angle with B_z, an angle that is determined by ℓ and m_ℓ. We say that **L** is **space quantized.** Space quantization is illustrated in Figure 3.26 for $\ell = 2$.

Since the energy of the electron does not depend on either ℓ or m_ℓ we can have a number of possible states for a given energy. For example, when the energy is E_2, then $n = 2$, which means that $\ell = 0$ or 1. For $\ell = 1$, we have $m_\ell = -1, 0, 1$, so that there are a total of three different orbitals for the electron.

Since the electron has a quantized orbital angular momentum, when an electron interacts with a photon, the electron must obey the law of the conservation of angular momentum, much as an ice skater does sudden fast spins by pulling in her

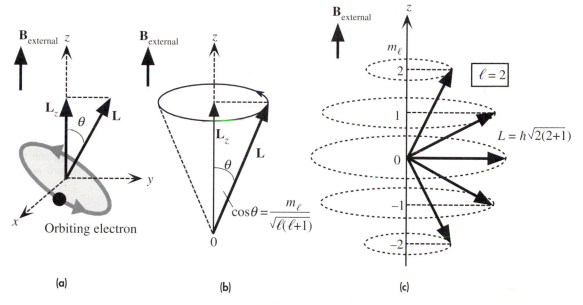

Figure 3.26
(a) The electron has an orbital angular momentum, which has a quantized component L along an external magnetic field $B_{external}$.
(b) The orbital angular momentum vector **L** rotates about the z axis. Its component L_z is quantized; therefore, the **L** orientation, which is the angle θ, is also quantized. **L** traces out a cone.
(c) According to quantum mechanics, only certain orientations (θ) for **L** are allowed, as determined by ℓ and m_ℓ.

arms. All experiments indicate that the photon has an intrinsic angular momentum with a constant magnitude given by $\hbar$. Therefore, when a photon of energy $h\upsilon = E_2 - E_1$ is absorbed, the angular momentum of the electron must change. This means that following photon absorption or emission, both the principal quantum number n and the orbital angular momentum quantum number ℓ must change.

The rules that govern which transitions are allowed from one state to another as a consequence of photon absorption or emission are called **selection rules.** As a result of photon absorption or emission, we must have

$$\Delta\ell = \pm 1 \quad \text{and} \quad \Delta m_\ell = 0, \pm 1 \qquad \text{[3.41]}$$

Selection rules for EM radiation

As an example, consider the excitation of the electron in the hydrogen atom from the ground energy E_1 to a higher energy level E_2. The photon energy $h\upsilon$ must be exactly $E_2 - E_1$. The wavefunction of the $1s$ ground state is $\psi_{1,0,0}$, whereas there are four wavefunctions at E_2: one $2s$ state, $\psi_{2,0,0}$; and three $2p$ states, $\psi_{2,1,-1}$, $\psi_{2,1,0}$, and $\psi_{2,1,1}$. The excited electron cannot jump into the $2s$ state, because $\Delta\ell$ must be ± 1, so it enters a $2p$ state corresponding to one of the orbitals $\psi_{2,1,-1}$, $\psi_{2,1,0}$, or $\psi_{2,1,1}$. Various allowed transitions for photon emission in the hydrogen atom are indicated in Figure 3.27.

Solving this for n, we find

$$n^2 = \frac{13.6}{(13.6 - 12.5)} = 12.36$$

so $n = 3.51$. But n can only be an integer; thus, the electron gets excited to the level $n = 3$ where its energy is $E_3 = -13.6/3^2 = -1.51$ eV.

The energy of the incoming electron after the collision is less by

$$(E_3 - E_1) = 13.6 - 1.51 = 12.09 \text{ eV}$$

Since the initial energy of the incoming electron was 12.5 eV, it leaves the collision with a kinetic energy of $12.5 - 12.09 = 0.41$ eV. From the E_3 level, the electron can undergo a transition from $n = 3$ to $n = 1$,

$$\Delta E_{31} = -1.51 \text{ eV} - (-13.6 \text{ eV}) = 12.09 \text{ eV}$$

The emitted radiation will have a wavelength λ given by $hc/\lambda = \Delta E$, so that

$$\lambda_{31} = hc/\Delta E_{31} = (6.626 \times 10^{-34} \text{ J s})(3 \times 10^8 \text{ m s}^{-1})/(12.09 \times 1.6 \times 10^{-19} \text{ J})$$

$$= 1.026 \times 10^{-7} \text{ m} \quad \text{or} \quad 102.6 \text{ nm} \quad \text{(in the ultraviolet region)}$$

Another possibility is the transition from $n = 3$ to $n = 2$, for which

$$\Delta E_{32} = -1.51 \text{ eV} - (-3.40 \text{ eV}) = 1.89 \text{ eV}$$

This will give a wavelength

$$\lambda_{32} = \frac{hc}{\Delta E_{32}} = 656 \text{ nm}$$

which is in the red region of the visible spectrum. For the transition from $n = 2$ to $n = 1$,

$$\Delta E_{21} = -3.40 \text{ eV} - (-13.6 \text{ eV}) = 10.2 \text{ eV}$$

which results in the emission of a photon of wavelength $\lambda_{21} = hc/\Delta E_{21} = 121.5$ nm. Note that each transition obeys $\Delta \ell = \pm 1$.

THE IONIZATION ENERGY OF He⁺   What is the energy required to further ionize He^+ ions to He^{++}? | **Example 3.14**

Solution

He^+ is a hydrogenic atom with one electron attracted by a nucleus with a $+2e$ charge. Thus $Z = 2$. The energy of the electron in a hydrogenic atom (in eV) is given by

$$E_n(\text{eV}) = -\frac{Z^2 \, 13.6}{n^2}$$

Since $Z = 2$, the energy required to ionize He^+ further is

$$|E_1| = |-(2^2)13.6| = 54.4 \text{ eV}$$

THE FRAUNHOFER LINES IN THE SUN'S SPECTRUM   The light from the sun includes extremely | **Example 3.15**
sharp "dark lines" at certain wavelengths, superimposed on a bright continuum at all other wavelengths, as discovered by Josef von Fraunhofer in 1829. One of these dark lines occurs

in the orange range and another in the blue. Fraunhofer measured their wavelengths to be 6563 Å and 4861 Å, respectively. With the aid of Figure 3.23, show that these are spectral lines from the hydrogen atom spectrum. (They are called the H_α and H_β Fraunhofer lines. Such lines provided us with the first clues to the chemical composition of the sun.)

Solution

The energy of the electron in a hydrogenic atom is

$$E_n = -\frac{Z^2 E_I}{n^2}$$

where $E_I = me^4/(8\varepsilon_0^2 h^2)$. Photon emission resulting from a transition from quantum number n_2 to n_1 has an energy

$$\Delta E = E_{n_2} - E_{n_1} = -Z^2 E_I \left(\frac{1}{n_2^2} - \frac{1}{n_1^2} \right)$$

From $h\upsilon = hc/\lambda = \Delta E$, we have

$$\frac{1}{\lambda} = \left(\frac{E_I}{hc} \right) Z^2 \left(\frac{1}{n_1^2} - \frac{1}{n_2^2} \right) = R_\infty Z^2 \left(\frac{1}{n_1^2} - \frac{1}{n_2^2} \right)$$

where $R_\infty = E_I/hc = 1.0974 \times 10^7$ m^{-1}. The equation for λ is called the **Balmer–Rydberg formula,** and R_∞ is called the **Rydberg constant.** We apply the Balmer–Rydberg formula with $n_1 = 2$ and $n_2 = 3$ to obtain

$$\frac{1}{\lambda} = (1.0974 \times 10^7 \text{ m}^{-1})(1^2)\left(\frac{1}{2^2} - \frac{1}{3^2} \right) = 1.524 \times 10^6 \text{ m}^{-1}$$

to get $\lambda = 6561$ Å. We can also apply the Balmer–Rydberg formula with $n_1 = 2$ and $n_2 = 4$ to get $\lambda = 4860$ Å.

Example 3.16   **GIANT ATOMS IN SPACE**   Radiotelescopic studies by B. Höglund and P.G. Mezger (*Science* vol. 150, p. 339, 1965) detected a 5009 MHz electromagnetic radiation in space. Show that this radiation comes from excited hydrogen atoms as they undergo transitions from $n = 110$ to 109. What is the size of such an excited hydrogen atom?

Solution

Since the energy of the electron is $E_n = -(Z^2 E_I/n^2)$, the energy of the emitted photon in the transition from n_2 to n_1 is

$$h\upsilon = E_{n_2} - E_{n_1} = Z^2 E_I(n_1^{-2} - n_2^{-2})$$

With $n_2 = 110$, $n_1 = 109$, and $Z = 1$, the frequency is

$$\upsilon = \frac{Z^2 E_I(n_1^{-2} - n_2^{-2})}{h}$$

$$= \frac{[(1.6 \times 10^{-19} \times 13.6)[(109^{-2} - 110^{-2})]}{(6.626 \times 10^{-34})}$$

$$= 5 \times 10^9 \text{ s}^{-1} \qquad \text{or} \qquad 5000 \text{ MHz}$$

The size of the atom from Equation (3.38) is on the order of

$$2r_{max} = 2n^2 a_o = 2(110^2)(52.918 \times 10^{-12} \text{ m}) = 1.28 \times 10^{-6} \text{ m} \quad \text{or} \quad 1.28 \ \mu\text{m}$$

A giant atom!

3.7.4 Electron Spin and Intrinsic Angular Momentum S

One aspect of electron behavior does not come from the simple Schrödinger equation. That is the spin of the electron about its own axis, which is analogous to the 24-hour spin of Earth around its axis.[5] Earth has an orbital angular momentum due to its motion around the sun, and an intrinsic or spin angular momentum due to its rotation about its own axis. Similarly, the electron has a **spin** or **intrinsic angular momentum,** denoted by **S**. In classical mechanics, in the absence of external torques, spin angular momentum is conserved. In quantum mechanics, this spin angular momentum is quantized, in a manner similar to that of orbital angular momentum. The magnitude of the spin has been found to be constant, with a quantized component S_z in the z direction along a magnetic field:

$$S = \hbar[s(s + 1)]^{1/2} \qquad s = \frac{1}{2} \qquad \text{[3.42]}$$

$$S_z = m_s \hbar \qquad m_s = \pm\frac{1}{2} \qquad \text{[3.43]}$$

where, in an analogy with ℓ and m_ℓ, we use the quantum numbers s and m_s, which are called the **spin** and **spin magnetic quantum numbers.** Contrary to our past experience with quantum numbers, s and m_s are not integers, but are $(1/2)$ and $\pm(1/2)$, respectively. The existence of electron spin was put forward by Goudsmit and Uhlenbeck in 1925 and derived by Dirac from relativistic quantum theory, which is beyond the scope of this book. Figure 3.28 illustrates the spin angular momentum of the electron and the two possibilities for S_z. When $S_z = +(1/2) \hbar$, using classical orbital motion as an analogy, we can label the spin of the electron as being in the clockwise direction, so $S_z = -(1/2) \hbar$ can be labelled as a counterclockwise spin. However, no such true clockwise or counterclockwise spinning of the electron can in reality[6] be identified. When $S_z = +(1/2) \hbar$, we could just as easily label the electron spin as "up," and call it "down" when $S_z = -(1/2) \hbar$. This terminology is used henceforth in this book.

Since the magnitude of the electron spin is constant, which is a remarkable fact, and is determined by $s = (1/2)$, we need not mention it further. It can simply be regarded as a fundamental property of the electron, in much the same way as its mass and charge. We do, however, need to specify whether $m_s = +(1/2)$ or

---

[5] Do not take the meaning of "spin" too literally, as in classical mechanics. Remember that the electron is assumed to have wave-like properties, which can have no classical spin.

[6] The explanation in terms of spin and its two possible orientational directions ("clockwise" and "counterclockwise") serve as mental aids in visualizing a quantum mechanical phenomenon. One question, however, is, "If the electron is a wave, what is spinning?"

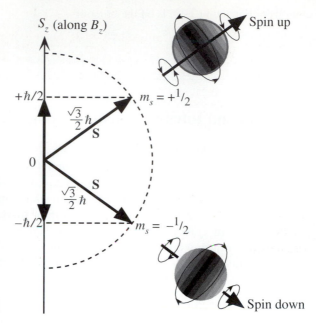

Figure 3.28   Spin angular momentum exhibits space quantization. Its magnitude along z is quantized, so the angle of **S** to the z axis is also quantized.

$-(1/2)$, since each of these selections gives the electron a different behavior. We therefore need four quantum numbers to specify what the electron is doing. Each state of the electron needs the spin magnetic quantum number m_s, in addition to n, ℓ, and m_ℓ. For each orbital $\psi_{n,\ell,m_\ell}(r, \theta, \phi)$, we therefore have two possibilities: $m_s = \pm(1/2)$. The quantum numbers n, ℓ, and m_ℓ determine the spatial extent of the electron by specifying the form of $\psi_{n,\ell,m_\ell}(r, \theta, \phi)$, whereas m_s determines the "direction" of the electron's spin. A full description of the

Table 3.3   The four quantum numbers for the hydrogenic atom

n	Principal quantum number	$n = 1, 2, 3, \ldots$	Quantizes the electron energy
ℓ	Orbital angular momentum quantum number	$\ell = 0, 1, 2, \ldots (n - 1)$	Quantizes the magnitude of orbital angular momentum L
m_ℓ	Magnetic quantum number	$m_\ell = 0, \pm 1, \pm 2, \ldots, \pm \ell$	Quantizes the orbital angular momentum component along a magnetic field B_z.
m_s	Spin magnetic quantum number	$m_s = \pm \dfrac{1}{2}$	Quantizes the spin angular momentum component along a magnetic field B_z.

behavior of the electron must therefore include all four quantum numbers n, ℓ, m, and m_s.

An **electronic state** is a wavefunction that defines both the spatial (ψ_{n,ℓ,m_ℓ}) and spin (m_s) properties of an electron. Frequently, an electronic state is simply denoted ψ_{n,ℓ,m_ℓ,m_s} which adds the spin quantum number to the orbital wavefunction.

The quantum numbers are extremely important, because they quantize the various properties of the electron: its total energy, orbital angular momentum, and the orbital and spin angular momenta along a magnetic field. Their significance is summarized in Table 3.3.

MAGNETIC DIPOLE MOMENT OF THE ELECTRON   Consider the electron orbiting the nucleus with an angular frequency ω, as illustrated in Figure 3.29a. The orbiting electron is equivalent to a current loop. The equivalent current i due to the orbital motion of the electron is

Example 3.17

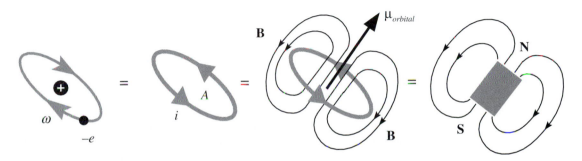

(a): The orbiting electron is equivalent to a current loop that behaves like a bar magnet.

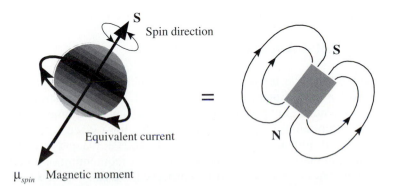

(b): The spinning electron can be imagined to be equivalent to a current loop as shown. This current loop behaves like a bar magnet, just as in the orbital case.

Figure 3.29

given by the charge flowing per unit time, or $i = charge/period = -e(\omega/2\pi)$. The negative sign indicates that the current i flows in the opposite direction to the electron motion. The magnetic field around the current loop is similar to that of a permanent magnet, as depicted in Figure 3.29a. The magnetic moment is defined as $\mu = iA$, the product of current and the area enclosed by the current loop. It is a vector normal to the surface A in a direction determined by the corkscrew rule applied to the circulation of the current i. Thus,

$$\mu = iA = \left(-\frac{e\omega}{2\pi}\right)(\pi r^2) = -\frac{e\omega r^2}{2}$$

Consider now the orbital angular momentum L, which is the linear momentum p multiplied by the radius r, or

$$L = pr = m_e vr = m_e \omega r^2$$

Using this, we can substitute for ωr^2 in $\mu = -\dfrac{e\omega r^2}{2}$ to obtain

$$\mu = -\frac{e}{2m_e}L$$

In vector notation, using the subscript *orbital* to identify the origin of the magnetic moment,

$$\boldsymbol{\mu}_{orbital} = -\frac{e}{2m_e}\mathbf{L}$$

This means that the direction of the orbital magnetic moment $\boldsymbol{\mu}_{orbital}$ is opposite to that of the orbital angular momentum $\mathbf{L}$ and is related to it by a constant $(e/2m_e)$.

Similarly, the spin angular momentum of the electron $\mathbf{S}$ leads to a spin magnetic moment $\boldsymbol{\mu}_{spin}$, which is in the opposite direction to $\mathbf{S}$ and is given by $\boldsymbol{\mu}_{spin} = -(e/m_e)\mathbf{S}$, as indicated in Figure 3.29b. Notice that there is no factor of 2 in the denominator. We see that, as a consequence of the orbital motion and also of the spin, the electron has two distinct magnetic moments. These moments act on each other, just like two magnets interact with each other. The result is a coupling of the orbital and spin angular momenta, $\mathbf{L}$ and $\mathbf{S}$, and their precession about the total angular momentum $\mathbf{J} = \mathbf{L} + \mathbf{S}$.

3.7.5   Total Angular Momentum J

The orbital angular momentum $\mathbf{L}$ and the spin angular momentum $\mathbf{S}$ add to give the electron a total angular momentum $\mathbf{J} = \mathbf{L} + \mathbf{S}$, as illustrated in Figure 3.30. There are a number of possibilities for the total angular momentum $\mathbf{J}$, based on the relative orientations of $\mathbf{L}$ and $\mathbf{S}$. For example, for a given $\mathbf{L}$, we can add $\mathbf{S}$ either in parallel or antiparallel, as depicted in Figures 3.30a and b, respectively.

Since in classical physics the total angular momentum of a body (not experiencing an external torque) must be conserved, we can expect J (the magnitude of $\mathbf{J}$) to be quantized. This turns out to be true. The magnitude of $\mathbf{J}$ and its z component along an external magnetic field are quantized via

$$J = \hbar[j(j + 1)]^{1/2} \tag{3.44}$$

$$J_z = m_j\hbar \tag{3.45}$$

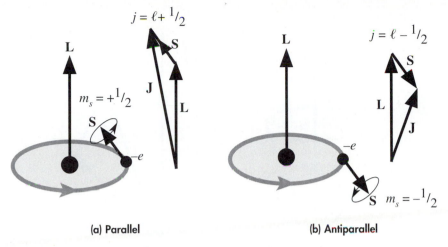

(a) Parallel   **(b) Antiparallel**

Figure 3.30   Orbital angular momentum vector **L** and spin angular momentum vector **S** can add either in parallel as in (a) or antiparallel, as in (b).

The total angular momentum vector $\mathbf{J} = \mathbf{L} + \mathbf{S}$, has a magnitude $J = \sqrt{[j(j+1)]}$, where in (a) $j = \ell + 1/2$ and in (b) $j = \ell - 1/2$.

where both j and m_j are quantum numbers[7] like ℓ and m_ℓ, but j and m_j can have fractional values. A rigorous theory of quantum mechanics shows that when $\ell > s$, the quantum numbers for the total angular momentum are given by $j = \ell + s$ and $\ell - s$ and $m_j = \pm j, \pm (j - 1)$. For example, for an electron in a p orbital, where $\ell = 1$, we have $j = 3/2$ and $1/2$, and $m_j = 3/2, 1/2, -1/2$, and $-3/2$. However, when $\ell = 0$ (as for all s orbitals), we have $j = s = 1/2$ and $m_j = m_s = \pm 1/2$, which are the only possibilities. We note from Equations 3.44 and 3.45 that $|J_z| < J$ and both are quantized, which means that **J** is space quantized; its orientation (or angle) with respect to the z axis is determined by j and m_j.

The spinning electron actually experiences a magnetic field $\mathbf{B}_{int}$ due to its orbital motion around the nucleus. If we were sitting on the electron, then in our reference frame, the positively-charged nucleus would be orbiting around us, which would be equivalent to a current loop. At the center of this current loop, there would be an "internal" magnetic field $\mathbf{B}_{int}$, which would act on the magnetic moment of the spinning electron to produce a torque. Since **L** and **S** add to give **J**, and since the latter quantity is space quantized (or conserved), then as a result of the internal torque on the electron, we must have **L** and **S** synchronously precessing about **J**, as illustrated in Figure 3.31a. If there is an external magnetic field **B** taken to be along z, this torque will act on the net magnetic moment due to **J** to cause this quantity to precess about **B**, as depicted in Figure 3.31b. Remember that the component along the z axis must be quantized and equal to $m_j \hbar$, so the torque can only cause precession. To understand the precession of the electron's angular momentum about the magnetic field **B**, think of a spinning top that precesses about the gravitational field of Earth.

---

[7] The quantum number j used here should not be confused with j for $\sqrt{-1}$.

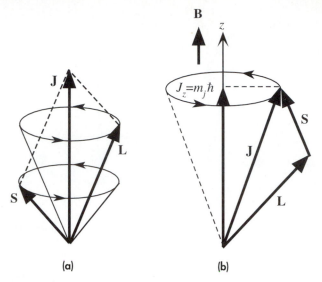

Figure 3.31

(a) The angular momentum vectors **L** and **S** precess around their resultant total angular momentum vector **J**.

(b) The total angular momentum vector is space quantized. Vector **J** precesses about the z axis, along which its component must be $m_j\hbar$.

3.8   THE HELIUM ATOM AND THE PERIODIC TABLE

3.8.1   He Atom and Pauli Exclusion Principle

In the He atom, there are two electrons in the presence of a nucleus of charge of $+2e$, as depicted in Figure 3.32. (Obviously, in higher-atomic-number elements, there will be Z electrons around a nucleus of charge $+Ze$.) The *PE* of an electron in the He atom consists of two interactions. The first is due to the coulombic attraction between itself and the positive nucleus; the second is due to the mutual repulsion between the two electrons. The *PE* function V of any one of the electrons, for example, that labelled as 1, therefore depends on both its distance from the nucleus r_1 and the separation of the two electrons, r_{12}. The *PE* of electron 1 thus depends on the locations of both the electrons, or

$$V(r_1,\, r_1) = \frac{-2e^2}{4\pi\varepsilon_o r_1} + \frac{e^2}{4\pi\varepsilon_o r_{12}}$$ [3.46]

When we use this *PE* in the Schrödinger equation for a single electron, we find the wavefunction and energy of one of the electrons in the He atom. We thus obtain the **one-electron wavefunction** and the **energy of one electron** within a many-electron atom.

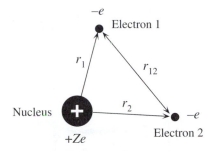

Figure 3.32   A helium-like atom.
The nucleus has a charge of $+Ze$, where
$Z = 2$ for He. If one electron is removed,
we have the He^+ ion, which is equivalent
to the hydrogenic atom with $Z = 2$.

One immediate and obvious result is that the energy of an electron now depends not only on n but also on ℓ, because the electron–electron potential energy term (the second term in Equation 3.46, which contains r_{12}) depends on the relative orientations of the electron orbitals, which change r_{12}. We therefore denote the electron energy by $E_{n,\ell}$. The dependence on ℓ is weaker than on n, as shown in Figure 3.33. As n and ℓ increase, $E_{n,\ell}$ also increases. Notice, however, that the energy of a $4s$ state is lower than that of a $3d$ state, and the same pattern also occurs at $4s$ and $5s$.

One of the most important theorems in quantum physics is the **Pauli exclusion principle,** which is based on experimental observations. This principle states that *no two electrons within a given system (e.g., an atom) may have all four identical quantum numbers, n, ℓ, m_ℓ and m_s.* Each set of values for n, ℓ, m_ℓ, and m_s represents a possible electronic state, that is, a wavefunction denoted by ψ_{n,ℓ,m_ℓ,m_s}, that the electron may (or may not) acquire. For example, an electron with the quantum numbers given by 2, 1, 1, 1/2 will have a definite wavefunction $\psi_{n,\ell,m_\ell,m_s} = \psi_{2,1,1,1/2}$, and it is said to be in the state $2p$, $m_\ell = 1$ and spin up. Its energy will be E_{2p}. The Pauli Exclusion Principle requires that no other electron be in this same state.

The orbital motion of an electron is determined by n, ℓ, and m_ℓ, whereas m_s determines the spin direction (up or down). Suppose two electrons are in the same orbital state, with identical n, ℓ, m_ℓ. By the Pauli Exclusion Principle, they would have to spin in opposite directions, as shown in Figure 3.34. One would have to spin "up" and the other "down." In this case we say that the electrons are **spin paired.** Two electrons can thus have the same orbitals (occupy the same region of space) if they pair their spins. However, the Pauli Exclusion Principal prevents a third electron from entering this orbital, since m_s can only have two values.

Using the Pauli exclusion principle, we can determine the electronic structure of many-electron atoms. For simplicity, we will use a box to represent and orbital state defined by a set of n, ℓ, m_ℓ values. Each box can take two electrons at most,

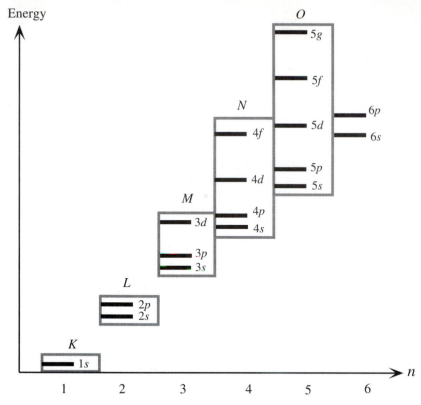

Figure 3.33   Energy of various one-electron states.
The energy depends on both *n* and ℓ.

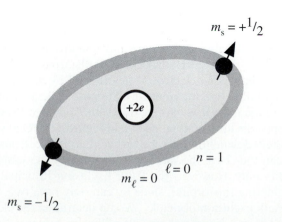

Figure 3.34   Paired spins in an orbital.

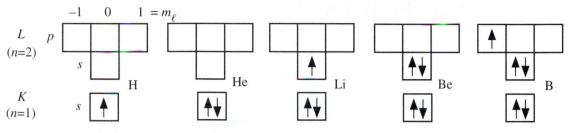

Figure 3.35 Electronic configurations for the first five elements. Each box represents an orbital $\psi(n, \ell, m_\ell)$.

with their spins paired. When we put an electron into a box, we are essentially assigning a wavefunction to that electron; that is, we are defining its orbital n, ℓ, m_ℓ. We use an arrow to show whether the electron is spinning up or down. As depicted in Figure 3.35, we arrange all of the boxes to correspond to the electronic subshells. As an example, consider boron, which has five electrons. The first electron enters the $1s$ orbital at the lowest energy. The second also enters this orbital by spinning in the opposite direction. The third goes into the $n = 2$ orbital. The lowest energy there is in the s orbitals corresponding to $\ell = 0$ and $m_\ell = 0$. The fourth electron can also enter the $2s$ orbital, provided that it spins in the opposite direction. Similarly, the fifth must go into another orbital, and the next nearest low-energy orbitals are those having $\ell = 1$ (p states) and $m_\ell = -1, 0, +1$. The final electronic structure of the B atom is shown in Figure 3.35.

We see that because the electron energy depends on n and ℓ, there are a number of states for a given energy $E_{n,\ell}$. Each of these states corresponds to different sets of m_ℓ and m_s. For example, the energy $E_{2,1}$ (or E_{2p}) corresponding to $n = 2$, $\ell = 1$ has six possible states, arising from $m_\ell = -1, 0, 1$ and $m_s = +1/2, -1/2$. Each m_ℓ state can have an electron spinning up or down, $m_s = +1/2$ or $m_s = -1/2$, respectively.

THE NUMBER OF STATES AT AN ENERGY LEVEL   Enumerate and identify the states corre- | **Example 3.18**
sponding to the energy level E_{3d}, or $n = 3$, $\ell = 2$.

Solution

When $n = 3$ and $\ell = 2$, m_ℓ and m_s can have these following values: $m_\ell = 2, 1, 0, 1, 2$, and $m_s = +1/2, -1/2$. This means there are ten combinations. The possible wavefunctions (electron states) are:

- $\psi_{3,2,2,1/2}$; $\psi_{3,2,1,1/2}$; $\psi_{3,2,0,1/2}$; $\psi_{3,2,-1,1/2}$; $\psi_{3,2,-2,1/2}$, all of which have spins up ($m_s = +1/2$)

- $\psi_{3,2,2,-1/2}$; $\psi_{3,2,1,-1/2}$; $\psi_{3,2,0,-1/2}$; $\psi_{3,2,-1,-1/2}$; $\psi_{3,2,-2,-1/2}$, all of which have spins down ($m_s = -1/2$)

3.8.2   Hund's Rule

In the many-electron atom, the electrons take up the lowest-energy orbitals and obey the Pauli Exclusion Principle. However, the Pauli Exclusion Principle does not determine how any two electrons distribute themselves among the many states of a given n and ℓ. For example, there are six $2p$ states corresponding to $m_\ell = -1$, 0, $+1$, with each m_ℓ having $m_s = \pm 1/2$. The two electrons could pair their spins and enter a given m_ℓ state, or they could align their spins (same m_s) and enter different m_ℓ states. An experimental fact deducted from spectroscopic studies shows that *electrons in the same n, ℓ orbitals prefer their spins to be parallel* (same m_s). This is known as **Hund's rule.**

The origin of Hund's Rule can be readily understood. If electrons enter the same m_ℓ state by pairing their spins (different m_s) their quantum numbers n, ℓ, m_ℓ will be the same and they will both occupy the same region of space (same ψ_{n,ℓ,m_ℓ} orbital). They will then experience a large coulombic repulsion and will have a large coulombic potential energy. On the other hand, if they parallel their spins (same m_s) they will each have a different m_ℓ and will therefore occupy different regions of space (different ψ_{n,ℓ,m_ℓ} orbitals), thereby reducing their coulombic repulsion.

The oxygen atom has eight electrons and its electronic structure is shown in Figure 3.36. The first two electrons enter the $1s$ box (orbital). The next two enter the $2s$ box. But p states can accommodate six electrons, so the remaining four electrons have a choice. Hund's Rule forces three of the four electrons to enter the boxes corresponding to $m_\ell = 1, 0, +1$, all with their spins parallel. The last electron can go into any of the $2p$ boxes, but it has no choice for spin. It must pair its spin with the electron already in the box. Thus, the oxygen atom has two unpaired electrons in half-occupied orbitals, as indicated in Figure 3.36. Since these two unpaired electrons spin in the same direction, they give the O atom a net angular momentum. An angular momentum due to charge rotation (i.e., spin) gives rise to a magnetic moment, $\boldsymbol{\mu}$. If there is an external magnetic field present, then $\boldsymbol{\mu}$ experiences a force given by $\boldsymbol{\mu} \cdot d\mathbf{B}/dx$. Oxygen atoms will therefore be deflected by a nonuniform magnetic field, as experimentally observed.

Following the Pauli Exclusion Principle and Hund's Rule, it is not difficult to build the electronic structure of various elements in the Periodic Table. There are

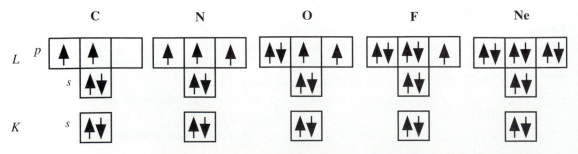

Figure 3.36   Electronic configurations for C, N, O, F, and Ne atoms.

Notice that in C, N, and O, Hund's Rule forces electrons to align their spins. For the Ne atom, all the K and L orbitals are full.

only a few instances of unusual behavior in the energy levels of the electronic states. The $4s$ state happens to be energetically lower than the $3d$ states, so the $4s$ state fills up first. Similarly, the $5s$ state is at a lower energy than the $4d$ states. These features are summarized in the energy diagram of Figure 3.33. There is a neat shorthand way of writing the electronic structure of any atom. To each $n\ell$ state, we attach a superscript to represent the number of electrons in those $n\ell$ states. For example, for oxygen, we write $1s^2 2s^2 2p^4$, or simply $[\text{He}]2s^2 2p^4$, since $1s^2$ is a full (closed) shell corresponding to He.

WHY IS THE Fe ATOM MAGNETIC?   The Fe atom has the electronic structure $[\text{Ar}]3d^6 4s^2$. | **Example 3.19**
Show that the Fe atom has four unpaired electrons and therefore a net angular momentum and a magnetic moment due to spin.

Solution

In a closed subshell, for example, $2p$ subshell with six states given by $m_\ell = -1, 0, +1$ and $m_s = +1/2$, all m_ℓ and m_s values have been taken up by electrons, so each m_ℓ orbital is occupied and has paired electrons. Each positive m_ℓ (or m_s) value assigned to an electron is canceled by the negative m_ℓ (or m_s) value assigned to another electron in the subshell. Therefore, *there is no net angular momentum from a closed subshell.* Only unfilled subshells contribute to the overall angular momentum. Thus, only the six electrons in the $3d$ subshell need be considered.

There are five d orbitals, corresponding to $m_\ell = -2, -1, 0, 1, 2$. Five of the six electrons obey Hund's Rule and align their spins, with each taking one of the m_ℓ values.

$$
m_\ell = -2 \quad -1 \quad 0 \quad 1 \quad 2
$$
$$
\uparrow \quad \uparrow \quad \uparrow \quad \uparrow \quad \uparrow
$$
$$
\downarrow
$$

The sixth must take the same m_ℓ as another electron. This is only possible if they pair their spins. Consequently, there are four electrons with unpaired spins in the Fe atom, which gives the Fe atom a net angular momentum. The Fe atom therefore possesses a magnetic moment as a result of four electrons having their charges spinning in the same direction.

3.9   STIMULATED EMISSION AND LASERS

3.9.1   Stimulated Emission and Photon Amplification

An electron can be excited from an energy level E_1 to a higher energy level E_2 by the absorption of a photon of energy $h\upsilon = E_2 - E_1$, as shown in Figure 3.37a. When an electron at a higher energy level transits down in energy to an unoccupied energy level, it emits a photon. There are essentially two possibilities for the emission process. The electron can spontaneously undergo the downward transition by itself, or it can be induced to do so by another photon.

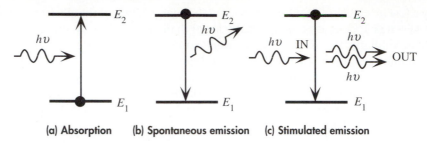

(a) Absorption   (b) Spontaneous emission   (c) Stimulated emission

Figure 3.37   Absorption, spontaneous emission, and stimulated emission.

In **spontaneous emission,** the electron falls in energy from level E_2 to E_1 and emits a photon of energy $h\upsilon = E_2 - E_1$, as indicated in Figure 3.37b. The transition is only spontaneous if the state with energy E_1 is not already occupied by another electron. In classical physics, when a charge accelerates and decelerates, as in an oscillatory motion, with a frequency υ, it emits an electromagnetic radiation also of frequency υ. The emission process during the transition of the electron from E_2 to E_1 appears as if the electron is oscillating with a frequency υ.

In **stimulated emission,** an incoming photon of energy $h\upsilon = E_2 - E_1$ stimulates the emission process by inducing the electron at E_2 to transit down to E_1. The emitted photon is in phase with the incoming photon, it is going in the same direction, and it has the same frequency, since it must also have the energy $E_2 - E_1$, as shown in Figure 3.37c. To get a feel for what is happening during stimulated emission, imagine the electric field of the incoming photon coupling to the electron and thereby driving it with the same frequency as the photon. The forced oscillation of the electron at a frequency $v = (E_2 - E_1)/h$ causes the electron to emit electromagnetic radiation, for which the electric field is totally in phase with that of the stimulating photon. When the incoming photon leaves the site, the electron can return to E_1, because it has emitted a photon of energy $h\upsilon = E_2 - E_1$.

Stimulated emission is the basis for photon amplification, since one incoming photon results in two outgoing photons, which are in phase. It is possible to achieve a practical light amplifying device based on this phenomenon. From Figure 3.37c, we see that to obtain stimulated emission, the incoming photon should not be absorbed by another electron at E_1. When we are considering using a collection of atoms to amplify light, we must therefore require that the majority of the atoms be at the energy level E_2. If this were not the case, the incoming photons would be absorbed by the atoms at E_1. When there are more atoms at E_2 than at E_1, we have what is called a **population inversion.** It should be apparent that with two energy levels, we can never achieve a population at E_2 greater than that at E_1, because, in the steady state, the incoming photon flux will cause as many upward excitations as downward stimulated emissions.

Let us consider the three-energy-level system shown in Figure 3.38. Suppose an external excitation causes the atoms[8] in this system to become excited to energy level E_3. This is called the **pump energy level,** and the process of exciting the atoms

---

[8] An atom is in an excited state when one (or more) of its electrons is excited from the ground energy to a higher energy level. The ground state of an atom has all the electrons in their lowest energy states consistent with the Pauli exclusion principle and Hund's rule.

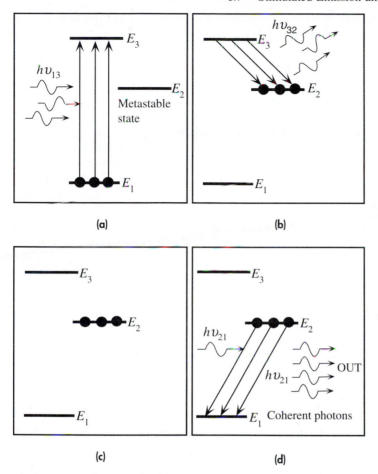

Figure 3.38   The principle of the LASER

(a) Atoms in the ground state are pumped up to energy level E_3 by incoming photons of energy $h\upsilon_{13} = E_3 - E_1$.

(b) Atoms at E_3 rapidly decay to the metastable state at energy level E_2 by emitting photons or emitting lattice vibrations: $h\upsilon_{32} = E_3 - E_2$.

(c) Since the states at E_2 are metastable, they quickly become populated, and there is a population inversion between E_2 and E_1.

(d) A random photon of energy $h\upsilon_{21} = E_2 - E_1$ can initiate stimulated emission. Photons from this stimulated emission can themselves further stimulate emissions, leading to an avalanche of stimulated emissions and coherent photons being emitted.

to E_3 is called **pumping.** In the present case, **optical pumping** is used, although this is not the only means of taking the atoms to E_3. Suppose further that the atoms in E_3 decay rapidly to energy level E_2, which happens to correspond to a state that does not rapidly and spontaneously decay to a lower-energy state. In other words, the state at E_2 is a **long-lived state.**[9] Quite often, the long-lived states are referred

[9] We will not examine what causes certain states to be long lived; we will simply accept that these states do not decay rapidly and spontaneously to lower-energy states.

to as **metastable states.** Since the atoms cannot decay rapidly from E_2 to E_1, they accumulate at this energy level, causing a population inversion between E_2 and E_1 as pumping takes more and more atoms to E_3 and hence to E_2.

When one atom at E_2 decays spontaneously, it emits a photon, which can go on to a neighboring atom and cause that to execute stimulated emission. The photons from the latter can then go on to the next atom at E_2 and cause that atom to emit by stimulated emission, and so on. The result is an avalanche effect of stimulated emission processes with all the photons in phase, so the light output is a large collection of coherent photons. This is the principle of the ruby laser in which the energy levels E_1, E_2, and E_3 are those of the Cr^{+3} ion in the Al_2O_3 crystal. At the end of the avalanche of stimulated emission processes, the atoms at E_2 will have returned to E_1 and can be pumped again to repeat the stimulated emission cycle again. The emission from E_2 to E_1 is called the **lasing emission.**

The system we have just described for photon amplification is a LASER, an acronym for light amplification by stimulated emission of radiation. In the ruby laser, pumping is achieved by using a xenon flash light. The lasing atoms are chromium ions (Cr^{3+}) in a crystal of alumina Al_2O_3 (saphire). The ends of the ruby crystal are silvered to reflect the stimulated radiation back and forward so that its intensity builds up, in much the same way we build up voltage oscillations in an electrical oscillator circuit. One of the mirrors is partially silvered to allow some of this radiation to be tapped out. What comes out is a highly coherent radiation with a high intensity. The coherency and the well-defined wavelength of this radiation are what make it distinctly different from a random stream of different-wavelength photons emitted from a tungsten bulb.

3.9.2   Helium–Neon Laser

With the Helium–Neon (HeNe) laser, the actual operation is not simple, since we need know such things as the energy states of the whole atom. We will therefore only consider the lasing emission at 632.8 nm, which gives the well-known red color to the laser light. The actual stimulated emission occurs from the Ne atoms; He atoms are used to excite the Ne atoms by atomic collisions.

Ne is an inert gas with a ground state ($1s^2 2s^2 2p^6$), which is represented as ($2p^6$) when the inner closed $1s$ and $2s$ subshells are ignored. If one of the electrons from the $2p$ orbital is excited to a $5s$ orbital, the excited configuration ($2p^5 5s^1$) is a state of the Ne atom that has higher energy. Similarly, He is an inert gas with the ground-state configuration of ($1s^2$). The state of He when one electron is excited to a $2s$ orbital can be represented as ($1s^1 2s^1$), which has higher energy.

The HeNe laser consists of a gaseous mixture of He and Ne atoms in a gas discharge tube, as shown schematically in Figure 3.39. The ends of the tube are mirrored to reflect the stimulated radiation and to build up the intensity within the cavity. If sufficient dc high voltage is used, electrical discharge is obtained within the tube, causing the He atoms to become excited by collisions with the drifting electrons. Thus,

$$He + e^- \rightarrow He* + e^-$$

where He* is an excited He atom.

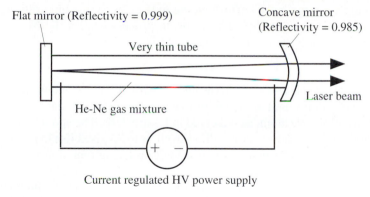

Figure 3.39   Schematic illustration of the HeNe laser.

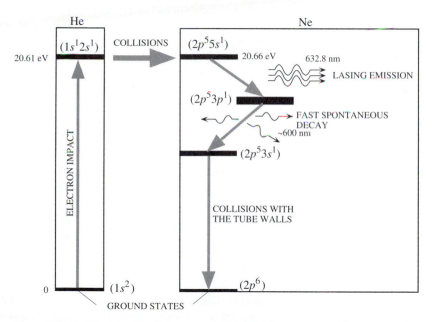

Figure 3.40   The principle of operation of the HeNe laser.
HeNe laser energy levels (for 632.8 nm emission).

The excitation of the He atom by an electron collision puts the second electron in He into a $2s$ state and changes its spin, so the excited He atom, He*, has the configuration $(1s^1 2s^1)$ with parallel spins. This atom is metastable (long lasting) with respect to the $(1s^2)$ state, as shown schematically in Figure 3.40. He* cannot spontaneously emit a photon and decay down to the $(1s^2)$ ground state because $\Delta \ell$ must be ± 1. Thus, a large number of He* atoms build up during the electrical discharge.

When an excited He atom collides with a Ne atom, it transfers its energy to the Ne atom by resonance energy exchange. This happens because, by good fortune,

Ne has an empty energy level, corresponding to the $(2p^55s^1)$ configuration, which matches that of $(1s^12s^1)$ of He*. The collision process excites the Ne atom and de-excites He* down to its ground energy, that is,

$$He* + Ne \rightarrow He + Ne*$$

With many He*–Ne collisions in the gaseous discharge, we end up with a large number of Ne* atoms and a population inversion between the $(2p^55s^1)$ and $(2p^53p^1)$ states of the Ne atom, as indicated in Figure 3.40. The spontaneous emission of a photon from one Ne* atom falling from $5s$ to $3p$ gives rise to an avalanche of stimulated emission processes, which leads to a lasing emission with a wavelength of 632.8 nm, in the red.

There are a few interesting facts about the HeNe laser, some of which are quite subtle. First, the $(2p^55s^1)$ and $(2p^53p^1)$ electronic configurations of the Ne atom actually have a spread of energies. For example for Ne$(2p^55s^1)$, there are four closely-spaced energy levels. Similarly, for Ne$(2p^53p^1)$, there are ten closely-separated energies. We can therefore achieve population inversion with respect to a number of energy levels. As a result, the lasing emissions from the HeNe laser contain a variety of wavelengths. The two lasing emissions in the visible spectrum, at 632.8 nm and 543 nm, can be used to build a red or green HeNe laser. Further, we should note that the energy of the Ne$(2p^54p^1)$ state (not shown) is above that of Ne$(2p^53p^1)$ but below that of Ne$(2p^55s^1)$. Consequently, there will also be stimulated transitions from Ne$(2p^55s^1)$ to Ne$(2p^54p^1)$, and hence a lasing emission at a wavelength of ~3.39 μm (IR). To suppress lasing emissions at the unwanted wavelengths (e.g., the infrared) and to obtain lasing only at the wavelength of interest, we can make the reflecting mirrors wavelength selective. This way the optical cavity builds up optical oscillations at the selected wavelength.

From $(2p^53p^1)$ energy levels, the Ne atoms decay rapidly to the $(2p^53s^1)$ energy levels by spontaneous emission. Most of the Ne atoms with the $(2p^53s^1)$ configuration, however, cannot simply return to the ground state $2p^6$, because the return of the electron in $3s$ requires that its spin be flipped to close the $2p$ subshell. An electromagnetic radiation cannot change the electron spin. Thus, the Ne$(2p^53s^1)$ energy levels are metastable. The only possible means of returning to the ground state (and for the next repumping act) is collisions with the walls of the laser tube. Therefore, we cannot increase the power obtainable from a HeNe laser simply by increasing the laser tube diameter, because that will accumulate more Ne atoms at the metastable $(2p^53s^1)$ states.

A typical HeNe laser, illustrated in Figure 3.39, consists of a narrow glass tube that contains the He and Ne gas mixture (typically, the He to Ne ratio is 5 : 1). The lasing emission intensity increases with tube length, since more Ne atoms are then used in stimulated emission. The intensity decreases with increasing tube diameter, since Ne atoms in the $(2p^53s^1)$ states can only return to the ground state by collisions with the walls of the tube. The ends of the tube are generally sealed with a flat mirror (99.9 percent reflecting) at one end and, for easy alignment, a concave mirror (98.5 percent reflecting) at the other end, to obtain an optical cavity within the tube. The outer surface of the concave mirror is ground to behave like a convergent lens, to compensate for the divergence in the beam arising from reflections from the concave mirror. The output radiation from the tube is typically a beam of diameter 0.5–2 mm and a divergence of 1 milliradians at a power of few milliwatts. In high-power HeNe lasers, the mirrors are external to the tube. In addition, Brewster windows are fused at the ends of the laser tube, to allow only polarized light to be transmitted and amplified within the cavity, so that the output radiation is polarized (that is, has electric field oscillations in one plane).

EFFICIENCY OF THE HeNe LASER   A typical low-power 2.5 mW HeNe laser tube operates at a dc voltage of 2 kV and carries a current of 5 mA. What is the efficiency of the laser?   **Example 3.20**

Solution

From the definition of efficiency,

$$Efficiency = \frac{(Output\ power)}{(Input\ power)}$$

$$= \frac{(2.5 \times 10^{-3}\ W)}{(5 \times 10^{-3}\ A)(2000\ V)} = 0.00025\ or\ 0.025\%$$

3.9.3   Laser Output Spectrum

The output radiation from a laser is not actually at one single well-defined wavelength corresponding to the lasing transition. Instead, the output covers a spectrum of wavelengths with a central peak. This is not a simple consequence of the Heisenberg Uncertainty Principle (which does broaden the output). Predominantly, it is a result of the broadening of the emitted spectrum by the **Doppler effect.** We recall from the kinetic molecular theory that gas atoms are in random motion, with an average translational kinetic energy of $(3/2)kT$. Suppose that these gas atoms emit radiation of frequency v_0 which we label as the source frequency. Then, due to the Doppler effect, when a gas atom moves toward an observer, the latter detects a higher frequency v_2, given by

$$v_2 = v_0\left(1 + \frac{v_x}{c}\right)$$

where v_x is the relative velocity of the atom with respect to the observer and c is the speed of light. When the atom moves away, the observer detects a smaller frequency, which corresponds to

$$v_2 = v_0\left(1 - \frac{v_x}{c}\right)$$

Since the atoms are in random motion, the observer will detect a range of frequencies, due to this Doppler effect. As a result, the frequency or wavelength of the output radiation from a gas (or a solid) laser will have a "linewidth" of $\Delta v = v_2 - v_1$, called a Doppler-broadened **linewidth** of a laser radiation. Other mechanisms also broaden the output spectrum, but we will ignore these at present.

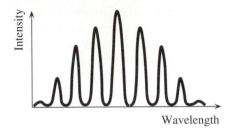

Figure 3.41   Typical output spectrum (intensity versus wavelength) for a laser

The reflections from the laser end mirrors give rise to traveling waves in opposite directions within the cavity. Since the waves are in phase, they interfere constructively, to set up a standing wave—in other words, stationary oscillations. Some of the energy in this wave is tapped by the 99 percent reflecting mirror to get an output, in much the same way that we tap the energy from an oscillating field in an LC circuit by attaching an antenna to it.

Only standing waves with certain wavelengths can be maintained within the optical cavity, just as only certain acoustic wavelengths can be obtained from musical instruments. Any standing wave in the cavity must have a half-wavelength $\lambda/2$ that fits into the cavity length L, or

$$n\left(\frac{\lambda}{2}\right) = L \qquad \text{[3.47]}$$

where n is an integer called the **mode number** of the standing wave. Each possible standing wave within the laser tube (cavity) satisfying Equation 3.47 is called a **cavity mode.** The laser output thus has a broad spectrum with peaks at certain wavelengths corresponding to various cavity modes existing within the Doppler-broadened emission curve. Figure 3.41 shows the expected output from a typical gas laser. At wavelengths satisfying Equation 3.47 that is, representing certain cavity modes, we have intensity spikes in the output. The net envelope of the output radiation is a Gaussian distribution, which is essentially due to the Doppler-broadened linewidth.

Even though we can try to get as parallel a beam as possible by lining the mirrors up perfectly, we will still be faced with diffraction effects at the output. When the output laser beam hits the end of the laser tube, it becomes diffracted, so the emerging beam is necessarily divergent. Simple diffraction theory can readily predict the divergence angle.

Example 3.21 | **DOPPLER-BROADENED LINEWIDTH**   Calculate the Doppler-broadened linewidths Δv and $\Delta \lambda$ for the HeNe laser transition $\lambda = 632.8$ nm, if the gas discharge temperature is about 127 °C. The atomic mass of Ne is 20.2 g mole^{-1}.

Solution

Due to the Doppler effect, the laser radiation from gas lasers is broadened around a central frequency v_o, which corresponds to the source frequency. The highest frequency detected will be due to the radiation emitted from atoms moving directly toward the observer, and the smallest frequency, detected will be result of emissions from atoms moving directly away from the observer. Therefore, the maximum width of the observed frequencies will be

$$\Delta v = v_o\left(1 + \frac{v_x}{c}\right) - v_o\left(1 - \frac{v_x}{c}\right) = \frac{2v_o v_x}{c}$$

From $\lambda = c/v$ we obtain the following by differentiation:

$$\frac{d\lambda}{dv} = -\frac{c}{v^2} = -\frac{\lambda}{v} = -\frac{\lambda^2}{c}$$

We need to know v_x, which is given by kinetic theory as $v_x^2 = kT/m$. For the HeNe laser, the Ne atoms lase, so

$$m = \frac{(20.2 \times 10^{-3} \text{ kg mol}^{-1})}{(6.023 \times 10^{23} \text{ mol}^{-1})} = 3.35 \times 10^{-26} \text{ kg}$$

Thus

$$v_x = \left[\frac{(1.38 \times 10^{-23} \text{ J K}^{-1})(127 + 273 \text{ K})}{(3.35 \times 10^{-26} \text{ kg})}\right]^{1/2} = 406 \text{ m s}^{-1}$$

The central frequency is

$$v_o = \frac{c}{\lambda_o} = \frac{(3 \times 10^8 \text{ m s}^{-1})}{(632.8 \times 10^{-9} \text{ m})} = 4.74 \times 10^{14} \text{ s}^{-1}$$

The frequency linewidth is

$$\Delta v_o = \frac{(2v_o v_x)}{c} = \frac{2(4.74 \times 10^{14} \text{ s}^{-1})(406 \text{ m s}^{-1})}{(3 \times 10^8 \text{ m s}^{-1})} = 1.285 \text{ GHz}$$

To get $\Delta\lambda$, we use $d\lambda/dv = -\lambda/v$, so that

$$\Delta\lambda = \Delta v\left|-\frac{\lambda}{v}\right| = \frac{(1.285 \times 10^9 \text{ Hz})(632.8 \times 10^{-9} \text{ m})}{(4.74 \times 10^{14} \text{ s}^{-1})}$$

$$= 1.71 \times 10^{-12} \text{ m} \qquad \text{or} \qquad 0.0017 \text{ nm}$$

ADDITIONAL TOPICS

3.10   TIME-DEPENDENT SCHRÖDINGER EQUATION

Under steady-state conditions, when the potential energy of a particle is independent of time, the wavefunction $\Psi(x, t)$ of the particle in one dimension can be written as

$$\Psi(x, t) = \psi(x) \exp\left(-\frac{jEt}{\hbar}\right)$$

where the time dependence has been separated out as a harmonic function. The spatial dependence $\psi(x)$ is then found by solving the time-independent Schrödinger equation. The electron wavefunctions $\psi_{n,\ell,m_\ell}(r, \theta, \phi)$ for the hydrogen atom are steady-state wavefunctions, or stationary states, because we assumed that

$$V = \frac{-e^2}{4\pi \varepsilon_o r}$$

was time independent. Suppose that for a certain duration of time, the electron interacts with an incoming electromagnetic (EM) radiation. During the interaction, the potential energy of the electron is time dependent inasmuch as the electric field oscillations in the radiation impose an oscillating force on the electron. The behavior of the electron cannot then be found by solving the time-independent Schrödinger equation; hence, no single wavefunction of the type $\psi_{n,\ell,m_\ell}(r, \theta, \phi)$ can describe the behavior during the interaction.

When the particle is in an environment where its *PE* has a time dependence, that is, $V = V(x, t)$, the general equation that describes the behavior of the particle through the wavefunction $\Psi(x, t)$ is the **time-dependent Schrödinger equation,** which is

Time-dependent
Schrödinger
equation

$$-\frac{\hbar^2}{2m} \frac{\partial^2 \Psi(x, t)}{\partial x^2} + V(x, t)\Psi(x, t) = j\hbar \frac{\partial \Psi(x, t)}{\partial t} \qquad \text{[3.48]}$$

This is the fundamental equation that we must solve to find the behavior of the particle, given its *PE* function, $V(x, t)$. This is not a trivial problem. Many scientists have devoted a lifetime formulating suitable methods for solving Equation 3.48 under various conditions. For example, with the appropriate *PE* and with a number of simplifying assumptions, Equation 3.48 can be solved to infer what happens when the electron in a hydrogenic atom interacts for a short duration with an incoming EM radiation. The result is that the transitions of the electron from one state to another due an absorption of radiation require certain conditions, such as $\Delta \ell = \pm 1$, and $\Delta m_\ell = 0, \pm 1$, which are called **selection rules.**

The time-independent Schrödinger equation is a special case of Equation 3.48. To verify this, suppose that $V(x, t) = V(x)$ and that we can separate out the time dependence and write

$$\Psi(x, t) = \psi(x) \exp\left(-\frac{jEt}{\hbar}\right)$$

Does the latter satisfy Equation 3.48? Substituting this expression for $\Psi(x, t)$ into Equation 3.48, we get

$$-\frac{\hbar^2}{2m} \exp\left(-\frac{jEt}{h}\right)\left[\frac{d^2\psi(x)}{dx}\right] + V(x)\psi(x) \exp\left(-\frac{jEt}{h}\right) = j\hbar\psi(x)\frac{d}{dt}\left[\exp\left(-\frac{jEt}{\hbar}\right)\right]$$

$$= E\psi(x) \exp\left(-\frac{jEt}{\hbar}\right)$$

After canceling out the exponential terms and collecting the terms with V and E, we find

$$\frac{\hbar^2}{2m} \frac{d^2 \psi(x)}{dx^2} + [E - V(x)]\psi(x) = 0 \qquad \textbf{[3.49]}$$

This is the time-independent Schrödinger equation that we used for determining the stationary states of a particle.

IMPORTANT TERMS

Angular momentum L about a point O is defined as $\mathbf{L} = \mathbf{p} \times \mathbf{r}$, where $\mathbf{p}$ is the linear momentum and $\mathbf{r}$ is the position vector of the body from O. For a circular orbit around O, the angular momentum is orbital and $L = pr = mvr$.

Complementarity principle suggests that the wave model and the particle model are complementary models in that one model alone cannot be used to explain all the observations in nature. For example, the electron diffraction phenomenon is best explained by the wave model, whereas in the Compton experiment, the electron is treated as a particle; that is, it is deflected by an impinging photon that imparts an additional momentum to the electron.

Compton effect is the scattering of a high-energy photon by a "free" electron. The effect is experimentally observed when an X-ray beam is scattered from a target that contains many conduction ("free") electrons, such as a metal or graphite.

De Broglie relationship relates the wave-like properties (e.g., wavelength λ) of matter to its particle-like properties (e.g., momentum p) via $\lambda = h/p$.

Doppler effect is the change in the measured frequency of a wave due to the motion of the source relative to the observer. In the case of electromagnetic radiation, if v is the relative velocity of the source object toward the observer and v_o is the source frequency, then the measured electromagnetic wave frequency is $v = v_o[1 + (v/c)]$ for $(v/c) \ll 1$.

Energy density ρ_E is the amount of energy per unit volume. In region where the electric field is $\mathcal{E}$, the energy stored per unit volume is $(1/2)\varepsilon_o \mathcal{E}^2$.

Flux is a term used to describe the rate of flow through a unit area. If ΔN is the number of particles flowing through an area A in time Δt, then particle flux Γ is defined as $\Gamma = \Delta N/(A \Delta t)$. If an amount of energy ΔE flows through an area A in time Δt, energy flux is $\Gamma_E = \Delta E/(A \Delta t)$, which defines the intensity ($\mathcal{I}$) of an electromagnetic wave.

Ground state is the state of the electron with the lowest energy.

Heisenberg's uncertainty principle states that the uncertainty Δx in the position of a particle and the uncertainty Δp_x in its momentum in the x direction obey $(\Delta x)(\Delta p_x) \gtrsim \hbar$. This is a consequence of the wave nature of matter and has nothing to do with the precision of measurement. If ΔE is the uncertainty in the energy of a particle during a time Δt, then according to the Uncertainty Principle, $(\Delta E)(\Delta t) \gtrsim \hbar$. To measure the energy of a particle without any uncertainty means that we would need an infinitely long time $\Delta t \to \infty$.

Hund's rule states that electrons in a given subshell $n\ell$ try to occupy separate orbitals (different m_ℓ) and keep their spins parallel (same m_s). In doing so, they achieve a lower energy than pairing their spins (different m_s) and occupying the same orbital (same m_ℓ).

Intensity ($\mathcal{I}$) is the flow of energy per unit area per unit time. It is equal to an energy flux.

LASER (light amplification by stimulated emission of radiation) is a device within which photon multiplication by stimulated emission produces an output radiation that is nearly monochromatic and coherent (vis-á-vis an incoherent stream of photons from a tungsten light bulb). Furthermore, the output beam has very little divergence.

Magnetic quantum number m_ℓ specifies the component of the orbital angular momentum L_z in the direction of a magnetic field along z so that $L_z = \pm\hbar m_\ell$, where m_ℓ can be a negative or positive integer from $-\ell$ to $+\ell$ including 0, that is, $-\ell$, $-(\ell - 1), \ldots, 0, \ldots, (\ell - 1), \ell$. The orbital ψ of the electron depends on m_ℓ, as well as on n and ℓ. The m_ℓ however, generally determines the angular variation of ψ.

Orbital is a region of space in an atom or molecule where an electron with a given energy may be found. Two electrons with opposite spins can occupy the same orbital. An orbit is a well-defined path for an electron, but it cannot be used to describe the whereabouts of the electron in an atom or molecule, because the electron has a probability distribution. The wavefunction $\psi_{n\ell m_\ell}(r, \theta, \phi)$ is often referred to as an orbital that represents the spatial distribution of the electron, since $|\psi_{n\ell m_\ell}(r, \theta, \phi)|^2$ is the probability of finding the electron per unit volume at (r, θ, ϕ).

Orbital (angular momentum) quantum number specifies the magnitude of the orbital angular momentum of the electron via $L = \hbar\sqrt{[\ell(\ell + 1)]}$, where ℓ is the orbital quantum number with values $0, 1, 2, 3, \ldots, n - 1$. The ℓ values 0, 1, 2, 3 are labelled the s, p, d, f states.

Orbital wavefunction describes the spatial dependence of the electron, not its spin. It is $\psi(r, \theta, \phi)$, which depends on n, ℓ, and m_ℓ, with the spin dependence m_s excluded. Generally, $\psi(r, \theta, \phi)$ is simply called an orbital.

Pauli exclusion principle requires that no two electrons in a given system may have the same set of quantum numbers, n, ℓ, m_ℓ, m_s. In other words, no two electrons can occupy a given state $\psi(n, \ell, m_\ell, m_s)$. Equivalently, up to two electrons with opposite spins can occupy a given orbital $\psi(n, \ell, m_\ell)$.

Photoelectric effect is the emission of electrons from a metal upon illumination with a frequency of light above a critical value which depends on the material. The kinetic energy of the emitted electron is independent of the light intensity and dependent on the light frequency v, via $KE = hv - \Phi$ where h is Planck's constant and Φ is a material-related constant called the **work function.**

Photon is a quantum of energy hv (where h is Planck's constant and v is the frequency) associated with electromagnetic radiation. A photon has a zero rest mass and a momentum p given by the de Broglie relationship $p = h/\lambda$, where λ is the wavelength. A photon does have a "moving mass" of hv/c^2, so it experiences gravitational attraction from other masses. For example, light from a star gets deflected as it passes by the sun.

Population inversion is the phenomenon of having more atoms occupy an excited energy level E_2, higher than a lower energy level, E_1, which means that the normal equilibrium distribution is reversed; that is, $N(E_2) > N(E_1)$. Population inversion occurs temporarily as a result of the excitation of a medium (pumping). If left on its own, the medium will eventually return to its equilibrium population distribution, with more atoms at E_1 than at E_2. For gas atoms, this means $N(E_2)/N(E_1) \approx \exp[-(E_2 - E_1)/kT]$.

Principal quantum number n is an integer quantum number with values 1, 2, 3, . . . that characterizes the total energy of an electron in an atom. The energy increases with n. With the other quantum numbers ℓ and m_ℓ, n determines the orbital of the electron in an atom, or $\psi_{n\ell m_\ell}(r, \theta, \phi)$. The values $n = 1, 2, 3, 4, \ldots$ are labelled the $K, L, M, N, \ldots$ shells, within each of which there may be subshells based on $\ell = 0, 1, 2, \ldots (n - 1)$ and corresponding to the $s, p, d, \ldots$ states.

Pumping means exciting atoms from their ground states to higher-energy states.

Schrödinger equation is a fundamental equation in nature, the solution of which describes the wave-like behavior of a particle. The equation cannot be derived from a more fundamental law. Its validity is based on its ability to predict any known physical phenomena. The solution requires as input the potential energy function $V(x, y, z, t)$ of the particle and the boundary and initial conditions. The *PE* function $V(x, y, z, t)$ describes the interaction of the particle with its environment. The time-independent Schrödinger equation describes the wave behavior of a particle under steady-state conditions, that is, when the *PE* is time-independent $V(x, y, z)$. If E is the total energy and $\nabla^2 = (\partial^2/\partial x^2 + \partial^2/\partial x^2 + \partial^2/\partial x^2)$, then

$$\nabla^2\psi - \left(\frac{2m}{\hbar^2}\right)[E - V(x, y, z)]\psi = 0$$

The solution of the time-independent Schrödinger equation gives the wavefunction $\psi(x, y, z)$ of the electron and its energy E. The interpretation of the wavefunction $\psi(x, y, z)$ is that $|\psi(x, y, z)|^2$ is the probability of finding the electron per unit volume at point $x, y\ z$.

Selection rules determine what values of ℓ and m_ℓ are allowed for an electron transition involving the emission and absorption of electromagnetic radiation, that is, a photon. In summary, $\Delta\ell = \pm 1$ and $\Delta m_\ell = 0, \pm 1$. The spin number m_s of the electron remains unchanged. Within an atom, the transition of the electron from one state $\psi(n, \ell, m_\ell, m_s)$ to another $\psi(n', \ell', m'_\ell, m'_s)$, due to collisions with other atoms or electrons, does not necessarily obey the selection rules.

Spin of an electron S is its intrinsic angular momentum (analogous to the spin of Earth around its own axis), which is space quantized to have two possibilities. The magnitude of the electron's spin is a constant, $\hbar\sqrt{3}/2$, but its component along a magnetic field in the z direction is $m_s\hbar$, where m_s is the spin magnetic quantum number, which is $+1/2$ or $-1/2$.

Spontaneous emission is the phenomenon in which a photon is emitted when an electron in a high-energy state $\psi(n, \ell, m_\ell, m_s)$ with energy E_2 spontaneously falls down to a lower, unoccupied energy

state $\psi(n', \ell', m'_\ell, m'_s)$ with energy E_1. The photon energy is $h\upsilon = (E_2 - E_1)$. Since the emitted photon has an angular momentum, the orbital quantum number ℓ of the electron must change, that is $\Delta\ell = \ell' - \ell = \pm 1$.

State is a possible wavefunction for the electron that defines its spatial (orbital) and spin properties. For example, $\psi(n, \ell, m_\ell, m_s)$ is a state of the electron. From the Schrödinger equation, each state corresponds to a certain electron energy E. We use the terms state of energy E, or *energy state*. There is generally more than one state ψ with the same energy E.

Stimulated emission is the phenomenon in which an incoming photon of energy $h\upsilon = E_2 - E_1$ interacts with an electron in a high-energy state $\psi(n, \ell, m_\ell, m_s)$ at E_2, and induces that electron to oscillate down to a lower, unoccupied energy state, $\psi(n', \ell', m'_\ell, m'_s)$ at E_1. The photon emitted by stimulation has the same energy and phase as the incoming photon, and it moves in the same direction. Consequently, stimulated emission results in two coherent photons, with the same energy, traveling in the same direction. The stimulated emission process must obey the selection rule $\Delta\ell = \ell' - \ell = \pm 1$, just as spontaneous emission must.

Tunneling is the penetration of an electron through a potential energy barrier by virtue of the electron's wave-like behavior. In classical mechanics, if the energy E of the electron is less than the *PE* barrier V_o, the electron cannot cross the barrier. In quantum mechanics, there is a distinct probability that the electron will "tunnel" through the barrier to appear on the other side. The probability of tunneling depends very strongly on the height and width of the *PE* barrier.

Wave is a periodically occurring disturbance, such as the displacement of atoms from their equilibrium positions in a solid carrying sound waves, or a periodic variation in a measurable quantity, such as the electric field $\mathcal{E}(x, t)$ in a medium or space. In a traveling wave, energy is transferred from one location to another by the oscillations. For example,

$\mathcal{E}_y(x, t) = \mathcal{E}_{yo} \sin(kx - \omega t)$ is a traveling wave in the x direction, where $k = 2\pi/\lambda$ and $\omega = 2\pi\upsilon$. The electric field in the y direction varies periodically along x, with a period λ called the wavelength. The field also varies with time, with a period $1/\upsilon$, where υ is the frequency. The wave propagates along the x direction with a velocity of propagation c. Electromagnetic waves are transverse waves in which the electric and magnetic fields $\mathcal{E}_y(x, t)$ and $B_z(x, t)$ are at right angles to each other, as well as to the direction of propagation x. A traveling wave in the electric field must be accompanied by a similar traveling wave in the associated magnetic field $B_z(x, t) = B_{zo} \sin(kx - \omega t)$. Typical wave-like properties are interference and diffraction.

Wave equation is a general partial differential equation in classical physics, of the form

$$v^2 \frac{\partial^2 u}{\partial x^2} - \frac{\partial^2 u}{\partial t^2} = 0$$

the solution of which describes the space and time dependence of the displacement $u(x, t)$ from equilibrium or zero, given the boundary conditions. The parameter v in the wave equation is the propagation velocity of the wave. In the case of electromagnetic waves in vacuum, the wave equation describes the variation of the electric (or magnetic) field $\mathcal{E}(x, t)$ with space and time, $(c^2 \partial^2 \mathcal{E}/\partial x^2) - (\partial^2 \mathcal{E}/t^2) = 0$, where c is the speed of light.

Wavefunction $\Psi(x, y, z, t)$ is a probability-based function used to describe the wave-like properties of a particle. It is obtained by solving the Schrödinger equation, which in turn requires a knowledge of the PE of the particle and the boundary and initial conditions. The term $|\Psi(x, y, z, t)|^2$ is the probability per unit volume of finding the electron at (x, y, z) at time t. In other words, $|\Psi(x, y, z, t)|^2 dxdydz$ is the probability of finding the electron in the small volume $dxdydz$ at (x, y, z) at time t. Under steady-state conditions, the wavefunction can be separated into a space-dependent component and a time-dependent component i.e., $\Psi(x, y, z, t) = \psi(z, y, z)\exp(-jEt/\hbar)$, where E is the energy of the particle and $\hbar = h/2\pi$. The spatial component, $\psi(x, y, z)$ satisfies the time-independent Schrödinger equation.

Wavenumber (or wavevector) k is the number of waves per 2π of length, that is, $k = 2\pi/\lambda$.

QUESTIONS AND PROBLEMS

3.1 Photons and photon flux

a. Consider a 1 kW AM radio transmitter at 700 kHz. Calculate the number of photons emitted from the antenna per second.

b. The average intensity of sunlight on Earth's surface is about 1 kW m^{-2}. The maximum intensity is at a wavelength around 800 nm Assuming that all the photons have an 800 nm wavelength, calculate the number of photons arriving on Earth's surface per unit time per unit area. What is the magnitude of the electric field in the sunlight?

c. Suppose that a solar cell device can convert each sunlight photon into an electron, which can then give rise to an external current. What is the maximum current that can be supplied per unit area (m^2) of this solar cell device?

3.2 **The human eye** Photons passing through the pupil are focused by the lens onto the retina and are detected by two types of photosensitive cells, called rods and cones. Rods are highly sensitive photoreceptors with a peak response at the wavelength 510 nm. They do not register color, but they are responsible for our vision under dimmed light conditions, which is termed **scotopic vision.** Cones are color sensitive and are responsible for our daytime vision, called **photopic vision.** There are three types of cone photoreceptors, which are sensitive to the blue, green, and red wavelengths 430 nm, 535 nm, and 575 nm, respectively. All three cones have an overall peak response of 555 nm.

a. Calculate the photon energy (in eV) for the peak responsivity of each photoreceptor in the eye.

b. The fovea is a retina region on the visual axis; images are focused onto this region. The density of the cones in the fovea is on the order of 150,000/mm². Below a light intensity of about 100 μW m^{-2}, cones are not functional and rods take over the vision.

    1. What is the minimum photon flux for color vision?

    2. If a visual sensation persists for a time (1/5)th of a second, how many photons does the eye need per cone for a visual color sensation?

    3. If the eye is 10 percent efficient overall, due to photon reflections, etc., how many photons are actually absorbed per cone to generate a colored visual sensation?

3.3 **The photoelectric effect** A photoelectric experiment indicates that green light of wavelength 5220 Å is the longest wavelength radiation that can cause the photoemission of electrons from a clean sodium surface.

a. What is the work function of sodium, in eV?

b. If a UV radiation of wavelength 2500 Å is incident upon the sodium surface, what will be the kinetic energy of the photoemitted electrons, in eV?

c. Given that the UV light of wavelength 2500 Å has an intensity of 20 mW/cm², if the emitted electrons are collected by applying a positive bias to the opposite electrode, what will be the photoelectric current density?

3.4 **The photoelectric effect** Consider the following metals and their work functions. For each metal, calculate the longest wavelength photon for photoemission. Which metal gives photoemission in the visible spectrum?

Metal	Al	Be	Cu	Cs	In	Rb	Sr
Work function (eV)	4.25	3.92	4.4	1.81	3.8	2.16	2.35

3.5   Diffraction by X-rays and an electron beam Diffraction studies on a polycrystalline Al sample using X-rays gives the smallest diffraction angle (2θ) of 29.5 degree corresponding to diffraction from the (111) planes. The lattice parameter a of Al (FCC), is 0.405 nm. If we wish to obtain the same diffraction pattern (same angle) using an electron beam, what should be the voltage needed to accelerate the electron beam? Note that the interplanar separation d for planes (h, k, ℓ) and the lattice parameter a for cubic crystals are related by $d = a/[h^2 + k^2 + \ell^2]^{1/2}$.

3.6   Heisenberg's uncertainty principle Show that if the uncertainty in the position of a particle is on the order of its de Broglie wavelength, then the uncertainty in its momentum is about the same as the momentum value itself.

3.7   Heisenberg's uncertainty principle An exicited electron in an Na atom emits radiation at a wavelength 589 nm and returns to the ground state. If the mean time for the transition is about 20 ns, calculate the inherent width in the emission line. What is the length of the photon emitted?

3.8   Tunneling

a.   Consider the phenomenon of tunneling through a potential energy barrier of height V_o and width a, as shown in Figure 3.16. What is the probability that the electron will be reflected? Given the transmission coefficient T, can you find the reflection coefficient R? What happens to R as a or V_o or both become very large?

b.   For a wide barrier ($\alpha a \gg 1$), show that T_o can at most be 4 and that $T_o = 4$ when $E = (1/2)V_o$.

3.9   Electron impact excitation

a.   A projectile electron of kinetic energy 12.2 eV collides with a hydrogen atom in a gas discharge tube. Find the n-th energy level to which the electron in the hydrogen atom gets excited.

b.   Calculate the possible wavelengths of radiation (in nm) that will be emitted from the excited H atom in the previous example as the electron returns to its ground state. Which one of these wavelengths will be in the visible spectrum?

c.   In neon street lighting tubes, gaseous discharge in the Ne tube involves electrons accelerated by the electric field impacting Ne atoms and exciting some of them of the $2p^5 3p^1$ states, as shown in Figure 3.40. What is the wavelength of emission? Can the Ne atom fall from the $2p^5 3p^1$ state to the ground state by spontaneous emission?

3.10   Line spectra of hydrogenic atoms Spectra of hydrogen-like atoms are classified in terms of electron transitions to a common lower energy level.

a.   All transitions from energy levels $n = 2, 3, \ldots$ to $n = 1$ (the K shell) are labelled K lines and constitute the **Lyman series.** The spectral line corresponding to the smallest energy difference ($n = 2$ to $n = 1$) is labelled the K_α line, next is labelled K_β, and so on. The transition from $n = \infty$ to $n = 1$ has the largest energy difference and defines the greatest photon energy (shortest wavelength) in the K series; hence it is called the absorption edge K_{ae}. What is the range of wavelengths for the K lines? What is K_{ae}? Where are these lines with respect to the visible spectrum?

b. All transitions from energy levels $n = 3, 4, \ldots$ to $n = 2$ (L shell) are labelled L lines and constitute the **Balmer series.** What is the range of wavelengths for the L lines (i.e., L_α and L_{ae})? Are these in the visible range?

c. All transitions from energy levels $n = 4, 5,$ to $n = 3$ (M shell) are labelled M lines and constitute the **Paschen series.** What is the range of wavelengths for the M lines? Are these in the visible range?

d. How would you expect the spectral lines to depend on the atomic number Z?

*3.11 X-rays and the Moseley relation X-rays are photons with wavelengths in the range 0.01 nm–10 nm, with typical energies in the range 100 eV to 100 keV. When an electron transition occurs in an atom from the L to the K shell, the emitted radiation is generally in the X-ray spectrum. For all atoms with atomic number $Z > 2$, the K shell is full. Suppose that one of the electrons in the K shell has been knocked out by an energetic projectile electron impacting the atom (the projectile electron would have been accelerated by a large voltage difference). The resulting vacancy in the K shell can then be filled by an electron in the L shell transiting down and emitting a photon. The emission resulting from the L to K shell transition is labelled the K_α line. The following table shows the K_α line data obtained for various materials.

Material	Mg	Al	S	Ca	Cr	Fe	Cu	Rb	W
Z	12	13	16	20	24	26	29	37	74
K_α line (nm)	0.987	0.834	0.537	0.335	0.229	0.194	0.154	0.093	0.021

a. If v is the frequency of emission, plot $v^{1/2}$ against the atomic number Z of the element.

b. H. G. Moseley, while still a graduate student of E. Rutherford in 1913, found the empirical relationship

$$v^{1/2} = B(Z - C)$$

where B and C are constants. What are B and C from the plot? Can you give a simple explanation as to why K_α absorption should follow this relationship?

3.12   **The He atom** Suppose that for the He atom, zero energy is taken to be the two electrons stationary at infinity (and infinitely apart) from the nucleus (He^{++}). Estimate the energy (in eV) of the electrons in the He atom by neglecting the electron–electron repulsion, that is, neglecting the potential energy due to the mutual coulombic repulsion between the electrons. How does this compare with the experimental value of -79 eV? How strong is the electron–electron repulsion energy?

3.13   **Hund's rule** For each of the following atoms and ions, sketch the electronic structure, using a box for an orbital wavefunction and an arrow (up or down) for an electron: (a) Manganese, $[Ar]3d^5 4s^2$; (b) Cobalt, $[Ar]3d^7 4s^2$; (c) Iron ion, Fe^{+2}, given that Fe is $[Ar]3d^6 4s^2$; (d) Neodymium ion, Nd^{+3}, given that Nd is $[Xe]4f^4 6s^2$.

3.14   **The HeNe Laser** A particular HeNe laser operating at 632.8 nm has a tube that is 40 cm long. The operating temperature is 130 °C

    a.   Calculate the Doppler-broadened linewidth $\Delta\lambda$ in the output spectrum.

    b.   What are the n values that satisfy the resonant cavity condition? How many modes are therefore allowed?

    c.   What is the separation in the frequencies of the modes? How does this change as the tube warms up during operation? Taking the linear expansion coefficient to be 10^{-6} K^{-1}, estimate the change in the mode frequency separation.

    d.   How can you increase the output intensity from the HeNe laser?

*3.15   **The Ar-ion laser** The argon-ion laser can provide powerful continuous wave (CW) visible coherent radiation of several watts. The laser operation is achieved as follows. The Ar atoms are ionized by electron collisions in a high-current electrical discharge. Further multiple collisions with electrons excite the argon ion Ar^+ to a group of $4p$ energy levels ~ 35 eV above the atomic ground state, as shown in Figure 3.42. Thus, a population inversion forms between the $4p$ levels and the $4s$ level, which is about 33.5 eV above the Ar atom ground level. Consequently, the stimulated radiation from the $4p$ levels down to the $4s$ level contains a series of wavelengths ranging from 351.1 nm   to   528.7 nm. Most of the power, however, is concentrated apporoximately equally in the 488 and 514.5 nm emissions. The Ar^+ ion at the lower laser level ($4s$) returns to its neutral atomic ground state via radiative decay, followed by recombination with an electron. The Ar atom is then ready for "pumping" again.

    a.   Calculate the energy drop in the excited Ar^+ ion when it is stimulated to emit the 514.5 nm radiation.

    b.   The Doppler-broadened linewidth Δv of the 514.5 nm radiation is about 3500 MHz. Δv is measured between half-power points in the output spectrum.

        1.   Calculate the Doppler-broadened width in the wavelength $\Delta\lambda$.

        2.   Estimate the operating temperature of the argon-ion gas, in °C.

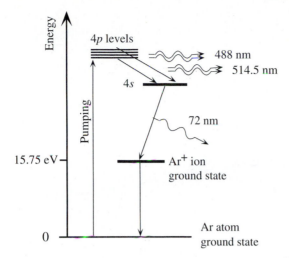

Figure 3.42   The Ar-ion laser energy diagram.

c.   In a particular argon-ion laser, the discharge tube, made of beryllia (beryllium oxide), is 30 cm long and has a bore 3 mm in diameter. When the laser is operated with a current of 40 A at 200 V dc, the total output power in the emitted radiation is 3 W. What is the efficiency of the laser?

4

Bonding, the Band Theory of Solids, and Statistics

One of the great successes of modern physics has been the application of quantum mechanics or the Schrödinger equation to the behavior of molecules and solids. For example, quantum mechanics explains the nature of the bond between atoms, and its consequences. How can carbon bond with four other carbon atoms? What determines the direction and strength of a bond? An intuitively obvious outcome from quantum mechanics is that the energy of the electron is still quantized in the molecule. In addition, the application of quantum mechanics to many atoms, as in a solid, leads to energy bands within which the electron energy levels are almost continuous. The electron energy falls within possible values in a band of energies. It is nearly impossible to comprehend the principles of operation of modern solid-state electronic devices without a good grasp of the band theory of solids. Since we are dealing with a large number of electrons in the solid, we must consider a statistical way of describing their behavior, just as we use the Maxwell distribution of velocities to explain the behavior of gas atoms. An equally important question, therefore, is "What is the probability that an electron is in a state with energy E within an energy band?"

4.1 HYDROGEN MOLECULE: MOLECULAR ORBITAL THEORY OF BONDING

Consider what happens when two hydrogen atoms approach each other to form the hydrogen molecule. This is the H–H (or H_2) system. Let us examine the energy levels of the H–H system as a function of the interatomic distance R. When the atoms are infinitely separated, each atom has its own set of energy levels, labeled

$1s$, $2s$, $2p$, etc. The electron energy in each atom is -13.6 eV with respect to the "free" state (electron infinitely separated from the parent nucleus). The energy of the two isolated hydrogen atoms is twice -13.6 eV.

As the atoms approach closer, the electrons interact both with each other and with the other nuclei. To obtain the wavefunctions and the new energy of the electrons, we need to find the new potential energy function *PE* for the electrons in this new environment and then solve the Schrödinger equation with this new *PE* function. The new energy is actually *lower* than twice -13.6 eV, which means that the H_2 formation is energetically favorable.

The bond formation between two H atoms can be easily explained by describing the behavior of the electron within the molecule. We use a **molecular orbital** ψ, which depends on the interaction of individual atomic wavefunctions and is regarded as an electron wavefunction within the molecule.

In the H_2 molecule, we cannot have two sets of identical atomic ψ_{1s} orbitals, for two reasons. First, this would violate the Pauli exclusion principle, which requires that, in a given system of electrons (those within the H_2 molecule), we cannot have two sets of identical quantum numbers. When the atoms were separated, we did not have this problem, because we had two isolated systems.

Second, as the two atoms approach each other, as shown in Figure 4.1, the atomic ψ_{1s} wavefunctions overlap. This overlap produces two new wavefunctions

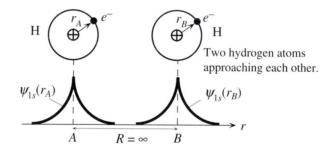

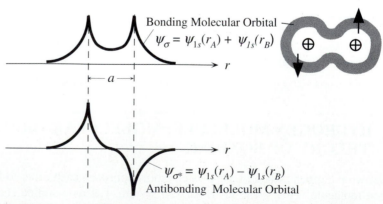

Figure 4.1   Formation of molecular orbitals, bonding, and antibonding (ψ_σ and ψ_{σ^*}) when two H atoms approach each other.

The two electrons pair their spins and occupy the bonding orbital ψ_σ.

with different energies and hence different quantum numbers. When the two atomic wavefunctions interfere, they can overlap either in phase (both positive, or both negative) or out of phase (one positive and the other negative), as a result of which two molecular orbitals are formed. These are conventionally labeled ψ_σ and $\psi_{\sigma*}$ as illustrated in Figure 4.1. Thus, two of the molecular orbitals in the H–H system are

$$\psi_\sigma = \psi_{1s}(r_A) + \psi_{1s}(r_B) \qquad [4.1]$$

$$\psi_{\sigma*} = \psi_{1s}(r_A) - \psi_{1s}(r_B) \qquad [4.2]$$

where the two hydrogen atoms are labeled A and B, and r_A and r_B are the respective distances of the electrons from their parent nucleus. In generating two separate molecular orbitals, ψ_σ and $\psi_{\sigma*}$, from a linear combination of two identical atomic orbitals, ψ_{1s}, we have used the **linear combination of atomic orbitals (LCAO)** method.

The first molecular orbital, ψ_σ, is symmetric and has considerable magnitude between the nuclei, whereas the second, $\psi_{\sigma*}$, is antisymmetric and has a node between the nuclei. The resulting electron probability distributions, $|\psi_\sigma|^2$ and $|\psi_{\sigma*}|^2$ are shown in Figure 4.2.

In an analogy to hydrogenic wavefunctions, since $\psi_{\sigma*}$ has a node, we would expect it to have a higher energy than the ψ_σ orbital and therefore a different energy quantum number, which means that the Pauli exclusion principle is no longer violated. We can also expect that because $|\psi_\sigma|^2$ has an appreciable electron concentration between the two nuclei, the electrostatic PE, and hence the total energy for the wavefunction ψ_σ, will be lower than that for $\psi_{\sigma*}$, as well as those for the individual atomic wavefunctions.

Of course, the true wavefunctions of the electrons in the H_2 system must be determined by solving the Schrödinger equation, but an intelligent guess is that these must look like ψ_σ and $\psi_{\sigma*}$. We can therefore use ψ_σ and $\psi_{\sigma*}$ in the Schrödinger equation, with the correct form of the PE term V, to evaluate the energies E_σ and

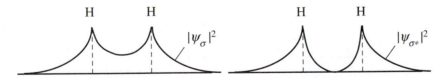

(a) Electron probability distributions for bonding and antibonding orbitals, ψ_σ and $\psi_{\sigma*}$.

(b) Lines representing contours of constant probability (darker lines represent greater relative probability).

Figure 4.2

$E_{\sigma*}$ of ψ_σ and $\psi_{\sigma*}$, respectively, as a function of R. The PE function V in the H–H system has positive PE contributions arising from electron–electron repulsions and proton–proton repulsions, but negative PE contributions arising from the attractions of the two electrons to the two protons.

The two energies, E_σ and $E_{\sigma*}$, are widely different, with E_σ below E_{1s} and $E_{\sigma*}$ above E_{1s}, as shown schematically in Figure 4.3a. As R decreases and the two H atoms get closer, the energy of the ψ_σ orbital state passes through a minimum at $R = a$. Each orbital state can hold two electrons with spins paired, and within the two hydrogen atoms, we have two electrons. If these enter the ψ_σ orbital and pair their spins, then this new configuration is energetically more favorable than two isolated H atoms. It corresponds to the hydrogen molecule H_2. The energy difference between that of the two isolated H atoms and the E_σ minimum energy at $R = a$ is the bonding energy, as illustrated in Figure 4.3a. When the two electrons in the H_2 molecule occupy the ψ_s orbital, their probability distribution (and hence, the negative charge distribution) is such that the negative PE, arising from the

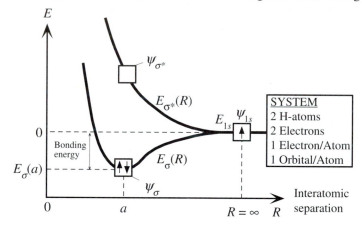

(a) Energy of ψ_σ and $\psi_{\sigma*}$ vs. the interatomic separation R.

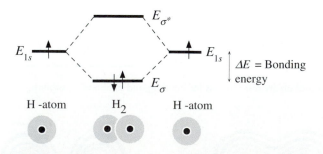

(b) Schematic diagram showing the changes in the electron energy as two isolated H atoms, far left and far right, come together to form a hydrogen molecule.

Figure 4.3   Electron energy in the system comprising two hydrogen atoms.

attractions of these two electrons to the two protons, is stronger in magnitude than the positive *PE*, arising from electron–electron repulsions and proton–proton repulsions and the kinetic energy of the two electrons. Therefore, the H_2 molecule is energetically stable.

The wavefunction ψ_σ corresponding to the lowest electron energy is called the **bonding orbital,** and $\psi_{\sigma*}$ is the **antibonding orbital.** When two atoms are brought together, the two identical atomic wavefunctions combine in two ways to generate two different molecular orbitals, each with a different energy. Effectively, then, an atomic energy level, such as E_{1s}, splits into two, E_σ and $E_{\sigma*}$. The splitting is due to the interaction (or overlap) between the atomic orbitals. Figure 4.3b schematically illustrates the changes in the electron energy levels as two isolated H atoms are brought together to form the H_2 molecule.

The splitting of a one-atom energy level when a molecule is formed is analogous to the splitting of the resonant frequency in an *RLC* circuit when two such circuits are brought together and coupled. Consider the *RLC* circuit shown in Figure 4.4a. The circuit is excited by an ac voltage source. The current peaks at the resonant frequency ω_0, as indicated in Figure 4.4a. When two such identical *RLC* circuits are coupled together and driven by an ac voltage source, the current develops, two peaks, at frequencies ω_1 and ω_2, below and above ω_0, as illustrated in Figure 4.4b. The two peaks at ω_1 and ω_2 are due to the mutual inductance that couples the two circuits, allowing them to interact. From this analogy, we can intuitively accept the energy splitting observed in Figure 4.3a.

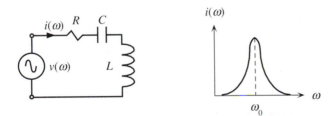

(a) There is one resonant frequency, ω_0, in an isolated *RLC* circuit.

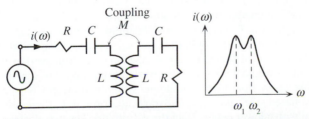

(b) There are two resonant frequencies in two coupled *RLC* circuits: one below and the other above ω_0.

Figure 4.4

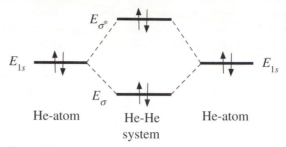

Figure 4.5
Two He atoms have four electrons. When He atoms come together, two of the electrons enter the E_σ level and two the $E_{\sigma*}$ level, so the overall energy is greater than two isolated He atoms.

Consider what happens when two He atoms come together. Recall that the $1s$ orbital has paired electrons and is full. The $1s$ atomic energy level will again split into two levels, E_σ and $E_{\sigma*}$, associated with the molecular orbitals ψ_σ and $\psi_{\sigma*}$, as illustrated in Figure 4.5. However, in the He–He system, there are four electrons, so two occupy the ψ_σ orbital state and two go to the $\psi_{\sigma*}$ orbital state. Consequently, the system energy is not lowered by bringing the two He atoms closer. Furthermore, quantum mechanical calculations show that the antibonding energy level $E_{\sigma*}$ shifts higher than the bonding level E_σ shifts lower. By the same token, although we could put an additional electron at $E_{\sigma*}$ in H_2 to make H_2^-, we could not make H_2^{2-} by placing two electrons at $E_{\sigma*}$.

From the He–He example, we can conclude that, as a general rule, the overlap of full atomic orbital states does not lead to bonding. In fact, full orbitals repel each other, because any overlap results in an increase in the system energy. To form a bond between two atoms, we essentially need an overlap of half-occupied orbitals, as in the H_2 molecule.

Example 4.1 | **AN ALKALI HALIDE MOLECULE (HF)**   We already know that H has a half-occupied $1s$ orbital, which can take part in bonding. Since the F atom has the electronic structure $1s^2 2s^2 p^5$, two of the p orbitals are full and one p orbital, p_x, is half full. This means that only the p_x orbital can participate in bonding. Figure 4.6 shows the electron orbitals in both H and F. When the H atom and the F atom approach each other to form an HF molecule, the ψ_{1s} orbital of H overlaps the p_x orbital of F. There are two possibilities for the overlap. First, ψ_{1s} and p_x can overlap in phase (both positive or both negative), to give a ψ_σ orbital that does not have a node between H and F, as shown in Figure 4.6. Second, they can overlap out of phase (one positive and the other negative), so that the overlap orbital $\psi_{\sigma*}$ has a node (similar to $\psi_{\sigma*}$ in Figure 4.1). We know from hydrogen atomic wavefunctions in Chapter 3 that orbitals with more nodes have higher energies. The molecular orbital ψ_σ therefore corresponds to a bonding orbital with a lower energy than the $\psi_{\sigma*}$ orbital. The two electrons, one from ψ_{1s} and the other from p_x, enter the ψ_σ orbital with spins paired, thereby forming a bond between H and F.

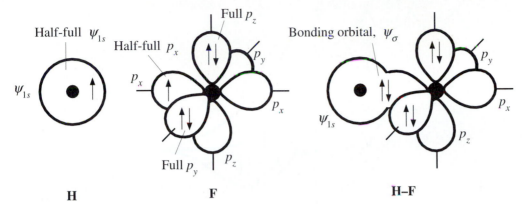

Figure 4.6
H has one half-empty ψ_{1s} orbital. F has one half-empty p_x orbital but full p_y and p_z orbitals. The overlap between ψ_{1s} and p_x produces a bonding orbital and an antibonding orbital. The two electrons fill the bonding orbital and thereby form a covalent bond between H and F.

4.2 BAND THEORY OF SOLIDS

We saw that when two hydrogen atoms approach each other, the atomic energy levels split into two. For the $1s$ case, these were labeled E_σ and $E_{\sigma*}$ (Figure 4.3). The split was due to the fact that the two identical atomic orbitals ψ_{1s} overlap within the molecule to generate two different molecular orbitals, ψ_σ and $\psi_{\sigma*}$, with different energies.

When we bring three hydrogen atoms (labeled A, B, and C) together, we generate three separate orbital states, ψ_a, ψ_b, and ψ_c, from three ψ_{1s} atomic states. Again, this occurs in three different ways, as illustrated in Figure 4.7a. The orbitals are:

$$\psi_a = \psi_{1s}(A) + \psi_{1s}(B) + \psi_{1s}(C) \qquad \textbf{[4.3a]}$$

$$\psi_b = \psi_{1s}(A) + \psi_{1s}(B) - \psi_{1s}(C) \qquad \textbf{[4.3b]}$$

$$\psi_c = \psi_{1s}(A) - \psi_{1s}(B) + \psi_{1s}(C) \qquad \textbf{[4.3c]}$$

where $\psi_{1s}(A)$, $\psi_{1s}(B)$, and $\psi_{1s}(C)$ are the $1s$ atomic wavefunctions centered around the atoms A, B, and C, respectively, as shown in Figure 4.7a. For example, the wavefunction $\psi_{1s}(A)$ represents $\psi_{1s}(r_A)$, which is centered around A and has the form $\exp(-r_A/a_o)$, where r_A is the distance from the nucleus of A, and a_o is the Bohr radius.

The energies E_a, E_b, and E_c of ψ_a, ψ_b, and ψ_c can be calculated from the Schrödinger equation by using the PE function of this system (the PE also includes proton–proton repulsions). It is clear that since ψ_a, ψ_b, and ψ_c are different, their energies E_a, E_b, and E_c are also different. Consequently, the $1s$ energy level splits into three separate levels, corresponding to the energies of ψ_a, ψ_b, and ψ_c, as

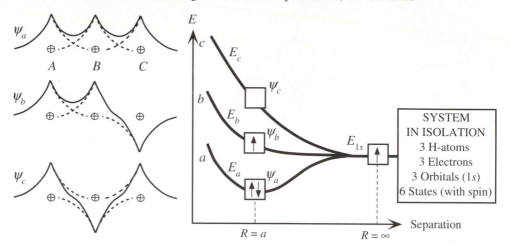

(a) Three molecular orbitals from three ψ_{1s} atomic orbitals overlapping in three different ways.

(b) The energies of the three molecular orbitals, labeled a, b, and c, in a system with three H atoms.

Figure 4.7

depicted by Figure 4.7b. By analogy with the electron wavefunctions in the hydrogen atom, we can argue that if the molecular wavefunction has more nodes, its energy is higher. Thus, ψ_a has the lowest energy E_a, ψ_b has the next higher energy E_b, and ψ_c has the highest energy E_c, as shown in Figure 4.7b. There are three electrons in the three-hydrogen system. The first two pair their spins and enter orbital ψ_a at energy E_a, and the third enters orbital ψ_b at energy E_b. Comparing Figures 4.7 and 4.3, we notice that although H_2 and H_3 both have two electrons in the lowest energy level, H_3 also has an extra electron at the higher energy level (E_b), which tends to increase the net energy of the atom. Thus, the H_3 molecule is much less stable than the H_2 molecule.[1]

Now consider the formation of a solid. Take N Li (lithium) atoms from infinity and bring them together to form the Li metal. Lithium has the electronic configuration $1s^2 2s^1$, which is somewhat like the hydrogen atom, since the K shell is closed and the third electron is alone in the $2s$ orbital.

Based on our previous discussions, we assume that the atomic energy levels will split into N separate energy levels. Since the $1s$ subshell is full and is close to the nucleus, it will not be affected much by the interatomic interactions; consequently, the energy of this state will experience only negligible splitting, if any. Since the $1s$ electrons will stay close to their parent nuclei, we will not consider them during formation of the solid.

In the system of N isolated Li atoms, we have N electrons in N ψ_{2s} orbitals at the energy E_{2s}, as illustrated in Figure 4.8 (at infinite interatomic separation). Let us assume that N is large (typically, $\sim 10^{23}$). As N atoms are brought together to

[1] The actual picture is not as simple as that sketched in Figure 4.7, which is an oversimplification. See G. Pimentel and R. Spratley, *Understanding Chemistry* (San Francisco: Holden-Day Inc., 1972), Section 18-6, pp. 682–687 for an excellent discussion.

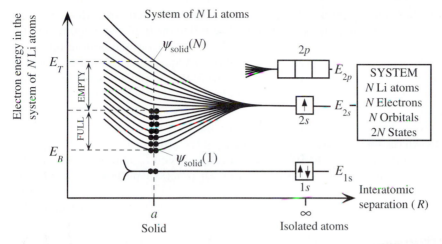

Figure 4.8   The formation of a 2s energy band from the 2s orbitals when N Li atoms come together to form the Li solid.

There are N 2s electrons, but 2N states in the band. The 2s band is therefore only half full. The atomic 1s orbital is close to the Li nucleus and remains undisturbed in the solid. Thus, each Li atom has a closed K shell (full 1s orbital).

form the solid, the energy level at E_{2s} splits into N finely-separated energy levels. The maximum width of the energy splitting depends on the closest interatomic distance (a) in the solid, as apparent in Figure 4.3a. The atoms separated by a distance greater than $R = a$ give rise to a lesser amount of energy splitting. The interatomic interactions between $N \psi_{2s}$ orbitals thus spread the N energy levels between the bottom and top levels, E_B and E_T, respectively, which are determined by the closest interatomic distance a. Put differently, E_B and E_T are determined by the distance between nearest neighbors. It is obvious that with N very large, the energy separation between two consecutive energy levels is very small; indeed, it is almost infinitesimal and not as exaggerated in the figure.

Remember that each energy level E_i in the Li metal of Figure 4.8 is the energy of an electron wavefunction $\psi_{solid}(i)$ in the solid, where $\psi_{solid}(i)$ is one particular combination of the N atomic wavefunctions ψ_{2s}. There are N different ways to combine N atomic wavefunctions ψ_{2s}, since each can be added in phase or out of phase, as is apparent in Equations 4.3a to c (see also Figures 4.7a and b). For example, when all $N \psi_{2s}$ are summed in phase, the resulting wavefunction $\psi_{solid}(1)$ is like ψ_a in Equation 4.3a, and it has the lowest energy. On the other hand, when $N \psi_{2s}$ are summed with alternating phases, $+ - + \ldots$, the resulting wavefunction $\psi_{solid}(N)$ is like ψ_c, and it has the highest energy. Other combinations of ψ_{2s} give rise to different energy values between E_B and E_T.

The single 2s energy level E_{2s} therefore splits into $N(\sim 10^{23})$ finely-separated energy levels, forming an **energy band,** as illustrated in Figure 4.8. Consequently, there are N separate energy levels, each of which can take two electrons with opposite spins. The N electrons fill all the levels up to and including the level at $N/2$. Therefore, the band is half full. We do not mean literally that the band is full to the half-energy point. The levels are not spread equally over the band from E_B to E_T, which means that the band cannot be full to the half-energy point. Half filled simply means half the states in the band are filled from the bottom up.

We have generated a half-filled band from a half-filled isolated $2s$ energy level. The energy band resulting from the splitting of the atomic $2s$ energy level is loosely termed the **$2s$ band.** By the same token, the atomic $1s$ levels are full, so any $1s$ band that forms from these $1s$ states will also be full. We can get an idea of the separation of energy levels in the $2s$ band by noting that the maximum separation, $E_T - E_B$, between the top and bottom of the band is on the order of 10 eV, but there are some 10^{23} atoms, giving rise to 10^{23} energy levels between E_B and E_T. Thus, the energy levels are finely separated, forming, for all practical purposes, a continuum of energy levels.

The $2p$ energy level, as well as the higher levels at $3s$ (and so on), also split into finely-separated energy levels, as shown in Figure 4.9. In fact, some of these energy levels overlap the $2s$ band; hence, they provide further energy levels and "extend" the $2s$ band into higher energy levels, as indicated in Figure 4.10, which shows how energy bands in metals are often represented. The vertical axis is the electron energy. The top of the $2s$ band, which is half full, overlaps the bottom of the $2p$ band which itself is overlapped near the top by the $3s$ band. We therefore have a band of energies that stretches from the bottom of the $2s$ band all the way to the vacuum level, as depicted in Figure 4.11. The reader may wonder what happened to the $3d$, $4s$ bands, etc. In the solid, the energies of these bands (including the top portion of the $3s$ band) are above the vacuum level, and the electron is free and far from the solid before it can acquire those energies.

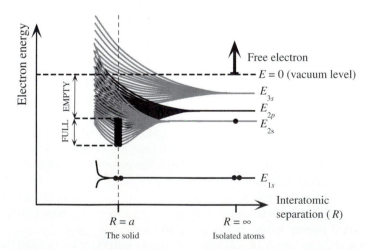

Figure 4.9   As Li atoms are brought together from infinity, the atomic orbitals overlap and give rise to bands. Outer orbitals overlap first. The $3s$ orbitals give rise to the $3s$ band, $2p$ orbitals to the $2p$ band, and so on. The various bands overlap to produce a single band in which the energy is nearly continuous.

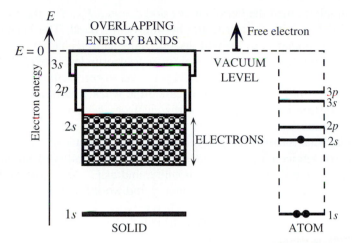

Figure 4.10
In a metal, the various energy bands overlap to give a single energy band that is only partially full of electrons. There are states with energies up to the vacuum level, where the electron is free.

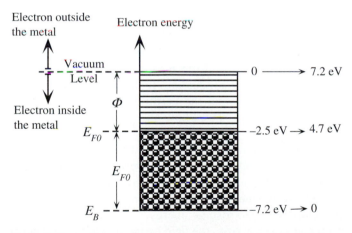

Figure 4.11 Typical electron energy band diagram for a metal.
All the valence electrons are in an energy band, which they only partially fill. The top of the band is the vacuum level, where the electron is free from the solid ($PE = 0$).

At a temperature of absolute zero, or nearly so, the thermal energy is insufficient to excite the electrons to higher energy levels, so all the electrons pair their spins and fill each energy level from E_B up to an energy level E_{FO} that we call the Fermi level at 0 K, as shown in Figure 4.11. The energy value for the Fermi level depends on where we take the reference energy. For example, if we take the vacuum level as the zero reference, then for the Li metal, E_{FO} is at -2.5 eV. The Fermi level is normally measured with respect to the bottom of the band, in which case, it is simply termed the Fermi energy and denoted E_{FO}. For the Li metal, E_{FO} is 4.7 eV which is with respect to the bottom of the band. The Fermi level has considerable significance, as we will discover later in this chapter.

At absolute zero, all the energy levels up to the Fermi level are full. The energy required to excite an electron from the Fermi level to the vacuum level, that is, to liberate the electron from the metal, is called the **work function** Φ of the metal. As the temperature increases, some of the electrons get excited to higher energy levels. To determine the probability of finding an electron at an energy level E, we must consider what is called "particle statistics," a topic that is key to understanding the behavior of semiconductor devices. Clearly, the probability of finding an electron at 0 K at some energy $E < E_{FO}$ is unity, and at $E > E_{FO}$, the probability is zero. Table 4.1 summarizes the Fermi energy and work function of a few selected metals.

Consider what happens when an electric field $\mathcal{E}_x$ is applied in the $-x$ direction by connecting a voltage source across a metal as shown in Figure 4.12. For the electron to accelerate under the force of the field $-e\mathcal{E}_x$, the electron must move to higher energy levels. Recall that classically, a force $F = -e\mathcal{E}_x$, will impart a KE given by

$$KE = \frac{p^2}{2m_e} = \frac{1}{2m_e}\left[\int_0^t F\,dt\right]^2$$

Thus, the electron gains energy from the electric field, provided it is able to do so, and the next higher energy level must be unoccupied. If it is occupied, the electron cannot move into it; therefore, the electron cannot be accelerated by the field and it cannot contribute to electrical conduction. It should be apparent that in metals, only the electrons near and around the Fermi energy play an important role in charge conduction. Figure 4.12 shows how an electron with an energy near E_{FO} contributes to electric current in the solid.

Notice that the application of the electric field bends the energy band, because the electrostatic PE of the electron is $-eV(x)$, where $V(x)$ is the voltage at position x. However, $V(x)$ changes linearly from 0 to V, by virtue of $dV/dx = -\mathcal{E}_x$. Since $PE = -eV(x)$ adds to the energy of the electron, the energy band must bend to account for the additional electrostatic energy.

The electron gains energy from the field by moving to higher energy levels and accelerating in the x direction from a to b to c, etc., until at some stage e it "collides" with a vibrating metal ion or an impurity. The electron then loses the energy gained from the field and returns to a lower energy level that is not already occupied by another electron. Once again, the electron starts to accelerate in the x direction and

Table 4.1   Fermi energy and work function of selected metals

Metal	Ag	Al	Au	Cs	Cu	Li	Mg	Na
Φ (eV)	4.26	4.28	5.1	2.14	4.65	2.3	3.7	2.75
E_{FO} (eV)	5.5	11.7	5.5	1.58	7.0	4.7	7.1	3.2

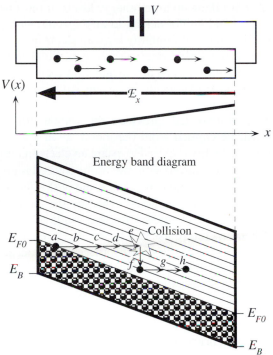

Figure 4.12
When an external field $\mathcal{E}_x$ is applied, electrons near E_{FO} can gain energy from the field, because there are empty energy levels above E_{FO}. These electrons can then accelerate along x under the action of the applied force $e\mathcal{E}_x$. As an electron moves from a to b to c and so on, its *KE* increases, but eventually it collides with a lattice vibration and loses the gained *KE*, dropping down to a lower level. Those electrons near the bottom of the band cannot gain energy from the field, because they do not have empty adjacent energy levels into which they can move.

climbs in energy, moving from f to g to h, and so on. The electrons in this partially empty band can therefore conduct electricity and are thus referred to as **conduction electrons.** Although the energy levels in Figure 4.12 have been drawn with highly exaggerated separations, they are in reality only very finely separated, with each one having a separate set of quantum numbers. We treat the band as an energy continuum.

The electrons in the energy band of a metal are the loosely bound valence electrons which become "free" in the metal and thereby form a kind of electron gas. This electron gas "holds" the metal ions together in the crystal structure and constitutes the metallic bond. This intuitive interpretation is quite apparent in Figure 4.9. When solid Li is formed from N atoms, the N electrons fill all the lower energy levels up to $N/2$. The energy of the system of N Li atoms, according to Figure 4.9, is therefore much less than that of N isolated Li atoms, by virtue of the N electrons taking up lower energy levels. It must be emphasized that the electrons within a band do not belong to any specific atom, but rather to the whole solid. All the $2s$ electrons essentially form an electron gas, and their energies fall within the energy band. These electrons are constantly moving around in the metal. In quantum mechanics terms, this means that their wavefunctions must be traveling waves, not the type that localizes the electron around a given atom (e.g., ψ_{n,ℓ,m_ℓ} in the hydrogen atom).

When a metal is illuminated, provided the wavelength of the radiation is correct, electrons will be emitted from the metal, as in the photoelectric effect. Since Φ is the minimum energy required to excite an electron into the vacuum level (out of the metal), the longest wavelength radiation required is given by $hc/\lambda = \Phi$.

The addition of heat to a metal can excite some of the electrons in the band to higher energy levels. Thus, heat can be absorbed by the metal's conduction electrons. The addition of heat also increases the amplitude of atomic vibrations. The heat capacity of a metal therefore has two terms, due to energy absorption by the lattice vibrations and energy absorption by the conduction electrons. At room temperature, the energy absorption by lattice vibrations dominates the heat capacity, whereas at the lowest temperatures, the electronic contribution is important.

Example 4.2 | **ELECTRICAL CONDUCTION BY A LIQUID METAL**   In the discussion on the formation of energy bands, we did not assume that the periodicity of atomic arrangements (crystal structure) was essential to the formation of an energy band. When N Li atoms are brought together, interatomic interactions cause the splitting of atomic energy levels and the formation of an energy band. The N valence electrons only partially fill this band. To form liquid Li, we again bring together almost as many Li atoms as we did in the crystal case, but we do not arrange the atoms in an orderly manner. The energy levels will still split into N, and we still get an energy band with N electrons partially filling the band. Consequently, when an electric field is applied, the electrons can gain energy from the field and move up to the empty levels within the band. Liquid metals are therefore good conductors, although their resistivity is higher than the crystal phase. In the liquid, the random arrangement of atoms causes the potential energy of the electron to fluctuate from site to site, and since force is $F = -d(PE)/dx$, the net effect is a random scattering force, which reduces the mean free time between scattering events and thus the mobility.

Example 4.3 | **METALLIC LIQUID HYDROGEN IN JUPITER AND ITS MAGNETIC FIELD**   The surface of Jupiter, as visualized schematically in Figure 4.13, mainly consists of a mixture of molecular hydrogen and He gases. Deep in the planet, however, the pressure is so tremendous that the hydrogen molecular bond breaks, leaving a dense ocean of hydrogen atoms. Hydrogen has only one electron in the $1s$ energy level. When atoms are densely packed, the $1s$ energy level forms an energy band, which is then only half filled. This is just like the Li metal, which means we can treat liquid hydrogen as a liquid metal, with electrical properties reminiscent of liquid mercury. Liquid hydrogen can sustain electrical currents, which in turn can give rise to the magnetic fields on Jupiter. The origin of the electrical currents are not known with

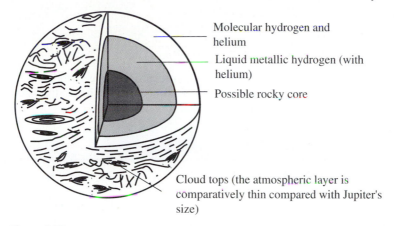

Molecular hydrogen and helium

Liquid metallic hydrogen (with helium)

Possible rocky core

Cloud tops (the atmospheric layer is comparatively thin compared with Jupiter's size)

Figure 4.13
The interior of Jupiter is believed to contain liquid hydrogen, which is metallic.
(Drawing adapted from: T. Hey, and P. Walters, *The Quantum Universe*, Cambridge, MA: Cambridge University Press, (1988), Figure 7.1, p. 96.)

certainty. We do know, however, that the core of the planet is hot and emanates heat, which causes convection currents. Temperature differences can readily give rise to electrical currents, by virtue of thermoelectric effects, as discussed later in this chapter.

WHAT MAKES A METAL?   The Be atom has an electronic structure of $1s^2 2s^2$. Although the Be atom has a full $2s$ energy level, solid Be is a metal. Why?

| **Example 4.4**

Solution

We will neglect the K shell ($1s$ state), which is full and very close to the nucleus, and consider only the higher energy states. In the solid, the $2s$ energy level splits into N levels, forming a $2s$ band. With $2N$ electrons, each level is occupied by spin-paired electrons. The $2s$ band is therefore full. However, the empty $2p$ band, from the empty $2p$ energy levels, over-laps the $2s$ band, thereby providing empty energy levels to these $2N$ electrons. Thus, the conduction electrons are in an energy band that is only partially filled; they can gain energy from the field to contribute to electrical conduction. Solid Be is therefore a metal.

FERMI SPEED OF CONDUCTION ELECTRONS IN A METAL   In copper, the Fermi energy of conduction electrons is 7.0 eV. What is the speed of the conduction electrons around this energy?

| **Example 4.5**

Solution

Since the conduction electrons are not bound to any one atom, their *PE* must be zero within the solid (but large outside), so all their energy is kinetic. For conduction electrons around the Fermi energy E_{FO} with a speed v_F, we have

$$\frac{1}{2} m v_F^2 = E_{FO}$$

so that

$$v_F = \sqrt{\frac{2E_{FO}}{m_e}} = \sqrt{\frac{2(1.6 \times 10^{-19}\ \text{C})(7.0\ \text{eV})}{(9.1 \times 10^{-31}\ \text{kg})}} = 1.6 \times 10^6\ \text{m s}^{-1}$$

Although the Fermi energy depends on the properties of the energy band, to a good approximation it is only weakly temperature dependent, so v_F will be relatively temperature insensitive, as we will show later in this chapter.

4.3   SEMICONDUCTORS

The Si atom has 14 electrons, which distribute themselves in the various atomic energy levels as shown in Figure 4.14. The inner shells ($n = 1$ and $n = 2$) are full and therefore "closed." Since these shells are near the nucleus, when Si atoms come together to form the solid, they are not much affected and they stay around the parent Si atoms. They can therefore be excluded from further discussion. The $3s$ and $3p$ subshells are farther away from the nucleus. When two Si atoms approach, these electrons strongly interact with each other. Therefore, in studying the formation of bands in the Si solid, we will only consider the $3s$ and $3p$ levels.

The first task is to examine why Si actually bonds with four neighbors, since the $3s$ orbital is full and there are only two electrons in the $3p$ orbitals. The full $3s$ orbital should not overlap a neighbor and become involved in bonding. Since only two $3p$ orbitals are half full, bonds should be formed with two neighboring Si

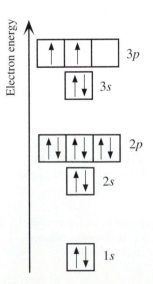

Figure 4.14   The electronic structure of Si

atoms. In reality, the $3s$ and $3p$ energy levels are quite close, and when four Si atoms approach each other, the interaction results in the four orbitals $\psi(3s)$, $\psi(3p_x)$, $\psi(3p_y)$, and $\psi(3p_x)$ mixing together to form four new **hybrid orbitals,** which are directed in tetrahedral directions; that is, each one is aimed as far away from the others as possible, as illustrated in Figure 4.15. We call this process sp^3 **hybridization,** since one s orbital and three p orbitals are mixed. (The superscript 3 on p has nothing to do with the number of electrons; it refers to the number of p orbitals used in the hybridization.)

The four sp^3 hybrid orbitals, ψ_{hyb}, each have one electron, so they are half occupied. This means that four Si atoms can have their orbitals ψ_{hyb} overlap to form bonds in tetrahedral directions, which is what actually happens; thus, one Si atom bonds with four other Si atoms.

In the same way, one Si atom bonds with four H atoms to form the important gas SiH_4, known as silane, which is widely used in semiconductor technology to fabricate Si devices. In SiH_4, four hybridized orbitals of the Si atom overlap with the $1s$ orbitals of four H atoms. In exactly the same way, one carbon atom bonds with four hydrogen atoms to form methane, CH_4.

There are two ways in which the hybrid orbital ψ_{hyb} can overlap with that of the neighboring Si atom to form two molecular orbitals. They can add in phase (both positive or both negative) or out of phase (one positive and the other negative) to produce a bonding or an antibonding molecular orbital, ψ_B and ψ_A, respectively, with energies E_B and E_A. Each Si–Si bond thus corresponds to two paired electrons in a bonding molecular orbital ψ_B. In the solid, there are N ($\sim 5 \times 10^{22}$ cm^{-3}) Si

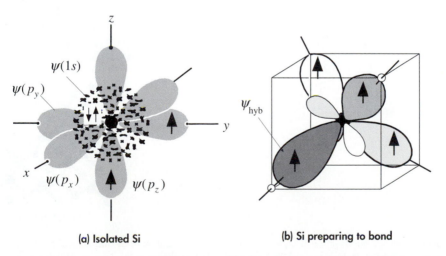

(a) Isolated Si	**(b) Si preparing to bond**

(a) Si is in Group IV in the Periodic Table. An isolated Si atom has 2 electrons in the $3s$ and 2 electrons in the $3p$ orbitals. (b) When Si is about to bond, the one $3s$ orbital and the three $3p$ orbitals become perturbed and mixed to form four hybridized orbitals, ψ_{hyb}, called sp^3 orbitals, which are directed towards the corners of a tetrahedron. The ψ_{hyb} orbital has a large major lobe and a small back lobe. Each ψ_{hyb} orbital takes one of the four valence electrons.

Figure 4.15

atoms, and there are nearly as many such ψ_B bonds. The interactions between the ψ_B orbitals (i.e., the Si–Si bonds) leads to the splitting of the E_B energy level to N levels, thereby forming an energy band labeled the valence band (VB) by virtue of the valence electrons it contains. Since the energy level E_B is full, so is the valence band. Figure 4.16 illustrates the formation of the VB from E_B.

In the solid, the interactions between the N number of ψ_A orbitals result in the splitting of the energy level E_A to N levels and the formation of an energy band that is completely empty and separated from the full valence band by a definite energy gap E_g. In this energy region, there are no states; therefore, the electron cannot have energy with a value within E_g. The energy band formed from N ψ_A orbitals is a conduction band (CB), as also indicated in Figure 4.16.

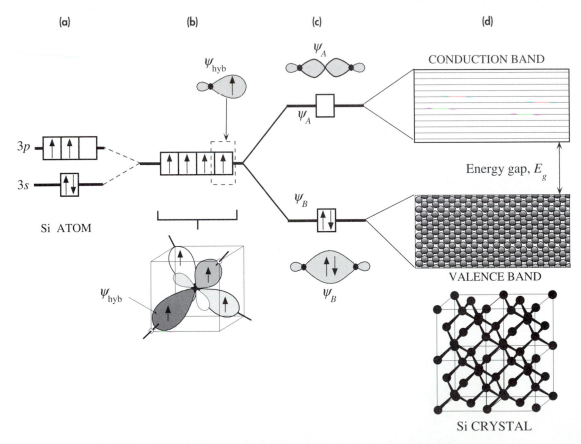

(a) Formation of energy bands in the Si crystal first involves hybridization of 3s and 3p orbitals to four identical ψ_{hyb} orbitals, which are at 109.5° to each other as shown in (b). (c) ψ_{hyb} orbitals on two neighboring Si atoms can overlap to form ψ_B or ψ_A. The first is a bonding orbital (full) and the second is an antibonding orbital (empty). In the crystal, ψ_B overlap to give the valence band (full) and ψ_A overlap to give the conduction band (empty).

Figure 4.16

The electronic states in the valence band (and also in the conduction band) extend throughout the whole solid, because they result from $N \, \psi_B$ orbitals interfering and overlapping each other. As before $N \, \psi_B$, orbitals can overlap in N different ways to produce N distinct wavefunctions ψ_{vb} that extend throughout the solid. We cannot relate a particular electron to a particular bond or site because the wavefunctions ψ_{vb} corresponding to the VB energies are not concentrated at a single location. The electrical properties of solids are based on the fact that in solids, such as semiconductors and insulators, there are certain bands of allowed energies for the electrons, and these bands are separated by energy gaps.

At temperatures above absolute zero, the atoms in a solid vibrate due to their thermal energy. Some of the atoms can acquire a sufficiently high energy from thermal fluctuations to strain and rupture their bonds. Physically, there is a possibility that the atomic vibration will impart sufficient energy to the electron for it to surmount the bonding energy and leave the bond. The electron must then enter a higher energy state. In the case of Si, this means entering a state in the CB, as shown in Figure 4.17. If there is an applied electric field $\mathcal{E}_x$ in the $+x$ direction then the excited electron will be acted on by a force $-e\mathcal{E}_x$ and it will try to move in the $-x$ direction. For it to do so, there must be empty higher energy levels, so that as the electron accelerates and gains energy, it moves up in the band. When an electron collides with a lattice vibration, it loses the energy acquired from the field and drops down within the CB. Again, it should be emphasized that states in an energy band are extended; that is, the electron is not localized to any one atom.

Note also that the thermal generation of an electron from the VB to the CB leaves behind a VB state with a missing electron. This unoccupied electron state has an apparent positive charge, because this crystal region was neutral prior to the removal of the electron. The VB state with the missing electron is called a hole and is denoted h^+. The hole can "move" in the direction of the field by exchanging places with neighboring valence electron; hence, it contributes to conduction, as will be discussed in Chapter 5.

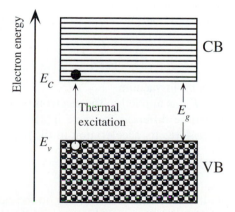

Figure 4.17   Energy band diagram of a semiconductor

Example 4.6

CUTOFF WAVELENGTH OF AN Si PHOTODETECTOR   What wavelengths of light can be absorbed by an Si photodetector given $E_g = 1.1$ eV? Can such a photodetector be used in fiberoptic communications at light wavelengths of 1.31 μm and 1.55 μm?

Solution

The energy bandgap E_g of Si is 1.1 eV. A photon must have at least this much energy to excite an electron from the VB to the CB, where the electron can drift. Excitation corresponds to the breaking of an Si–Si bond. A photon of less energy does not get absorbed, because its energy will put the electron in the bandgap where there are no states. Thus, $hc/\lambda > E_g$ gives

$$\lambda < \frac{hc}{E_g} = \frac{(6.6 \times 10^{-34} \text{ J s})(3 \times 10^8 \text{ m s}^{-1})}{(1.1 \text{ eV})(1.6 \times 10^{-19} \text{ C})}$$

$$= 1.13 \times 10^{-6} \text{ m} \qquad \text{or} \qquad 1.1 \ \mu\text{m}$$

Since optical communications networks use wavelengths of ~1.31 and ~1.55 μm, these light waves will not be absorbed by Si and thus cannot be detected by an Si photodetector.

4.4   ELECTRON EFFECTIVE MASS

When an electric field $\mathcal{E}_x$ is applied to a metal, an electron near the Fermi level can gain energy from the field and move to higher energy levels, as shown in Figure 4.12. The external force $F_{ext} = e\mathcal{E}_x$ is in the x direction, and it drives the electron along x. The acceleration of the electron is still given by $a = F_{ext}/m_e$, where m_e is the mass of the electron in vacuum.

The law $F_{ext} = m_e a$ cannot strictly be valid for the electron inside a solid, because the electron interacts with the host ions and experiences internal forces F_{int} as it moves around, as depicted in Figure 4.18. The electron therefore has a *PE* that varies with distance. Recall that we interpret mass as inertial resistance against acceleration per unit applied force. When an external force F_{ext} is applied to an electron in vacuum, the electron will accelerate by an amount

$$a_{vac} = \frac{F_{ext}}{m_e} \qquad\qquad \textbf{[4.4]}$$

as determined by its mass m_e in vacuum.

When the same force F_{ext} is applied to the electron inside a crystal, the acceleration of the electron will be different, because it will also experience internal forces, as shown in Figure 4.18. Its acceleration in the crystal will be

$$a_{\text{cryst}} = \frac{F_{ext} + F_{int}}{m_e} \qquad\qquad \textbf{[4.5]}$$

where F_{int} is the sum of all the internal forces acting on the electron, which is quite different than Equation 4.4. To the outside agent applying the force F_{ext}, the electron will appear to be exhibiting a different inertial mass, since its acceleration

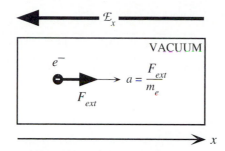

(a) An external force F_{ext} applied to an electron in vacuum results in an acceleration $a_{vac} = F_{ext}/m_e$.

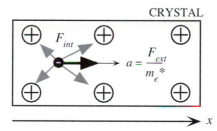

(b) An external force F_{ext} applied to an electron in a crystal results in an acceleration $a_{cryst} = F_{ext}/m_e^*$.

Figure 4.18

will be different. It would be most useful for the external agent if the effect of the internal forces in F_{int} could be accounted for in a simple way, and if the acceleration could be calculated from the external force F_{ext} alone, through something like Equation 4.4. This is indeed possible.

In a crystalline solid, the atoms are arranged periodically, and the variation of F_{int}, and hence the *PE*, or $V(x)$, of the electron with distance along x, is also periodic. In principle, then, the effect on the electron motion can be predicted and accounted for. When we solve the Schrödinger equation with the periodic *PE*, or $V(x)$, we essentially obtain the effect of these internal forces on the electron motion. It has been found that when the electron is in a band that is not full, we can still use Equation 4.4, but instead of the mass in vacuum m_e, we must use the effective mass m_e^* of the electron in that particular crystal. The effective mass is a quantum mechanical quantity that behaves in the same way as the inertial mass in classical mechanics. The acceleration of the electron in the crystal is then simply

$$a_{cryst} = \frac{F_{ext}}{m_e^*} \qquad [4.6]$$

Table 4.2   Effective mass m_e^* of electrons in some metals

Metal	Ag	Au	Bi	Cu	K	Li	Na	Ni	Pt	Zn
$\dfrac{m_e^*}{m_e}$	0.99	1.10	0.047	1.01	1.12	1.28	1.2	28	13	0.85

The effects of all internal forces are incorporated into m_e^*. It should be emphasized that m_e^*, is obtained theoretically from the solution of the Schrödinger equation for the electron in a particular crystal, a task that is by no means trivial. However, the effective mass can be readily measured. For some of the familiar metals, m_e^* is very close to m_e. For example, in copper, $m_e^* = m_e$ for all practical purposes, whereas in lithium $m_e^* = 1.28m_e$, as shown in Table 4.2. On the other hand, m_e^* for many metals and semiconductors is appreciably different than electron mass in vacuum and can even be negative. (m_e^* depends on the properties of the band that contains the electron. This is further discussed at the end of Chapter 5.)

4.5   ENERGY BANDS: DENSITY OF STATES

Although we know there are many energy levels (some $\sim 10^{23}$) in a given band, we have not yet considered how many states (or electron wavefunctions) there are per unit energy per unit volume in that band. We can appreciate the distribution of energy states from a simple intuitive consideration. We know that the amount of separation between the split energy levels increases as atoms are brought together from infinity (Figure 4.3). Put differently, energy splitting increases with the degree of interaction between the atoms.

A given atom in a solid has a fixed number of nearest neighbors, but many distant neighbors, as illustrated in Figure 4.19. If we consider just a single plane and an arbitrary atom, there may be four nearest neighbors, as shown in Figure 4.19a. Isolated from others, this system should give rise to four split energy levels, with the widest energy separation; the interaction occurs between nearest neighbors. As we move away from an atom, there are more and more neighbors. For example, an atom may have eight 4th-neighbors in one plane, as shown in Figure 4.19b. In isolation from the rest, the energy would have split into nine narrowly-separated levels, since the atoms are further isolated. In the crystal, there are hundreds and thousands of distant neighbors, so the number of narrowly-split energy levels will correspondingly be very large. We therefore expect the number of energy levels (each corresponding to an electron wavefunction in the crystal) in the central regions of the band to be very large, as depicted in Figure 4.19c.

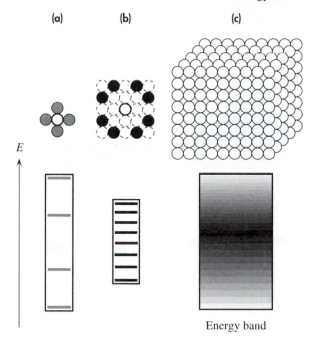

(a) An arbitrary atom in a single plane has four nearest neighbors. Isolated from others, this system would have four split energy levels with the widest energy separation; the interaction is between nearest neighbors. (b) There are eight 4th-neighbors on this plane. In isolation from the rest, there would be nine narrowly-separated split energy levels as the atoms are further isolated. (c) In the crystal, there are hundreds and thousands of distant neighbors, so the number of energy levels, which are narrowly split in energy, will be correspondingly very large.

Figure 4.19

Figure 4.20 illustrates how the energy and volume density of electronic states change across an energy band. We define the **density of states** $g(E)$ such that $g(E)\, dE$ is the number of states (i.e., wavefunctions) in the energy interval E to $(E + dE)$ per unit volume of the sample. Thus, the number of states per unit volume up to some energy E' is

$$S_v(E') = \int_0^{E'} g(E)\, dE \qquad\qquad \text{[4.7]}$$

which is called the total number of states per unit volume with energies less than E'. This is denoted $S_v(E')$.

To determine the density of states function $g(E)$, we must first determine the number of states with energies less than E' in a given band. This is tantamount to calculating $S_v(E')$ in Equation 4.7. Instead, we will improvise and use the energy

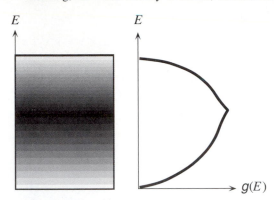

Figure 4.20   The density of states $g(E)$ versus E in an energy band

levels for an electron in a three-dimensional potential well. Recall that the energy of an electron in a cubic *PE* well of size L is given by

$$E = \frac{h^2}{8m_e L^2}(n_1^2 + n_2^2 + n_3^2) \qquad [4.8]$$

where n_1, n_2, and n_3 are integers 1, 2, 3, ... The spatial dimension L of the well now refers to the size of the entire solid, as the electron is confined to be somewhere inside that solid. Thus, L is very large compared to atomic dimensions, which means that the separation between the energy levels is very small.

Each combination of n_1, n_2, and n_3 is one electron orbital state. For example, $\psi_{n_1,n_2,n_3} = \psi_{1,1,2}$ is one possible orbital state. Suppose that in Equation 4.8 E is given as E'. We need to determine how many combinations of n_1, n_2, n_3, (i.e., how many ψ) have energies, less than E', as given by Equation 4.8. Assume that $(n_1^2 + n_2^2 + n_3^2) = n'^2$. The object is to enumerate all possible choices of integers for n_1, n_2, and n_3 that satisfy $n_1^2 + n_2^2 + n_3^2 \le n'^2$.

The two-dimensional case is easy to solve. Consider $n_1^2 + n_2^2 \le n'^2$ and the two-dimensional n- space where the axes are n_1 and n_2, as shown in Figure 4.21. The two-dimensional space is divided by lines drawn at $n_1 = 1, 2, 3, \ldots$ and $n_2 = 1, 2, 3, \ldots$ into infinitely many boxes (squares), each of which has a unit area and represents a possible state ψ_{n_1,n_2}. For example, the state $n_1 = 1$, $n_2 = 3$ is shaded, as is that for $n_1 = 2$, $n_2 = 2$.

Clearly, the area contained by n_1, n_2 and the circle defined by $n'^2 = n_1^2 + n_2^2$ (just like $r^2 = x^2 + y^2$) is the number of states that satisfy $n_1^2 + n_2^2 \le n'^2$. This area is $(1/4)(\pi n'^2)$.

In the three-dimensional case, $n_1^2 + n_2^2 + n_3^2 \le n'^2$ is required, as indicated in Figure 4.22. This is the volume contained by the positive n_1, n_2, and n_3 axes and the surface of a sphere of radius n'. Each state has a unit volume, and within the sphere, $n_1^2 + n_2^2 + n_3^2 \le n'^2$ is satisfied. Therefore, the number of orbital states $S_{orb}(n')$ within this volume is given by

$$S_{orb}(n') = \frac{1}{8}\left(\frac{4}{3}\pi n'^3\right) = \frac{1}{6}\pi n'^3$$

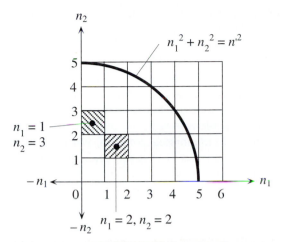

Figure 4.21   Each state, or electron wavefunction in the crystal, can be represented by a box at n_1, n_2.

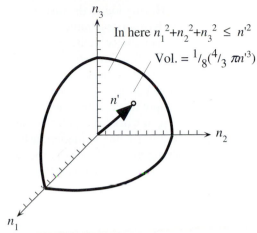

Figure 4.22   In three dimensions, the volume defined by a sphere of radius n' and the positive axes n_1, n_2 and n_3, contains all the possible combinations of positive n_1, n_2, and n_3 values that satisfy $n_1^2 + n_2^2 + n_3^2 \leq n'^2$.

Each orbital state can take two electrons with opposite spins, which means that the number of states, including spin, is given by

$$S(n') = 2S_{orb}(n') = \frac{1}{3}\pi n'^3$$

We need this expression in terms of energy. Substituting $n'^2 = 8m_e L^2 E'/h^2$ from Equation 4.8 in $S(n')$, we get

$$S(E) = \frac{\pi L^3 (8m_e E')^{3/2}}{3h^3}$$

Since L^3 is the physical volume of the solid, the number of states per unit volume $S_v(E')$ with energies $E \leq E'$ is

$$S_v(E) = \frac{\pi (8m_e E')^{3/2}}{3h^3} \qquad [4.9]$$

Furthermore, from Equation 4.7, $dS_v/dE = g(E)$. By differentiating Equation 4.9 with respect to energy, we get

$$g(E) = (8\pi 2^{1/2}) \left(\frac{m_e}{h^2}\right)^{3/2} E^{1/2} \qquad [4.10] \qquad \textit{Density of states}$$

Equation 4.10 shows that the density of states increases proportionately to $E^{1/2}$ with energy from the bottom of the energy band. As we approach the top of the band, according to our understanding in Figure 4.20, $g(E)$ decreases with the energy as $(E_T - E)^{1/2}$, so that as $E \to E_T$, $g(E) \to 0$. The variation of $g(E)$ with energy in a band can therefore be described as a square-root dependence on the energy from the band edge.

Having found the distribution of the electron energy states, Equation 4.10, we now wish to determine the number of states that actually contain electrons; that is, the probability of finding an electron at an energy level E. This is given by the Fermi–Dirac statistics.

As an example, one convenient way of calculating the population of a city is to find the density of houses in that city (i.e., the number of houses per unit area), multiply that by the probability of finding a human in a house, and finally, integrate the result over the area of the city. The problem is working out the chances of actually finding someone at home, using a mathematical formula. For those who like analogies, if $g(A)$ is the density of houses and $f(A)$ is the probability that a house is occupied, then the population of the city is

$$n = \int_{City} f(A)g(A)\, dA$$

where the integration is done over the entire area of the city. This equation can be used to find the number of electrons per unit volume within a band. If E is the electron energy and $f(E)$ is the probability that a state with energy E is occupied, then

$$n = \int_{Band} f(E)g(E)\, dE$$

where the integration is done over all the energies of the band.

Example 4.7

DENSITY OF STATES IN A BAND   Given that the width of an energy band is typically ~10 eV, calculate the following, in per cm^3 and per eV units:

a.   The density of states at the center of the band.

b.   The number of states per unit volume within a small energy range kT about the center.

c.   The density of states at kT above the bottom of the band.

d.   The number of states per unit volume within a small energy range of kT to $2kT$ from the bottom of the band.

Solution

The density of states, or the number of states per unit energy range per unit volume, $g(E)$, is given by

$$g(E) = (8\pi\, 2^{1/2})\left(\frac{m_e}{h^2}\right)^{3/2} E^{1/2}$$

which gives the number of states per cubic meter per Joule of energy. Substituting $E = 5$ eV, we have

$$g_{center} = (8\pi\, 2^{1/2})\left(\frac{9.1 \times 10^{-31}}{(6.626 \times 10^{-34})^2}\right)^{3/2} (5 \times 1.6 \times 10^{-19})^{1/2} = 9.50 \times 10^{46}\ m^{-3}\ J^{-1}$$

Converting to cm^{-3} and eV^{-1}, we get

$$g_{center} = (9.50 \times 10^{46}\ m^{-3}\ J^{-1})(10^{-6}\ m^3\ cm^{-3})(1.6 \times 10^{-19}\ J\ eV^{-1})$$

$$= 1.52 \times 10^{22}\ cm^{-3}\ eV^{-1}$$

To find the number of states per unit volume within kT at the center of the band, we multiply g_{center} by kT or $(1.52 \times 10^{22} \text{ cm}^{-3} \text{ eV}^{-1})(0.026 \text{ eV})$ to get $3.9 \times 10^{20} \text{ cm}^{-3}$. This is not a small number!

At kT above the bottom of the band, at 300 K($kT = 0.026$ eV), we have

$$g_{0.026} = (8\pi 2^{1/2}) \left(\frac{9.1 \times 10^{-31}}{(6.626 \times 10^{-34})^2} \right)^{3/2} (0.026 \times 1.6 \times 10^{-19})^{1/2}$$

$$= 6.84 \times 10^{45} \text{ m}^{-3} \text{ J}^{-1}$$

Converting to cm^{-3} and eV^{-1} we get

$$g_{0.026} = (6.84 \times 10^{45} \text{ m}^{-3} \text{ J}^{-1})(10^{-6} \text{ m}^3 \text{ cm}^{-3})(1.6 \times 10^{-19} \text{ J eV}^{-1})$$

$$= 1.10 \times 10^{21} \text{ cm}^{-3} \text{ eV}^{-1}$$

Within kT, the volume density of states is

$$(1.10 \times 10^{21} \text{ cm}^{-3} \text{ eV}^{-1})(0.026 \text{ eV}) = 2.8 \times 10^{19} \text{ cm}^{-3}$$

This is very close to the bottom of the band and is still very large.

TOTAL NUMBER OF STATES IN A BAND | **Example 4.8**

a.   Based on the overlap of atomic orbitals to form the electron wavefunction in the crystal, how many states should there be in a band?

b.   Consider the density of the states function

$$g(E) = (8\pi 2^{1/2}) \left(\frac{m_e}{h^2} \right)^{3/2} E^{1/2}$$

By integrating $g(E)$, estimate the total number of states in a band per unit volume, and compare this with the atomic concentration for silver. For silver, we have $E_{FO} = 5.5$ eV and $\Phi = 4.5$ eV. (Note that "state" means a distinct wavefunction, including spin.)

Solution

a.   We know that when N atoms come together to form the solid; N atomic orbitals can overlap N different ways to produce N orbitals or $2N$ states in the crystal, since each orbital has two states, spin up and spin down. These states form the band.

b.   For silver, $E_{FO} = 5.5$ eV and $\Phi = 4.5$ eV, so the width of the energy band is 10 eV. To estimate the total volume density of states, we assume that the density of states $g(E)$ reaches its maximum at the center of the band $E = E_{ctr}$. Integrating $g(E)$ from the bottom of the band, $E = 0$, to the center, $E = E_{ctr}$, yields the number of states per unit volume up to the center of the band. This is half the total number of states in the whole band, that is, $(1/2) S_{band}$, where S_{band} is the number of states per unit volume in the band and is determined by

$$\frac{1}{2} S_{band} = \int_0^{E_{ctr}} g(E) \, dE = \frac{16\pi 2^{1/2}}{3} \left(\frac{m_e}{h^2} \right)^{3/2} E_{ctr}^{3/2}$$

or

$$\frac{1}{2} S_{band} = \frac{16\pi 2^{1/2}}{3} \left(\frac{9.1 \times 10^{-31} \text{ kg}}{(6.626 \times 10^{-34} \text{ J s})^2} \right)^{3/2} (5 \text{ eV} \times 1.6 \times 10^{-19} \text{ C})^{3/2}$$

$$= 5.08 \times 10^{28} \text{ m}^{-3} = 5.08 \times 10^{22} \text{ cm}^{-3}$$

Thus

$$S_{band} = 10.16 \times 10^{22} \text{ states cm}^{-3}$$

We must now calculate the number of atoms per unit volume in silver. Given the density $d = 10.5 \text{ g cm}^{-3}$ and the atomic mass $M_{at} = 107.9$ of silver, the atomic concentration is

$$n_{Ag} = \frac{dN_A}{M_{at}} = 5.85 \times 10^{22} \text{ atoms cm}^{-3}$$

As expected, the density of states is almost twice the atomic concentration, even though we used a crude approximation to estimate the density of states.

4.6 STATISTICS: COLLECTIONS OF PARTICLES

4.6.1 Boltzmann Classical Statistics

Given a collection of particles in random motion and colliding with each other,[2] we need to determine the concentration of particles in the energy range E to $(E + dE)$. Consider the process shown in Figure 4.23, in which two electrons with energies E_1 and E_2 interact and then move off in different directions, with energies E_3 and E_4. Let the probability of an electron having an energy E be $P(E)$, where $P(E)$ is the fraction of electrons with an energy E. Assume there are no restrictions to the electron energies, that is, that we can ignore the Pauli Exclusion Principle. The probability of this event is then $P(E_1)P(E_2)$. The probability of the reverse process, in which electrons with energies E_3 and E_4 interact, is $P(E_3)P(E_4)$. Since we have thermal equilibrium, that is, the system is in equilibrium, the forward process must be just as likely as the reverse process, so that

$$P(E_1)P(E_2) = P(E_3)P(E_4) \qquad [4.11]$$

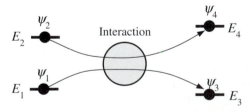

Figure 4.23
Two electrons with initial wavefunctions ψ_1 and ψ_2 at E_1 and E_2 interact and end up at different energies E_3 and E_4. Their corresponding wavefunctions are ψ_3 and ψ_4.

[2] From Chapter 1, we can associate this with the Kinetic Theory of Gases. The energies of the gas molecules, which are moving around randomly, are distributed according to the Maxwell–Boltzmann statistics.

Furthermore, the energy in this collision must be conserved, so we also need

$$E_1 + E_2 = E_3 + E_4 \qquad [4.12]$$

We therefore need to find the $P(E)$ that satisfies both Equations 4.11 and 4.12. Based on our experience with the distribution of energies among gas molecules, we can guess that the solution for Equations 4.11 and 4.12 would be

$$P(E) = A \exp\left(-\frac{E}{kT}\right) \qquad [4.13]$$

where k is the Boltzmann constant, T is the temperature, and A is a constant. We can show that Equation 4.13 is a solution to Equations 4.11 and 4.12 by a simple substitution. Equation 4.13 is the **Boltzmann probability function** and is shown in Figure 4.24. The probability of finding a particle at an energy E therefore decreases exponentially with energy. We assume, of course, that any number of particles may have a given energy E. In other words, there is no restriction such as permitting only one particle per state at an energy E, as in the Pauli Exclusion Principle.

Suppose that we have N_1 particles at energy level E_1 and N_2 particles at a higher energy E_2. Then, by Equation 4.13, we have

$$\frac{N_2}{N_1} = \exp\left(-\frac{E_2 - E_1}{kT}\right) \qquad [4.14] \qquad \textit{Boltzmann statistics}$$

If $E_2 - E_1 \gg kT$, then N_2 can be orders of magnitude smaller than N_1. As the temperature increases, N_2/N_1 also increases. Therefore, increasing the temperature populates the higher energy levels.

Classical particles obey the Boltzmann statistics. Whenever there are many more states (by orders of magnitude) than the number of particles, the likelihood of two particles having the same set of quantum numbers is negligible and we do not have to worry about the Pauli Exclusion Principle. In these cases, we can use the Boltzmann statistics. An important example is the statistics of electrons in the conduction band of a semiconductor where, in general, there are many more states than electrons.

4.6.2   Fermi–Dirac Statistics

Now consider the interaction for which no two electrons can be in the same quantum state, which is essentially obedience to the Pauli Exclusion Principle, as shown in Figure 4.23. We assume that we can have only one electron in a particular quantum state ψ (including spin) associated with the energy value E. We therefore need those states that have energies E_3 and E_4 to be not occupied. Let $f(E)$ be the probability that an electron is in such a state, with energy E in this new interaction environment. The probability of the forward event in Figure 4.24 is

$$f(E_1)\, f(E_2)[1 - f(E_3)][1 - f(E_4)]$$

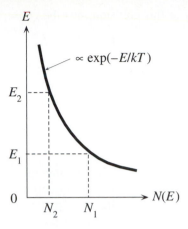

Figure 4.24
The Boltzmann energy distribution describes the statistics of particles, such as electrons, when there are many more available states than the number of particles.

The square brackets represent the probability that the states with energies E_3 and E_4 are empty. In thermal equilibrium, the reverse process, the electrons with E_3 and E_4 interacting to transfer to E_1 and E_2, has just as equal a likelihood as the forward process. Thus, $f(E)$ must satisfy the equation

$$f(E_1)\, f(E_2)[1 - f(E_3)][1 - f(E_4)]$$
$$= f(E_3)\, f(E_4)[1 - f(E_1)][1 - f(E_2)] \qquad [4.15]$$

In addition, for energy conservation, we must have

$$E_1 + E_2 = E_3 + E_4 \qquad [4.16]$$

By an "intelligent guess," the solution to Equations 4.15 and 4.16 is

$$f(E) = \frac{1}{1 + A \exp\left(\dfrac{E}{kT}\right)} \qquad [4.17]$$

where A is a constant. You can check that this is a solution by substituting Equation 4.17 into 4.15 and using Equation 4.16. Letting $A = \exp(-E_F/kT)$, we can write Equation 4.17 as

Fermi–Dirac statistics

$$f = \frac{1}{1 + \exp\left[\dfrac{E - E_F}{kT}\right]} \qquad [4.18]$$

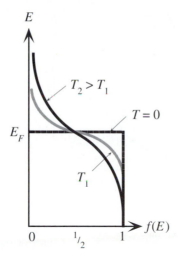

Figure 4.25
The Fermi–Dirac function $f(E)$, describes the statistics of electrons in a solid. The electrons interact with each other and the environment, obeying the Pauli exclusion Principle.

where E_F is a constant called the **Fermi energy.** The probability of finding an electron in a state with energy E is given by Equation 4.18, which is called the **Fermi–Dirac function.**

The behavior of the Fermi–Dirac function is shown in Figure 4.25. Note the effect of temperature. As T increases, $f(E)$ extends to higher energies. At energies of a few $kT(0.026 \text{ eV})$ above E_F, $f(E)$ behaves almost like the Boltzmann function

$$f(E) = \exp\left[-\frac{(E - E_F)}{kT}\right] \qquad (E - E_F) \gg kT \qquad \textbf{[4.19]}$$

Above absolute zero, at $E = E_F$, $f(E_F) = 1/2$. We define the Fermi energy as that energy for which the probability of occupancy $f(E_F)$ equals $1/2$. The approximation to $f(E)$ in Equation 4.19 at high energies is often referred to as the **Boltzmann tail** to the Fermi–Dirac function.

4.7   MODERN THEORY OF METALS

We know that the number of states $g(E)$ for an electron, per unit energy per unit volume increases with energy as $g(E) \propto E^{1/2}$. We have also calculated that the probability of an electron being in a state with an energy E is the Fermi–Dirac function $f(E)$. Consider the energy band diagram for a metal and the density of states $g(E)$ for that band, as shown in Figure 4.26a and b, respectively.

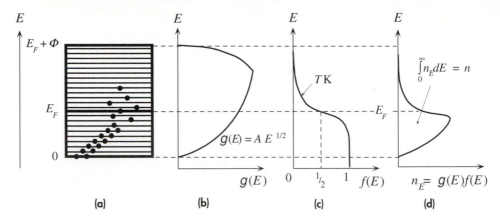

(a) Above 0 K, due to thermal excitation, some of the electrons are at energies above E_F. (b) The density of states, $g(E)$ versus E in the band. (c) The probability of occupancy of a state at an energy E is $f(E)$. (d) The product $g(E)f(E)$ is the number of electrons per unit energy per unit volume, or the electron concentration per unit energy. The area under the curve on the energy axis is the concentration of electrons in the band.

Figure 4.26

At absolute zero, all the energy levels up to E_F are full. At 0 K, $f(E)$ has the step form at E_F (Figure 4.25). This clarifies why E_F in $f(E)$ is termed the Fermi energy. At 0 K, $f(E) = 1$ for $E < E_F$, and $f(E) = 0$ for $E > E_F$, so at 0 K, E_F separates the empty and full energy levels. This explains why we restricted ourselves to 0 K or thereabouts when we introduced E_F in the band theory of metals.

At some finite temperature, $f(E)$ is *not* zero beyond E_F, as indicated in Figure 4.26c. This means that some of the electrons are excited to, and thereby occupy, energy levels above E_F. If we multiply $g(E)$, by $f(E)$, we obtain the number of electrons per unit energy per unit volume, denoted n_E. The distribution of electrons in the energy levels is described by $n_E = g(E)f(E)$.

Since $f(E) = 1$ for $E \ll E_F$, the states near the bottom of the band are all occupied; thus, $n_E \propto E^{1/2}$ initially. As E passes through E_F, $f(E)$ starts decreasing sharply. As a result, n_E takes a turn and begins to decrease sharply as well, as depicted in Figure 4.26d.

In the small energy range E to $(E + dE)$, there are $n_E dE$ electrons per unit volume. When we sum all $n_E dE$ from the bottom to the top of the band ($E = 0$ to $E = E_F + \Phi$), we get the total number of valence electrons per unit volume, n, in the metal, as follows:

$$n = \int_0^{\text{Top of band}} n_E \, dE = \int_0^{\text{Top of band}} g(E)f(E) \, dE \qquad \text{[4.20]}$$

Since $f(E)$ falls very sharply when $E > E_F$, we can carry the integration to $E = \infty$, rather than to $(E_F + \Phi)$, because $f \to 0$ when $E \gg E_F$. Putting in the functional forms of $g(E)$ and $f(E)$ (e.g., from Equations 4.10 and 4.18), we obtain

$$n = \frac{8\pi 2^{1/2} m_e^{3/2}}{h^3} \int_0^\infty \frac{E^{1/2} \, dE}{1 + \exp\left(\dfrac{E - E_F}{kT}\right)} \qquad [4.21]$$

If we could integrate this, we would obtain an expression relating n and E_F. At 0 K, however, $E_F = E_{FO}$ and the integrand exists only for $E < E_{FO}$. If we integrate at 0 K, Equation 4.21 yields

$$E_{FO} = \left(\frac{h^2}{8m_e}\right)\left(\frac{3n}{\pi}\right)^{2/3} \qquad [4.22] \qquad \textit{Fermi energy at } T = 0 \text{ K}$$

It may be thought that E_F is temperature independent, since it was sketched that way in Figure 4.25. However, in our derivation of the Fermi–Dirac statistics, there was no restriction that demanded this. Indeed, since the number of electrons in a band is fixed, E_F at a temperature T is implicitly determined by Equation 4.22, which can be solved to express E_F in terms of n and T. It turns out that at 0 K, E_F is given by Equation 4.22, and it changes very little with temperature. In fact, by utilizing various mathematical approximations, it is not too difficult to integrate Equation 4.21 to obtain the Fermi energy at a temperature T, as follows:

$$E_F(T) = E_{FO}\left[1 - \frac{\pi^2}{12}\left(\frac{kT}{E_{FO}}\right)^2\right] \qquad [4.23] \qquad \textit{Fermi energy at } T \text{ K}$$

which shows that $E_F(T)$ is only weakly temperature dependent, since $E_{FO} \gg kT$.

The Fermi energy has an important significance in terms of the average energy E_{av} of the conduction electrons in a metal. In the energy range E to $(E + dE)$, there are $n_E dE$ electrons with energy E. The average energy of an electron will therefore be

$$E_{av} = \frac{\int E n_E \, dE}{\int n_E \, dE} \qquad [4.24]$$

If we substitute $g(E)f(E)$ for n_E and integrate, the result at 0 K is

$$E_{av}(0) = \frac{3}{5} E_{FO} \qquad [4.25] \qquad \textit{Average energy per electron at } 0 \text{ K}$$

Above absolute zero, the average energy is approximately

$$E_{av}(T) = \frac{3}{5} E_{FO}\left[1 + \frac{5\pi^2}{12}\left(\frac{kT}{E_{FO}}\right)^2\right] \qquad [4.26] \qquad \textit{Average energy per electron at } T \text{ K}$$

Since $E_{FO} \gg kT$, the second term in the square brackets is much smaller than unity, and $E_{av}(T)$ shows only a very weak temperature dependence. Furthermore, in our model of the metal, the electrons are free to move around within the metal, where their potential energy PE is zero, whereas outside the metal, it is $E_F + \Phi$ (Figure 4.11). Therefore, their energy is purely kinetic. Thus, Equation 4.26 gives the average KE of the electrons in a metal

$$\frac{1}{2} m_e v_e^2 = E_{av} \approx \frac{3}{5} E_{FO}$$

where v_e is the root mean square (rms) speed of the electrons, which is simply called the **effective speed.** The effective speed v_e depends on the Fermi energy E_{FO} and is relatively insensitive to temperature. Compare this with the behavior of molecules in an ideal gas. In that case, the average $KE = (3/2)kT$, so $(1/2)mv^2 = (3/2)kT$. Clearly, the average speed of molecules in a gas increases with temperature.

The relationship $(1/2)mv_e^2 \approx (3/2)E_{FO}$ is an important conclusion that comes from the application of quantum mechanical concepts, ideas that lead to $g(E)$ and $f(E)$ and so on. It cannot be proved without invoking quantum mechanics.

The fact that the average electronic speed is nearly constant is the only way to explain the observation that the resistivity of a metal is proportional to T (and not $T^{3/2}$), as we saw in Chapter 2. In a metal, a conduction electron is essentially scattered by its collisions with thermally vibrating metal ions. Recall that the mean frequency of such collisions is proportional to the cross-sectional area of the thermally vibrating ion, or $S = \pi a^2$ (where a is the amplitude of the thermal vibrations), the speed of the electron (approximately v_e), and the concentration of scatters (N_s). The effective speed v_e has very little temperature dependence, and N_s is constant because all atoms in the metal are executing thermal vibrations at sufficiently high temperatures (e.g., > 100 K). The average KE of the thermal vibrations of the atoms along one coordinate is[3] $(1/4)M\omega^2 a^2$, where M is the mass of the atom and ω is its frequency of oscillation. By Maxwell's equipartition of energy theorem, this average KE must be $(1/2)kT$, so $a^2 \propto T$. Thus, the mean free time between collisions is

$$\tau = \frac{1}{(\pi a^2)v_e N_s} \propto \frac{1}{(\pi a^2)} \propto \frac{1}{T}$$

Since the drift mobility μ_d of the electrons is determined by the mean free time between collisions τ, we have

$$\mu_d = \frac{e\tau}{m_e} \propto \frac{1}{T}$$

The conductivity σ is $en\mu_d$, where n is the concentration of conduction electrons (which is constant), and this leads to the desired result:

$$\sigma \propto \frac{1}{T}$$

Note that in a metal, there is a large number of free electrons moving around, with a large temperature-independent effective speed v_e. When an electric field is applied, the electrons drift in the opposite direction to the field, with a velocity $v_d = \mu_d \mathcal{E}_x$, where μ_d is the drift mobility. The drift velocity v_d is much smaller than the effective speed v_e.

One final aspect of metals is that we assumed *all* the valence electrons constituting the electron gas could contribute to charge conduction. We also mentioned

---

[3] We assume that the atomic oscillations are simple harmonic motion.

that when an electron is accelerated by an applied field, it gains energy and must therefore move to higher energy levels. This is only possible if the higher energy levels are empty. As is obvious from Figure 4.26, the energy levels around E_F are not very populated, so the electrons with energies around E_F make the major contribution to the electric current. The electrons at the Fermi energy have velocities given by $v_F = \sqrt{[2E_F/m_e]}$, which is called the Fermi speed. Since the electrons near E_F contribute to conduction, when calculating the mean free path and the drift mobility, we should take the mean electronic speed as being close to v_F, which, like v_e, has very little temperature dependence.

MEAN SPEED OF CONDUCTION ELECTRONS IN A METAL   Calculate the Fermi energy E_{FO} at 0 K for copper and estimate the average speed of the conduction electrons in Cu. The density of Cu is 8.93 g cm^{-3} and the relative atomic mass (atomic weight) is 63.5. | **Example 4.9**

Solution

Assuming each Cu atom donates one free electron, we can find the concentration of electrons from the density d, atomic mass M_{at}, and Avogadro's number N_A, as follows:

$$n = \frac{dN_A}{M_{at}} = \frac{8.93 \times 6.02 \times 10^{23}}{63.5}$$

$$= 8.47 \times 10^{22} \text{ cm}^{-3} \quad \text{or} \quad 8.47 \times 10^{28} \text{ m}^{-3}$$

The Fermi energy at 0 K is given by Equation 4.22:

$$E_{FO} = \left(\frac{h^2}{8m_e}\right)\left(\frac{3n}{\pi}\right)^{2/3}$$

Substituting $n = 8.47 \times 10^{28}$ m^{-3} and the values for h and m_e, we obtain

$$E_{FO} = 1.127 \times 10^{-18} \text{ J} \quad \text{or} \quad 7.03 \text{ eV}$$

To estimate the mean speed of the electrons, we calculate the rms speed v_e from $(1/2)m_e v_e^2 = (3/5)E_{FO}$. The mean speed will be close to the rms speed. Thus, $v_e = (6E_{FO}/5m_e)^{1/2}$. Substituting for E_{FO} and m_e, we find $v_e = 1.22 \times 10^6$ m s^{-1}.

MEAN FREE PATH OF AN ELECTRON IN A METAL   If the electron mobility in Cu is 43 cm^2 V^{-1} s^{-1}, what is the mean free path of electrons between collisions with the lattice? | **Example 4.10**

Solution

Apply the mobility formula $\mu_d = e\tau/m_e$ to obtain the mean free time, as follows:

$$\tau = \frac{\mu_d m_e}{e} = \frac{(43 \times 10^{-4} \text{ m}^2 \text{ V}^{-1} \text{ s}^{-1})(9.1 \times 10^{-31} \text{ kg})}{(1.6 \times 10^{-19} \text{ C})} = 2.45 \times 10^{-14} \text{ s}$$

The effective mean free path is

$$\ell_e = v_e\tau = (1.22 \times 10^6 \text{ m s}^{-1})(2.45 \times 10^{-14} \text{ s})$$

$$= 2.99 \times 10^{-8} \text{ m} \quad \text{or} \quad 29.9 \text{ nm}$$

4.8   FERMI ENERGY SIGNIFICANCE

4.8.1   Metal–Metal Contacts: Contact Potential

Suppose that two metals, platinum (Pt) with a work function 5.36 eV and molybdenum (Mo) with a work function 4.20 eV, are brought together, as shown in Figure 4.27a. We know that in metals, all the energy levels up to the Fermi level are full. Since the Fermi level is higher in Mo (due to a smaller Φ), the electrons in Mo are more energetic. They therefore immediately go over to the Pt surface (by tunneling), where there are empty states at lower energies, which they can occupy. This electron transfer from Mo to the Pt surface reduces the total energy of the electrons in the Pt–Mo system, but at the same time, the Pt surface becomes negatively charged with respect to the Mo surface. Consequently, a contact voltage (or a potential difference) develops at the junction between Pt and Mo, with the Mo side being positive.

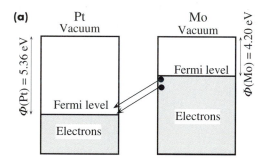

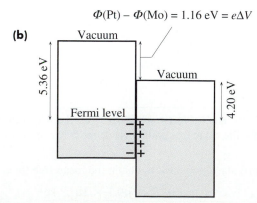

(a) Electrons are more energetic in Mo, so they tunnel to the surface of Pt.   (b) Equilibrium is reached when the Fermi levels are lined up.

Figure 4.27

When two metals are brought together, there is a contact potential ΔV.

   The electron transfer from Mo to Pt continues until the contact potential is large enough to prevent further electron transfer: the system reaches equilibrium. It should be apparent that the transfer of energetic electrons from Mo to Pt continues until the two Fermi levels are lined up, that is, until the Fermi level is uniform and the same in both metals, so that no part of the system has more (or less) energetic electrons, as illustrated in Figure 4.27b. Otherwise, the energetic electrons in one part of the system will flow towards a region with lower-energy states. Under these conditions, the Pt–Mo system is in equilibrium. The contact voltage ΔV is determined by the difference in the work functions, that is,

$$e\,\Delta V = 5.36 \text{ eV} - 4.20 \text{ eV} = 1.16 \text{ eV}$$

   We should note that away from the junction on the Mo side, we must still provide an energy of $\Phi = 4.20$ eV to free an electron, whereas away from the junction on the Pt side, we must provide $\Phi = 5.36$ eV to free an electron. This means that the vacuum energy level going from Mo to Pt has a step $\Delta\Phi$ at the junction. Since we must do work equivalent to $\Delta\Phi$ to get a free electron (e.g., on the metal surface) from the Mo surface to the Pt surface, this represents a voltage of $\Delta\Phi/e$ or 1.16 V.

   From the Second Law of Thermodynamics,[4] this contact voltage cannot do work, that is, it cannot drive current in an external circuit. To see this, we can close the Pt metal–Mo metal circuit to form a ring, as depicted in Figure 4.28. As soon as we close the circuit, we create another junction with a contact voltage that is equal and opposite to that of the first junction. Consequently, going around the circuit, the net voltage is zero and the current is therefore zero.

   There is a deep significance of the Fermi energy E_F, which should at least be mentioned. For a given metal the Fermi energy represents the free energy per

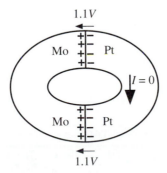

Figure 4.28
There is no current when a closed circuit is formed by two different metals, even though there is a contact potential at each contact. The contact potentials oppose each other.

---

[4] By the way, the Second Law of Thermodynamics simply says that you cannot extract heat from a system in thermal equilibrium and do work (i.e., charge × voltage).

electron called the **electrochemical potential** μ. In other words, the Fermi energy is a measure of the potential of an electron to do electrical work ($V \Delta Q$) or nonmechanical work, through chemical or physical processes.[5] In general, when two metals are brought into contact, the Fermi level (with respect to vacuum) in each will be different. This difference means a difference in the chemical potential $\Delta\mu$, which in turn means that the system will do external work, which is obviously not possible. Instead, electrons are immediately transferred from one metal to the other, until the free energy per electron μ for the whole system is minimized and is uniform across the two metals, so that $\Delta\mu = 0$. We can guess that if the Fermi level in one metal could be maintained at a higher level than the other, by using an external energy source (e.g., light or heat), for example, then the difference could be used to do electrical work.

4.8.2   The Seebeck Effect and the Thermocouple

Consider a copper rod that is heated at one end and cooled at the other end, as depicted in Figure 4.29. The electrons in the hot region are more energetic and therefore have greater velocities than those in the cold region.[6] Consequently, there

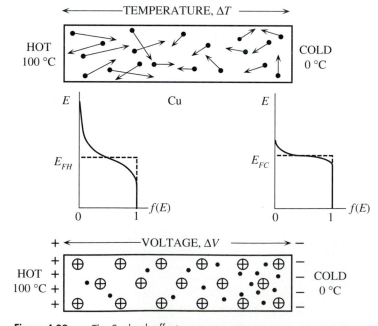

Figure 4.29   The Seebeck effect
A temperature gradient along a conductor gives rise to a potential difference.

[5] A change in any type of PE can, in principle, be used to do work, that is, $\Delta(PE)$ = Work done. Chemical PE is the potential to do nonmechanical work (e.g., electrical work) by virtue of physical or chemical processes. The chemical PE per electron is E_F and ΔE_F = Electrical work.

[6] Even though the mean speed has only a small temperature dependence, as discussed in section 4.7, this small change is important in understanding the thermoelectric effect.

is a net diffusion of electrons from the hot end toward the cold end, which leaves exposed, positive metal ions behind in the hot region and accumulates electrons in the cold region. This situation prevails until the electric field developed between the positive ions in the hot region and the excess electrons in the cold region prevents further electron motion from the hot to the cold. A voltage therefore develops between the hot and cold ends, with the hot end at positive potential. The potential difference ΔV across a piece of metal due to a temperature difference ΔT is called the **Seebeck effect.**[7] To gauge the magnitude of this effect, we introduce a special coefficient defined as the potential difference developed per unit temperature difference, or

$$S = \frac{dV}{dT}$$ [4.27]   *Thermoelectric power, or Seebeck coefficient*

The coefficient S is widely referred to as the **thermoelectric power,** even though this is misleading since it refers to a voltage difference rather than power. However, the term has stuck. A recent alternative and more appropriate term is the **Seebeck coefficient.** Here, S is a material property that depends on temperature, $S = S(T)$, and is tabulated for many materials as a function of temperature. Given the Seebeck coefficient $S(T)$ for a material, Equation 4.27 yields the voltage difference between two points where temperatures are T_o and T, as follows:

$$\Delta V = \int_{T_o}^{T} S\ dT$$ [4.28]

This intuitive (and correct) line of thinking is not apparent from the Fermi energy, because the Fermi energy E_F in the hot region is actually very slightly lowered by the increase in temperature, by virtue of Equation 4.23. However, this is only the Fermi energy, whereas what we need is the average energy per electron, $E_{av}(T)$, which we have already evaluated as Equation 4.26,

$$E_{av}(T) = \frac{3}{5} E_{FO} \left[1 + \frac{5\pi^2}{12} \left(\frac{kT}{E_{FO}} \right)^2 \right]$$ [4.29]

Although E_F may be slightly lower, the Fermi–Dirac distribution actually extends to much higher energies when the temperature is raised, as depicted in Figure 4.29, so the average energy per electron, as determined by Equation 4.29, is actually greater in the hot end. Consequently, more energetic electrons in the hot end diffuse toward the cold region until a potential difference ΔV that prevents further diffusion is built up. We should also note that the average energy per electron, as determined by Equation 4.29, also depends on the material, by virtue of E_{FO}.

Suppose that a small temperature difference δT results in a voltage difference δV between the accumulated electrons and exposed, positive metal ions, as depicted in Figure 4.30. Suppose that one electron manages to diffuse from the hot region to the cold region. It must do work $e\delta V$ against the potential difference δV.

---

[7] Thomas Seebeck observed the effect in 1821, using two different metals, as in the thermocouple, which is the only way to observe the phenomenon. It was Thomson (Lord Kelvin) who explained the observed effect.

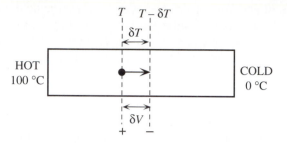

Figure 4.30   Consider a small length δx, over which the temperature difference is δT and voltage difference is δV

This work decreases the average energy of the electron by δE_{av} from E_{av}(hot) to E_{av}(cold), or

$$e\,\delta V = E_{av}(T + \delta T) - E_{av}(T)$$

Substituting for $E_{av}(T)$ from Equation 4.29, expanding $(T + \delta T)$, and neglecting the δT^2 term, we obtain

$$e\,\delta V \approx \frac{\pi^2 k^2 T\,\delta T}{2E_{FO}}$$

Since $S = \dfrac{\delta V}{\delta T}$, the Seebeck coefficient is given by

Seebeck coefficient for metals

$$S \approx \frac{\pi^2 k^2 T}{2eE_{FO}} \qquad\qquad [4.30]$$

A more rigorous theory considers the actual motion of the electrons and their interactions with the lattice vibrations in the presence of a temperature gradient. This approach makes S somewhat larger. We will consider Equation 4.30 to be a good approximation of the Seebeck coefficient for metals. For example, for Cu, $E_{FO} = 7.0$ eV. Therefore, at $T = 300$ K (27 °C), Equation 4.30, predicts 1.57 μV K^{-1}, which is on the order of the experimentally inferred value of about 1.84 μV K^{-1}. Table 4.3 summarizes some typical experimental values for the Seebeck coefficient of a few metals. The values are in the "microvolt per Kelvin" range.

Suppose we try to measure the voltage difference ΔV across the copper rod by using copper wires connected to a voltmeter, as indicated in Figure 4.31. The same temperature difference now also exists across the copper connecting wires; therefore, an identical voltage also develops across the connecting wires, opposing that across the copper rod. Consequently, no net voltage will be registered by the voltmeter. It is possible, however, to read a net voltage difference, if the connecting wires are of a different material, that is, have a different Seebeck coefficient than that of copper. Then, the thermally induced voltage across this material will be different than that across the copper rod.

Table 4.3   Seebeck coefficient of a few pure metals at 0 °C and 27 °C (S is in $\mu V\ K^{-1}$)

Metal	S(0 °C)	S(27 °C)
Ag	1.38	1.51
Au	1.79	1.94
Cu	1.70	1.84
Li	14.00	
Mo	4.71	5.57

(N. Cusack and P. Kendall, *Proceedings of the Royal Society* 72, no. 289, (1958); and A.H. Wilson, *The Theory of Metals*, 2nd ed. Cambridge, UK; Cambridge University Press, 1958, p. 207)

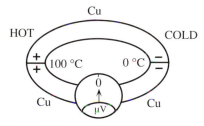

Figure 4.31

If Cu wires are used to measure the Seebeck voltage across the Cu rod, then the net emf is zero.

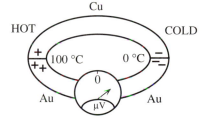

Figure 4.32   The Cu-Au thermocouple The cold end is maintained at 0 °C, which is the reference temperature. The other junction is used to sense the temperature. In this example, it is heated to 100 °C.

The Seebeck effect is fruitfully utilized in the thermocouple (TC), shown in Figure 4.32, which uses two different metals. One junction is maintained at a reference temperature T_o, and the other is used to sense the temperature T. The voltage across each metal element depends on its Seebeck coefficient. The potential difference between the two wires will depend on $(S_A - S_B)$. By virtue of Equation 4.28, the emf between the two wires, $V_{AB} = \Delta V_A - \Delta V_B$, is then given by

$$V_{AB} = \int_{T_o}^{T} (S_A - S_B)\, dT = \int_{T_o}^{T} S_{AB}\, dT \qquad \textbf{[4.31]}$$

where $S_{AB} = S_A - S_B$ is defined as the thermoelectric power for the thermocouple pair AB. For the chromel–alumel (K-type) TC, for example, $S_{AB} \approx 40\ \mu V\ K^{-1}$ at 300 K.

The output voltage from a TC pair obviously depends on the two metals used. Instead of tabulating the emf from all possible pairs of materials in the world, which

Although the Cu–Au pair can be used to measure temperatures, its output emf is far too small. A more practical pair that gives higher voltages, such as chromel–alumel, iron–constantan, or copper–constantan, etc., would be preferable. It would also be so useful to have thermocouple voltages for various metals listed against some reference metal. The reference is usually Pt, with the reference junction at 0 °C. From Table 4.4, we can read Cu–Pt and Au–Pt emfs as $V_{Cu-Pt} = 0.76$ mV   and $V_{Au-Pt} = 0.78$ mV at 100 °C, with the experimental error being around ±0.01 mV. For the Cu–Au pair, we have

$$V_{Cu-Au} = V_{Cu-Pt} - V_{Au-Pt} = 0.78 \text{ mV} - 0.76 \text{ mV} = 0.02 \text{ mV} \qquad \text{or} \qquad 20 \ \mu V$$

In order of magnitude, this is close to the value estimated in this example, within experimental errors and within the limit of the theory.

Example 4.12   **THE THERMOCOUPLE EQUATION**   We can only measure differences between the thermoelectric powers of different materials. When two different metals, A and B, form a thermocouple, as in Figure 4.32, the net emf is the voltage difference between the two elements. From Equation 4.28, we have

$$V_{AB} = \Delta V_A - \Delta V_B = \int_{T_o}^{T} S_A \, dT - \int_{T_o}^{T} S_B \, dT$$

From Equation 4.30, we can substitute $S = \pi^2 k^2 T / 2e E_{FO}$ to obtain

$$V_{AB} = \left(\frac{\pi^2 k^2}{2e E_{FAO}}\right) \int_{T_o}^{T} T \, dT - \left(\frac{\pi^2 k^2}{2e E_{FBO}}\right) \int_{T_o}^{T} T \, dT$$

where E_{FAO} and E_{FBO} refer to the E_{FO} for A and B, respectively. Integrating from T_o to T, the emf is

$$V_{AB} = A(T^2 - T_o^2) - B(T^2 - T_o^2) = C(T^2 - T_o^2)$$

where $A = \pi^2 k^2 / 4e E_{FAO}$, $B = \pi^2 k^2 / 4e E_{FBO}$,   and $C = A - B$. According to this result, the emf is parabolic in T.

We can now expand V_{AB} about T_o by using Taylor's expansion

$$F(T) = F(T_o) + \Delta T \frac{dF}{dT} + \frac{1}{2}(\Delta T)^2 \left(\frac{d^2 F}{dT^2}\right)$$

where $F = V_{AB}$, $\Delta T = T - T_o$, and the derivatives are evaluated at T_o. The result is the thermocouple equation

$$V_{AB}(T) = a(\Delta T) + b(\Delta T)^2$$

The coefficients a and b are $2CT_o$ and C, respectively.

It is clear that the magnitude of the emf produced depends on $C = A - B$, or $S_A - S_B$ which we can label S_{AB}. The greater the thermoelectric power difference S_{AB} for the TC, the larger the emf produced. For the copper-constantan TC, S_{AB} is about 43 μV K^{-1}. On the other hand, for the Cu–Au TC of Example 4.10, with $E_{FOA} = 5.51$ eV (Au) and $E_{FOB} = 7.01$ eV (Cu) at 300 K, we have

$$S_{AB} = \frac{\pi^2 k^2 T}{2e} \left(\frac{1}{E_{FAO}} - \frac{1}{E_{FBO}}\right) = (2.00 - 1.57) \ \mu V \ K^{-1} = 0.43 \ \mu V \ K^{-1}$$

which is 50 times less than that for the K-type TC.

4.9   THERMIONIC EMISSION AND VACUUM TUBE DEVICES

4.9.1   Thermionic Emission: Richardson–Dushman Equation

Even though most of us view vacuum tubes as electrical antiques, their basic principle of operation (electrons emitted from a heated cathode) still finds application in the cathode ray tube and various microwave tubes, such as the UHF triode, klystron, and magnetron. Therefore, it is useful to examine how electrons are emitted when a metal is heated.

When a metal is heated, the electrons become more energetic as the Fermi–Dirac function extends to higher temperatures. Some of the electrons have sufficiently large energies to leave the metal and become free. This situation is self-limiting because as the electrons accumulate outside the metal, they prevent more electrons from leaving the metal. (Put differently, emitted electrons leave a net positive charge behind, which pulls the electrons in.) Consequently, we need to replenish the "lost" electrons and collect the emitted ones, which is done most conveniently using the vacuum tube arrangement in a closed circuit, as shown in Figure 4.34. The cathode, heated by a filament, emits electrons. A battery connected between the cathode and the anode replenishes the cathode electrons and provides a positive bias to the anode to collect the thermally-emitted electrons from the cathode. The vacuum inside the tube ensures that the electrons do not collide with the air molecules and become dispersed, with some even being returned to the cathode by collisions. Therefore, the vacuum is essential.

We know that only those electrons with energies greater than $E_F + \Phi$ (Fermi energy + work function) which are moving toward the surface can leave the metal. Their number depends on the temperature, by virtue of the Fermi–Dirac statistics. Figure 4.35 shows how the concentration of conduction electrons with energies above $E_F + \Phi$ increases with temperature. We know that conduction electrons behave as if they are free within the metal. We can therefore take the *PE*

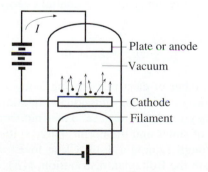

Figure 4.34   Thermionic electron emission in a vacuum tube

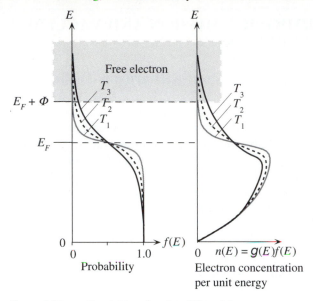

Figure 4.35   Fermi–Dirac function $f(E)$ and the energy density of electrons $n(E)$ (electrons per unit energy and per unit volume) at three different temperatures.

The electron concentration extends more and more to higher energies as the temperature increases. Electrons with energies in excess of $E_F + \Phi$ can leave the metal (thermionic emission).

to be zero within the metal, but $E_F + \Phi$ outside the metal. The energy E of the electron within the metal is then purely kinetic, or

$$E = \frac{1}{2}m_e v_x^2 + \frac{1}{2}m_e v_y^2 + \frac{1}{2}m_e v_z^2 \qquad \text{[4.33]}$$

Suppose that the surface of the metal is perpendicular to the direction of emission, say along x. For an electron to be emitted from the surface, its $KE = (1/2)\,mv_x^2$ along x must be greater than the potential energy barrier $E_F + \Phi$, that is,

$$\frac{1}{2}mv_x^2 > E_F + \Phi \qquad \text{[4.34]}$$

Let $dn(v_x)$ be the number of electrons moving along x with velocities in the range v_x to $(v_x + dv_x)$, with v_x satisfying emission in Equation 4.34. These electrons will be emitted when they reach the surface. Their number $dn(v_x)$ can be determined from the density of states and the Fermi–Dirac statistics, since energy and velocity are related through Equation 4.33. Close to $E_F + \Phi$, the Fermi-Dirac function will approximate the Boltzman distribution, $f(E) = \exp[-E - E_F/kT]$. The number $dn(v_x)$ is therefore at least proportional to this exponential energy factor.

The emission of $dn(v_x)$ electrons will give a thermionic current density $dJ_x = ev_x \, dn(v_x)$. This must be integrated (summed) for all velocities satisfying Equation 4.34 to obtain the total current density J_x, or simply J. Since $dn(v_x)$ includes an exponential energy function, the integration also leads to an exponential. The final result is

$$J = B_o T^2 \exp\left(-\frac{\Phi}{kT}\right)$$ [4.35]

where $B_o = 4\pi em_e k^2/h^3$. Equation 4.35 is called the **Richardson–Dushman equation,** and B_o is the Richardson–Dushman constant, whose value is 1.20×10^6 A m^{-2} K^{-2}. We see from Equation 4.35 that the emitted current from a heated cathode varies exponentially with temperature and is sensitive to the work function Φ of the cathode material. Both factors are apparent in Figure 4.35.

The wave nature of electrons means that when an electron approaches the surface, there is a probability that it may be reflected back into the metal, instead of being emitted over the potential barrier. As the potential energy barrier becomes very large, $\Phi \rightarrow \infty$, the electrons are totally reflected and there is no emission. Taking into account that waves can be reflected, the thermionic emission equation is appropriately modified to

$$J = B_e T^2 \exp\left(-\frac{\Phi}{kT}\right)$$ [4.36]   *Thermionic emission*

where $B_e = (1 - R)B_o$ is the emission constant and R is the reflection coefficient. The value of R will depend on the material and the surface conditions. For most metals, B_e is about half of B_o, whereas for some oxide coatings on Ni cathodes used in thermionic tubes, B_e can be as low as 1×10^2 A m^{-2} K^{-2}.

Equation 4.35 was derived by neglecting the effect of the applied field on the emission process. Since the anode is positively-biased with respect to the cathode, the field will not only collect the emitted electrons (by drifting them to the anode), but will also enhance the process of thermal emission by lowering the potential energy barrier Φ.

VACUUM TUBES   It is clear from the Richardson–Dushman equation that to obtain an efficient thermionic cathode, we need high temperatures and low work functions. Metals such as tungsten (W) and tantalum (Ta) have high melting temperatures but high work functions. For example, for W, the melting temperature T_m is 3680 °C and its work function is about 4.5 eV. Some metals have low work functions, but also low melting temperatures, a typical example being Cs with $\Phi = 1.8$ eV and $T_m = 28.5$ °C. If we use a thin film coating of a low Φ material, such as ThO or BaO, on a high-melting-temperature base metal such as W, we can maintain the high melting properties and obtain a lower Φ. For example, Th on W has a $\Phi = 2.6$ eV and $T_m = 1845$ °C. Most vacuum tubes use indirectly heated cathodes that consist of the oxides of B, Sr, and Ca on a base metal of Ni. The operating temperatures for these cathodes are typically 800 °C.

A certain transmitter-type vacuum tube has a cylindrical Th-coated W (thoriated tungsten) cathode, which is 4 cm long and 2 mm in diameter. Estimate the saturation current

Example 4.13

if the tube is operated at a temperature of 1600 °C, given that the emission constant is $B_e = 3.0 \times 10^4$ A m^{-2} K^{-2} for Th on W.

Solution

We apply the Richardson–Dushman equation with $\Phi = 2.6$ eV, $T = (1600 + 273)$ K = 1873 K, and $B_e = 3.0 \times 10^4$ A m^{-2} K^{-2}, to find the maximum current density that can be obtained from the cathode at 1873 K, as follows:

$$J = (3.0 \times 10^4 \text{ A m}^{-2} \text{ K}^{-2})(1873 \text{ K})^2 \exp\left[-\frac{(2.6 \times 1.6 \times 10^{-19})}{(1.38 \times 10^{-23} \times 1873)} \right]$$

$$= 1.08 \times 10^4 \text{ A m}^{-2}$$

The emission surface area is

$$A = \pi(\text{diameter})(\text{length}) = \pi(2 \times 10^{-3})(4 \times 10^{-2}) = 2.5 \times 10^{-4} \text{ m}^2$$

so the saturation current, which is the maximum current obtainable (i.e., the thermionic current), is

$$I = JA = (1.08 \times 10^4 \text{ A m}^{-2})(2.5 \times 10^{-4} \text{ m}^2) = 2.7 \text{ A}$$

4.9.2   Field-Assisted Emission: The Schottky Effect

When a positive voltage is applied to the anode with respect to the cathode, the electric field at the cathode helps the thermionic emission process by lowering the *PE* barrier Φ. This is called the **Schottky effect.** Consider the *PE* of the electron just outside the surface of the metal. The electron is pulled in by the effective positive charge left in the metal. To represent this attractive *PE* we use the **theorem of image charges** in electrostatics,[8] which says that an electron at a distance x from the surface of a conductor possesses a potential energy that is

$$PE_{\text{image}}(x) = \frac{-e^2}{16\pi\varepsilon_o x} \qquad [4.37]$$

where ε_o is the absolute permittivity.

This equation is valid for x much greater than the atomic separation a; otherwise, we must consider the interaction of the electron with the individual ions. Further, Equation 4.37 has a reference level of zero *PE* at infinity ($x = \infty$), but we defined *PE* = 0 to be inside the metal. We must therefore modify Equation 4.37 to conform to our definition of zero *PE* as a reference. Figure 4.36a shows how this "image *PE*" varies with x in this system. In the region $x < x_o$, we artificially bring $PE_{\text{image}}(x)$ to zero at $x = 0$, so our definition *PE* = 0 within the metal is maintained. Far away from the surface, the *PE* is expected to be $(E_F + \Phi)$ (and not zero, as in Equation 4.37), so we modify Equation 4.37 to read

$$PE_{\text{image}}(x) = (E_F + \Phi) - \frac{e^2}{16\pi\varepsilon_o x} \qquad [4.38]$$

---

[8] An electron at a distance x from the surface of a conductor experiences a force as if there were a positive charge of $+e$ at a distance $2x$ from it. The force is $e^2/[4\pi\varepsilon_o(2x)^2]$ or $e^2/[16\pi\varepsilon_o x^2]$. The result is called the image charge theorem. Integrating the force gives the potential energy in Equation 4.37.

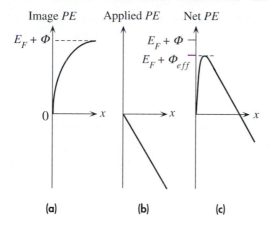

(a) PE of the electron near the surface of a conductor.
(b) Electron PE due to an applied field, that is between cathode and anode. (c) The overall PE is the sum.

Figure 4.36

The present model, which takes $PE(x)$ from 0 to $(E_F + \Phi)$ along Equation 4.38, is in agreement with the thermionic emission analysis, since the electron must still overcome a *PE* barrier of $E_F + \Phi$ to escape.

From the definition of potential, which is potential energy per unit charge, when a voltage difference is applied between the anode and cathode, there is a *PE*, gradient just outside the surface of the metal, given by $eV(x)$, or

$$PE_{\text{applied}}(x) = -ex\mathcal{E}$$ [4.39]

where $\mathcal{E}$ is the applied field and is assumed, for all practical purposes, to be uniform. The variation of $PE_{\text{applied}}(x)$ with x is depicted in Figure 4.36b. The total $PE(x)$ of the electron outside the metal is the sum of Equations 4.38 and 4.39, as sketched in Figure 4.36c,

$$PE(x) = (E_F + \Phi) - \frac{e^2}{16\pi\varepsilon_o x} - ex\mathcal{E}$$ [4.40]

Note that the $PE(x)$ outside the metal no longer goes up to $(E_F + \Phi)$, and the *PE* barrier against thermal emission is effectively reduced to $(E_F + \Phi_{eff})$, where Φ_{eff} is a new effective work function that takes into account the effect of the applied field. The new barrier $(E_f + \Phi_{eff})$ can be found by locating the maximum of $PE(x)$, that is, by differentiating Equation 4.40 and setting it to zero. The effective work function in the presence of an applied field is therefore

$$\Phi_{eff} = \Phi - \left[\frac{e^3\mathcal{E}}{4\pi\varepsilon_o}\right]^{1/2}$$ [4.41]

This lowering of the work function by the applied field, as predicted by Equation 4.41, is the Schottky effect. The current density is given by the Richardson–Dushman equation, but with Φ_{eff} instead of Φ,

Field-assisted thermionic emission

$$J = B_e T^2 \exp\left[\frac{-(\Phi - \beta_S \mathcal{E}^{1/2})}{kT}\right]$$

[4.42]

where $\beta_S = [e^3/4\pi\varepsilon_o]^{1/2}$ is the **Schottky coefficient,** whose value is 3.79×10^{-5} $(\text{eV}/\sqrt{(\text{V m}^{-1})})$.

When the field becomes very large, for example, $\mathcal{E} > 10^7$ V cm^{-1}, the $PE(x)$ outside the metal surface may bend sufficiently steeply to give rise to a narrow PE barrier. In this case, there is a distinct probability that an electron at an energy E_F will tunnel through the barrier and escape into vacuum, as depicted in Figure 4.37. The likelihood of tunneling depends on the effective height Φ_{eff} of the PE barrier above E_F, as well as the width x_F of the barrier at energy level E_F. Since tunneling is temperature independent, the emission process is termed **field emission.** The tunneling probability P depends on Φ_{eff} and x_F via[9]

$$P \approx \exp\left[\frac{-2(2m_e\Phi_{eff})^{1/2}x_F}{\hbar}\right]$$

[4.43]

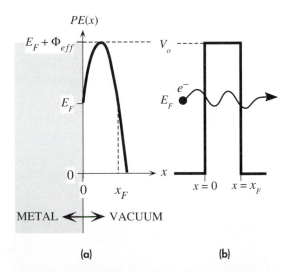

(a) Field emission is the tunneling of an electron at an energy E_F through the narrow PE barrier induced by a large applied field. (b) For simplicity, we take the barrier to be rectangular.

Figure 4.37

---

[9] In Chapter 3 we showed that the transmission probability $T = T_o \exp(-2\alpha a)$ where $\alpha^2 = 2m(V_o - E)/\hbar^2$ and a is the barrier width. The pre-exponential constant T_o can be taken to be ~ 1. Clearly $V_o - E = \Phi_{eff}$ since electrons with $E = E_F$ are tunneling and $a = x_F$.

We can easily find x_F by noting that when $x = x_F$, $PE(x_F)$ is level with E_F, as shown in Figure 4.37. From Equation 4.40, when the field is very strong, then around $x \approx x_F$ the second term is negligible compared to the third, so putting $x = x_F$ and $PE(x_F) = E_F$ in Equation 4.40 yields $\Phi = e\mathcal{E}x_F$. Substituting for Φ_{eff} from Equation 4.41 and $x_F = \Phi/e\mathcal{E}$ in Equation 4.43, we can obtain the tunneling probability P

$$P \approx \exp\left[\frac{-2(2m_e\Phi_{eff})^{1/2}\Phi}{e\hbar\mathcal{E}}\right]$$

which can be rewritten

$$P \approx \exp\left(-\frac{\mathcal{E}_C}{\mathcal{E}}\right) \qquad \mathcal{E}_C = \frac{2(2m_e\Phi_{eff})^{1/2}\Phi}{e\hbar} \qquad \text{[4.44]}$$

where $\mathcal{E}_C$ is a constant with the dimensions of an electric field and is often termed the **critical field for tunneling.**

We take the x direction to represent the normal to the surface from which field emission occurs. In the metal, the electron flux along x to the surface is nv_x, where v_x is the mean speed of the electrons along x and n is the electron concentration. Of these electrons, $(nv_x)P$ tunnel through to the vacuum, so the current density is $e(nv_x)P$. Therefore,

$$J_{\text{field emission}} \approx env_x \exp\left(-\frac{\mathcal{E}_C}{\mathcal{E}}\right) \qquad \text{[4.45]}$$

Field-assisted tunneling: Fowler–Nordhein equation

Equation 4.45 is a simplified **Fowler–Nordheim equation** for field emission. The equation shows that the current increases dramatically with the applied field. Since the thermionic emission contribution is negligible at room temperature, the tunneling current in the scanning tunneling microscope is given by Equation 4.45.

FIELD-ASSISTED THERMIONIC EMISSION   Consider the vacuum tube of Example 4.12. It has a cathode with a work function of $\Phi = 2.6$ eV. Suppose that the anode is separated from the cathode by a distance of 1 mm and that the anode voltage is 4 kV. What is the theoretical saturation current?

| **Example 4.14**

Solution

The field at the cathode is

$$\mathcal{E} = \frac{(4 \times 10^3 \text{ V})}{(10^{-3} \text{ m})} = 4 \times 10^6 \text{ V m}^{-1}$$

The reduction in the work function will be $\beta_S\sqrt{\mathcal{E}}$, where $\beta_S = 3.79 \times 10^{-5}$ (eV/$\sqrt{\text{V m}^{-1}}$) is the Schottky coefficient. Thus,

$$\Phi_{eff} = \Phi - \beta_S\sqrt{\mathcal{E}}$$

$$= 2.6 - (3.79 \times 10^{-5})\sqrt{4 \times 10^6} = 2.52 \text{ eV}$$

Even though the applied field is quite large, the change in Φ is not drastic. The new current will be

$$I_2 = I_o \exp\left(-\frac{\Phi_{eff}}{kT}\right)$$

where $I_o = AB_eT^2$. Originally, we had

$$I_1 = I_o \exp\left(-\frac{\Phi}{kT}\right)$$

So,

$$I_2 = I_1 \exp\left(\frac{\Phi - \Phi_{eff}}{kT}\right) = (2.7) \exp\left[\frac{0.0758 \times 1.6 \times 10^{-19}}{1.38 \times 10^{-23} \times 1873}\right] = 4.3 \text{ A}$$

This is a notable change in the current.

ADDITIONAL TOPICS

4.10   BAND THEORY OF METALS: ELECTRON DIFFRACTION IN CRYSTALS

A rigorous treatment of the band theory of solids involves extensive quantum mechanical analysis and is beyond the scope of this book. However, we can attain a satisfactory understanding through a semiquantitative treatment.

We know that the wavefunction of the electron moving freely along x in space is a traveling wave of the spatial form $\psi_k(x) = \exp(jkx)$, where k is the wavevector $k = 2\pi/\lambda$ of the electron and $\hbar k$ is its momentum. Here, $\psi_k(x)$ represents a traveling wave because it must be multiplied by $\exp(-j\omega t)$ (where $\omega = E/\hbar$) to get the total wavefunction $\Psi(x, t) = \exp[j(kx - \omega t)]$.

We will assume that an electron moving freely within the crystal and within a given energy band should also have a traveling-wave type of wavefunction,

$$\psi_k(x) = A \exp(jkx) \qquad \text{[4.46]}$$

where k is the electron wavevector in the crystal and A is the amplitude. This is a reasonable expectation, since, to a first order, we can take the *PE* of the electron inside a solid as zero, $V = 0$. Yet, the *PE* must be large outside, so the electron is contained within the crystal. When the *PE* is zero, Equation 4.46 is a solution to the Schrödinger equation. The momentum of the electron described by the traveling wave Equation 4.46 is then $\hbar k$ and its energy is

$$E_k = \frac{(\hbar k)^2}{2m_e} \qquad \text{[4.47]}$$

The electron, as a traveling wave, will freely propagate through the crystal. However, not all traveling waves, can propagate in the lattice. The electron cannot have any k value in Equation 4.46 and still move through the crystal. Waves can be reflected and diffracted, whether they are electron waves, X-rays, or visible light. Diffraction occurs when reflected waves interfere constructively. Certain k values will cause the electron wave to be diffracted, preventing the wave from propagating.

The simplest illustration that certain k values will result in the electron wave being diffracted is shown in Figure 4.38 for a hypothetical linear lattice in which diffraction is simply a reflection (what we call diffraction becomes Bragg reflection). The electron is assumed to be propagating in the forward direction along x with a traveling wave function of the type in Equation 4.46. At each atom, some of this wave will be reflected. At A, the reflected wave is A' and has a magnitude A'. If the reflected waves A', B', and C', will reinforce each other, a full reflected wave will be created, traveling in the backward direction. The reflected waves A', B', C', ... will reinforce each other if the path difference between A', B', C', ... is $n\lambda$, where λ is the wavelength and $n = 1, 2, 3, \ldots$ is an integer. When wave B' reaches A', it has traveled an additional distance of $2a$. The path difference between A' and B' is therefore $2a$. For A' and B' to reinforce each other, we need

$$2a = n\lambda \qquad n = 1, 2, 3, \ldots$$

Substituting $\lambda = \dfrac{2\pi}{k}$, we obtain the condition in terms of k

$$k = \frac{n\pi}{a} \qquad n = 1, 2, 3, \ldots \qquad \text{[4.48]}$$

Thus, whenever k is such that it satisfies the condition in Equation 4.48, all the reflected waves reinforce each other and produce a backward-traveling, reflected wave of the following form (with a negative k value):

$$\psi_{-k}(x) = A \exp(-jkx) \qquad \text{[4.49]}$$

This wave will also probably suffer a reflection, since its k satisfies Equation 4.48, and the reflections will continue. The crystal will then contain waves

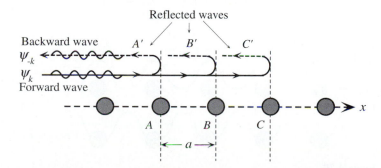

Figure 4.38   An electron wave propagation through a linear lattice.

For certain k values, the reflected waves at successive atomic planes reinforce each other, giveing rise to a reflected wave traveling in the backward direction. The electron cannot then propagate through the crystal.

traveling in the forward and backward directions. These waves will interfere to give **standing waves** inside the crystal. Hence, whenever the k value satisfies Equation 4.48, traveling waves cannot propagate through the lattice. Instead, there can only be standing waves. For k satisfying Equation 4.48, the electron wavefunction consists of waves ψ_k and ψ_{-k} interfering in two possible ways to give two possible standing waves:

$$\psi_c(x) = A \exp(jkx) + A \exp(-jkx) = A_c \cos\left(\frac{n\pi x}{a}\right) \qquad [4.50]$$

$$\psi_s(x) = A \exp(jkx) - A \exp(-jkx) = A_s \sin\left(\frac{n\pi x}{a}\right) \qquad [4.51]$$

The probability density distributions, $|\psi_c(x)|^2$ and $|\psi_s(x)|^2$, for the two standing waves are shown in Figure 4.39. The first standing wave $\psi_c(x)$ is at a maximum on the ion cores, and the other $\psi_s(x)$ is at a maximum between the ion cores. Note also that both the standing waves, $\psi_c(x)$ and $\psi_s(x)$, are solutions to the Schrödinger equation.

The closer the electron is to a positive nucleus, the lower is its electrostatic PE, by virtue of $-e^2/4\pi\varepsilon_o r$. The PE of the electron distribution in $\psi_c(x)$ is lower than that in $\psi_s(x)$, because the maxima for $\psi_c(x)$ are nearer the positive ions. Therefore, the energy of the electron in $\psi_c(x)$ is lower than that of the electron in $\psi_s(x)$, or $E_c < E_s$.

It is not difficult to evaluate the energies E_c and E_s. The kinetic energy of the electron is the same in both $\psi_c(x)$ and $\psi_s(x)$, because these wavefunctions have the same k value and KE is given by $(\hbar k)^2/2m_e$. However, there is an electrostatic PE arising from the interaction of the electron with the ion cores, and this PE is different for the two wavefunctions. Suppose that $V(x)$ is the electrostatic PE of the electron at position x. We then must find the average, using the probability density distribution. Given that $|\psi_c(x)|^2\, dx$ is the probability of finding the electron at x in dx, the potential energy V_c of the electron is simply $V(x)$ averaged over the entire linear length L of the crystal. Thus, the potential energy V_c for $\psi_c(x)$ is

$$V_c = \frac{1}{L} \int_0^L V(x)|\psi_c(x)|^2\, dx = -V_n \qquad [4.52]$$

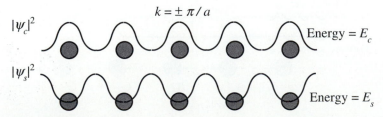

Figure 4.39

Forward and backward waves in the crystal with $k = \pm\pi/a$ give rise to two possible standing waves, ψ_c and ψ_s. Their probability density distributions, $|\psi_c|^2$ and $|\psi_s|^2$, have maxima either at the ions or between the ions respectively.

where V_n is the numerical result of the integration, which depends on $k = n\pi/a$ or n, by virtue of Equation 4.50. The integration in Equation 4.52 is a negative number that depends on n. We do not need to evaluate the integral, as we only need its final numerical result.

Using $|\psi_s(x)|^2$, we can also find V_s, the *PE* associated with $\psi_s(x)$. The result is that V_s is a positive quantity given by $+V_n$, where V_n is again the numerical result of the integration in Equation 4.52, which depends on n. The energies of the wavefunctions ψ_c and ψ_s whenever $k = n\pi/a$ are

$$E_c = \frac{(\hbar k)^2}{2m_e} - V_n \qquad k = \frac{n\pi}{a} \qquad\qquad \text{[4.53]}$$

$$E_s = \frac{(\hbar k)^2}{2m_e} + V_n \qquad k = \frac{n\pi}{a} \qquad\qquad \text{[4.54]}$$

Clearly, whenever k has the critical values $n\pi/a$, there are only two possible values for the energies, E_c and E_s as determined by Equations 4.53 and 4.54; no other energies are allowed in between. These two energies are separated by $2V_n$.

Away from the critical k values determined by $k = n\pi/a$, the electron simply propagates as a traveling wave; the wave does not get reflected. The energy is then given by the free-running wave solution to the Schrödinger equation, that is, Equation 4.47,

$$E_k = \frac{(\hbar k)^2}{2m_e} \qquad\qquad \text{Away from } k = \frac{n\pi}{a} \qquad\qquad \text{[4.55]}$$

It seems that the energy of the electron increases parabolically with k along Equation 4.55 and then suddenly, at $k = n\pi/a$, it suffers a sharp discontinuity and increases parabolically again. Although the discontinuities at the critical points $k = n\pi/a$ are expected, by virtue of the Bragg reflection of waves, reflection effects will still be present to a certain extent, even within a small region around $k = n\pi/a$. The individual reflections shown in Figure 4.38 do not occur exactly at the origins of the atoms at $x = a, 2a, 3a, \ldots$. Rather, they occur over some distance, since the wave must interact with the electrons in the ion cores to be reflected. We therefore expect E–k behavior to deviate from Equation 4.55 in the neighborhood of the critical points, even if k is not exactly $n\pi/a$. Figure 4.40 shows the E–k behavior we expect, based on these arguments.

In Figure 4.40, we notice that there are certain energy ranges occurring at $k = \pm(n\pi/a)$ in which there are no allowed energies for the electron. As we saw previously, the electron cannot posses an energy between E_c and E_s at $k = \pi/a$. These energy ranges form energy gaps at the critical points $k = \pm(n\pi/a)$.

The range of k values from zero to the first energy gap at $k = \pm(\pi/a)$ defines a zone of k values called the **first Brillouin zone**. The zone between the first and

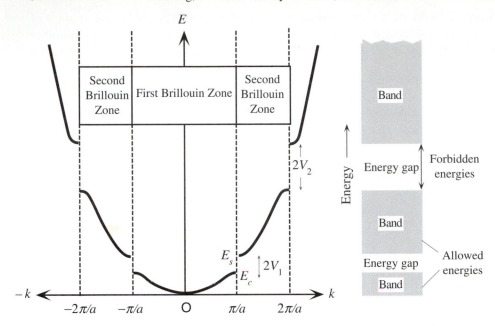

Figure 4.40   The energy of the electron as a function of its wavevector k inside a one-dimensional crystal.

There are discontinuities in the energy at $k = \pm n\pi/a$, where the waves suffer Bragg reflections in the crystal. For example, there can be no energy value for the electron between E_c and E_s. Therefore, $E_c - E_s$ is an energy gap at $k = \pm\pi/a$. Away from the critical k values, the E–k behavior is like that of a free electron, with E increasing with k as $E = (\hbar k)^2/2m_e$. In a solid, these energies fall within an energy band.

second energy gap defines the second Brillouin zone, and so on. The Brillouin zone boundaries therefore identify where the energy discontinuities, or gaps, occur along the k axis.

Electron motion in the three-dimensional crystal can be readily understood based on the concepts described here. For simplicity, we consider an electron propagating in a two-dimensional crystal, which is analogous, for example, to propagation in the xy plane of a crystal, as depicted in Figure 4.41. For certain k values and in certain directions, the electron will suffer diffraction and will be unable to propagate in the crystal.

Suppose that the electron's k vector along x is k_1. Whenever $k_1 = n\pi/a$, the electron will be diffracted by the planes perpendicular to x, that is, the (10) planes.[10] Similarly, it will be diffracted by the (01) planes whenever its k vector along y is $k_2 = n\pi/a$. The electron can also be diffracted by the (11) planes, whose separation is $a/\sqrt{2}$. If the component of k perpendicular to the (11) plane is k_3, then whenever $k_3 = \pm n\pi(\sqrt{2}/a)$, the electron will experience diffraction. These

---

[10] We use Miller indices in two dimensions by dropping the third digit but keeping the same interpretation. The direction along x is [10] and the plane perpendicular to x is (10).

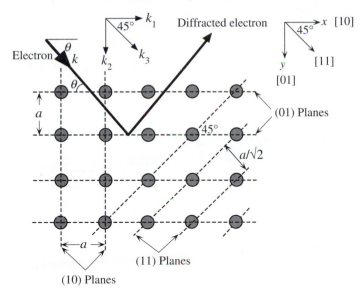

Figure 4.41   Diffraction of the electron in a two-dimensional crystal. Diffraction occurs whenever k has a component satisfying $k_1 = \pm n\pi/a$, $k_2 = \pm n\pi/a$ or $k_3 = \pm n\pi \sqrt{2}/a$. In general terms, diffraction occurs when $k \sin \theta = n\pi/a$.

diffraction conditions can all be expressed through the Bragg diffraction condition $2a \sin \theta = n\lambda$, or

$$k \sin \theta = \frac{n\pi}{a} \qquad \text{[4.56]}$$

Bragg diffraction condition

When we plot the energy of the electron as a function of k, we must consider the direction of k, since the diffraction behavior in Equation 4.56 depends on $\sin \theta$. Along x, at $\theta = 0$, the energy gap occurs at $k = \pm (n\pi/a)$. Along $\theta = 45°$, it is at $k = \pm n\pi (\sqrt{2}/a)$, which is farther away. The E–k behavior for the electron in the two-dimensional lattice is shown in Figure 4.42 for the [10] and [11] directions. The figure shows that the first energy gap along x, in the [10] direction, $k = \pi/a$. Along the [11] direction, which is at 45° to the x axis, the first gap is at $k = \pi \sqrt{2}/a$.

When we consider the overlap of the energy bands along [10] and [11], in the case of a metal, there is no apparent energy gap. The electron can always find any energy simply by changing its direction.

The effects of overlap between energy bands and of energy gaps in different directions are illustrated in Figure 4.43. In the case of a semiconductor, the energy gap along [10] overlaps that along [11], so there is an overall energy gap. The electron in the semiconductor cannot have an energy that falls into this energy gap.

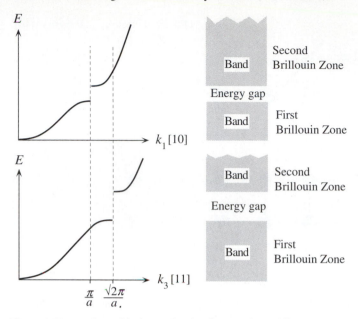

Figure 4.42   The *E–k* behavior for the electron along different directions in the two-dimensional crystal

The energy gap along [10] is at π/a whereas it is at $\sqrt{2}\pi/a$ along [11].

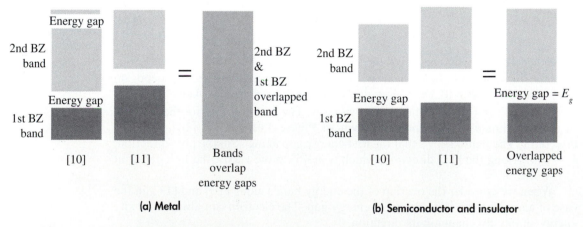

(a) Metal

(b) Semiconductor and insulator

Figure 4.43   (a) For the electron in a metal, there is no apparent energy gap because the second BZ (Brillouin zone) along [10] overlaps the first BZ along [11]. Bands overlap the energy gaps. Thus, the electron can always find any energy by changing its direction. (b) For the electron in a semiconductor, there is an energy gap arising from the overlap of the energy gaps along [10] and [11] directions. The electron can never have an energy within this energy gap E_g.

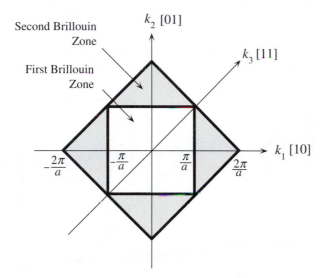

Figure 4.44   The Brillouin zones in two dimensions for the cubic lattice.

The Brillouin zones identify the boundaries where there are discontinuities in the energy, there are energy gaps.

The first and second Brillouin zones for the two-dimensional lattice of Figure 4.41 are shown in Figure 4.44. The zone boundaries mark the occurrences of energy gaps in k space (space defined by k axes along the x and y directions). When we look at the $E–k$ behavior, we must consider the crystal directions. This is most conveniently done by plotting energy contours in k space, as in Figure 4.45. Each contour connects all those values of k that possess the same energy. A point such as P on an energy contour gives the value of k for that energy along the direction OP. Initially, the energy contours are circles, as the energy follows $(\hbar k)^2/2m_e$ behavior, whatever the direction of k. However, near the critical values, that is, near the Brillouin zone boundaries, E increases more slowly than the parabolic relationship, as is apparent in Figure 4.40. Therefore, the circles begin to bulge as critical k values are approached. In Figure 4.45, the high-energy contours are concentrated in the corners of the zone, simply because the critical value is reached last along [11]. The energy contours do not continue smoothly across the zone boundary, because of the energy discontinuity in the $E–k$ relationship at the boundary. Indeed, Figure 4.42 shows that the lowest energy in the second Brillouin zone may be lower than the highest energy in the first Brillouin zone.

There are two cases of interest. In the first, there is no apparent energy gap, as in Figure 4.45a, which corresponds to Figure 4.43a. The electron can have any energy value. In the second case, there is a range of energies that are not allowed, as shown in Figure 4.45b, which corresponds to Figure 4.43b.

In three dimensions, the $E–k$ energy contour in Figure 4.45 becomes a surface in three-dimensional k space. To understand the use of such $E–k$ contours or surfaces, consider that an $E–k$ contour (or a surface) is made of many finely-

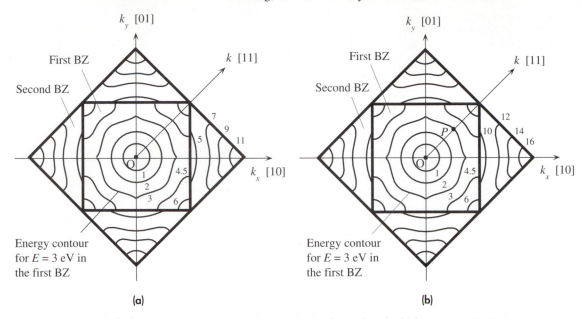

(a) In a metal, the lowest energy in the second zone (5 eV) is lower than the highest energy (6 eV) in the first zone. There is an overlap of energies between the Brillouin zones. (b) In a semiconductor or an insulator, there is an energy gap between the highest energy contour (6 eV) in the first zone and the lowest energy contour (10 eV) in the second zone.

Figure 4.45   Energy contours in k space (space defined by k_x, k_y).

Each contour represents the same energy value. Any point P on the contour gives the values of k_x and k_y for that energy in that direction from O. For point P, $E = 3$ eV and OP along [11] is k.

separated individual points, each representing a possible electron wavefunction ψ_k with a possible energy E. At absolute zero, all the energies up to the Fermi energy are taken by the valence electrons. In k space, the energy surface, corresponding to the Fermi energy is termed the **Fermi surface.** The shape of this Fermi surface provides a means of interpreting the electrical and magnetic properties of solids.

For example, Na has one $3s$ electron per atom. In the solid, the $3s$ band is half full. The electrons take energies up to E_F, which corresponds to a spherical Fermi surface within the first Brillouin zone, as indicated in Figure 4.46a. We can then say that all the valence electrons (or nearly all) in this alkali solid exhibit an $E = (\hbar k)^2 / 2m_e$ type of behavior, as if they were free. When an external force is applied, such as an electric or magnetic field, we can treat the electron behavior as if it were free inside the metal with a constant mass. This is a desirable simplification for studying such metals. We can illustrate this desirability with an example. The Hall coefficient R_H derived in Chapter 2 was based on treating the electron as if were a free particle inside the metal, or

$$R_H = -\frac{1}{en}$$ [4.57]

For Na, the experimental value of R_H is -2.50×10^{-10} m³ C⁻¹. Using the density (0.97 g cm⁻³) and atomic mass (23) of Na and one valence electron per atom, we can calculate $n = 2.54 \times 10^{28}$ m⁻³ and $R_H = -2.46 \times 10^{-10}$ m³ C⁻¹, which is very close to the experimental value.

In the case of Cu, Ag, and Au (the IB metals in the Periodic Table), the Fermi surface is inside the first Brillouin zone, but it is not spherical as depicted in Figure 4.46b. Also, it touches the centers of the zone boundaries. Some of those electrons near the zone boundary behave quite differently than $E = (\hbar k)^2/2m_e$, although the majority of the electrons in the sphere do exhibit this type of behavior. To an extent, we can expect the free electron derivations to hold. The experimental value of R_H for cu is -0.55×10^{-10} m³ C⁻¹, whereas the expected value, based on Equation 4.57 with one electron per atom, is -0.73×10^{-10} m³ C⁻¹, which is noticeably greater than the experimental value.

The divalent metals Be, Mg, and Ca have closed outer s subshells and should have a full s band in the solid. Recall that electrons in a full band cannot respond to an applied field and drift. We also know that there should be an overlap between the s and p bands, forming one partially-filled continuous energy band, so these metals are indeed conductors. In terms of Brillouin zones, their structure is based on Figure 4.43a, which has the second zone overlapping the first Brillouin zone. The Fermi surface extends into the second zone and the corners of the first zone are empty, as depicted in Figure 4.46c. Since there are empty energy levels next to the Fermi surface, the electrons can gain energy and drift in response to an applied field. But the surface is not spherical; indeed, near the corners of the first zone, it even has the wrong curvature. Therefore, it is no longer possible to describe these electrons on the Fermi surface as obeying $E = (\hbar k)^2/2m_e$. When a magnetic field is applied to a drifting electron to bend its trajectory, its total behavior changes unpredictably from that expected when it is acting as a free particle. The external force changes the momentum $\hbar k$ and the corresponding change in the energy

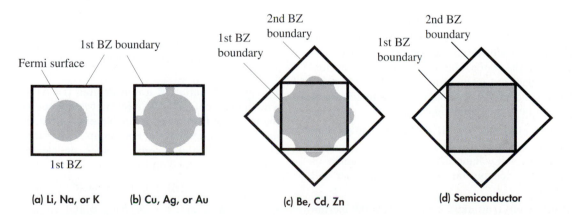

(a) Li, Na, or K   (b) Cu, Ag, or Au   (c) Be, Cd, Zn   (d) Semiconductor

(a) Monovalent group IA metals. (b) Group IB metals. (c) Be (group IIA), Zn, and Cd (Group IIB). (d) A semiconductor.

Figure 4.46   Schematic sketches of Fermi surfaces in two dimensions, representing various materials qualitatively.

depends on the Fermi surface and can be quite complicated. To finish the example on the Hall coefficient, we note that based on two valence electrons per atom (Group IIA), the Hall coefficient for Be should be -0.25×10^{-10} m^3 C^{-1}, but the measured value is a positive coefficient of $+2.44 \times 10^{-10}$ m^3 C^{-1}. Equation 4.57 is therefore useless. It seems that the electrons moving at the Fermi surface of Be are equivalent to the motion of positive charges (like holes), so the Hall effect registers a positive coefficient.

The Fermi surface of a semiconductor is simply the boundary of the first Brillouin zone, because there is an energy gap between the first and the second Brillouin zones, as depicted in Figure 4.43b. In a semiconductor, all the energy levels up to the energy gap are taken up by the valence electrons. The first Brillouin zone forms the valence band and the second forms the conduction band.

IMPORTANT TERMS

Average energy E_{av} of an electron in a metal is determined by the Fermi–Dirac statistics and the density of states. It increases with the Fermi energy and also with the temperature.

Boltzmann statistics describes the behavior of a collection of particles (e.g., gas atoms) in terms their energy distribution. It specifies the number of particles $N(E)$ with given energy, though $N(E) \propto \exp(-E/kT)$, where k is the Boltzmann constant. The description is nonquantum mechanical in that there is no restriction on the number of particles that can have same state (the same wavefunction) with an energy E. Also, it applies when there are only a few particles compared to the number of possible states, so that the likelihood of two particles having the same state becomes negligible. This is generally the case for thermally excited electrons in the conduction band of a semiconductor, where there are many more states than electrons. The kinetic energy distribution of gas molecules in a tank obeys the Boltzmann statistic.

Density of states $g(E)$ is the number of electron states (e.g., wave functions, $\psi(n, \ell, m_\ell, m_s)$) per unit energy per unit volume. Thus, $g(E)\,dE$ is number of states in the energy range E to $(E + dE)$ per unit volume.

Effective electron mass m_e^* represents the inertial resistance of an electron inside a crystal against an acceleration imposed by an external force, such as the applied electric field. If $F_{ext} = eE_x$ is the external applied force due to the applied field $\mathcal{E}_x$, then the effective mass m_e^* determines the acceleration a of the electron by $eE_x = m_e^* a$. This takes into account the effect of the internal fields on the motion of the electron. In vacuum, where there are no internal fields, m_e^* is the mass in vacuum m_e.

Fermi–Dirac statistics determines the probability of an electron occupying a state at an energy level E. This takes into account that a collection of electrons must obey the Pauli exclusion principle. The Fermi–Dirac function quantifies this probability via $f(E) = 1/\{1 + \exp[(E - E_F)/kT]\}$, where E_F is the Fermi energy.

Fermi energy is the maximum energy of the electrons in a metal at 0 K.

Field emission is the tunneling of an electron from the surface of a metal into vaccum, due to the application of a strong electric field (typically $\mathcal{E} > 10^9$ V m^{-1}).

Linear combination of atomic orbitals (LCAO) is a method for obtaining the electron wavefunction in the molecule from a linear combination of individual atomic wavefunctions. For example, when three H atoms A, B, and C come together, the electron wavefunctions, based on LCAO, are

$$\psi_a = \psi_{1s}(A) + \psi_{1s}(B) + \psi_{1s}(C)$$
$$\psi_b = \psi_{1s}(A) + \psi_{1s}(B) - \psi_{1s}(C)$$
$$\psi_c = \psi_{1s}(A) - \psi_{1s}(B) + \psi_{1s}(C)$$

where $\psi_{1s}(A)$, $\psi_{1s}(B)$, and $\psi_{1s}(C)$ are atomic wavefunctions centered around the H atoms A, B, and C, respectively. The ψ_a, ψ_b, and ψ_c represent molecular orbital wavefunctions for the electron; they reflect the behavior of the electron, or its probability distribution, in the molecule.

Molecular orbital wavefunction, or simply molecular orbital, is a wavefunction for an electron within a system of two or more nuclei (e.g., molecule). A molecular orbital determines the probability distribution of the electron within the molecule, just as the atomic orbital determines the electron's probability distribution within the atom. The molecular orbital can take two electrons with opposite spins.

Orbital is a region of space in an atom or molecule where an electron with a given energy may be found. An orbit, which is a well-defined path for an electron, cannot be used to describe the whereabouts of the electron in an atom or molecule because the electron has a probability distribution. Orbitals are generally represented by a surface within which the total probability is high, for example, 90 percent.

Orbital wavefunction, or simply orbital, describes the spatial dependence of the electron. The orbital is $\psi(r, \theta, \phi)$, which depends on n, ℓ, and m_ℓ, and the spin dependence m_s is excluded.

Peltier effect is the phenomenon of heat absorption or production at a junction between two dissimilar materials when a current passes through the junction. The direction of the current determines whether heat is absorbed or ejected at the junction.

Seebeck effect is the development of a built-in potential difference across a material as a result of a temperature gradient. If dV is the built-in potential across a temperature difference dT, then the Seebeck coefficient S is defined as $S = dV/dT$. The coefficient gauges the magnitude of the Seebeck effect. Only the net Seebeck voltage difference between different metals can be measured. The principle of the thermocouple is based on the Seebeck effect.

State is a possible wavefunction for the electron that defines its spatial (orbital) and spin properties, for example, $\psi(n, \ell, m_\ell \, m_s)$ is a state of the electron. From the Schrödinger equation, each state corresponds to a certain electron energy E. We thus speak of a state with energy E, state of energy E, or even an energy state. Generally there may be more than one state, ψ, with the same energy E.

Thermionic emission is the emission of electrons from the surface of a heated metal.

Work function is the minimum energy needed to free an electron from the metal at a temperature of absolute zero. It is the energy separation of the Fermi level from the vacuum level.

QUESTIONS AND PROBLEMS

4.1   Phase of an atomic orbital

   a.   What is the functional form of a 1s wavefunction $\psi(r)$? Sketch schematically the atomic wavefunction $\psi_{1s}(r)$ as a function of distance from the nucleus.

   b.   What is the total wavefunction $\Psi_{1s}(r, t)$?

   c.   What is meant by two wavefunctions $\Psi_{1s}(A)$ and $\Psi_{1s}(B)$ that are out of phase?

   d.   Sketch schematically the two wavefunction $\Psi_{1s}(A)$ and $\Psi_{1s}(B)$ at one instant.

4.2   Molecular orbitals and atomic orbitals   Consider a ring of seven atoms representing a hypothetical molecule. Suppose that each atomic wavefunction is a 1s wavefunction.

a. Using a circle to represent the atomic $1s$ orbital and $+$ and $-$ to represent its phase, show that there are seven different possible molecular wavefunctions resulting from overlaps of atomic wavefunctions of various phases.

b. Sketch linearly the probability distributions $|\psi|^2$.

c. If more nodes in the wavefunction lead to greater energies, sketch schematically the energy vs. interatomic separation behavior of the various molecular orbitals for this molecule.

d. Suppose that the molecule is not a ring, but a linear molecule. How many distinct molecular orbitals are there? How would you extend this finding to N atoms in the solid, where $N \sim 10^{22}$ cm^{-3}?

4.3 Diamond and tin   Germanium, silicon, and diamond have the same crystal structure, that of diamond. Bonding in each case involves sp^3 hybridization. The bonding energy decreases as we go from C to Si to Ge, as noted in Table 4.5.

a. What would you expect for the bandgap of diamond? How does it compare with the experimental value of 5.5 eV?

b. Tin has a tetragonal crystal structure, which makes it different than its group members, diamond, silicon, and germanium.

    1. Is it a metal or a semiconductor?
    2. What experiments do you think would expose its semiconductor properties?

4.4 Compound III–V Semiconductors   Indium as an element is a metal. It has a valency of III. Sb as an element is a metal and has a valency of V. InSb is a semiconductor, with each atom bonding to four neighbors, just like in silicon. Explain how this is possible and why InSb is a semiconductor and not a metal alloy. [Consider the electronic structure and sp^3 hybridization for each atom.]

Table 4.5

Property	Diamond	Silicon	Germanium	Tin (α–Sn)
Melting temperature, °C	3800	1417	937	232
Covalent radius, nm	0.077	0.117	0.122	0.146
Bond energy, eV	3.60	1.84	1.7	1.2
First ionization energy, eV	11.26	8.15	7.88	7.33
Bandgap, eV	?	1.12	0.67	?

4.5 Compound II–VI Semiconductors   CdSe is a semiconductor, with each atom bonding to four neighbours, just like in silicon. In terms of covalent bonding and the positions of Cd and Se in the Periodic Table, explain how this is possible. Would you expect the bonding in CdSe to have more ionic character than that in III–V semiconductors?

*4.6 Density of states for a two-dimensional electron gas   Consider a two-dimensional electron gas in which the electrons are restricted to move freely within a square area a^2 in the xy plane. Following the procedure in Section 4.5, show that the density of states $g(E)$ is constant (independent of energy).

4.7 Fermi energy of Cu   The Fermi energy of electrons in copper at room temperature is 7.0 eV. The electron drift mobility in copper, from Hall effect measurements, is 33 cm^2 V^{-1} s^{-1}.

  a.   What is the speed v_F of conduction electrons with energies around E_F in copper? By how many times is this larger than the average thermal speed $v_{thermal}$ of electrons, if they behaved like an ideal gas (Maxwell–Boltzmann statistics)? Why is v_F much larger than $v_{thermal}$?

  b.   What is the De Broglie wavelength of these electrons? Will the electrons get diffracted by the lattice planes in copper, given that interplanar separation in Cu $= 2.09$Å?

  c.   Calculate the mean free path of electrons in copper and comment. [Solution guide: Diffraction of waves occurs when $2d \sin \theta = \lambda$, which is the Bragg condition. Find the relationship between λ and d that results in $\sin \theta > 1$ and hence no diffraction]

4.8 Fermi energy and electron concentration   Consider the metals in Table 4.6 from groups I, II and III in the Periodic Table. Calculate the Fermi energies at absolute zero, and compare the values with the experimental values. What is your conclusion?

4.9 Temperature dependence of the Fermi energy

  a.   Given that the Fermi energy for Cu is 7.0 eV at absolute zero, calculate the E_F at 300 K. What is the percentage change in E_F and what is your conclusion?

Table 4.6

Metal	Group	M_{at}	Density (g cm^{-3})	E_F(eV) [Calculated]	E_F(eV) [Experiment]
Cu	I	63.55	8.96	—	6.5
Zn	II	65.38	7.14	—	11.0
Al	III	27	2.70	—	11.8

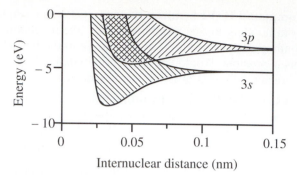

Figure 4.47   The energy bands in sodium

b.   Given the Fermi energy for Cu at absolute zero, calculate the average energy and mean speed per conduction electron at absolute zero and 300 K, and comment.

4.10   **Energy bands in Na**   Figure 4.47 shows the formation of the 3s and 3p energy bands in Na as a function of internuclear separation. The structure of the Na atom is [Ne]$3s^1$.

a.   Where is the equilibrium separation (estimate) if some electrons in the 3s band spill over into the states in the 3p band?

b.   What is the width of the energy band at the equilibrium internuclear separation?

c.   Given that the Fermi energy E_F for Na is about 3.2 eV, estimate the work function for Na and compare that with the experimental value of 2.75 eV.

d.   Sketch schematically the density of states, $g(E)$ as a function of E.

4.11   **Thermoelectric effects and E_F**   Consider a thermocouple pair that consists of copper and aluminum. One junction is at 100 °C and the other is at 0 °C. A voltmeter (with a very large input resistance) is inserted into the aluminum wire. Use the properties of Cu and Al in Table 4.7 to estimate the emf registered by the voltmeter and identify the positive end.

4.12   **The thermocouple equation**   Although inputting the measured emf for V in the thermocouple equation $V = a\,\Delta T + b(\Delta T)^2$ leads to a quadratic equation, which in principle can be solved for ΔT, in general ΔT is related to the measured emf via

$$\Delta T = a_1 V + a_2 V^2 + a_3 V^3 + \ldots$$

with the coefficients a_1, a_2, etc., determined for each pair of TCs. By carrying out a Taylor's expansion of TC equation, find the first two coefficients a_1 and a_2. Using an emf table for the K-type thermocouple or Figure 4.33, evaluate a_1 and a_2.

Table 4.7

	Cu	**Al**
Atomic mass, M_{at}	63.55	27
Density, g	8.96	2.7
Conduction electrons per atom	1	3

Table 4.8

	B_e (A m^{-2} K^{-2})	Φ(eV)
Th on W	3×10^4	2.6
Oxide coating	100	1

4.13 **Thermionic emission** A vacuum tube is required to have a cathode operating at 800 °C and providing an emission current of 10 A. What should be the surface area of the cathode for the two materials in Table 4.8? What should be the operating temperature for the Th on W cathode, if it is to have the same surface area as the oxide-coated cathode?

4.14 **Field-assisted thermal emission** Consider the field-assisted emission and tunneling currents through an SiO_2 insulating layer on the surface of an Si wafer. Suppose there is an Al electrode of area 50 μm $\times$ 50 μm on the SiO_2 and there is a voltage of 10 V with respect of the Si wafer. Consider two thicknesses for the SiO_2, (a) 20 Å, and (b) 100 Å, where (1 Å = 1^{-10} m). The work function of Al is 4.2 eV, but this refers to electron emission into vacuum, whereas in this case, the electron is emitted into the oxide. Given that the potential energy barrier Φ_B between Al and SiO_2 is about 3.1 eV, and $\varepsilon_r = 3.9$ for SiO_2, estimate the field-assisted emission and tunneling currents through these two oxide thicknesses? What is your conclusion? (Note: replace ε_o with $\varepsilon_o \varepsilon_r$ in β_s).

*4.15 **Overlapping bands** Consider Cu and Ni with their density of states as schematically sketched in Fig. 4.48. Both have overlapping 3d and 4s bands, but the 3d band is very narrow compared to the 4s band. In the case of Cu the band is full, whereas in Ni, it is only partially filled.

a. In Cu, do the electrons in the 3d band contribute to electrical conduction? Explain.

b. In Ni, do electrons in both bands contribute to conduction? Explain.

c. Do electrons have the same effective mass in the two bands? Explain.

d. Can an electron in the 4s band with energy around E_F become scattered into the 3d band as a result of a scattering process? Consider both metals.

e. Scattering of electrons from the 4s band to the 3d band and vice versa can be viewed as an additional scattering process. How would you expect the resistivity of Ni to compare with that of Cu, even though Ni has 2 valence electrons and nearly the same density as Cu? In which case would you expect a stronger temperature dependence for the resistivity?

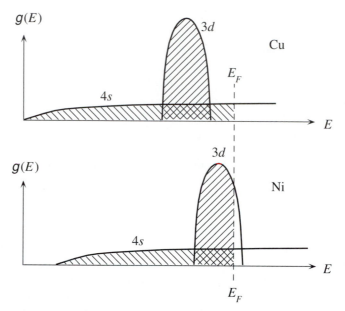

Figure 4.48   Density of states and electron filling in Cu and Ni

chapter

5

Semiconductors

In this chapter we develop a basic understanding of the properties of intrinsic and extrinsic semiconductors. Although most of our discussions and examples will be based on Si, the ideas are applicable to Ge and to the compound semiconductors such as GaAs, InP, and others. By intrinsic Si we mean an ideal perfect crystal of Si that has no impurities and crystal defects such as dislocations and grain boundaries. The crystal thus consists of Si atoms perfectly bonded to each other in the diamond structure. At temperatures above absolute zero, we know that the Si atoms in the crystal lattice will be vibrating with a distribution of energies. Even though the average energy of the vibrations is at most $3kT$ and incapable of breaking the Si−Si bond, a few of the lattice vibrations in certain crystal regions may nonetheless be sufficiently energetic to "rupture" a Si−Si bond. When a Si−Si bond is broken, a "free" electron is created that can wander around the crystal and also contribute to electrical conduction in the presence of an applied field. The broken bond has a missing electron that causes this region to be positively charged. The vacancy left behind by the missing electron in the bonding orbital is called a **hole.** An electron in a neighboring bond can readily tunnel into this broken bond and fill it, thereby effectively causing the hole to be displaced to the original position of the electron. By electron tunneling from a neighboring bond, holes are therefore also free to wander around the crystal and also contribute to electrical conduction in the presence of an applied field. In an intrinsic semiconductor, the number of thermally generated electrons is equal to the number of holes (broken bonds). In an extrinsic semiconductor, impurities are added to the semiconductor that can contribute either excess electrons or excess holes. For example, when an impurity such as arsenic is added to Si, each As atom acts as a donor and contributes a free electron to the crystal. Since these electrons do not come from broken bonds, the numbers of electrons and holes are not equal in an extrinsic semiconductor, and the As-doped Si in this example will have excess electrons. It will be an n-type Si since electrical conduction will be mainly due to the motion of electrons. It is also possible to obtain a p-type Si crystal in which hole concentration is in excess of the electron concentration due to, for example, boron doping.

5.1   INTRINSIC SEMICONDUCTORS

5.1.1   Silicon Crystal and Energy Band Diagram

The electronic configuration of an isolated Si atom is $[Ne]3s^2p^2$. However, in the vicinity of other atoms, the $3s$ and $3p$ energy levels are so close that the interactions result in the *four* orbitals $\psi(1s)$, $\psi(3p_x)$, $\psi(3p_y)$, and $\psi(3p_z)$ mixing together to form *four* new hybrid orbitals (called ψ_{hyb}) that are symmetrically directed as farther away from each other as possible (towards the corners of a tetrahedron). In two dimensions, we can simply view the orbitals pictorially as in Figure 5.1a. The four hybrid orbitals, ψ_{hyb}, each have one electron so that they are half-occupied. Therefore, a ψ_{hyb} orbital of one Si atom can overlap a ψ_{hyb} orbital of a neighboring Si atom to form a covalent bond with two spin-paired electrons. In this manner one Si atom bonds with four other Si atoms by overlapping the half-occupied ψ_{hyb} orbitals, as illustrated in Figure 5.1b. Each Si–Si bond corresponds to a bonding orbital, ψ_B,

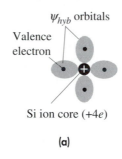

(a)

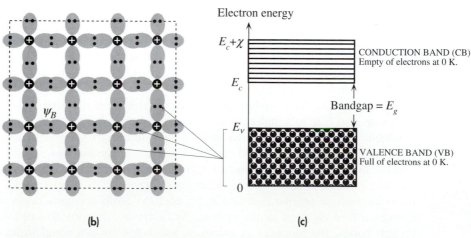

(b)   (c)

Figure 5.1

(a) A simplified two-dimensional illustration of a Si atom with four hybrid orbitals, ψ_{hyb}. Each orbital has one electron.
(b) A simplified two-dimensional view of a region of the Si crystal showing covalent bonds.
(c) The energy band diagram at absolute zero of temperature.

obtained by overlapping two neighboring ψ_{hyb} orbitals. Each bonding orbital (ψ_B) has two spin-paired electrons and is therefore *full*. Neighboring Si atoms can also form covalent bonds with other Si atoms, thus forming a three-dimensional network of Si atoms. The resulting structure is the Si crystal in which each Si atom bonds with four Si atoms in a tetrahedral arrangement. The crystal structure is that of *a diamond,* which was described in Chapter 1. We can imagine the Si crystal in two dimensions as depicted in Figure 5.1b. The electrons in the covalent bonds are the valence electrons.

The energy band diagram of the silicon crystal is shown in Figure 5.1c.[1] The vertical axis is the electron energy in the crystal. The valence band (VB) contains those electronic states that correspond to the overlap of bonding orbitals (ψ_B). Since all the bonding orbitals (ψ_B) are full with valence electrons in the crystal, the VB is also full with these valence electrons at absolute zero of temperature. The conduction band (CB) contains electronic states that are at higher energies, those corresponding to the overlap of antibonding orbitals. The CB is separated from the VB by an energy gap E_g, called the **band gap.** The width of the CB is called the **electron affinity,** χ. The energy level E_v marks the top of the VB and E_c marks the bottom of the CB. The general energy band diagram in Figure 5.1c applies to all crystalline semiconductors with appropriate changes in the energies.

The electrons shown in the VB in Figure 5.1c are those in the covalent bonds between the Si atoms in Figure 5.1b. An electron in the VB, however, is not localized to an atomic site but extends throughout the whole solid. Although the electrons appear localized in Figure 5.1b, at the bonding orbitals between the Si atoms this is not, in fact, true. In the crystal, the electrons can tunnel from one bond to another and exchange places. If we were to work out the wavefunction of a valence electron in the Si crystal, we would find that it extends throughout the whole solid. This means that the electrons in the covalent bonds are indistinguishable. We cannot label an electron from the start and say that this electron is in this covalent bond between these two atoms.

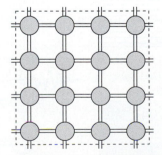

Figure 5.2   A two-dimensional pictorial view of the Si crystal showing covalent bonds as two lines where each line is a valence electron.

[1] The formation of energy bands in the silicon crystal was described in detail in Chapter 4.

We can crudely represent the silicon crystal in two dimensions as shown in Figure 5.2. Each covalent bond between Si atoms is represented by two lines corresponding to two spin-paired electrons. Each line represents a valence electron.

5.1.2   Electrons and Holes

The only empty electronic states in the silicon crystal are in the CB (Figure 5.1c). An electron placed in the CB is free to move around the crystal and also respond to an applied electric field because there are plenty of neighboring empty energy levels. An electron in the CB can easily gain energy from the field and move to higher energy levels because these states are empty. Generally we can treat an electron in the CB as if it were free within the crystal with certain modifications to its mass, as explained later.

Since the only empty states are in the CB, the excitation of an electron from the VB requires a minimum energy of E_g. Figure 5.3a shows what happens when a photon of energy $h\upsilon > E_g$ is incident on an electron in the VB. This electron absorbs the incident photon and gains sufficient energy to surmount the energy gap E_g and reach the CB. Consequently, a free electron and a "hole," corresponding to a missing electron in the VB, are created. In some semiconductors such as Si and Ge, the photon absorption process also involves lattice vibrations (vibrations of the Si atoms), which we have not shown in Figure 5.3b.

Although in this specific example a photon of energy $h\upsilon > E_g$ creates an electron–hole pair, this is not necessary. In fact, in the absence of radiation, there is an electron–hole generation process going on in the sample as a result of

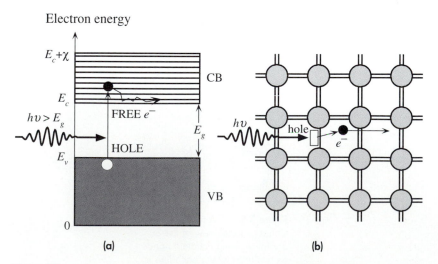

Figure 5.3

(a) A photon with an energy greater than E_g can excite an electron from the VB to the CB.
(b) When a photon breaks a Si–Si bond, a free electron and a hole in the Si–Si bond are created.

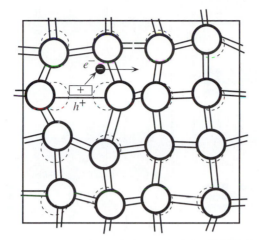

Figure 5.4   Thermal vibrations of atoms can break bonds and thereby create electron–hole pairs.

thermal generation. Due to thermal energy, the atoms in the crystal are constantly vibrating, which corresponds to the bonds between the Si atoms being periodically deformed. In a certain region, the atoms, at some instant, may be moving in such a way that a bond becomes overstretched, as pictorially depicted Figure 5.4. This will result in the overstretched bond rupturing and hence releasing an electron into the CB (the electron effectively becomes "free"). The empty electronic state of the missing electron in the bond is what we call a **hole** in the valence band. The free electron, which is in the CB, can wander around the crystal and contribute to the electrical conduction when an electric field is applied. The region remaining around the hole in the VB is positively charged because a charge of $-e$ has been removed from an otherwise neutral region of the crystal. This hole, denoted as h^+, can also wander around the crystal as if it were free. This is because an electron in a neighboring bond can "jump," that is, tunnel, into the hole to fill the vacant electronic state at this site and thereby create a hole at its original position. This is effectively equivalent to the hole being displaced in the opposite direction, as illustrated in Figure 5.5a. This single step can reoccur, causing the hole to be further displaced. As a result, the hole moves around the crystal as if it were a free positively charged entity, as pictured in Figure 5.5a to d. Its motion is quite independent from that of the original electron. When an electric field is applied, the hole will drift in the direction of the field and hence contribute to electrical conduction. It is now apparent that there are essentially two types of charge carriers in semiconductors: *electrons* and *holes*. A hole is effectively an empty electronic state in the VB that behaves as if it were a positively charged "particle" free to respond to an applied electric field.

When a wandering electron in the CB meets a hole in the VB, the electron has found an empty state of lower energy and therefore occupies the hole. The electron falls from the CB to the VB to fill the hole, as depicted in Figure 5.5e and f. This is called **recombination** and results in the annihilation of an electron in the

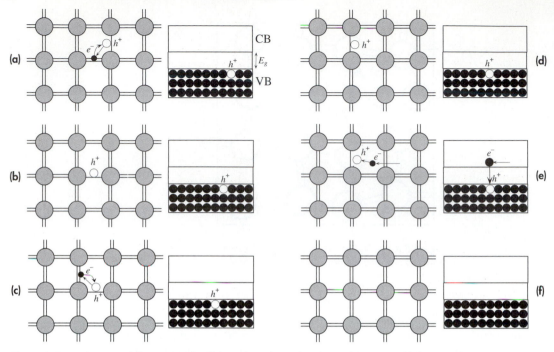

Figure 5.5   A pictorial illustration of a hole in the valence band wandering around the crystal due to the tunneling of electrons from neighboring bonds.

CB and a hole in the VB. The excess energy of the electron falling from CB to VB in certain semiconductors such as GaAs and InP is emitted as a photon. In Si and Ge the excess energy is lost as lattice vibrations (heat).

It must be emphasized that the illustrations in Figure 5.5 are pedagogical pictorial visualizations of hole motion based on classical notions and cannot be taken too seriously, as discussed in more advanced texts (see also Section 5.11). We should remember that the electron has a wavefunction in the crystal that is extended and not localized, as the pictures in Figure 5.5 imply. Further, the hole is a concept that corresponds to an empty valence band wavefunction that normally has an electron. Again, we cannot localize the hole to a particular site, as the pictures in Figure 5.5 imply.

5.1.3   Conduction in Semiconductors

When an electric field is applied across a semiconductor as shown in Figure 5.6, the energy bands bend. The total electron energy E is $KE + PE$, but now there is an additional electrostatic PE contribution that is not constant in an applied electric field. A uniform electric field $\mathcal{E}_x$ implies a linearly decreasing potential, $V(x)$, by virtue of $(dV/dx) = -\mathcal{E}_x$, that is, $V = -Ax + B$. This means that the PE, $-eV(x)$, of the electron is now $eAx - eB$, which increases linearly across the sample. All the energy levels and hence the energy bands must therefore tilt up in the x direction, as shown in Figure 5.6, in the presence of an applied field.

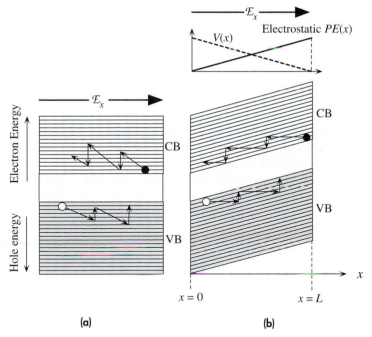

Figure 5.6   When an electric field is applied, electrons in the CB and holes in the VB can drift and contribute to the conductivity.

(a) A simplified illustration of drift in $\mathcal{E}_x$.

(b) Applied field bends the energy bands since the electrostatic *PE* of the electron is $-eV(x)$ and $V(x)$ decreases in the direction of $\mathcal{E}_x$, whereas *PE* increases.

Under the action of $\mathcal{E}_x$, the electron in the CB moves to the left and immediately starts gaining energy from the field. When the electron collides with a thermal vibration of a Si atom, it loses some of this energy and thus "falls" down in energy in the CB. After the collision the electron starts to accelerate again, until the next collision, and so on. We recognize this process as the drift of the electron in an applied field, as illustrated in Figure 5.6. The drift velocity v_{de} of the electron is $\mu_e \mathcal{E}_x$ where μ_e is the drift mobility of the electron. In a similar fashion, the holes in the VB also drift in an applied field, but here the drift is along the field. Notice that when a hole gains energy, it moves "down" in the VB because the potential energy of the hole is of opposite sign to that of the electron.

Since both electrons and holes contribute to electrical conduction, we may write the current density, *J*, from its definition, as

$$J = env_{de} + epv_{dh} \qquad \text{[5.1]}$$

where *n* is the electron concentration in the CB, *p* is the hole concentration in the VB, and v_{de} and v_{dh} are the drift velocities of electrons and holes in response to an applied electric field $\mathcal{E}_x$, Thus,

$$v_{de} = \mu_e \mathcal{E}_x \qquad \text{and} \qquad v_{dh} = \mu_h \mathcal{E}_x \qquad \text{[5.2]}$$

where μ_e and μ_h are the electron and hole drift mobilities. In Chapter 2 we derived the drift mobility μ_e of the electrons in a conductor as

$$\mu_e = \frac{e\tau_e}{m_e} \tag{5.3}$$

where τ_e is the mean free time between scattering events and m_e is the electronic mass. The ideas on electron motion in metals can also be applied to the electron motion in the CB of a semiconductor to rederive Equation 5.3. We must, however, use an effective mass m_e^* for the electron in the crystal rather than the mass m_e in free space. A "free" electron in a crystal is not entirely free because as it moves it interacts with the potential energy (PE) of the ions in the solid and therefore experiences various internal forces. The effective mass, m_e^*, accounts for these internal forces in such a way that we can relate the acceleration, a, of the electron in the CB to an external force F_{ext} (e.g., $-e\mathcal{E}_x$) by $F_{ext} = m_e^* a$ just as we do for the electron in vacuum by $F_{ext} = m_e a$. In applying the $F_{ext} = m_e^* a$ type of description to the motion of the electron, we are assuming, of course, that the effective mass of the electron can be calculated or measured experimentally. It is important to remark that the true behavior is governed by the solution of the Schrödinger equation in a periodic lattice (crystal) from which it can be shown we can indeed describe the inertial resistance of the electron to acceleration in terms an effective mass, m_e^*. The effective mass depends on the interaction of the electron with its environment within the crystal.

We can now speculate on whether the hole can also have a mass. As long as we view mass as resistance to acceleration, that is, inertia, there is no reason why the hole should not have a mass. Accelerating the hole means accelerating electrons tunneling from bond to bond in the opposite direction. Therefore it is apparent that the hole will have a nonzero finite inertial mass because otherwise the smallest external force will impart an infinite acceleration to it. If we represent the effective mass of the hole in the VB by m_h^*, then the hole drift mobility will be

$$\mu_h = \frac{e\tau_h}{m_h^*} \tag{5.4}$$

where τ_h is the mean free time between scattering events for holes.

Taking Equation 5.1 for the current density further, we can write the conductivity of a semiconductor as

Conductivity of a semiconductor

$$\sigma = en\mu_e + ep\mu_h \tag{5.5}$$

where n and p are the electron and hole concentrations in the CB and VB respectively. This is a general equation valid for all semiconductors.

5.1.4   Electron and Hole Concentrations

The general equation for the conductivity of a semiconductor, Equation 5.5, depends on n, the electron concentration, and p, the hole concentration. How do we determine these quantities? We follow the procedure schematically shown in Figure 5.7a to d in which the density of states is multiplied by the probability of a

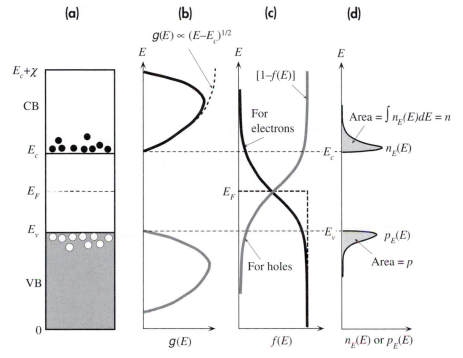

Figure 5.7
(a) Energy band diagram.
(b) Density of states (number of states per unit energy per unit volume).
(c) Fermi–Dirac probability function (probability of occupancy of a state).
(d) The product of $g(E)$ and $f(E)$ is the energy density of electrons in the CB (number of electrons per unit energy per unit volume). The area under $n_E(E)$ versus E is the electron concentration.

state being occupied and integrated over the entire CB for n and over the entire VB for p.

We define $g_{cb}(E)$ as the density of states in the CB, that is, the number of states per unit energy per unit volume. The probability of finding an electron in a state with energy E is given by the Fermi–Dirac function, $f(E)$, which is discussed in Chapter 4. Then $g_{cb}(E)f(E)$ is the actual number of electrons per unit energy per unit volume, $n_E(E)$, in the CB. Thus,

$$n_E \, dE = g_{cb}(E)f(E) \, dE$$

is the number of electrons in the energy range E to $E + dE$. Integrating this from the bottom (E_c) to the top ($E_c + \chi$) of the CB gives the electron concentration n, number of electrons per unit volume, in the CB. In other words,

$$n = \int_{E_c}^{E_c+\chi} n_E(E) \, dE = \int_{E_c}^{E_c+\chi} g_{cb}(E)f(E) \, d(E)$$

We will assume that $(E_c - E_F) \gg kT$ (i.e., E_F is at least a few kT below E_c) so that

$$f(E) \approx \exp[-(E - E_F)/kT]$$

We are thus replacing Fermi–Dirac statistics by Boltzmann statistics and thereby inherently assuming that the number of electrons in the CB is far less than the number of states in this band.

Further, we will take the upper limit to be $E = \infty$ rather than $E_c + \chi$ since $f(E)$ decays rapidly with energy so that $g_{cb}(E)f(E) \rightarrow 0$ near the top of the band. Furthermore, since $g_{cb}(E)f(E)$ is significant only close to E_c, we can use

$$g_{cb}(E) = \frac{(\pi 8\sqrt{2})m_e^{*3/2}}{h^3}(E - E_c)^{1/2}$$

for an electron in a three-dimensional PE well without having to consider the exact form of $g_{cb}(E)$ across the whole band. Thus

$$n \approx \frac{(\pi 8\sqrt{2})m_e^{*3/2}}{h^3}\int_0^\infty (E - E_c)^{1/2}\exp\left[-\frac{(E - E_F)}{kT}\right]dE$$

which leads to

Electron concentration in CB

$$n = N_c\exp\left[-\frac{(E_c - E_F)}{kT}\right] \qquad [5.6]$$

where

Effective density of states at CB edge

$$N_c = 2\left[\frac{2\pi m_e^* kT}{h^2}\right]^{3/2}$$

The result of the integration in Equation 5.6 seems to be simple, but it is an approximation as it assumes that $(E_c - E_F) \gg kT$. N_c is a temperature-dependent constant, called the **effective density of states at the CB edge.** Equation 5.6 can be interpreted as follows. If we take all the states in the conduction band and replace them with an effective concentration, N_c (number of states per unit volume), at E_c and then multiply this simply by the Boltzmann probability function, $f(E) = \exp[-(E_c - E_F)]$, we obtain the concentration of electrons at E_c, that is, in the conduction band. N_c is thus an effective density of states at the CB band edge.

We can carry out a similar analysis for the concentration of holes in the VB. Multiplying the density of states $g_{vb}(E)$ in the VB with the probability of occupancy by a hole, $[1 - f(E)]$, that is, the probability that an electron is absent, gives p_E, the hole concentration per unit energy. Integrating this over the VB gives the hole concentration

$$p = \int_0^{E_v} p_E\, dE = \int_0^{E_v} g_{vb}(E)[(1 - f(E)]\, dE$$

With the assumption that E_F is a few kT above E_v, the integration simplifies to

Hole concentration in VB

$$p \approx N_v\exp\left[-\frac{(E_F - E_v)}{kT}\right] \qquad [5.7]$$

where N_v is the effective density of states at the VB edge and is given by

$$N_v = 2\left[\frac{2\pi m_h^* kT}{h^2}\right]^{3/2}$$

Effective density of states at VB edge

We can now see the virtues of studying the density of states $g(E)$ as a function of energy E and the Fermi–Dirac function $f(E)$. Both were central factors in deriving the expressions for n and p. There are no specific assumptions in our derivations above, except for E_F being a few kT away from the band edges, which means that Equations 5.6 and 5.7 are generally valid.

The general equations that determine the free electron and hole concentrations are thus given by Equations 5.6 and 5.7. It is interesting to consider the product np,

$$np = N_c \exp\left[-\frac{(E_c - E_F)}{kT}\right]N_v \exp\left[-\frac{(E_F - E_v)}{kT}\right] = N_c N_v \exp\left[-\frac{(E_c - E_v)}{kT}\right]$$

or

$$np = N_c N_v \exp\left(-\frac{E_g}{kT}\right) \qquad [5.8]$$

where $E_g = E_c - E_v$ is the band gap energy. First, we note that this is a general expression in which the right-hand side, $N_c N_v \exp(-E_g/kT)$, is a constant that depends on the temperature and the material properties, for example, E_g, and not on the position of the Fermi level. In the special case of an intrinsic semiconductor, $n = p$, which we can denote as n_i (the intrinsic concentration), so that $N_c N_v \exp(-E_g/kT)$ must be n_i^2. From Equation 5.8 we therefore have

$$np = n_i^2 = N_c N_v \exp\left(-\frac{E_g}{kT}\right) \qquad [5.9] \quad \textit{Mass action law}$$

This is a general equation that is valid as long as we have thermal equilibrium and external excitation, such as photogeneration, is excluded. It states that the product np is a temperature-dependent constant. If we somehow increase the electron concentration, then we inevitably reduce the hole concentration. The constant n_i has a special significance because it represents the free electron and hole concentrations in the intrinsic material.

An intrinsic semiconductor is a pure semiconductor crystal in which the electron and hole concentrations are equal. By pure we mean virtually no impurities in the crystal. We should also exclude crystal defects that may capture carriers of one sign and thus result in unequal electron and hole concentrations. Clearly in a pure semiconductor, electrons and holes are generated in pairs by thermal excitation across the band gap. It must be emphasized that Equation 5.9 is generally valid and therefore applies to both intrinsic and nonintrinsic ($n \neq p$) semiconductors.

When an electron and hole meet in the crystal, they "recombine." The electron falls in energy and occupies the empty electronic state that the hole represents. Consequently, the broken bond is "repaired," but we lose two free charge carriers. Recombination of an electron and hole results in their annihilation. In a semiconductor we therefore have thermal generation of electron–hole pairs by thermal excitation from the VB to the CB, and we also have recombination of electron–hole pairs that removes them from their conduction and valence bands respectively. The rate of recombination R will be proportional to the number of electrons and also to the number of holes. Thus

$$R \propto np$$

The rate of generation G will depend on how many electrons are available for excitation at E_v, that is, N_v; how many empty states are available at E_c, that is, N_c; and the probability that the electron will make the transition, that is, $\exp(-E_g/kT)$, so that

$$G \propto N_c N_v \, \exp\left(-\frac{E_g}{kT} \right)$$

Since in thermal equilibrium we have no continuous increase in n or p, we must have the rate of generation equal to the rate of recombination, that is, $G = R$. This is equivalent to Equation 5.9.

In sketching the diagrams in Figure 5.7a to d to illustrate the derivation of the expressions for n and p (in Equations 5.6 and 5.7), we assumed that the Fermi level E_F is somewhere around the middle of the energy band gap. This was not an assumption in the mathematical derivations but only in the sketches. From Equations 5.6 and 5.7 we also note that the position of Fermi level is important in determining the electron and hole concentrations. It serves as a "mathematical crank" to determine n and p.

We first consider an intrinsic semiconductor, $n = p = n_i$. Setting $p = n_i$ in Equation 5.7, we can solve for the Fermi energy in the intrinsic semiconductor, E_{Fi}, that is,

$$N_v \, \exp\left[-\frac{(E_{Fi} - E_v)}{kT} \right] = (N_c N_v)^{1/2} \, \exp\left(-\frac{E_g}{2kT} \right)$$

which leads to

$$E_{Fi} = E_v + \frac{1}{2}E_g - \frac{1}{2}kT \, \ln\left(\frac{N_c}{N_v} \right) \qquad \text{[5.10]}$$

Furthermore, substituting the proper expressions for N_c and N_v we get

$$E_{Fi} = E_v + \frac{1}{2}E_g - \frac{3}{4}kT \, \ln\left(\frac{m_e^*}{m_h^*} \right) \qquad \text{[5.11]}$$

It is apparent from these equations that if $N_c = N_v$ or $m_e^* = m_h^*$, then

$$E_{Fi} = E_v + \frac{1}{2}E_g$$

that is, E_{Fi} is right in the middle of the energy gap. Normally, however, the effective masses will not be equal and the Fermi level will be slightly shifted down from midgap by an amount $(3/4) \, kT \, \ln(m_e^*/m_h^*)$, which is quite small compared with $(1/2) \, E_g$. For Si and Ge, the hole effective mass is slightly greater than the electron effective mass, so E_{Fi} is slightly above the mid-gap.

The condition $np = n_i^2$ means that if we can somehow increase the electron concentration in the CB over the intrinsic value—for example, by adding impurities into the Si crystal that donate additional electrons to the CB—we will then have $n > p$. The semiconductor is then called n-type. The Fermi level must be closer to E_c than E_v, so that

$$E_c - E_F < E_F - E_v$$

and Equations 5.6 and 5.7 yield $n > p$. The np product always yields n_i^2 in thermal equilibrium in the absence of external excitation, for example, illumination.

It is also possible to have an excess of holes in the VB over electrons in the CB, for example, by adding impurities that remove electrons from the VB and thereby generate holes. In that case E_F is closer to E_v than to E_c. A semicoductor in which $p > n$ is called a p-type semiconductor. The general band diagrams with the appropriate Fermi levels for intrinsic, n, and p-type semiconductors (e.g., i-Si, n-Si, and p-Si respectively) are illustrated in Figure 5.8a to c.

It is apparent that if we know where E_F is, then we have effectively determined n and p by virtue Equations 5.6 and 5.7. We can view E_F as a material property that is related to the concentration of charge carriers that contribute to electrical conduction. Its significance, however, goes beyond n and p. It also determines the energy needed to remove an electron from the semiconductor. The energy difference between the vacuum level (where the electron is free) and E_F is the work function, Φ, of the semiconductor, the energy required to remove an electron even though there are no electrons at E_F in a semiconductor.

The Fermi level can also be interpreted in terms of the potential energy per electron for electrical work similar to the interpretation of electrostatic PE, V. Just as $e\Delta V$ is the electrical work involved in taking a charge e across a potential difference, ΔV, any difference in E_F in going from one end of a material (or system) to another is available to do an amount ΔE_F of external work. A corollary to this is that if electrical work is done on the material, for example, by passing a current through it, then the Fermi level is not uniform in the material. ΔE_F then represents the work done per electron. For a material in thermal equilibrium and not subject to any external excitation such as illumination or connections to a voltage supply, the Fermi level in the material must therefore be uniform, $\Delta E_F = 0$.

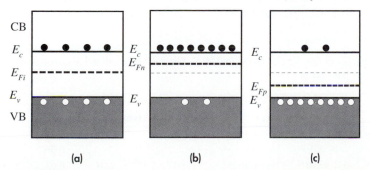

(a)　　　　　　　(b)　　　　　　　(c)

Figure 5.8   Energy band diagrams for (a) intrinsic, (b) n-type, and (c) p-type semiconductors.

In all cases, $np = n_i^2$.

What is the average energy of an electron in the conduction band of a semiconductor? Also, what is the mean speed of an electron in the conduction band? We note that the concentration of electrons with energies E to $E + dE$ is $n_E(E)\,dE$ or $g_{cb}(E)f(E)\,dE$. Thus the average energy of elections in the CB, by definition of the mean, is

$$\overline{E}_{CB} = \frac{1}{n} \int_{CB} E g_{cb}(E) f(E)\,dE$$

where the integration is over the CB. This must also be the average kinetic energy since we assumed that an electron in the CB is "free." Substituting the proper expressions for $g_{cb}(E)$ and $f(E)$ in the above integrand, and carrying out the integration from E_c to the top of the band, we find the very simple result that

$$\overline{E}_{CB} = \frac{3}{2}kT \qquad [5.12]$$

This is just like the average kinetic energy of gas atoms (such as He atoms) in a tank assuming that the atoms (or the "particles") do not interact, that is, they are independent. We know from the kinetic theory that the statistics of a collection of independent gas atoms obeys the classical Maxwell–Boltzmann description with an average energy given by Equation 5.12. We should also recall that the description of electron statistics in a metal involves the Fermi–Dirac function, which is based on the Pauli exclusion principle. In a metal the average energy of the conduction electron is $(3/5)E_F$ and, for all practical purposes, temperature independent. We see that the collective electron behavior is completely different in the two solids. We can explain the difference by noting that the conduction band in a semiconductor is only scarcely populated by electrons, which means that there are many more electronic states than electrons and thus the likelihood of two electrons trying to occupy the same electronic state is practically nil. We can then neglect the Pauli exclusion principle and use the Boltzmann statistics. This is not the case for metals where the number of conduction electrons and the number of states are comparable in magnitude.

Table 5.1 is a comparative table of some of the properties of the important semiconductors, Ge, Si, and GaAs.

Table 5.1   Selected typical properties of Ge, Si, and GaAs at 300 K. Effective mass related to conductivity (labeled a) is different than that for density of states (labeled b).

	E_g (eV)	χ (eV)	N_c (cm^{-3})	N_v (cm^{-3})	n_i (cm^{-3})	μ_e (cm^2 V^{-1} s^{-1})	μ_h (cm^2 V^{-1} s^{-1})	m_e^*/m_e	m_h^*/m_e	ε_r
Ge	0.66	4.13	1.04×10^{19}	6.0×10^{18}	2.4×10^{13}	3900	1900	0.12 a 0.56 b	0.23a 0.40b	16
Si	1.10	4.01	2.8×10^{19}	1.04×10^{19}	1.45×10^{10}	1350	450	0.26 a 1.08 b	0.38a 0.56b	11.9
GaAs	1.42	4.07	4.7×10^{17}	7×10^{18}	1.8×10^6	8500	400	0.067 a,b	0.40a 0.50b	13.1

INTRINSIC CONCENTRATION AND CONDUCTIVITY OF Si   Given that the density of states | **Example 5.1**
related effective masses of electrons and holes in Si are approximately $1.08m_e$ and $0.56m_e$
respectively and the electron and hole drift mobilities at room temperature are 1350 and
450 cm^2 V^{-1} s^{-1}, calculate the intrinsic concentration and intrinsic resistivity of Si.

Solution

We simply calculate the effective density of states N_c and N_v by

$$N_c = 2\left[\frac{2\pi m_e^* kT}{h^2}\right]^{3/2} \quad \text{and} \quad N_v = 2\left[\frac{2\pi m_h^* kT}{h^2}\right]^{3/2}$$

Thus

$$N_c = 2\left[\frac{2\pi(1.08 \times 9.1 \times 10^{-31}\text{ kg})(1.38 \times 10^{-23}\text{ J K}^{-1})(300\text{ K})}{(6.63 \times 10^{-34}\text{ J s})^2}\right]^{3/2}$$

$$= 2.8 \times 10^{25}\text{ m}^{-3} \quad \text{or} \quad 2.8 \times 10^{19}\text{ cm}^{-3}$$

and

$$N_v = 2\left[\frac{2\pi(0.56 \times 9.1 \times 10^{-31}\text{ kg})(1.38 \times 10^{-23}\text{ J K}^{-1})(300\text{ K})}{(6.63 \times 10^{-34}\text{ J s})^2}\right]^{3/2}$$

$$= 1.05 \times 10^{25}\text{ m}^{-3} \quad \text{or} \quad 1.05 \times 10^{19}\text{ cm}^{-3}$$

The intrinsic concentration is

$$n_i = (N_c N_v)^{1/2} \exp\left(\frac{-E_g}{2kT}\right)$$

so that

$$n_i = [(2.8 \times 10^{19}\text{ cm}^{-3})(1.05 \times 10^{19}\text{ cm}^{-3})]^{1/2} \exp\left[\frac{(-1.10\text{ eV})}{2(300\text{ K})(8.62 \times 10^{-5}\text{ eV K}^{-1})}\right]$$

$$n_i = 1.0 \times 10^{10}\text{ cm}^{-3}$$

The conductivity is

$$\sigma = en\mu_e + ep\mu_h = en_i(\mu_e + \mu_h)$$

that is,

$$\sigma = (1.6 \times 10^{-19}\text{C})(1.0 \times 10^{10}\text{ cm}^{-3})(1350 + 450\text{ cm}^2\text{ V}^{-1}\text{ s}^{-1})$$

$$= 2.9 \times 10^{-6}\ \Omega^{-1}\text{ cm}^{-1}$$

The resistivity is

$$\rho = \frac{1}{\sigma} = 3.5 \times 10^5\ \Omega\text{ cm}$$

MEAN SPEED OF ELECTRONS IN THE CB   Estimate the mean speed of electrons in the conduc- | **Example 5.2**
tion band of Si and thus predict the temperature dependence of the drift mobility due to
scattering from thermal vibrations. The effective mass of an electron in the conduction band
is $0.26m_e$.

Solution

The fact that the average $KE, (1/2)\, m_e^* v_e^2$, of an electron in the CB of a semiconductor is
$(3/2)\, kT$ means that the effective mean speed, v_e, must be

$$v_e = \left[\frac{3kT}{m_e^*}\right]^{1/2} = \left[\frac{(3 \times 1.38 \times 10^{-23} \times 300)}{(0.26 \times 9.1 \times 10^{-31})}\right]^{1/2} = 2.3 \times 10^5 \text{ m s}^{-1}$$

The effective mean speed v_e is called the thermal velocity v_{th} of the electron.

The mean free time τ between scattering events due to atomic thermal vibrations is inversely proportional to both the mean speed v_e of the electron and the scattering cross section of the thermal vibrations, that is,

$$\tau \propto \frac{1}{v_e(\pi a^2)}$$

where a is the amplitude of the atomic thermal vibrations. But, $v_e \propto T^{1/2}$ and $(\pi a^2) \propto kT$, so that $\tau \propto T^{-3/2}$ and consequently $\mu_e \propto T^{-3/2}$.

NOTE: The effective mass used in the density of states calculations is actually different than that used in transport calculations such as the mean speed, drift mobility, and so on.

5.2   EXTRINSIC SEMICONDUCTORS

By introducing small amounts of impurities into an otherwise pure Si crystal, it is possible to obtain a semiconductor in which the concentration of carriers of one polarity is much in excess of the other type. Such semiconductors are referred to as **extrinsic semiconductors** vis-à-vis the intrinsic case of a pure and perfect crystal. For example, by adding pentavalent impurities, such as arsenic, which have a valency of more than four, we can obtain a semiconductor in which the electron concentration is much larger than the hole concentration. In this case we will have an n-type semiconductor. If we add trivalent impurities, such as boron, which have a valency of less than four, then we find that we have an excess of holes over electrons. We now have a p-type semiconductor. How do impurities change the concentrations of holes and electrons in a semiconductor?

5.2.1 n-Type Doping

Consider what happens when small amounts of a pentavalent (valency of 5) element from group V in the periodic table, such as As, P, Sb, are introduced into a pure Si crystal. We only add small amounts (e.g., one impurity atom for every million host atoms) because we wish to surround each impurity atom by millions of Si atoms, thereby forcing the impurity atoms to bond with Si atoms in the same diamond crystal structure. Arsenic has five valence electrons, whereas Si has four. Thus when an As atom bonds with four Si atoms, it has one electron left unbonded. It cannot find a bond to go into, so it is left orbiting around the As atom, as illustrated in Figure 5.9. The As$^+$ ionic center with an electron, e^-, orbiting it is just like a hydrogen atom in a silicon environment. We can easily calculate how much energy is required to free this electron away from the As site, thereby ionizing the As impurity. Had this been a hydrogen atom in free space, the energy required to

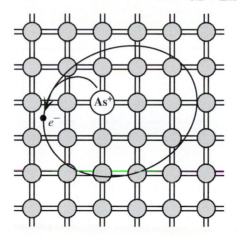

Figure 5.9 Arsenic-doped Si crystal. The four valence electrons of As allow it to bond just like Si, but the fifth electron is left orbiting the As site. The energy required to release the free fifth electron into the CB is very small.

remove the electron from its ground state (at $n = 1$) to far away from the positive center would have been given by $-E_n$ with $n = 1$. The binding energy of the electron in the H-atom is thus

$$E_b = -E_1 = \frac{m_e e^4}{8\varepsilon_o^2 h^2} = 13.6 \text{ eV}$$

If we wish to apply this to the electron around an As$^+$ core in the Si crystal environment, we must use $\varepsilon_r \varepsilon_o$ instead of ε_o, where ε_r is the relative permittivity of silicon, and also the effective mass of the electron m_e^* in the silicon crystal. Thus, the binding energy of the electron to the As$^+$ site in the Si crystal is

$$E_b^{Si} = \frac{m_e^* e^4}{8\varepsilon_r^2 \varepsilon_o^2 h^2} = (13.6 \text{ eV})\left(\frac{m_e^*}{m_e}\right)\left(\frac{1}{\varepsilon_r^2}\right) \qquad \text{[5.13]}$$

With $\varepsilon_r = 11.9$ and $m_e^* \approx (1/3) m_e$ for silicon, we find $E_b^{Si} = 0.032$ eV, which is comparable with the average thermal energy of atomic vibrations at room temperature, $\sim 3kT (\sim 0.07 \text{ eV})$. Thus, the fifth valence electron can be readily freed by thermal vibrations of the Si lattice. The electron will then be "free" in the semiconductor, or, in other words, it will be in the CB. The energy required to excite the electron to the CB is therefore 0.032 eV. The addition of As atoms introduces localized electronic states at the As sites because the fifth electron has a localized wavefunction, of the hydrogenic type, around As$^+$. The energy E_d of these states is 0.032 eV below E_c because this is how much energy is required to take the electron away into the CB. Thermal excitation by the lattice vibrations at room temperature is sufficient to ionize the As atom, that is, excite the electron from E_d into the CB. This process creates free electrons but immobile As$^+$ ions, as

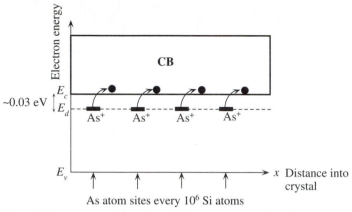

Figure 5.10 Energy band diagram for an *n*-type Si doped with 1 ppm As. There are donor energy levels just below E_c around As$^+$ sites.

shown in the energy band diagram of an *n*-type semiconductor in Figure 5.10. Because the As atom donates an electron into the CB, it is called a **donor** atom. E_d is the electron energy around the donor atom. E_d is close to E_c so that the spare fifth electron from the dopant can be readily donated to the CB. If N_d is the donor atom concentration in the crystal, then provided that $N_d \gg n_i$, at room temperature the electron concentration in the CB will be nearly equal to N_d, that is $n \approx N_d$. The hole concentration will be $p = n_i^2/N_d$, which is less than the intrinsic concentration because a few of the large number of electrons in the CB recombine with holes in the VB so as to maintain $np = n_i^2$. The conductivity will then be

$$\sigma = eN_d\mu_e + e\left(\frac{n_i^2}{N_d}\right)\mu_h \approx eN_d\mu_e \qquad \text{[5.14]}$$

At low temperatures, however, not all the donors will be ionized and we need to know the probability, denoted as $f_d(E_d)$, of finding an electron in a state with energy E_d as a donor. This probability function is similar to the Fermi–Dirac function $f(E_d)$ except that it has a factor of $1/2$ multiplying the exponential term,

$$f_d(E_d) = \frac{1}{1 + \dfrac{1}{2}\exp\left[\dfrac{(E_d - E_F)}{kT}\right]} \qquad \text{[5.15]}$$

The factor $1/2$ is due to the fact that the electron state at the donor can take an electron with spin either up or down but not both[2] (once the donor has been occupied, a second electron cannot enter this site). Thus, the number of ionized

---

| [2] The proof can be found in advanced solid state physics texts.

donors at a temperaure T is given by

$$N_d^+ = N_d \times (\text{probability of not finding an electron at } E_d) \qquad \text{[5.16]}$$

$$= N_d[1 - f_d(E_d)]$$

$$N_d^+ = \frac{N_d}{1 + 2\exp\left[\dfrac{(E_F - E_d)}{kT}\right]}$$

5.2.2 *p*-Type Doping

We saw that introducing a pentavalent atom into a Si crystal results in *n*-type doping because the fifth electron cannot go into a bond and escapes from the donor into the CB by thermal excitation. By similar arguments, we should anticipate that doping a Si crystal with a trivalent atom (valency of 3) such as B, Al, Ga, or In will result in a *p*-type Si crystal. We consider doping Si with small amounts of B as shown in Figure 5.11a. Because B has only three valence electrons, when it shares them with four neighboring Si atoms, one of the bonds has a missing electron, which of course is a hole. A nearby electron can tunnel into this hole and displace the hole further away from the Boron atom. As the hole moves away, it gets attracted by the negative charge left behind on the Boron atom and therefore takes an orbit around the B^- ion, as shown in Figure 5.11b. The binding energy of this hole to the B^- ion can be calculated using the hydrogenic atom analogy as in the *n*-type Si case. This binding energy turns out to be very small, ~ 0.05 eV, so that

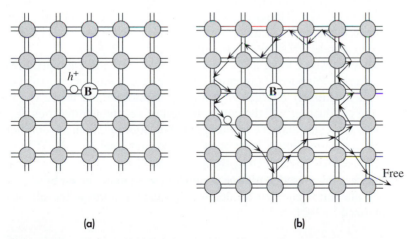

(a)   (b)

Figure 5.11 Boron-doped Si crystal.
B has only three valence electrons. When it substitutes for a Si atom, one of its bonds has an electron missing and therefore a hole, as shown in (a). The hole orbits around the B^- site by the tunneling of electrons from neighboring bonds, as shown in (b). Eventually, thermally vibrating Si atoms provide enough energy to free the hole from the B^- site into the VB, as shown.

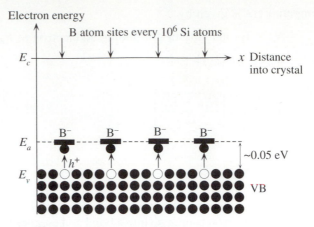

Figure 5.12   Energy band diagram for a *p*-type Si doped with 1 ppm B.

There are acceptor energy levels E_a just above E_v around B⁻ sites. These acceptor levels accept electrons from the VB and therefore create holes in the VB.

at room temperature the thermal vibrations of the lattice can free the hole away from the B⁻ site. A free hole, we recall, exists in the VB. The escape of the hole from the B⁻ site involves the B atom *accepting* an electron from a neighboring Si–Si bond (from the VB), which effectively results in the hole being displaced away and its eventual escape to freedom in the VB. The B atom introduced into the Si crystal therefore acts as an electron acceptor and, because of this, it is called an **acceptor impurity.** The electron accepted by the B atom comes from a nearby bond. On the energy band diagram, an electron leaves the VB and gets accepted by a B atom, which becomes negatively charged. This process leaves a hole in the VB that is free to wander away, as illustrated in Figure 5.12.

It is apparent that doping a silicon crystal with a trivalent impurity results in a *p*-type material. We have many more holes than electrons for electrical conduction since the negatively charged B atoms are immobile and hence cannot contribute to the conductivity. If the concentration of acceptor impurities, N_a, in the crystal is much greater than the intrinsic concentration, n_i, then at room temperature all the acceptors would have been ionized and thus $p \approx N_a$. The electron concentration is then determined by the mass action law, $n = n_i^2/N_a$, which is much smaller than p, and consequently the conductivity is simply given by $\sigma = eN_a\mu_h$.

Typical ionization energies for donor and acceptor atoms in the silicon crystal are summarized in Table 5.2.

Table 5.2   Examples of donor and acceptor ionization energies (eV) in Si

Donors			Acceptors		
P	**As**	**Sb**	**B**	**Al**	**Ga**
0.045	0.054	0.039	0.045	0.057	0.072

5.2.3   Compensation Doping

What happens when a semiconductor contains both donors and acceptors? **Compensation doping** is a term used to describe the doping of a semiconductor with both donors and acceptors to control the properties. For example, a p-type semiconductor doped with N_a acceptors can be converted to an n-type semiconductor by simply adding donors until the concentration N_d exceeds N_a. The effect of donors compensates for the effect of acceptors and vice versa. The electron concentration is then given by $N_d - N_a$ provided the latter is larger than n_i. When both acceptors and donors are present, what essentially happens is that electrons from donors recombine with the holes from the acceptors so that the mass action law $np = n_i^2$ is obeyed. Remember that we cannot simultaneously increase the electron and hole concentrations because that leads to an increase in the recombination rate that returns the electron and hole concentrations to satisfy $np = n_i^2$. When an acceptor atom accepts a valence band electron, a hole is created in the VB. This hole then recombines with an electron from the CB. Suppose that we have more donors than acceptors. If we take the initial electron concentration as $n = N_d$, then the recombination between the electrons from the donors and N_a holes generated by N_a acceptors results in the electron concentration reduced by N_a to $n = N_d - N_a$. By a similar argument, if we have more acceptors than donors, the hole concentration becomes $p = N_a - N_d$, with electrons from N_d donors recombining with holes from N_a acceptors. Thus there are two compensation effects:

1.   More donors: $N_d - N_a \gg n_i$ $n = (N_d - N_a)$ and $p = \dfrac{n_i^2}{(N_d - N_a)}$

2.   More acceptors: $N_a - N_d \gg n_i$ $p = (N_a - N_d)$ and $n = \dfrac{n_i^2}{(N_a - N_d)}$

The above arguments assume that the temperature is sufficiently high for donors and acceptors to have been ionized. This will be the case at room temperature. At low temperatures, we have to consider donor and acceptor statistics and the charge neutrality of the whole crystal, as in Example 5.8.

RESISTIVITY OF INTRINSIC AND DOPED Si   Find the resistance of a 1 cm^3 pure silicon crystal. | **Example 5.3**
What is the resistance when the crystal is doped with arsenic if the doping is 1 in 10^9, that is, 1 part per billion (ppb) (note that this doping corresponds to one foreigner living in China). Given data: Atomic concentration in silicon is 5×10^{22} cm^{-3}, $n_i = 1.45 \times 10^{10}$ cm^{-3}, $\mu_e = 1350$ cm^2 V^{-1} s^{-1}, and $\mu_h = 450$ cm^2 V^{-1} s^{-1}.

Solution

For the intrinsic case, we apply

$$\sigma = en\mu_e + ep\mu_h = en(\mu_e + \mu_h)$$

so that

$$\sigma = (1.6 \times 10^{-19} \text{ C})(1.45 \times 10^{10} \text{ cm}^{-3})(1350 + 450 \text{ cm}^2 \text{ V}^{-1} \text{ s}^{-1})$$

$$= 4.18 \times 10^{-6} \ \Omega^{-1} \text{ cm}^{-1}$$

Since $L = 1$ cm and $A = 1$ cm^{-2}, the resistance is

$$R = \frac{L}{\sigma A} = \frac{1}{\sigma} = 2.39 \times 10^5 \ \Omega \ \text{or } 239 \ \text{k}\Omega$$

When the crystal is doped with 1 in 10^9, then

$$N_d = \frac{N_{si}}{10^9} = \frac{5 \times 10^{22}}{10^9} = 5 \times 10^{13} \ \text{cm}^{-3}$$

At room temperature all the donors are ionized, so that

$$n = N_d = 5 \times 10^{13} \ \text{cm}^{-3}$$

The hole concentration is

$$p = \frac{n_i^2}{N_d} = \frac{(1.45 \times 10^{10})^2}{(5 \times 10^{13})} = 4.2 \times 10^6 \ \text{cm}^{-3} \ll n_i$$

Therefore,

$$\sigma = en\mu_e = (1.6 \times 10^{-19} \ \text{C})(5 \times 10^{13} \ \text{cm}^{-3})(1350 \ \text{cm}^2 \ \text{V}^{-1} \ \text{s}^{-1})$$
$$= 1.08 \times 10^{-2} \ \Omega^{-1} \ \text{cm}^{-1}$$

Further,

$$R = \frac{L}{\sigma A} = \frac{1}{\sigma} = 92.6 \ \Omega$$

Notice the drastic fall in the resistance when the crystal is doped with only 1 in 10^9 atoms.

Doping the silicon crystal with boron instead of arsenic, but still in amounts of 1 in 10^9, means that $N_a = 5 \times 10^{13}$ cm^{-3}, which results in a conductivity of

$$\sigma = ep\mu_h = (1.6 \times 10^{-19} \ \text{C})(5 \times 10^{13} \ \text{cm}^{-3})(450 \ \text{cm}^2 \ \text{V}^{-1} \ \text{s}^{-1})$$
$$= 3.6 \times 10^{-3} \ \Omega^{-1} \ \text{cm}^{-1}$$

Therefore,

$$R = \frac{1}{\sigma} = 278 \ \Omega$$

The reason for a higher resistance with p-type doping compared with the same amount of n-type doping is that $\mu_h < \mu_e$.

Example 5.4 | **COMPENSATION DOPING**   An n-type Si semiconductor containing 10^{16} phosphorus (donor) atoms cm^{-3} has been doped with 10^{17} boron (acceptor) atoms cm^{-3}. Calculate the electron and hole concentrations in this semiconductor.

Solution

This semiconductor has been compensation doped with excess acceptors over donors so that

$$N_a - N_d = 10^{17} - 10^{16} = 9 \times 10^{16} \text{ cm}^{-3}$$

This is much larger than the intrinsic concentration, $n_i = 1.45 \times 10^{10} \text{ cm}^{-3}$ at room temperature, so that

$$p = N_a - N_d = 9 \times 10^{16} \text{ cm}^{-3}$$

The electron concentration

$$n = \frac{n_i^2}{p} = \frac{(1.45 \times 10^{10} \text{ cm}^{-3})^2}{(9 \times 10^{16} \text{ cm}^{-3})} = 2.34 \times 10^3 \text{ cm}^{-3}$$

Clearly, the electron concentration and hence its contribution to electrical conduction is completely negligible compared with the hole concentration. Thus, by excessive boron doping, the n-type semiconductor has been converted to a p-type semiconductor.

THE FERMI LEVEL IN n AND p TYPE Si   An n-type Si wafer has been doped uniformly with 10^{16} antimony (Sb) atoms cm^{-3}. Calculate the position of the Fermi energy with respect to the Fermi energy, E_{Fi}, in intrinsic Si. The above n-type Si sample is further doped with 2×10^{17} boron atoms cm^{-3}. Calculate the position of the Fermi energy with respect to the Fermi energy, E_{Fi}, in intrinsic Si, and hence with respect to the Fermi energy in the n-type case above. | **Example 5.5**

Solution

Sb gives n-type doping with $N_d = 10^{16} \text{ cm}^{-3}$, and since $N_d \gg n_i (= 1.45 \times 10^{10} \text{ cm}^{-3})$, we have

$$n = N_d = 10^{16} \text{ cm}^{-3}$$

For intrinsic Si,

$$n_i = N_c \exp\left[-\frac{(E_c - E_{Fi})}{kT}\right]$$

whereas for doped Si,

$$n = N_c \exp\left[-\frac{(E_c - E_{Fn})}{kT}\right] = N_d$$

where E_{Fi} and E_{Fn} are the Fermi energies in the intrinsic and n-type Si. Dividing the two expressions,

$$\frac{N_d}{n_i} = \exp\left[\frac{(E_{Fn} - E_{Fi})}{kT}\right]$$

so that

$$E_{Fn} - E_{Fi} = kT \ln\left(\frac{N_d}{n_i}\right) = (0.0259 \text{ eV}) \ln\left(\frac{10^{16}}{1.45 \times 10^{10}}\right) = 0.348 \text{ eV}$$

When the wafer is further doped with boron, the acceptor concentration is

$$N_a = 2 \times 10^{17} \text{ cm}^{-3} > N_d = 10^{16} \text{ cm}^{-3}$$

The semiconductor is compensation doped and compensation converts the semconductor to *p*-type Si. Thus

$$p = N_a - N_d = (2 \times 10^{17} - 10^{16}) = 1.9 \times 10^{17} \text{ cm}^{-3}$$

For intrinsic Si,

$$p = n_i = N_v \exp\left[\frac{-(E_{Fi} - E_v)}{kT}\right]$$

whereas for doped Si,

$$p = N_v \exp\left[\frac{-(E_{Fp} - E_v)}{kT}\right] = N_a - N_d$$

where E_{Fi} and E_{Fp} are the Fermi energies in the intrinsic and *p*-type Si respectively. Dividing the two expressions,

$$\frac{p}{n_i} = \exp\left[\frac{-(E_{Fp} - E_{Fi})}{kT}\right]$$

so that

$$E_{Fp} - E_{Fi} = -kT \ln\left(\frac{p}{n_i}\right) = -(0.0259 \text{ eV}) \ln\left(\frac{1.9 \times 10^{17}}{1.45 \times 10^{10}}\right)$$

$$= -0.42 \text{ eV}$$

Example 5.6

ENERGY BAND DIAGRAM OF AN *n*-TYPE SEMICONDUCTOR CONNECTED TO A VOLTAGE SUPPLY Consider the energy band diagram for an *n*-type semiconductor that is connected to a voltage supply of *V* and is carrying a current. The applied voltage drops uniformly along the semiconductor so that the electrons in the semiconductor now also have an imposed electrostatic potential energy that decreases toward the positive terminal, as depicted in Figure 5.13. The whole band structure, the CB and the VB, therefore tilts. When an electron drifts from *A* toward *B*, its *PE* decreases because it is approaching the positive terminal. The Fermi level, E_F, is above that for the intrinsic case, E_{Fi}.

We should remember that an important property of the Fermi level is that a change in E_F within a system is available externally to do electrical work. As a corollary we note that when electrical work is done on the system, for example, when a battery is connected to a semiconductor, then E_F is not uniform throughout the whole system. A change in E_F within a system ΔE_F, is equivalent to electrical work per electron or *eV*. E_F therefore follows the electrostatic *PE* behavior and the change in E_F from one end to the other, $E_F(A) - E_F(B)$, is just *eV*, the energy expended in taking an electron through the semiconductor, as shown in Figure 5.13. Electron concentration in the semiconductor is uniform so that $E_c - E_F$ must be constant from one end to the other. Thus the CB, VB, and E_F all bend by the same amount.

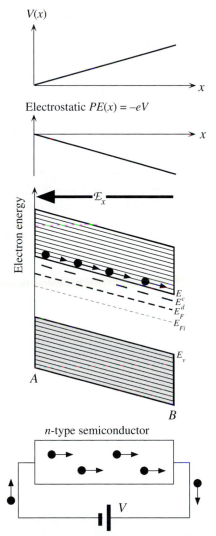

Figure 5.13   Energy band diagram of an *n*-type semiconductor connected to a voltage supply of *V* volts.

The whole energy diagram tilts because the electron now also has an electrostatic potential energy.

5.3 TEMPERATURE DEPENDENCE OF CONDUCTIVITY

So far we have been calculating conductivities and resistivities of doped semiconductors at room temperature by simply assuming that $n \approx N_d$ for n-type and $p \approx N_a$ for p-type doping, with the proviso that the concentration of dopants is much greater than the intrinsic concentration n_i. To obtain the conductivity at other temperatures we have to consider two factors: the temperature dependence of the carrier concentration and the drift mobility.

5.3.1 Carrier Concentration Temperature Dependence

Consider an n-type semiconductor doped with N_d donors per unit volume where $N_d \gg n_i$. We take the semiconductor down to very low temperatures until its conductivity is practically nil. At this temperature, the donors will *not* be ionized because the thermal vibrational energy is insufficiently small. As the temperature is increased, some of the donors become ionized and donate their electrons to the CB, as shown in Figure 5.14a. The Si–Si bond breaking, that is, thermal excitation from E_v to E_c, is unlikely because it takes too much energy. Since the donor ionization energy $\Delta E = E_c - E_d$ is very small ($\ll E_g$), thermal generation involves exciting electrons from E_d to E_c. The electron concentration at low temperatures is given by the expression

$$n = \left(\frac{1}{2} N_c N_d\right)^{1/2} \exp\left(-\frac{\Delta E}{2kT}\right) \qquad [5.17]$$

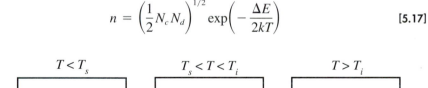

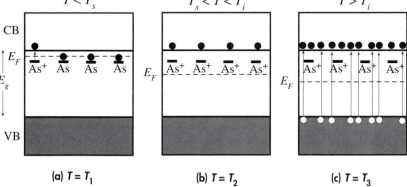

(a) $T = T_1$ (b) $T = T_2$ (c) $T = T_3$

Figure 5.14

(a) Below T_s, the electron concentration is controlled by the ionization of the donors.
(b) Between T_s and T_i, the electron concentration is equal to the concentration of donors since they would all have ionized.
(c) At high temperatures, thermally generated electrons from the VB exceed the number of electrons from ionized donors and the semiconductor behaves as if intrinsic.

similar to the intrinsic case, that is,

$$n = (N_c N_v)^{1/2} \exp\left(-\frac{E_g}{2kT}\right) \qquad [5.18]$$

Equation 5.18 is valid when thermal generation occurs across the band gap E_g from E_v to E_c. Equation 5.17 is the counterpart of Equation 5.18 taking into account that at low temperatures the excitation is from E_d to E_c (across ΔE) and that instead of N_v, we have N_d as the number of available electrons. The numerical factor $1/2$ in Equation 5.17 arises because donor occupation statistics is different by this factor from the usual Fermi–Dirac function, as mentioned earlier.

As the temperature is increased further, eventually all the donors become ionized and the electron concentration is equal to the donor concentration, that is, $n = N_d$, as depicted in Figure 5.14b. This state of affairs remains unchanged until very high temperatures are reached, when thermal generation across the band gap begins to dominate. At very high temperatures, thermal vibrations of the atoms will be so strong that many Si–Si bonds will be broken and thermal generation across E_g will dominate. The electron concentration in the CB will then be mainly due to thermal excitation from the VB to the CB, as illustrated in Figure 5.14c. But this process also generates an equal concentration of holes in the VB. Accordingly, the semiconductor behaves as if it were intrinsic. The electron concentration at these temperatures will therefore be equal to the intrinsic concentration n_i, which is given by Equation 5.18.

The dependence of the electron concentration on temperature thus has three regions:

1. **Low Temperature Range $(T < T_s)$.**   The increase in temperature at these low temperatures ionizes more and more donors. The donor ionization continues until we reach a temperature T_s, called the **saturation temperature,** when all donors have been ionized and we have saturation in the concentration of ionized donors. The electron concentration is given by Equation 5.17. This temperature range is often referred to as the **ionization range.**

2. **Medium Temperature Range $(T_s < T < T_i)$.**   Since all the donors have been ionized in this range, $n = N_i$. This condition remains unchanged until $T = T_i$ when n_i, which is temperature dependent, becomes equal to N_d. It is this temperature range $T_s < T < T_i$ that utilizes the n-type doping properties of the semiconductor in pn junction device applications. This temperature range is often referred to as the **extrinsic range.**

3. **High Temperature Range $(T > T_i)$.**   The concentration of electrons generated by thermal excitation across the band gap, n_i, is now much larger than N_d, so that the electron concentration $n = n_i(T)$. Furthermore, as excitation occurs from the VB to the CB, the hole concentration $p = n$. This temperature range is referred to as the **intrinsic range.**

Figure 5.15 shows the behavior of the electron concentration with temperature in an n-type semiconductor. By convention we plot $\ln(n)$ versus the reciprocal temperature, T^{-1}. At low temperatures, $\ln(n)$ versus T^{-1} is almost a straight line

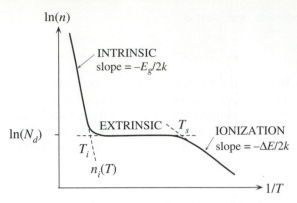

Figure 5.15   The temperature dependence of the electron concentration in an *n*-type semiconductor.

with a slope $-(\Delta E/2k)$, since the temperature dependence of $N_c^{1/2}\,(\propto T^{3/4})$ is negligible compared with the $\exp(-\Delta E/2kT)$ part in Equation 5.17. In the high temperature range, however, the slope is quite steep and almost $-E_g/2k$ since Equation 5.18 implies that

$$n \propto T^{3/2} \exp\left(-\frac{E_g}{2kT}\right)$$

and the exponential part again dominates over the $T^{3/2}$ part. In the intermediate range, n is equal to N_d and practically independent of the temperature.

Figure 5.16 displays the temperature dependence of the intrinsic concentration in Ge, Si, and GaAs as $\log(n_i)$ versus $1/T$ where the slope of the lines is, of course, a measure of the band gap energy, E_g. The $\log(n_i)$ versus $1/T$ graphs can be used to find, for example, whether the dopant concentration at a given temperature is more than the intrinsic concentration. As we will find out in Chapter 6, the reverse saturation current in a *pn* junction diode depends on n_i^2, so Figure 5.16 also indicates how this saturation current varies with temperature.

Example 5.7 | **SATURATION AND INTRINSIC TEMPERATURES**   An *n*-type Si sample has been doped with 10^{15} phosphorus atoms cm^{-3}. The donor energy level for P in Si is 0.045 eV below the conduction band edge energy. (*a*) Estimate the temperature above which the sample behaves as if intrinsic. (*b*) Estimate the lowest temperature above which all the donors are ionized.

Solution

Remember that $n_i(T)$ is highly temperature dependent, as shown in Figure 5.16 so that as T increases, eventually at $T \approx T_i$, n_i becomes comparable to N_d. Beyond T_i, $n_i(T > T_i) \gg N_d$. Thus we need to solve

$$n_i(T_i) = N_d = 10^{15} \text{ cm}^{-3}.$$

From the $\log(n_i)$ versus $10^3/T$ graph for Si in Figure 5.16, when $n_i = 10^{15}$ cm^{-3}, $(10^3/T_i) = 1.8$, giving $T_i = 556$ K   or   283 °C.

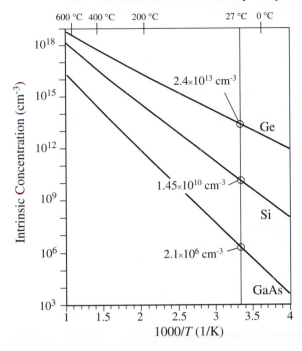

Figure 5.16   The temperature dependence of the intrinsic concentration.

We will assume that all donors are ionized, say at $T = T_s$, when

$$n = \left(\frac{1}{2}N_c N_d\right)^{1/2} \exp\left(-\frac{\Delta E}{2kT_s}\right) \approx N_d$$

This is the temperature at which the ionization behavior intersects the extrinsic region. In the above equation, $N_d = 10^{15}$ cm^{-3}, $\Delta E = 0.045$ eV, and $N_c \propto T^{3/2}$, that is,

$$N_c(T_s) = N_c(300 \text{ K})\left(\frac{T_s}{300}\right)^{3/2}$$

Clearly, then, the equation can only be solved numerically. Similar equations occur in a wide range of physical problems where one term has the strongest temperature dependence. Here, $\exp(-\Delta E/kT_s)$ dominates over $N_c = AT_s^{3/2}$. First assume N_c is that at 300 K, $N_c = 2.8 \times 10^{19}$ cm^{-3}, and evaluate T_s,

$$T_s = \frac{\Delta E}{k \ln\left(\dfrac{N_c}{2N_d}\right)}$$

$$T_s = \frac{0.045 \text{ eV}}{(8.62 \times 10^{-5} \text{ eV K}^{-1}) \ln\left[\dfrac{2.8 \times 10^{19} \text{ cm}^{-3}}{2(1.0 \times 10^{15} \text{ cm}^{-3})}\right]} = 54.7 \text{ K}$$

At $T = 54.7$ K,

$$N_c(54.7 \text{ K}) = N_c(300 \text{ K})\left(\frac{54.7}{300}\right)^{3/2} = 2.18 \times 10^{18} \text{ cm}^{-3}$$

With this new N_c at a lower temperature, the improved T_s is 74.6 K. You can of course repeat the iteration to obtain an even better value for T_s. At 74.6 K, $N_c = 3.47 \times 10^{18} \text{ cm}^{-3}$, so a better T_s is 70 K. The extrinsic range of this semiconductor is therefore from about 70 K to 556 K   or -203 °C to about 283 °C.

Example 5.8 | **TEMPERATURE DEPENDENCE OF THE ELECTRON CONCENTRATION**   By considering the mass action law, charge neutrality within the crystal, and occupation statistics of electronic states, we can show that at the lowest temperatures the electron concentration in an n-type semiconductor is given by

$$n = \left(\frac{1}{2}N_c N_d\right)^{1/2} \exp\left(-\frac{\Delta E}{2kT}\right)$$

where $\Delta E = E_c - E_d$. Furthermore, at the lowest temperatures, the Fermi energy is midway between E_d and E_c.

There are only a few physical principles that must be considered to arrive at the effect of doping on the electron and hole concentrations. For an n-type semiconductor, these are:

1. **Charge Carrier Statistics.**

$$n = N_c \exp\left[-\frac{(E_c - E_F)}{kT}\right]$$

2. **Mass Action Law.**

$$np = n_i^2$$

3. **Electrical Neutrality of the Crystal.**   We must have the same number of positive and negative charges:

$$p + N_d^+ = n$$

where N_d^+ is the concentration of *ionized* donors.

4. **Statistics of Ionization of the Dopants.**

$$N_d^+ = N_d \times \text{(probability of not finding an electron at } E_d) = N_d[1 - f_d(E_d)]$$

$$N_d^+ = \frac{N_d}{1 + 2\exp\left[\dfrac{(E_F - E_d)}{kT}\right]}$$

Solving Equations 1 to 4 for n will give the dependence of n on T and N_d. For example, from the mass action law, Equation 2, and the charge neutrality condition, Equation 3, we get

$$\frac{n_i^2}{n} + N_d^+ = n$$

This is a quadratic equation in n. Solving this equation gives

$$n = \frac{1}{2}(N_d^+) + \left[\frac{1}{4}(N_d^+)^2 + n_i^2\right]^{1/2}$$

Clearly, the above should give the behavior of n as a function of T and N_d when we also consider the statistics in Equation 4. In the low temperature region ($T < T_s$) discussed above, n_i^2 is negligible in the expression for n and we have

$$n = N_d^+ = \frac{N_d}{1 + 2\exp\left[\frac{(E_F - E_d)}{kT}\right]} \approx \frac{1}{2}N_d \exp\left[-\frac{(E_F - E_d)}{kT}\right]$$

But the statistical description in Equation 1 is generally valid so that multiplying the above by Equation 1 and taking the square root eliminates E_F from the expression, giving

$$n = \left(\frac{1}{2}N_c N_d\right)^{1/2} \exp\left[-\frac{(E_c - E_d)}{2kT}\right]$$

To find the location of the Fermi energy, consider the general expression

$$n = N_c \exp\left[-\frac{(E_c - E_F)}{kT}\right]$$

which must now correspond to n at low temperatures. Equating the two and rearranging to obtain E_F we find

$$E_F = \frac{E_c + E_d}{2} + \frac{1}{2}kT \ln\left(\frac{N_d}{2N_c}\right)$$

which puts the Fermi energy near the middle of $\Delta E = E_c - E_d$ at low temperatures.

5.3.2   Drift Mobility: Temperature and Impurity Dependence

The temperature dependence of the drift mobility follows two distinctly different temperature variations. In the high temperature region, it is observed that the drift mobility is limited by scattering from lattice vibrations. As the magnitude of atomic vibrations increases with temperature, the drift mobility decreases in the fashion $\mu \propto T^{-3/2}$. However, at low temperatures the lattice vibrations are not sufficiently strong to be the major limitation to the mobility of the electrons. It is observed that at low temperatures the scattering of electrons by ionized impurities is the major mobility limiting mechanism and $\mu \propto T^{3/2}$, as we will show below. We recall from Chapter 2 that the electron drift mobility, μ, depends on the mean free time, τ, between scattering events via

$$\mu = \frac{e\tau}{m_e^*} \qquad\qquad \text{[5.19]}$$

in which

$$\tau = \frac{1}{Sv_{th}N_s} \qquad\qquad \text{[5.20]}$$

where S is the cross-sectional area of the scatterer; v_{th} is the mean speed of the electrons, called the **thermal velocity;** and N_s is the number of scatterers per unit volume. If a is the amplitude of the atomic vibrations about the equilibrium, then $S = \pi a^2$. As the temperature increases, so does the amplitude, a, of the lattice vibrations following $a^2 \propto T$ behavior, as shown in Chapter 2. An electron in the CB is free to wander around and therefore has only KE. We also know that the mean energy per electron in the CB is $(3/2)\,kT$, just as if the kinetic molecular theory could be applied to all those electrons in the CB. Therefore

$$\frac{1}{2}m_e^* v_{th}^2 = \frac{3}{2}kT$$

so that $v_{th} \propto T^{1/2}$. Thus the mean time, τ_L, between scattering events from lattice vibrations is

$$\tau_L = \frac{1}{(\pi a^2)v_{th}N_s} \propto \frac{1}{(T)(T^{1/2})} \propto T^{-3/2}$$

which leads to a lattice vibration scattering limited mobility, denoted as μ_L, of the form

Lattice scattering limited mobility

$$\mu_L \propto T^{-3/2} \qquad\qquad \textbf{[5.21]}$$

At low temperatures, scattering of electrons by thermal vibrations of the lattice will not be as strong as the electron scattering brought about by ionized donor impurities. As an electron passes by an ionized donor As^+, it is attracted and thus deflected from its straight path, as schematically shown in Figure 5.17. This type of scattering of an electron is what limits the drift mobility at low temperatures.

The PE of an electron at a distance r from an As^+ ion is due to the Coulombic attraction, and its magnitude is given by

$$|PE| = \frac{e^2}{4\pi\varepsilon_o\varepsilon_r r}$$

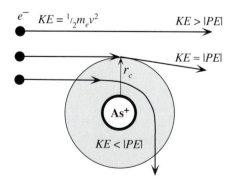

Figure 5.17   Scattering of electrons by an ionized impurity.

If the *KE* of the electron approaching an As$^+$ ion is larger than its *PE* at distance *r* from As$^+$, then the electron will essentially continue without feeling the *PE* and therefore without being deflected, and we can say that it has not been scattered. Effectively, due to its high *KE*, the electron does not feel the coulombic pull of the donor. On the other hand, if the *KE* of the electron is less than its *PE* at *r* from As$^+$, then the *PE* of the coulombic interaction will be so strong that the electron will be strongly deflected. This is illustrated in Figure 5.17. The critical radius, r_c, corresponds to the case when the electron is just scattered, which is when $KE \approx |PE(r_c)|$. But average $KE = (3/2) kT$, so that at $r = r_c$

$$\frac{3}{2} kT = |PE(r_c)| = \frac{e^2}{4\pi \varepsilon_o \varepsilon_r r_c}$$

from which $r_c = e^2/[6\pi \varepsilon_o \varepsilon_r kT]$. As the temperature increases, the scattering radius decreases. The scattering cross section, $S = \pi r_c^2$, is thus given by

$$S = \frac{\pi e^4}{(6\pi \varepsilon_o \varepsilon_r kT)^2} \propto T^{-2}$$

Incorporating $v_{th} \propto T^{1/2}$ as well, the temperature dependence of the mean scattering time τ_I between impurities, from Equation 5.20, must be

$$\tau_I = \frac{1}{S v_{th} N_I} \propto \frac{1}{(T^{-2})(T^{1/2}) N_I} \propto \frac{T^{3/2}}{N_I}$$

where N_I is the concentration of ionized impurities (all ionized impurities including donors and acceptors). Consequently, the ionized impurity scattering limited mobility from Equation 5.19 is

$$\mu_I \propto \frac{T^{3/2}}{N_I} \qquad \text{[5.22]}$$

Ionized impurity scattering limited mobility

Note also that μ_I decreases with increasing ionized dopant concentration, N_I, which itself may be temperature dependent. Indeed, at the lowest temperatures, below the saturation temperature T_s, N_I will be strongly temperature dependent because not all the donors would have been fully ionized.

The overall temperature dependence of the drift mobility is then, simply, the reciprocal additions of the μ_I and μ_L by virtue of Mathiessen's rule, that is,

$$\frac{1}{\mu_e} = \frac{1}{\mu_I} + \frac{1}{\mu_L} \qquad \text{[5.23]}$$

Effective mobility

so that the scattering process having the lowest mobility determines the overall (effective) drift mobility.

The experimental temperature dependence of the electron drift mobility in both Ge and Si is shown in Figure 5.18 as a log–log plot for various donor concentrations. The slope on this plot corresponds to the index *n* in $\mu_e \propto T^n$. The simple theoretical sketches in the insets show how μ_I and μ_L from Equations 5.21 and 5.22 depend on the temperature. For Ge, at low doping concentrations (e.g., $N_d = 10^{13}$ cm^{-3}), the experiments indicate a $\mu_e \propto T^{-1.5}$ type of behavior, which is in

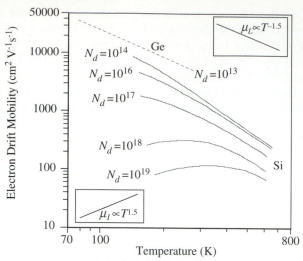

Figure 5.18   Log–log plot of drift mobility versus temperature for *n*-type Ge and *n*-type Si samples.

Various donor concentrations for Si are shown. N_d are in cm^{-3}. The upper right inset is the simple theory for lattice limited mobility, whereas the lower left inset is the simple theory for impurity scattering limited mobility.

agreement with μ_e determined by μ_L in Equation 5.21. Curves for Si at low-level doping (μ_I negligible) at high temperatures, however, evince a $\mu_e \propto T^{-2.5}$ type of behavior rather than $T^{-1.5}$, which can be accounted for in a more rigorous theory. As the donor concentration increases, the drift mobility decreases by virtue of μ_I getting smaller. At the highest doping concentrations and at low temperatures, the electron drift mobility in Si exhibits almost a $\mu_e \propto T^{3/2}$ type of behavior. Similar arguments can be extended to the temperature dependence of the hole drift mobility.

The dependences of the room temperature electron and hole drift mobilities on the dopant concentration for Si are shown in Figure 5.19 where, as expected, past a certain amount of impurity addition, the drift mobility is overwhelmingly controlled by μ_I in Equation 5.23.

5.3.3   Conductivity Temperature Dependence

The conductivity of an extrinsic semiconductor doped with donors depends on the electron concentration and the drift mobility, both of which have been determined above. At the lowest temperatures in the ionization range, the electron concentration depends exponentially on the temperature by virtue of

$$n = \left(\frac{1}{2}N_c N_d\right)^{1/2} \exp\left[\frac{-(E_c - E_d)}{2kT}\right]$$

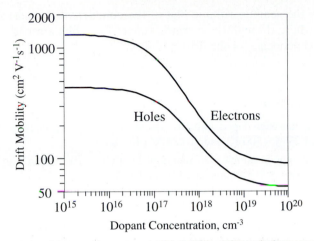

Figure 5.19   The variation of the drift mobility with dopant concentration in Si for electrons and holes.

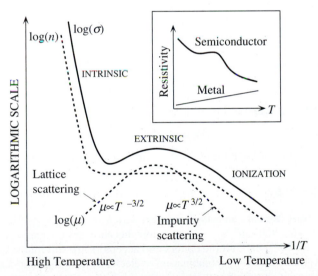

Figure 5.20   Temperature dependence of electrical conductivity for a doped (n-type) semiconductor.

which then also dominates the temperature dependence of the conductivity. In the intrinsic range at the highest temperatures, the conductivity is dominated by the temperature dependence of n_i since

$$\sigma = en_i(\mu_e + \mu_h)$$

and n_i is an exponential function of temperature in contrast to $\mu \propto T^{-3/2}$. In the extrinsic temperature range, $n = N_d$ and is constant, so that the conductivity follows the temperature dependence of the drift mobility. Figure 5.20 shows schematically the semilogarithmic plot of the conductivity against the reciprocal temperature where through the extrinsic range σ exhibits a broad "S" due to the temperature dependence of the drift mobility.

Example 5.9 | **COMPENSATION-DOPED Si**

a.   A Si sample has been doped with 10^{17} arsenic atoms cm^{-3}. Calculate the conductivity of the sample at 27 °C (300 K) and at 127 °C (400 K).

b.   The above *n*-type Si sample is further doped with 9×10^{16} boron atoms cm^{-3}. Calculate the conductivity of the sample at 27 °C and 127 °C.

Solution

a.   The arsenic dopant concentration, $N_d = 10^{17}$ cm^{-3}, is much larger than the intrinsic concentration, n_i, which means that $n = N_d$ and $p = (n_i^2/N_d) \ll n$ and can be neglected. Thus $n = 10^{17}$ cm^{-3} and the electron drift mobility at $N_d = 10^{17}$ cm^{-3} is 800 cm^2 V^{-1} s^{-1} from the drift mobility versus dopant concentration graph in Figure 5.19 so that

$$\sigma = en\mu_e + ep\mu_h = eN_d\mu_e$$
$$= (1.6 \times 10^{-19} \text{ C})(10^{17} \text{ cm}^{-3})(800 \text{ cm}^2 \text{ V}^{-1} \text{ s}^{-1}) = 12.8 \; \Omega^{-1} \text{ cm}^{-1}$$

At $T = 127$ °C $= 400$ K,

$$\mu_e \approx 420 \text{ cm}^2 \text{ V}^{-1} \text{ s}^{-1}$$

(from the μ_e versus T graph in Figure 5.18). Thus,

$$\sigma = eN_d\mu_e = 6.72 \; \Omega^{-1} \text{ cm}^{-1}$$

b.   With further doping we have $N_a = 9 \times 10^{16}$ cm^{-3}, so that from the compensation effect

$$N_d - N_a = 1 \times 10^{17} - 9 \times 10^{16} = 10^{16} \text{ cm}^{-3}$$

Since $N_d - N_a \gg n_i$, we have an *n*-type material with $n = N_d - N_a = 10^{16}$ cm^{-3}. But the drift mobility is still 800 cm^2 V^{-1} s^{-1} as before because, even though $N_d - N_a$ is now 10^{16} cm^{-3} and not 10^{17} cm^{-3}, all the donors and acceptors are still ionized and hence still scatter the charge carriers. The recombination of electrons from the donors and holes from the acceptors does not alter the fact that at room temperature all the dopants will be ionized. Effectively, the compensation effect is as if all electrons from the donors were being accepted by the acceptors. Although with compensation doping the net electron concentration is $n = N_d - N_a$, the drift mobility scattering is determined by $(N_d + N_a)$, which in this case is $10^{17} + 9 \times 10^{16}$ cm$^{-3} = 1.9 \times 10^{17}$ cm^{-3}, which gives an electron drift mobility of ~800 cm^2 V^{-1} s^{-1} at 300 K and ~420 cm^2 V^{-1} s^{-1} at 400 K. Then, neglecting the hole concentration, $p = n_i^2/(N_d - N_a)$, we have:

At 300 K, $\sigma = e(N_d - N_a)\mu_e \approx (1.6 \times 10^{-19} \text{ C})(10^{16} \text{ cm}^{-3})(800 \text{ cm}^2 \text{ V}^{-1} \text{ s}^{-1})$
$$= 1.28 \; \Omega^{-1} \text{ cm}^{-1}$$

At 400 K, $\sigma = e(N_d - N_a)\mu_e \approx (1.6 \times 10^{-19} \text{ C})(10^{16} \text{ cm}^{-3})(420 \text{ cm}^2 \text{ V}^{-1} \text{ s}^{-1})$
$$= 0.67 \; \Omega^{-1} \text{ cm}^{-1}$$

5.3.4 Degenerate and Nondegenerate Semiconductors

The general exponential expression for the concentration of electron in the CB,

$$n \approx N_c \exp\left[-\frac{(E_c - E_F)}{kT}\right] \qquad [5.24]$$

is based on replacing Fermi–Dirac statistics with Boltzmann statistics, which is only valid when E_c is several kT above E_F. In other words, we assumed that the number of states in the CB far exceeds the number of electrons there, so the likelihood of two electrons trying to occupy the same state is almost nil. This means that the Pauli exclusion principle can be neglected and the electron statistics can be described by the Boltzmann statistics. N_c is a measure of the density of states in the CB. The Boltzmann expression for n is valid only when $n \ll N_c$. Those semiconductors for which $n \ll N_c$ and $p \ll N_v$ are termed **nondegenerate semiconductors.** They essentially follow all the discussions above and exhibit all the normal semiconductor properties outlined above.

When the semiconductor has been excessively doped with donors, then n may be so large, typically 10^{19}–10^{20} cm^{-3}, that it may be comparable to or greater than N_c. In that case the Pauli exclusion principle becomes important in the electron statistics and we have to use the Fermi–Dirac statistics. Equation 5.24 for n is then no longer valid. Such a semiconductor exhibits properties that are more metal-like than semiconductor-like, for example, the resistivity follows $\rho \propto T$. Semiconductors that have $n > N_c$ or $p > N_v$ are called **degenerate semiconductors.**

The large carrier concentration in a degenerate semiconductor is due to its heavy doping. For example, as the donor concentration in an n-type semiconductor is increased, at sufficiently high doping levels, the donor atoms become so close to each other that their orbitals overlap to form a narrow energy band that overlaps and becomes part of the conduction band. The valence electrons from the donors fill the band from E_c. This situation is reminiscent of the valence electrons filling overlapping energy bands in a metal. In a degenerate n-type semiconductor, the Fermi level is therefore within the CB, or above E_c just like E_F is within the band in a metal. The majority of the states between E_c and E_F are full of electrons as indicated in Figure 5.21. In the case of a p-type degenerate semiconductor, the

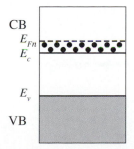

(a) Degenerate *n*-type semiconductor

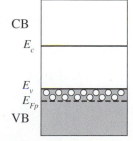

(b) Degenerate *p*-type semiconductor

Figure 5.21

Fermi level lies in the VB below E_v. It should be emphasized that one cannot simply assume that $n = N_d$ or $p = N_a$ in a degenerate semiconductor because the dopant concentration is so large that they interact with each other. Not all dopants are able to become ionized, and the carrier concentration eventually reaches a saturation typically around $\sim 10^{20}$ cm^{-3}. Furthermore, the mass action law, $np = n_i^2$, is not valid for degenerate semiconductors.

Degenerate semiconductors have many important uses. For example, they are used in laser diodes, zener diodes, and ohmic contacts in ICs, and as metal gates in many microelectronic MOS devices.

5.4   RECOMBINATION AND MINORITY CARRIER INJECTION

5.4.1   Direct and Indirect Recombination

Above absolute zero of temperature, the thermal excitation of electrons from the VB to the CB continuously generates free electron–hole pairs. It should be apparent that in equilibrium there should be some annihilation mechanism that returns the electron from the CB down to an empty state (a hole) in the VB. When a free electron, wandering around in the CB of a crystal, "meets" a hole, it falls into this low-energy empty electronic state and fills it. This process is called **recombination.** Intuitively, recombination corresponds to the free electron finding an incomplete bond with a missing electron. The electron then enters and completes this bond. The free electron in the CB and the free hole in the VB are consequently annihilated. On the energy band diagram, the recombination process is represented by returning the electron from the CB (where it is free) into a hole in the VB (where it is in a bond). Figure 5.22 shows a direct recombination mechanism, for example, as it occurs in GaAs, in which a free electron recombines with a free hole when they

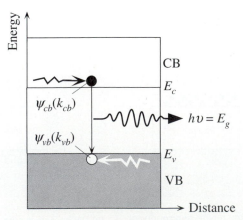

Figure 5.22   Direct recombination in GaAs.
$k_{cb} = k_{vb}$ so that momentum conservation is satisfied.

meet at one location in the crystal. The excess energy of the electron is lost as a photon of energy $hv = E_g$. In fact, it is this type of recombination that results in the emitted light from light emitting diodes (LEDs).

The recombination process between an electron and a hole, like every other process in nature, must obey the momentum conservation law. The wavefunction of an electron in the CB, $\psi_{cb}(k_{cb})$, will have a certain momentum, $\hbar k_{cb}$, associated with the wavevector k_{cb} and, similarly, the electron wavefunction, $\psi_{vb}(k_{vb})$, in the VB will have a momentum $\hbar k_{cb}$ associated with the wavevector k_{vb}. Conservation of linear momentum during recombination requires that when the electron drops from the CB to the VB, its wavevector should remain the same, $k_{vb} = k_{cb}$. For the elemental semiconductors, Si and Ge, the electronic states $\psi_{vb}(k_{vb})$ with $k_{vb} = k_{cb}$ are right in the middle of the VB and are therefore fully occupied. Consequently, there are no empty states in the VB that can satisfy $k_{vb} = k_{cb}$, and so direct recombination in Si and Ge is next to impossible. For some compound semiconductors, such as GaAs and InSb, for example, the states with $k_{vb} = k_{cb}$ are right at the top of the valence band, so they are essentially empty (contain holes). Consequently, an electron in the CB of GaAs can drop down to an empty electronic state at the top of the VB and maintain $k_{vb} = k_{cb}$. Thus **direct recombination** is highly probable in GaAs, and it is this very reason that makes GaAs an LED material.

In elemental semiconductor crystals, for example, in Si and Ge, electrons and holes usually recombine through recombination centers. A recombination center increases the probability of recombination because it can "take up" any momentum difference between a hole and electron. The process essentially involves a third body, which may be an impurity atom or a crystal defect. The electron is captured by the recombination center and thus becomes localized at this site. It is "held" at the center until some hole arrives and recombines with it. In the energy band diagram picture shown in Figure 5.23a, the recombination center provides a localized electronic state below E_c in the bandgap, which is at a certain location in the crystal. When an electron approaches the center, it is captured. The electron is then localized and bound to this center and "waits" there for a hole with which it can recombine. In this recombination process, the energy of the electron is usually lost to lattice vibrations (as "sound") via the "recoiling" of the third body. Emitted lattice vibrations are called phonons. A **phonon** is a quantum of energy associated with atomic vibrations in the crystal analogous to the photon.

Typical recombination centers, besides the donor and acceptor impurities, might be metallic impurities and crystal defects such as dislocations, vacancies, or interstitials. Each has its own peculiar behavior in aiding recombination, which will not be described here.

It is instructive to mention briefly the phenomenon of charge carrier **trapping** since in many devices this can be the main limiting factor on the performance. An electron in the conduction band can be captured by a localized state, just like a recombination center, located in the bandgap, as shown in Figure 5.23b. The electron falls into the trapping center at E_t and becomes temporarily removed from the CB. At a later time, due to an incident energetic lattice vibration, it becomes excited back into the CB and is available for conduction again. Thus trapping involves the temporary removal of the electron from the CB, whereas in the case of recombination, the electron is permanently removed from the CB since the capture is followed by recombination with a hole. We can view a trap as essentially

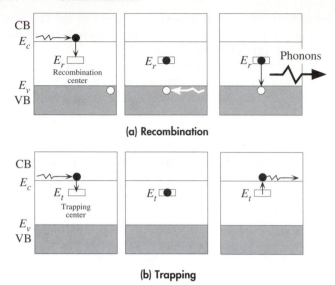

(a) Recombination

(b) Trapping

Figure 5.23   Recombination and trapping

(a) Recombination in Si via a recombination center that has a localized energy level at E_r in the bandgap, usually near the middle.

(b) Trapping and detrapping of electrons by trapping centers. A trapping center has a localized energy level in the band gap.

being a flaw in the crystal that results in the creation of a localized electronic state, around the flaw site, with an energy in the band gap. A charge carrier passing by the flaw can be captured and lose its freedom. The flaw can be an impurity or a crystal imperfection in the same way as a recombination center. The only difference is that when a charge carrier is captured at a recombination site, it has no possibility of escaping again because the center aids recombination. Although Figure 5.23b illustrates an electron trap, similar arguments also apply to hole traps, which are normally closer to E_v. In general, flaws and defects that give localized states near the middle of the band gap tend to act as recombination centers.

5.4.2   Minority Carrier Lifetime

Consider what happens when an n-type semiconductor, doped with $5 \times 10^{13}\,\text{cm}^{-3}$ donors, is uniformly illuminated with appropriate wavelength light to photogenerate electron–hole pairs (EHPs), as shown in Figure 5.24. We will now define thermal equilibrium majority and minority carrier concentrations in an extrinsic semiconductor. In general, the subscript n or p is used to denote the type of semiconductor, and o to refer to thermal equilibrium in the dark.

In an n-type semiconductor, electrons are the majority carriers and holes are the minority carriers

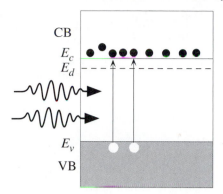

Figure 5.24 Low level photoinjection into an *n*-type semiconductor in which $\Delta n_n < n_{no}$

n_{no} is defined as the **majority carrier concentration** (electron concentration in an *n*-type semiconductor) in thermal equilibrium in the dark. These electrons, constituting the majority carriers, are thermally ionized from the donors.

p_{no} is termed the **minority carrier concentration** (hole concentration in an *n*-type semiconductor) in thermal equilibrium in the dark. These holes that constitute the minority carriers are thermally generated across the band gap.

In both cases the subscript *no* refers to an *n*-type semiconductor and thermal equilibrium conditions respectively. Thermal equilibrium means that the mass action law is obeyed and $n_{no}p_{no} = n_i^2$.

When we illuminate the semiconductor we create *excess* EHPs by photogeneration. Suppose that the electron and hole concentrations at any instant are denoted by n_n and p_n, which are defined as the *instantaneous* majority (electron) and minority (hole) concentrations respectively. At any instant and at any location in the semiconductor, we define the departure from the equilibrium by **excess concentrations** as

Δn_n is the *excess* electron (majority carrier) concentration: $\Delta n_n = n_n - n_{no}$

and

Δp_n is the *excess* hole (minority carrier) concentration: $\Delta p_n = p_n - p_{no}$

Under illumination, at any instant, therefore

$$n_n = n_{no} + \Delta n_n \quad \text{and} \quad p_n = p_{no} + \Delta p_n$$

Photoexcitation creates EHPs or an equal number of electrons and holes, as shown in Figure 5.24, which means that

$$\Delta p_n = \Delta n_n$$

and obviously the mass action law is not obeyed: $n_n p_n \neq n_i^2$. It is worth remembering that

$$\frac{dn_n}{dt} = \frac{d\Delta n_n}{dt} \quad \text{and} \quad \frac{dp_n}{dt} = \frac{d\Delta p_n}{dt}$$

since n_{no} and p_{no} depend only on temperature.

Let us assume that we have "weak" illumination, which causes, say, only a 10% change in n_{no}, that is,

$$\Delta n_n = 0.1 n_{no} = 0.5 \times 10^{13} \text{ cm}^{-3}$$

Then

$$\Delta p_n = \Delta n_n = 0.5 \times 10^{13} \text{ cm}^{-3}$$

Figure 5.25 shows a single-axis plot of the majority (n_n) and minority (p_n) concentrations in the dark and in light. The scale is logarithmic to allow large orders of magnitude changes to be recorded. Under illumination, the minority carrier concentration is

$$p_n = p_{no} + \Delta p_n = 4.5 \times 10^6 + 0.5 \times 10^{13} \approx 0.5 \times 10^{13} = \Delta p_n$$

That is, $p_n \approx \Delta p_n$, which shows that although n_n changes by only 10%, p_n changes *drastically,* that is, by a factor $\sim 10^{13}$.

Figure 5.26 shows a pictorial view of what is happening inside an n-type semiconductor when light is switched on at a certain time and then later switched off again. Obviously when the light is switched off, the condition $p_n = \Delta p_n$ (state B in Figure 5.26) must eventually revert back to the dark case (state A) where $p_n = p_{no}$. In other words, the excess minority carriers, Δp_n, and excess majority carriers, Δn_n, must be removed. This removal occurs by recombination. Excess holes recombine with the electrons available and disappear. This, however, takes

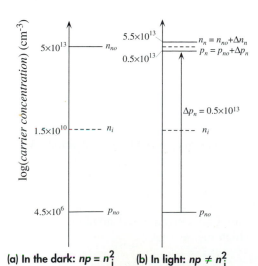

(a) In the dark: $np = n_i^2$ (b) In light: $np \neq n_i^2$

Figure 5.25   Low level injection in an n-type semiconductor does not significantly affect n_n but drastically affects the minority carrier concentration p_n.

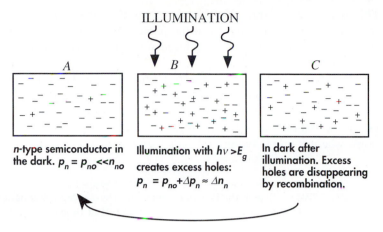

Figure 5.26 Illumination of an *n*-type semiconductor results in excess electron and hole concentrations.

After the illumination, the recombination process restores equilibrium; the excess electrons and holes simply recombine.

time because the electrons and holes have to find each other. In order to describe the rate of recombination, we introduce a temporal quantity, denoted by τ_h and called the **minority carrier lifetime (mean recombination time),** which is defined as follows: τ_h is the average time a hole exists in the VB from its generation to its recombination, that is, the mean time the hole is free before recombining with an electron. An alternative and equivalent definition is that $1/\tau_h$ is the average probability per unit time that a hole will recombine with an electron. We must remember that the recombination process occurs through recombination centers, so the recombination time, τ_h, will depend on the concentration of these centers and their effectiveness in capturing the minority carriers. Once a minority carrier has been captured by a recombination center, there are many majority carriers available to recombine with it, so τ_h in an indirect process is independent of the majority carrier concentration. This is the reason for defining the recombination time as a minority carrier lifetime.

If the minority carrier recombination time is, say, 10 seconds and if there are some 1,000 holes, then it is clear that these excess holes will be disappearing at a rate of $1000/10$ seconds $= 100$ per second. The rate of recombination of excess minority carriers is simply $\Delta p_n/\tau_h$. At any instant, therefore

$$
\begin{array}{ccc}
\text{Rate of increase in excess} & \text{Rate of} & \text{Rate of recombination} \\
\text{hole concentration} & = & \text{photogeneration} & - & \text{of excess holes}
\end{array}
$$

If G_{ph} is the rate of photogeneration, then clearly the net rate of change of Δp_n is

$$
\frac{d\,\Delta p_n}{dt} = G_{ph} - \frac{\Delta p_n}{\tau_h}
$$

[5.25] *Excess minority carrier contration*

This is a general expression that describes the time evolution of the excess minority carrier concentration given the photogeneration rate, G_{ph}; the minority carrier lifetime, τ_h; and the initial condition at $t = 0$. The only assumption is weak injection ($\Delta p_n < n_{no}$).

We should note that the recombination time τ_h depends on the semiconductor material, impurities, crystal defects, temperature, and so forth, and there is no typical value to quote. It can be anywhere from nanoseconds to seconds. Later it will be shown that certain applications require a short τ_h, as in fast switching of pn junctions, whereas others require a long τ_h, for example, persistent luminescence.

Example 5.10 | **PHOTORESPONSE TIME**   Sketch the hole concentration when a step illumination is applied to an n-type semiconductor at time $t = 0$ and switched off at time $t = t_{off}(\gg \tau_h)$.

Solution

We use Equation 5.25 with $G_{ph} = $ constant in $0 \leq t \leq t_{off}$. Since Equation 5.25 is a first order differential equation, integrating it we simply find

$$\ln\left[G_{ph} - \left(\frac{\Delta p_n}{\tau_h} \right) \right] = \frac{-t}{\tau_h} + C_1$$

where C_1 is the integration constant. At $t = 0$, $\Delta p_n = 0$, so that $C_1 = G_{ph}$. Therefore the solution is

$$\Delta p_n(t) = \tau_h G_{ph}\left[1 - \exp\left(-\frac{t}{\tau_h} \right) \right] \qquad 0 \leq t < t_{off} \qquad \text{[5.26]}$$

We see that as soon as the illumination is turned on, the minority carrier concentration rises exponentially toward its steady-state value $\Delta p_n(\infty) = \tau_h G_{ph}$. This is reached after a time $t > \tau_h$.

At the instant the illumination is switched off, we assume that $t_{off} \gg \tau_h$ so that from Equation 5.26,

$$\Delta p_n(t_{off}) = \tau_h G_{ph}$$

We can define t' to be the time measured from $t = t_{off}$, that is, $t' = t - t_{off}$. Then

$$\Delta p_n(t' = 0) = \tau_h G_{ph}$$

Solving Equation 5.25 with $G_{ph} = 0$ in $t > t_{off}$ or $t' > 0$, we get

$$\Delta p_n(t') = \Delta p_n(0) \exp\left(-\frac{t'}{\tau_h} \right)$$

where $\Delta p_n(0)$ is actually an integration constant that is equivalent to the boundary condition on Δp_n at $t' = 0$. Putting $t' = 0$ and $\Delta p_n = \tau_h G_{ph}$ gives

$$\Delta p_n(t') = \tau_h G_{ph} \exp\left(-\frac{t'}{\tau_h} \right) \qquad \text{[5.27]}$$

We see that the excess minority carrier concentration decays exponentially from the instant the light is switched off with a time constant equal to the minority carrier recombination time. The time evolution of the minority carrier concentration is sketched in Figure 5.27.

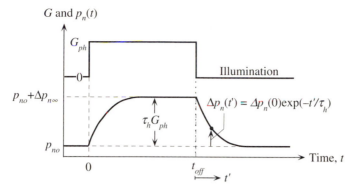

Figure 5.27   Illumination is switched on at time $t = 0$ and then off at $t = t_{off}$.

The excess minority carrier concentration, $\Delta p_n(t)$, rises exponentially to its steady-state value with a time constant τ_h. From t_{off}, the excess minority carrier concentration decays exponentially to its equilibrium value.

PHOTOCONDUCTIVITY   Suppose that a direct band gap semiconductor with no traps is illuminated with light of intensity $I(\lambda)$ and wavelength λ that will cause photogeneration as shown in Figure 5.28. The area of illumination is $A = (L \times W)$, and the thickness (depth) of the semiconductor is D. If η is the quantum efficiency (number of free EHPs generated per absorbed photon) and τ the recombination lifetime of the photogenerated carriers, show that the steady-state photoconductivity, defined as

$$\Delta\sigma = \sigma(\text{in light}) - \sigma(\text{in dark})$$

is given by

$$\Delta\sigma = \frac{e\eta\, I\lambda\tau(\mu_e + \mu_h)}{hcD} \qquad \textbf{[5.28]}$$

| Example 5.11

A photoconductive cell has a CdS crystal 1 mm long, 1 mm wide, and 0.1 mm thick with electrical contacts at the end, so that the receiving area of radiation is 1 mm², whereas the area of each contact is 0.1 mm². The cell is illuminated with a blue radiation of wavelength 450 nm and intensity 1 mW/cm². For unity quantum efficiency and an electron recombination time of 1 ms, calculate (a) the number of EHPs generated per second, (b) the photoconductivity of the sample, and (c) the photocurrent produced if 50 V is applied to the sample.

Note that CdS photoconductor is a direct band gap semiconductor with an energy gap $E_g = 2.6$ eV, electron mobility $\mu_e = 0.034$ m² V⁻¹ s⁻¹, and hole mobility $\mu_h = 0.0018$ m² V⁻¹ s⁻¹.

Solution

If Γ_{ph} is the number of photons arriving per unit area per unit second (the photon flux), then $\Gamma_{ph} = I/h\upsilon$ where I is the light intensity (energy flowing per unit area per second) and $h\upsilon$ is the energy per photon. The quantum efficiency η is defined as the number of free EHPs generated per absorbed photon. Thus, the number EHPs generated *per unit volume per*

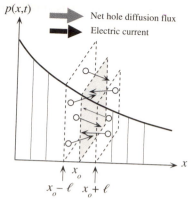

Figure 5.30   Arbitrary hole concentration $p(x, t)$ profile in a semiconductor. There is a net diffusion (flux) of holes from higher to lower concentrations. There are more holes crossing x_o coming from the left $(x_o - \ell)$ than coming from the right $(x_o + \ell)$.

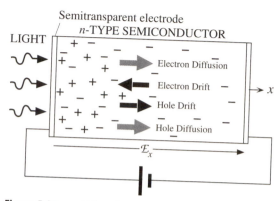

Figure 5.31   When there is an electric field and also a concentration gradient, charge carriers move by both diffusion and drift.

Suppose that there is also a positive electric field, $\mathcal{E}_x$, acting along $+x$ in Figures 5.29 and 5.30. A practical example is shown in Figure 5.31 in which a semiconductor is sandwiched between two electrodes, the left one semitransparent. By connecting a battery to the electrodes, an applied field of $\mathcal{E}_x$ is set up in the semiconductor along $+x$. The left electrode is continuously illuminated, so excess EHPs are generated at this surface that give rise to concentration gradients in n and p. The applied field imposes an electrical force on the charges, which then try to drift. Holes drift toward the right and electrons toward the left. Charge motion then involves both drift and diffusion. The total current density due to the electrons drifting, driven by $\mathcal{E}_x$, and also diffusing, driven by dn/dx, is then given by adding Equation 5.33 to the usual electron drift current density,

Total electron current due to drift and diffusion

$$J_e = en\mu_e \mathcal{E}_x + eD_e \frac{dn}{dx} \qquad [5.35]$$

We note that as $\mathcal{E}_x$ is along x, as is the drift current (first term), but the diffusion current (second term) is actually in the opposite direction by virtue of a negative dn/dx.

Similarly, the hole current due to holes drifting and diffusing, Equation 5.34, is given by

Total hole current due to drift and diffusion

$$J_h = ep\mu_h \mathcal{E}_x - eD_h \frac{dp}{dx} \qquad [5.36]$$

In this case the drift and diffusion currents are in the same direction.

We mentioned that the diffusion coefficient is a measure of the ease with which the diffusing charge carriers move in the medium. But drift mobility is also a measure of the ease with which the charge carriers move in the medium. The two quantities are related through the **Einstein relation,**

$$\frac{D_e}{\mu_e} = \frac{kT}{e} \quad \text{and} \quad \frac{D_h}{\mu_h} = \frac{kT}{e} \qquad \text{[5.37]} \quad \textit{Einstein relation}$$

In other words, the diffusion coefficient is proportional to the temperature and mobility. This is a reasonable expectation since increasing the temperature will increase the mean speed and thus accelerate diffusion. The randomizing effect against diffusion in one particular direction is introduced by the scattering of the carriers from lattice vibrations, impurities, and so forth, so that the longer the mean free path between scattering events, the larger the diffusion coefficient. This is examined in Example 5.12.

We equated the diffusion coefficient D to $\ell^2/2\tau$ in Equation 5.32. Our analysis, as represented in Figure 5.29, is oversimplified because we simply assumed that all electrons move a distance ℓ before scattering and all are free for a time τ. We essentially assumed that all those at a distance ℓ from x_o and moving towards x_o cross the plane exactly in time τ. This assumption is not entirely true because scattering is a stochastic process and consequently not all electrons moving towards x_o will cross it even in the segment of thickness ℓ. A rigorous statistical analysis shows that the diffusion coefficient is given by

$$D = \frac{\ell^2}{\tau} \qquad \text{[5.38]} \quad \textit{Diffusion coefficient}$$

THE EINSTEIN RELATION   Using the relation between the drift mobility and the mean free time, τ, between scattering events and the expression for the diffusion coefficient, $D = \ell^2/\tau$, derive the Einstein relation for electrons. | **Example 5.12**

Solution

In one dimension, for example, along x, the diffusion coefficient for electrons is given by $D_e = \ell^2/\tau$ where ℓ is the mean free path along x and τ is the mean free time between scattering events for electrons. The mean free path $\ell = v_x\tau$, where v_x is the mean (or effective) speed of the electrons along x. Thus,

$$D_e = v_x^2 \, \tau$$

In the conduction band and in one dimension, the mean *KE* of electrons is $(1/2)kT$, so $(1/2)kT = (1/2)m_e^* v_x^2$ where m_e^* is the effective mass of the electron in the CB. This gives

$$v_x^2 = \frac{kT}{m_e^*}$$

Substituting for v_x in D_e above we get,

$$D_e = \frac{kT\tau}{m_e^*} = \frac{kT}{e}\left(\frac{e\tau}{m_e^*}\right)$$

Further, we know from Chapter 2 that the electron drift mobility, μ_e, is related to the mean free time τ via $\mu_e = e\tau/m_e^*$ so that we can substitute for τ to obtain,

$$D_e = \frac{kT}{e}\mu_e$$

which is the Einstein relation. We assumed that Boltzmann statistics, that is, $v_x^2 = kT/m_e^*$ is applicable, which, of course, is true for the conduction band electrons in a semiconductor but not for the conduction electrons in a metal. Thus, the Einstein relation is only valid for electrons and holes in a nondegenerate semiconductor and certainly not valid for electrons in a metal.

Example 5.13 | **DIFFUSION COEFFICIENT OF ELECTRONS IN Si**   Calculate the diffusion coefficient of electrons at 27 °C in n-type Si doped with 10^{15} As atoms cm^{-3}.

Solution

From the μ_e versus dopant concentration graph, the electron drift mobility μ_e with 10^{15} cm^{-3} of dopants is about 1300 cm^2 V^{-1} s^{-1}, so

$$D_e = \frac{\mu_e kT}{e} = (1300 \text{ cm}^2 \text{ V}^{-1} \text{ s}^{-1})(0.0259 \text{ V}) = 33.7 \text{ cm}^2 \text{ s}^{-1}$$

Example 5.14 | **BUILT-IN POTENTIAL DUE TO DOPING VARIATION**   Suppose that due to a variation in the amount of donor doping in a semiconductor, the electron concentration is nonuniform across the semiconductor, that is, $n = n(x)$. What will be the potential difference between two points in the semiconductors where the electron concentrations are n_1 and n_2? If the donor profile in an n-type semiconductor is $N(x) = N_o\exp(-x/b)$, evaluate the built-in field $\mathcal{E}_x$. What is your conclusion?

Solution

Consider a nonuniformly doped n-type semiconductor in which immediately after doping the donor concentration, and hence the electron concentration, decreases toward the right. Initially, the sample is neutral everywhere. The electrons will immediately diffuse from higher to lower concentration regions. But this diffusion accumulates *excess* electrons in the right region and exposes the positively charged donors in the left region, as depicted in Figure 5.32. The electric field between the accumulated negative charges and the exposed donors prevents further accumulation. Equilibrium is reached when the diffusion toward the right is just balanced by the drift of electrons toward the left. The total current in the sample must be zero (it is an open circuit),

$$J_e = en\mu_e \mathcal{E}_x + eD_e\frac{dn}{dx} = 0$$

But the field is related to the potential difference by $\mathcal{E}_x = -(dV/dx)$ so that

$$-en\mu_e\frac{dV}{dx} + eD_e\frac{dn}{dx} = 0$$

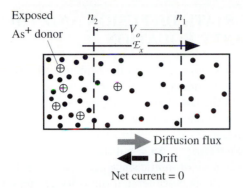

Net current = 0

Figure 5.32   Nonuniform doping profile results in electron diffusion toward the less concentrated regions.

This exposes positively charged donors and sets up a built-in field $\mathcal{E}_x$. In the steady state, the diffusion of electrons toward the right is balanced by their drift toward the left.

We can now use the Einstein relation $D_e/\mu_e = kT/e$ to eliminate D_e and μ_e and then cancel dx and integrate the equation,

$$\int_{V_1}^{V_2} dV = \frac{kT}{e} \int_{n_1}^{n_2} \frac{dn}{n}$$

Integrating

$$V_2 - V_1 = \frac{kT}{e} \ln\left(\frac{n_2}{n_1}\right) \qquad [5.39]$$

To find the built-in field, we will assume that (and this is a reasonable assumption) the diffusion of electrons toward the right has not drastically upset the original $n(x) = N_d(x)$ variation because the field builds up quickly to establish equilibrium. Thus

$$n(x) \approx N_d(x) = N_o \exp\left(-\frac{x}{b}\right)$$

Substituting into the equation for $J_e = 0$ above, and again using the Einstein relation, we obtain $\mathcal{E}_x$ as

$$\mathcal{E}_x = \frac{kT}{be} \qquad [5.40]$$

Note: As a result of the fabrication process, the base region of a bipolar transistor has nonuniform doping, which can be approximated by $N_d(x)$ as above. The resulting electric field $\mathcal{E}_x$ in Equation 5.40 acts to drift minority carriers faster and therefore speeds up the transistor operation as discussed in Chapter 6.

5.6   STEADY-STATE DIFFUSION AND THE CONTINUITY EQUATION

Many semiconductor devices operate on the principle that when minority carriers are injected into a semiconductor, they can move across the specimen due to diffusion alone, even in the absence of an applied field. Consider, for example, the continuous illumination of one end of an n-type semiconductor bar by light that is absorbed near the surface, as depicted in Figure 5.33. Both excess electrons and holes are photogenerated ($\Delta p_n = \Delta n_n$), but the percentage increase in the concentration of holes is much more dramatic since $p_{no} \ll n_{no}$. We therefore will focus our attention on the fate of the photogenerated minority carriers, that is, holes.

Suppose that illumination is such that it causes the hole concentration at $x = 0$ to be $p_n(0)$. Then as holes diffuse toward the right, they meet with electrons and recombine, as a result of which the hole concentration $p_n(x)$ decays rapidly with distance into the semiconductor. If the bar is very long, then far away from the injection end, we would expect p_n to be equal to the thermal equilibrium concentration p_{no}. We may intelligently guess that the excess hole concentration $\Delta p_n(x)$ decays exponentially as

$$\Delta p_n(x) = \Delta p_n(0) \exp\left(-\frac{x}{L_h}\right) \qquad \text{[5.41]}$$

where

$$\Delta p_n = p_n(x) - p_{no}$$

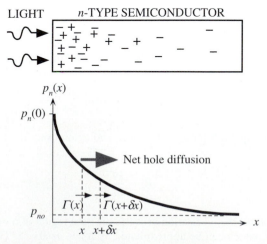

Figure 5.33   Steady-state minority carrier concentration profile in an n-type semiconductor that is illuminated at one end.

and L_h is some "length" constant. L_h is called the minority carrier **diffusion length** and represents the average distance diffused by minority carriers before recombination. It is given by

$$L_h = (D_h \tau_h)^{1/2}$$

where τ_h is the mean recombination time and D_h is the diffusion coefficient of holes. This equation is similar to Equation 5.38 in which the diffusion coefficient D was related to the mean free path, ℓ, and the mean free time, τ, between scattering events for charge carriers, $\ell^2 = D\tau$.

We need to establish the correct methodology that we must use to predict the minority carrier concentration profile under steady-state injection conditions and in the absence of an applied field. Consider an arbitrary hole concentration profile $p_n(x)$, which has been established as a steady-state response to an excitation (e.g., photogeneration above) as illustrated in Figure 5.33. Consider a small section (of unit cross-sectional area) of the bar between x and $x + \delta x$ as shown in Figure 5.33. Let the hole flux into this section be $\Gamma_h(x)$ and that out from the section be $\Gamma_h(x + \delta x)$. The difference in the two fluxes is the rate at which the holes are being lost in this section due to recombination. From elementary calculus, the change in the flux can be written as

$$\delta\Gamma_h(x) = \Gamma_h(x + \delta x) - \Gamma_h(x) \approx \left(\frac{d\Gamma_h}{dx}\right)\delta x$$

This change in the flux $\delta\Gamma_h$ represents the rate at which the excess holes are recombining between x and $x + \delta x$. If we divide $\delta\Gamma_h$ by the elementary volume $A\,\delta x$ (or δx) we get the rate of change in the number of holes per unit volume (hole concentration) in this section due to recombination. Thus

$$\frac{d\Gamma_h}{dx} = \frac{\Delta p_n}{\tau_n}$$

We can now substitute for the flux, Γ_h, from Equation 5.32 (Fick's first law)

$$\Gamma_h = -D_h \frac{dp_n}{dx} = -D_h \frac{d\Delta p_n}{dx}$$

we obtain

$$D_h \frac{d^2\Delta p_n}{dx^2} = \frac{\Delta p_n}{\tau_h}$$

By defining $D_h \tau_h = L_h^2$ and naming it the *diffusion length,* we can express this as

$$\frac{d^2\Delta p_n}{dx^2} = \frac{\Delta p_n}{L_h^2} \qquad \textbf{[5.42]}$$

Steady-state continuity equation with $\mathcal{E} = 0$

Equation 5.42 describes the steady-state behavior of minority carrier concentration in a semiconductor under steady-state excitation. When the appropriate boundary conditions are also included, its solution gives the spatial dependence of the minority carrier concentration in the semiconductor. Since Equation 5.42 is a differential equation, its exact solution necessarily requires the boundary conditions of the problem, one of which is usually $p_n(0)$. We should note that Equation 5.42 is a special case of a more general equation called the **continuity equation,** which may be found in more advanced textbooks on semiconductor devices.

It is apparent that Equation 5.41 for $\Delta p_n(x)$ is a particular solution of Equation 5.42 given the boundary conditions that at $x = 0$, $\Delta p_n = \Delta p_n(0)$ and the semiconductor bar is very long, $\Delta p_n = 0$ as $x \to \infty$.

Example 5.15 | **INFINITELY LONG SEMICONDUCTOR ILLUMINATED AT ONE END**   Find the minority carrier concentration profile $p_n(x)$ in an infinite n-type semiconductor that is illuminated continuously at one end as in Figure 5.33. Assume that photogeneration occurs near the surface. Show that the mean distance diffused by the minority carriers before recombination is L_h.

Solution

Continuous illumination means that we have steady-state conditions and thus Equation 5.42 can be used. The general solution of this second-order differential equation is

$$\Delta p_n(x) = A \exp\left(-\frac{x}{L_h}\right) + B \exp\left(\frac{x}{L_h}\right) \qquad \text{[5.43]}$$

where A and B are constants that have to be found from the boundary conditions. For an infinite bar, at $x = \infty$, $\Delta p_n(\infty) = 0$ gives $B = 0$. At $x = 0$, $\Delta p_n = \Delta p_n(0)$ so $A = \Delta p_n(0)$. Thus the excess (photoinjected) hole concentration at position x is

$$\Delta p_n(x) = \Delta p_n(0) \exp\left(-\frac{x}{L_h}\right) \qquad \text{[5.44]}$$

which is shown in Figure 5.33. To find the mean position of the photoinjected holes, we use the definition of the "mean", that is,

$$\bar{x} = \frac{\int_0^\infty x\, \Delta p_n(x)\, dx}{\int_0^\infty \Delta p_n(x)\, dx}$$

Substituting for $\Delta p_n(x)$ from Equation 5.44 and carrying out the integration gives $\bar{x} = L_h$. We conclude that the diffusion length L_h is the average distance diffused by the minority carriers before recombination. As a corollary, we should infer that $1/L_h$ is the mean probability per unit distance that the hole recombines with an electron.

5.7   OPTICAL ABSORPTION

We have already seen that a photon of energy $h\upsilon$ greater than E_g can be absorbed in a semiconductor, resulting in the excitation of an electron from the valence band to the conduction band, as illustrated in Figure 5.34. The average energy of electrons in the conduction band is $(3/2)\,kT$ above E_c (average kinetic energy is $(3/2)\,kT$), which means that the electrons are very close to E_c. If the photon energy is much larger than the bandgap energy E_g, then the excited electron is not near E_c and has to lose the extra energy $h\upsilon - E_g$ to reach thermal equilibrium. The excess energy $h\upsilon - E_g$ is lost to lattice vibrations as heat as the electron is scattered from

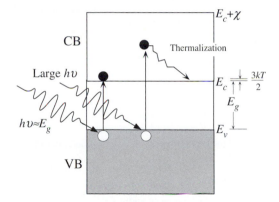

Figure 5.34   Optical absorption generates electron–hole pairs.

Energetic electrons must lose their excess energy to lattice vibrations until their average energy is (3/2) kT in the CB.

one atomic vibration to another. This process is called **thermalization.** If, on the other hand, the photon energy hv is less than the bandgap energy, the photon will not be absorbed and we can say that the semiconductor is transparent to wavelengths longer than hc/E_g provided that there are no energy states in the bandgap. There, of course, will be reflections occurring at the air/semiconductor surface due to the change in the refractive index.

Suppose that I_o is the intensity of a beam of photons incident on a semiconductor material. Thus, I_o is the energy incident per unit area per unit time. If Γ_{ph} is the photon flux, then

$$I_o = hv\Gamma_{ph}$$

When the photon energy is greater than E_g, photons from the incident radiation will be absorbed by the semiconductor. The absorption of photons requires the excitation of valence band electrons, and there are only so many of them with the right energy *per unit volume*. Consequently, absorption depends on the thickness of the semiconductor. Suppose that $I(x)$ is the light intensity at x and δI is the change in the light intensity in the small elemental volume of thickness δx at x due to photon absorption, as illustrated in Figure 5.35. Then δI will depend on the number of photons arriving at this volume, $I(x)$, and the thickness δx. Thus

$$\delta I = -\alpha I \, \delta x$$

where α is a proportionality constant that depends on the photon energy and hence wavelength, that is, $\alpha = \alpha(\lambda)$. The negative sign ensures that δI is a reduction. The constant α as defined by this equation is called the **absorption coefficient** of the semiconductor. It is therefore defined by

$$\alpha = -\frac{\delta I}{I \, \delta x} \qquad\qquad \textbf{[5.45]}$$

Definition of absorption coefficient

which has the dimensions of length^{-1} (m^{-1}).

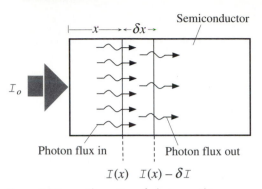

Figure 5.35 Absorption of photons within a small elemental volume of width δx

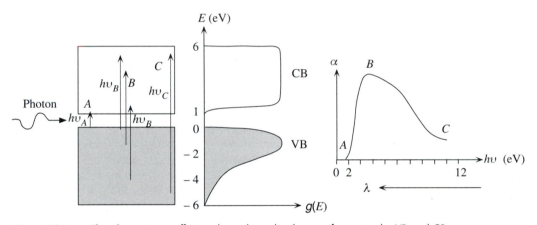

Figure 5.36 The absorption coefficient depends on the density of states in the VB and CB.
Photon energies that involve maximum $g(E)$ transitions as in B generally lead to large absorption coefficients. The absorption process shown would be typical of an amorphous semiconductor such as hydrogenated amorphous silicon (silicon containing hydrogen atoms) in which $g(E)$ for the CB does not exhibit distinct peaks and troughs.

When we integrate Equation 5.45 for illumination with constant wavelength light, we get the **Beer–Lambert law,** the transmitted intensity decreases exponentially with the thickness,

Beer–Lambert law

$$I(x) = I_o \exp(-\alpha x) \qquad [5.46]$$

The absorption coefficient depends on the photon absorption processes occurring in the semiconductor. In the case of a band-to-band (interband) absorption, α rises sharply for photon energies above E_g. The general features of the photon energy dependence of the absorption coefficient, shown in Figure 5.36, can be understood from the density-of-states diagram also illustrated in the same figure.

Density of states, $g(E)$, represents the number of states per unit energy per unit volume. We assume that the VB states are filled and the CB states are empty since the number of electrons in the CB is much smaller than the number of states in this band ($n \ll N_c$). The photon absorption process increases when there are more VB states available since more electrons can be excited. We also need available CB states into which the electrons can be excited, otherwise the electrons cannot find empty states to fill. The probability of photon absorption depends on both the density of VB states *and* the density of CB states. For photons of energy $h\nu_A = E_g$, the absorption can only occur from E_v to E_c where the density of states is low and thus the absorption coefficient is small, which is illustrated as A in Figure 5.36. For photon energies $h\nu_B$, which can take electrons from the middle region of the VB to the middle of the CB, the densities of states are large and α is also large, as indicated by B in Figure 5.36. Furthermore, there are more choices of excitation for the $h\nu_B$ photon, as illustrated by the three arrows in the figure. If the photon energy $h\nu_C$ is such that it can only excite electrons from around the bottom of the VB to the top of the CB, depicted as C in Figure 5.36, then the densities of states are small and α is accordingly also small. The absorption coefficient therefore typically increases with $h\nu$, reaches a maximum, and then decreases again.

In reality, the density of states, $g(E)$, of a real crystalline semiconductor material is much more complicated, with various sharp peaks and troughs on the density of states function, particularly away from the band edges. In addition, the absorption process has to satisfy the conservation of momentum, which means that certain transitions from the CB to the VB will be more favorable than others. The resulting α versus $h\nu$ characteristic of a crystalline semiconductor has distinct peaks and sharp features. The simplest example in Figure 5.36 is likely to be more representative of an amorphous (noncrystalline) semiconductor such as amorphous silicon. These semiconductor materials generally have smoothly varying density of states functions and exhibit α versus $h\nu$ behavior with no distinct sharp features, as depicted in Figure 5.36.

PHOTOCONDUCTIVITY OF A THIN SLAB   Modify the photoconductivity expression | **Example 5.16**

$$\Delta\sigma = \frac{e\eta I_o \lambda \tau (\mu_e + \mu_h)}{hcD},$$

derived for a direct band gap semiconductor in Figure 5.28 to take into account that some of the light intensity is transmitted through the material.

Solution

If we assume that all the photons are absorbed (there is no transmitted light intensity), then the photoconductivity expression is

$$\Delta\sigma = \frac{e\eta I_o \lambda \tau (\mu_e + \mu_h)}{hcD}$$

But, in reality, $I_o \exp(-\alpha D)$ is the transmitted intensity through the specimen with thickness D so that absorption is determined by the intensity lost in the material, $I_o[1 - \exp(-\alpha D)]$, which means that $\Delta\sigma$ must be accordingly scaled down to

$$\Delta\sigma = \frac{e\eta I_o[1 - \exp(-\alpha D)]\lambda \tau (\mu_e + \mu_h)}{hcD}$$

Example 5.17 | **PHOTOGENERATION IN GaAs**   Suppose that a GaAs sample is illuminated with a 50 mW HeNe laser beam (wavelength 632.8 nm) on its surface. Calculate how much power is dissipated as heat in the sample. Give your answer as mW. The energy band gap, E_g, of GaAs is 1.42 eV.

Solution

Suppose P_L is the power in the laser beam; then $P_L = IA$, where I is the intensity of the beam and A is the area of incidence. The photon flux, photons arriving per unit area per unit time, is

$$\Gamma_{ph} = \frac{I}{hv} = \frac{P_L}{Ahv}$$

so that the number of EHPs generated per unit time is

$$\frac{dN}{dt} = \Gamma_{ph} A = \frac{P_L}{hv}$$

These carriers thermalize—lose their excess energy as lattice vibrations (heat) via collisions with the lattice—so that eventually their average kinetic energy becomes $(3/2)\,kT$ above E_g. Remember that we assume that electrons in the CB are nearly free, so they must obey the kinetic theory and hence have an average kinetic energy of $(3/2)\,kT$. The average energy of the electron is then $E_g + (3/2)\,kT \approx 1.46$ eV. The excess energy

$$\Delta E = hv - \left(E_g + \frac{3}{2}kT \right)$$

is lost to the lattice as heat, that is, lattice vibrations. Since each electron loses ΔE amount of energy as heat, the heat power generated is

$$P_H = \left(\frac{dN}{dt} \right) \Delta E = \left(\frac{P_L}{hv} \right)(\Delta E)$$

The incoming photon has an energy $hv = \dfrac{hc}{\lambda} = 1.96$ eV, so that

$$P_H = \frac{(50 \text{ mW})(1.96 \text{ eV} - 1.46 \text{ eV})}{1.96 \text{ eV}} = 12.76 \text{ mW}$$

5.8   LUMINESCENCE

When an excited electron in the conduction band of a direct bandgap semiconductor recombines with a hole in the valence band, the recombination process may lead to the emission of a photon. Emission of light by an electron–hole recombination processes is generally referred to as **luminescence**. In contrast, light emitted from an ordinary light bulb is due to the heating of the metal filament. The emission of radiation from a heated object is called **incandescence.**

In luminescence, emission of radiation requires the initial excitation of electrons. If the electron excitation is due to photon absorption, then the process is identified as **photoluminescence.** The direct electron–hole recombination mechanism generally occurs very quickly. For example, typical minority carrier lifetimes are in the range of nanoseconds, so light emission from a semiconductor stops within nanoseconds after the removal of excitation. Such luminescence processes are normally identified as **fluorescence.** The emission of light from a fluorescent tube is actually a fluorescence process. The tube contains a gas mixture of argon and mercury. The Ar and Hg gas atoms become excited by the electrical discharge process and emit light mainly in the ultraviolet region, which is absorbed by the fluorescent coating on the tube. The excited electrons in the fluorescent coating material then recombine, emitting light in the visible spectrum.

There are also materials, called **phosphors,** from which light emission may continue for milliseconds to hours after the cessation of excitation. These slow luminescence processes are normally referred to as **phosphorescence.**

To understand the mechanism that allows the emission of light many minutes after the cessation of excitation, we consider a semiconductor that has localized states in the bandgap at an energy level E_t, as depicted in Figure 5.37. These localized states are electron traps and temporarily capture an electron from the conduction band and thereby immobilize it. After a while a strong lattice vibration returns the electron back into the conduction band (by thermal excitation). The traps may be due to crystal defects or they may be added impurities. The time the electron spends trapped at E_t depends on the energy depth of the trap from the conduction band, $E_c - E_t$. As indicated in Figure 5.37, initially the incident photon excites a valence band electron to the conduction band. The electron then thermalizes, that is, loses the excess energy as it is collides with lattice vibrations, and falls close to E_c. As the electron wanders in the conduction band, it becomes captured by a trap at E_t. It remains captured until a strong lattice vibration excites it back

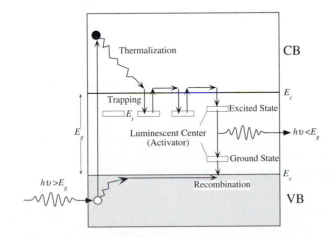

Figure 5.37   Optical absorption generates an EHP.

The electron thermalizes and then becomes trapped at a local center and thereby removed from the CB. Later it becomes detrapped and wanders in the CB again. Eventually it is captured by a luminescent center, where it recombines with a hole, emitting a photon. Traps therefore delay recombination.

into the conduction band. The electron wanders around again in the conduction band and eventually it becomes trapped in an excited state of a luminescent center or an activator. Typically, luminescent centers are intentionally added impurities or crystal defects such as interstitials or vacancies. The electron then falls down in energy to the ground state of the activator, releasing a photon. Later the electron at the ground state recombines with a wandering hole in the VB. Thus the activator acts as a radiative recombination center. With some impurities, the energy that is released when the electron falls down to the ground state is in the form of lattice vibrations so that the impurity acts as a nonradiative recombination center.

The time interval between photogeneration and recombination can be quite long if the electron remains captured at E_t for a considerable length of time. In fact, the electron may become trapped and detrapped many times before it finally recombines, so the emission of light can persist for a relatively long time after the cessation of excitation.

There are many examples of phosphors with various activators. For example, ZnS is a typical phosphor material. Small amounts of Cu in the ZnS phosphor act as an activator with luminescence occurring in the green region. Mn, on the other hand, is an activator that gives luminescence in the red region.

It is also possible to excite electrons into the CB by bombarding the material with a high energy electron beam. If these electrons recombine with holes and emit light, then the process is called **cathodoluminescence**. This is the mechanism that allows us to view the electron beam trace on the screen of a CRT (cathode ray tube). The electron beam excites EHPs in the phosphor coating on the CRT screen. In the case of color CRT displays, typically the screen is coated uniformly with three sets of phosphor dots that exhibit cathodoluminescence in the blue, red, and green wavelengths.

In **electroluminescence** an electric current, either ac or dc, is used to excite electrons into the CB. They then recombine with holes and emit light. For example, passing a current through certain semiconducting phosphors such as ZnS doped with Mn causes light emission by electroluminescence. The emission of light from a light-emitting diode (LED) is an example of injection electroluminescence in which the applied voltage causes charge carrier injection and recombination in a device (diode) that has a junction between a *p*-type and an *n*-type semiconductor.

5.9   SCHOTTKY JUNCTION

5.9.1   Schottky Diode

We consider what happens when a metal and an *n*-type semiconductor are brought into contact. In practice, this process is frequently carried out by the evaporation of a metal onto the surface of a semiconductor crystal in vacuum.

The energy band diagrams for the metal and the semiconductor are shown in Figure 5.38. The work function, denoted as Φ, is the energy difference between the

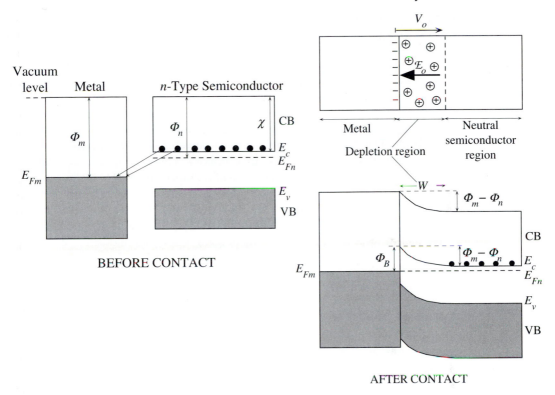

Figure 5.38   Formation of a Schottky junction between a metal and an *n*-type semiconductor

vacuum level and the Fermi level. Vacuum level defines the energy where the electron is free from that particular solid and where the electron has zero *KE*.

For the metal, the work function Φ_m is the minimum energy required to remove an electron from the solid. In the metal there are electrons at the Fermi level, E_{Fm}, but in the semiconductor there are none at E_{Fn}. Nonetheless, the semiconductor work function Φ_n still represents the energy required to remove an electron from the semiconductor. It may be thought the minimum energy required to remove an electron from the semiconductor is simply the electron affinity, χ, but this is not so. Thermal equilibrium requires that only a certain fraction of all the electrons in the semiconductor should be in the CB at a given temperature. When an electron is removed from the conduction band, then thermal equilibrium can be maintained only if an electron is excited from the VB to CB, which involves absorbing heat (energy) from the environment; thus it takes more energy than simply χ. We will not derive the effective energy required to remove an electron but state that, as for the metal, this is equal to Φ_n, even though there are no electrons at E_{Fn}.

We assume that $\Phi_m > \Phi_n$, the work function of the metal is greater than that of the semiconductor. When the two solids come into contact, the more energetic electrons in the CB of the semiconductor can readily tunnel into the metal in search of lower empty energy levels (just above E_{Fm}) and accumulate near the surface of the metal, as illustrated in Figure 5.38. Electrons tunneling from the semiconductor leave behind an electron-depleted region of width W in which there are exposed positively charged donors, in other words, net positive space charge. The contact potential, called the **built-in potential** V_o, therefore develops between the metal and the semiconductor. There is obviously also a built-in electric field $\mathcal{E}_o$ from the positive charges to the negative charges on the metal surface. Eventually this built-in potential reaches a value that prevents further accumulation of electrons at the metal surface and an equilibrium is reached. The value of the built-in voltage V_o is the same as that in the metal–metal junction case in Chapter 4, namely, $(\Phi_m - \Phi_n)/e$. The depletion region has been depleted of free carriers (electrons) and hence contains the exposed positive donors. This region thus constitutes a space charge layer (SCL) in which there is a nonuniform internal field directed from the semiconductor to the metal surface. The maximum value of this built-in field is denoted as $\mathcal{E}_o$ and occurs right at the metal–semiconductor junction (this is where there are a maximum number of field lines from positive to negative charges).

The Fermi level throughout the whole solid, the metal and semiconductor in contact, must be uniform in equilibrium. Otherwise, a change in the Fermi level, ΔE_F, going from one end to the other end will be available to do external (electrical) work. Thus, E_{Fm} and E_{Fn} line up. The W region, however, has been depleted of electrons, so in this region $E_c - E_{Fn}$ must increase so that n decreases. The bands must bend to increase $E_c - E_{Fn}$ towards the junction, as depicted in Figure 5.38. Far away from the junction, we, of course, still have an n-type semiconductor. The bending is just enough for the vacuum level to be continuous and changing by $\Phi_m - \Phi_n$ from the semiconductor to the metal, as this much energy is needed to take an electron across from the semiconductor to the metal. The PE barrier for electrons moving from the metal to the semiconductor is the called **Schottky barrier height,** Φ_B which is given by

$$\Phi_B = \Phi_m - \chi = eV_o + (E_c - E_{Fn})$$

which is greater than eV_o.

Under open circuit conditions, there is no net current flowing through the metal–semiconductor junction. The number of electrons thermally emitted over the PE barrier Φ_B from the metal to the semiconductor is equal to the number of electrons thermally emitted over eV_o from the semiconductor to the metal. Emission probability depends on the PE barrier for emission through the Boltzmann factor. There are two current components due to electrons flowing through the junction. The current due to electrons being thermally emitted from the metal to the CB of the semiconductor is

$$J_1 = A_1 \exp\left(-\frac{\Phi_B}{kT}\right)$$
[5.47]

where A_1 is some constant, whereas the current due to electrons being thermally emitted from the CB of the semiconductor to the metal is

$$J_2 = A_2 \exp\left(-\frac{eV_o}{kT}\right)$$
[5.48]

where A_2 is some other constant different than A_1.

In equilibrium, that is, open circuit conditions in the dark, the currents are equal but in the reverse directions:

$$J_{\text{open circuit}} = J_2 - J_1 = 0$$

Under forward bias conditions, the semiconductor side is connected to the negative terminal, as depicted schematically in Figure 5.39a. Since the depletion region, W, has a much larger resistance than the neutral n-region (outside W) and the metal side, nearly all the voltage drop is across the depletion region. The applied bias is in the opposite direction to the built-in voltage V_o. Thus V_o is reduced to $V_o - V$. Φ_B remains unchanged. The semiconductor band diagram outside the depletion region has been effectively shifted up with respect to the metal side by an amount eV because

$$PE = \text{Charge} \times \text{Voltage}$$

The charge is negative but so is the voltage connected to the semiconductor, as shown in Figure 5.39a.

The PE barrier for thermal emission of electrons from the semiconductor to the metal is now $e(V_o - V)$. The electrons in the CB can now readily overcome the PE barrier to the metal.

The current, J_2^{for}, due the electron emission from the semiconductor to the metal, is now

$$J_2^{\text{for}} = A_2 \exp\left[-\frac{e(V_o - V)}{kT}\right] \qquad \text{[5.49]}$$

Since Φ_B is the same, J_1 remains unchanged. The net current is then

$$J = J_2^{\text{for}} - J_1 = A_2 \exp\left[-\frac{e(V_o - V)}{kT}\right] - A_2 \exp\left(-\frac{eV_o}{kT}\right)$$

or

$$J = A_2 \exp\left(-\frac{eV_o}{kT}\right)\left[\exp\left(\frac{eV}{kT}\right) - 1\right]$$

giving

$$J = J_o\left[\exp\left(\frac{eV}{kT}\right) - 1\right] \qquad \text{[5.50]} \qquad \textit{Schottky Junction}$$

where J_o is a constant that depends on the material and surface properties of the two solids. In fact, examination of the above steps shows that J_o is also J_1 in Equation 5.47.

When the Schottky junction is reverse biased, then the positive terminal is connected to the semiconductor, as illustrated in Figure 5.39b. The applied voltage, V_r, drops across the depletion region since this region has very few carriers and is highly resistive. The built-in voltage V_o thus increases to $V_o + V_r$. Effectively, the

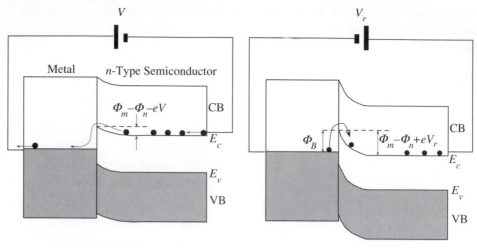

(a) Forward-biased Schottky junction. Electrons in the CB of the semiconductor can readily overcome the small *PE* barrier to enter the metal.

(b) Reverse-biased Schottky junction. Electrons in the metal cannot easily overcome the *PE* barrier Φ_B to enter the semiconductor.

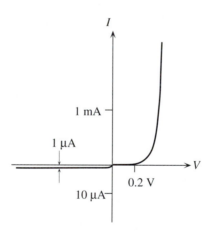

(c) *I–V* characteristics of a Schottky junction exhibit rectifying properties (negative current axis is in microamps).

Figure 5.39   The Schottky junction

semiconductor band diagram is shifted down with respect to the metal side because the charge is negative but the voltage is positive and $PE = \text{Charge} \times \text{Voltage}$. The *PE* barrier for thermal emission of electrons from the CB to the metal becomes $e(V_o + V_r)$, which means that the corresponding current component becomes

$$J_2^{\text{rev}} = A_2 \exp\left[-\frac{e(V_o + V_r)}{kT} \right] \ll J_1 \tag{5.51}$$

Since generally V_o is typically a fraction of a volt and the reverse bias is more than a few volts, $J_2^{rev} \ll J_1$ and the reverse bias current is essentially limited by J_1 only and is very small. Thus, under reverse bias conditions, the current is primarily due to the thermal emission of electrons over the barrier Φ_B from the metal to the CB of the semiconductor. Figure 5.39c illustrates the I–V characteristics of a typical Schottky junction. The I–V characteristics exhibit rectifying properties, and the device is called a Schottky diode.

Equation 5.50, which is derived for forward bias conditions, is also valid under reverse bias by making V negative, that is, $V = -V_r$. Furthermore, it turns out to be applicable not only to Schottky type metal–semiconductor junctions but also to junctions between a p-type and an n-type semiconductor, pn junctions, as we will show in Chapter 6. Under a forward bias, V_f, which is greater than 25 mV at room temperature, the forward current is simply

$$J_f = J_o \exp\left(\frac{eV_f}{kT}\right) \qquad V_f > \frac{kT}{e}$$

It should be mentioned that it is also possible to obtain a Schottky junction between a metal and a p-type semiconductor. This arises when $\Phi_m < \Phi_p$, where the Φ_p is the work function for the p-type semiconductor.

5.9.2   Schottky Junction Solar Cell

The built-in field in the depletion region of the Schottky junction allows this type of device to function as a photovoltaic device and also as a photodetector. We consider a Schottky device that has a thin metal film (usually Au) deposited onto an n-type semiconductor. The energy band diagram is shown in Figure 5.40. The metal is sufficiently thin (~ 10 nm) to allow light to reach the semiconductor.

For photon energies greater than E_g, EHPs are generated in the depletion region in the semiconductor, as indicated in Figure 5.40. The field in this region separates the EHPs and drifts the electrons toward the semiconductor and holes toward the metal. When an electron reaches the neutral n-region, there is now one extra electron there and therefore an additional negative charge. This end therefore becomes more negative with respect to the situation in the dark or the equilibrium situation. When a hole reaches the metal, it recombines with an electron and reduces the effective charge there by one electron, thus making it more positive relative to its dark state. Under open circuit conditions, therefore, a voltage develops across the Schottky junction device with the metal end positive and semiconductor end negative.

The photovoltaic explanation in terms of the energy band diagram is simple. At the point of photogeneration, the electron finds itself at a PE slope as E_c is decreasing toward the semiconductor, as shown in Figure 5.40. It has no option but to roll down the slope just as a ball that is let go on a slope would roll down the slope to decrease its gravitational PE. Recall that there are many more empty states in the CB than electrons, so there is nothing to prevent the electron from rolling down the

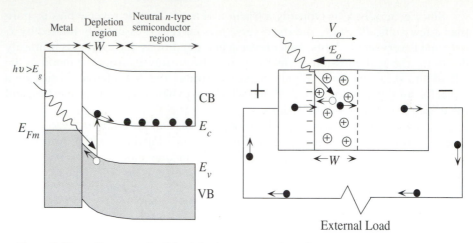

Figure 5.40   The principle of the Schottky junction solar cell.

CB in search of lower energy. When the electron reaches the neutral region (flat E_c region), it upsets the equilibrium there. There is now an additional electron in the CB and this side acquires a negative charge. If we remember that hole energy increases downwards on the energy band diagram, then similar arguments also apply to the photogenerated hole in the VB, which rolls down its own PE slope to reach the surface of the metal and recombine with an electron there.

If the device is connected to an external load, then the extra electron in the neutral n-region is conducted through the external leads, through the load, toward the metal side, where it replenishes the lost electron in the metal. As long as photons are generating EHPs, the flow of electrons around the external circuit will continue and there will be photon energy to electrical energy conversion. It is useful to think of the neutral n-type semiconductor region as a conductor, an extension of the external wire (except that the n-type semiconductor has a higher resistivity). As soon as the photogenerated electron crosses the depletion region, it reaches a conductor and is conducted around the external circuit to the metal side to replenish the lost electron there.

For photon energies less than E_g, the device can still respond, providing that the $h\upsilon$ can excite an electron from E_{Fm} in the metal over the PE barrier Φ_B into the CB, from where the electron will roll down toward the neutral n-region. In this case, $h\upsilon$ must only be greater than Φ_B.

If the Schottky junction diode is reverse biased, as shown in Figure 5.41, then the reverse bias V_r increases the built-in potential V_o to $V_o + V_r$ ($V_r \gg V_o$). The internal field increases to substantially high values. This has the advantage of increasing the drift velocity of the EHPs ($\upsilon_d = \mu_d \mathcal{E}$) in the depletion region and therefore shortening the transit time required to cross the depletion width. The device responds faster and is useful as a fast photodetector. The photocurrent i_{photo} in the external circuit is due to the drift of photogenerated carriers in the depletion region and can be readily measured.

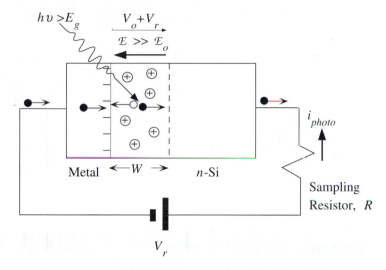

$h\upsilon > E_g$

$\dfrac{V_o + V_r}{\mathcal{E} \gg \mathcal{E}_o}$

i_{photo}

Metal ←W→ n-Si

Sampling
Resistor, R

V_r

Figure 5.41   Reverse-biased Schottky photodiodes are frequently used as fast photodetectors.

THE SCHOTTKY DIODE   The Schottky junction current in Equation 5.50 is the same as the Richardson–Dushman equation for thermionic emission over a potential barrier Φ_B derived in Chapter 4. The quantity J_o, which represents the reverse saturation current density J_1 in Equation 5.47, is given by

$$J_o = B_e T^2 \exp\left(-\frac{\Phi_B}{kT}\right)$$

where B_e is the effective Richardson constant. For many metal-semiconductor Schottky junctions B_e is typically $10 - 100 \text{A K}^{-2} \text{ cm}^{-2}$.

a.   Consider a Schottky junction diode between Al and n-Si, doped with 10^{16} donors cm^{-3}. The cross-sectional area is 1 mm^2. Given that the electron affinity (χ) of Si is 4.01 eV and the work function of Al is 4.28 eV, what is the theoretical barrier height, Φ_B, from the metal to the semiconductor?

b.   Given that the experimental barrier height Φ_B is about 0.5 eV, what is the reverse saturation current and the current when there is a forward bias of 0.1 V across the diode?

Solution

a.   From Figure 5.38, it is clear that the barrier height Φ_B is

$$\Phi_B = \Phi_{Fm} - \chi = 4.28 - 4.01 = 0.27 \text{ eV}$$

The experimental value is typically 0.5 eV, greater than the theoretical value due to various effects at the metal–semiconductor interface arising from defects, impurities, and so forth.

b.   If A is the cross-sectional area, 0.01 cm^2, taking B_e to be typically 30 A K^{-2} cm^{-2}, the saturation current is

Example 5.18

$$I_o = AB_e T^2 \exp\left(-\frac{\Phi_B}{kT}\right) = (0.01)(30)(300^2) \exp\left(-\frac{0.5 \text{ eV}}{0.026 \text{ eV}}\right)$$

$$= 1.20 \times 10^{-4} \text{ A} \quad \text{or} \quad 120 \ \mu\text{A}$$

The forward current is

$$I_f = I_o\left[\exp\left(\frac{eV_f}{kT}\right) - 1\right] = (0.12 \times 10^{-3} \text{ mA})\left[\exp\left(\frac{0.1 \text{ eV}}{0.026 \text{ eV}}\right) - 1\right] = 5.5 \text{ mA}$$

5.10   OHMIC CONTACTS AND THERMOELECTRIC COOLERS

An ohmic contact is a junction between a metal and a semiconductor that does not limit the current flow. The current is essentially limited by the resistance of the semiconductor outside the contact region rather than the thermal emission rate of carriers across a potential barrier at the contact. In the Schottky diode, the I–V characteristics were determined by the thermal emission rate of carriers across the contact. It should be mentioned that, contrary to intuition, when we talk about an ohmic contact, we do not generally infer a linear I–V characteristic for the ohmic contact itself. We only imply that the contact does not limit the current flow.

Figure 5.42 shows the formation of an ohmic contact between a metal and an n-type semiconductor. The work function of the metal, Φ_m, is smaller than the work function, Φ_n of the semiconductor. There are more energetic electrons in the metal than in the CB, which means that the electrons (around E_{Fm}) tunnel into the semiconductor in search of lower energy levels, which they find around E_c, as indicated in the figure. Consequently many electrons pile in the CB of the semiconductor near the junction. Equilibrium is reached when the accumulated electrons in the CB of the semiconductor prevent further electrons tunneling from the metal. Put more rigorously, equilibrium is reached when the Fermi level is uniform across the whole system from one end to the other.

The semiconductor region near the junction in which there are excess electrons is called the **accumulation region.** To show the increase in n, we draw the semiconductor energy bands bending downwards to decrease $E_c - E_{Fn}$, which increases n. Going from the far end of the metal to the far end of the semiconductor, there are always conduction electrons. In sharp contrast, the depletion region of the Schottky junction separates the conduction electrons in the metal from those in the semiconductor. It can be seen from the contact in Figure 5.42 that the conduction electrons immediately on either side of the junction (at E_{Fm} and E_c) have about the same energy and therefore there is no barrier involved when they cross the junction in either direction under the influence of an applied field.

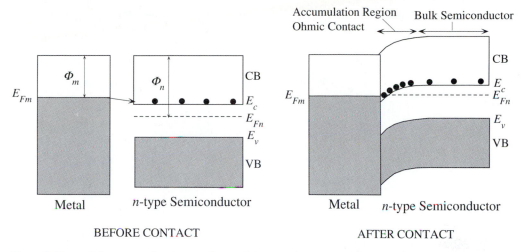

Figure 5.42   When a metal with a smaller work function than an n-type semiconductor is put into contact with the n-type semiconductor, the resulting junction is an ohmic contact in the sense that it does not limit the current flow.

It is clear that the excess electrons in the accumulation region increase the conductivity of the semiconductor in this region. When a voltage is applied to the structure, the voltage drops across the higher resistance region, which is the bulk semiconductor region. Both the metal and the accumulation region have comparatively high concentrations of electrons compared with the bulk of the semiconductor. The current is therefore determined by the resistance of the bulk region. The current density is then simply $J = \sigma \mathcal{E}$ where σ is the conductivity of the semiconductor in the bulk and $\mathcal{E}$ is the applied field in this region.

One of the interesting and important applications of semiconductors is in **thermoelectric,** or **Peltier,** devices, which enable small volumes to be cooled by direct currents. Whenever a dc current flows through a contact between two dissimilar materials, heat is either released or absorbed in the contact region, depending on the direction of the current. Suppose that there is a dc current flowing from an n-type semiconductor to a metal through an ohmic contact, as depicted in Figure 5.43a. Then electrons are flowing from the metal to the CB of the semiconductor. We only consider the contact region where the Peltier effect occurs. Current is carried by electrons near the Fermi level, E_{Fm}, in the metal. These electrons then cross over into the CB of the semiconductor and when they reach the end of the contact region, their energy is E_c plus average KE (which is $(3/2)kT$). There is therefore an increase in the average energy ($PE + KE$) per electron in the contact region. The electron must therefore absorb heat from the environment (lattice vibrations) to gain this energy as it drifts through the junction. Thus, the passage of an electron from the metal to the CB of an n-type semiconductor involves the absorption of heat at the junction.

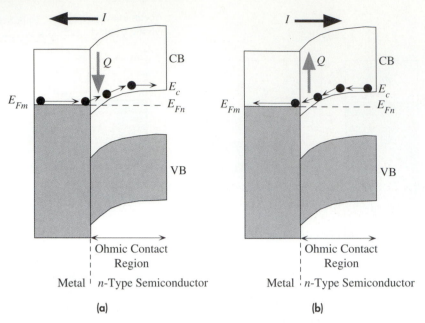

Figure 5.43

(a) Current from an *n*-type semiconductor to the metal results in heat absorption at the junction.

(b) Current from the metal to an *n*-type semiconductor results in heat release at the junction.

When the current direction is from the metal to the *n*-type semiconductor, the electrons flow from the CB of the semiconductor to the Fermi level of the metal as they pass through the contact. Since E_{Fm} is lower than E_c the passing electron has to lose energy, which it does to lattice vibrations as heat. Thus, the passage of a CB electron from the *n*-type semiconductor to the metal involves the release of heat at the junction, as indicated in Figure 5.43b.

It is apparent that depending on the direction of the current flow through a junction between a metal and an *n*-type semiconductor, heat is either absorbed or released at the junction. Although we considered current flow between a metal and an *n*-type semiconductor through an ohmic contact, this thermoelectric effect is a general phenomenon that occurs at a junction between any two dissimilar materials. It is called the **Peltier effect** after its discoverer. In the case of metal–*p*-type semiconductor junctions, heat is absorbed for current flowing from the metal to the *p*-type semiconductor and heat is released in the other direction. Thermoelectric effects occurring at metal–semiconductor junctions are summarized in Figure 5.44. It is important not to confuse the Peltier effect with the Joule heating of the semiconductor and the metal. Joule heating, which we simply call I^2R (or $J^2\rho$) heating, arises from the finite resistivity of the material. It is due to the conduction electrons losing their energy gained from the field to lattice vibrations when they become scattered by such vibrations, as discussed in Chapter 2.

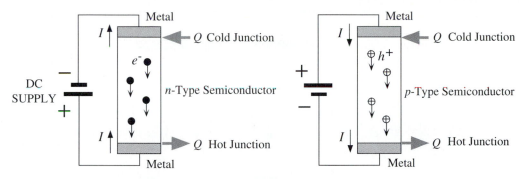

Figure 5.44   When a dc current is passed through a semiconductor to which metal contacts have been made, one junction absorbs heat and cools (the cold junction) and the other releases heat and warms (the hot junction).

It is self-evident that when a current flows through a semiconductor sample with metal contacts at its ends, as depicted in Figure 5.44, one of the contacts will always absorb heat and the other will always release heat. The contact where heat is absorbed will be cooled and is called the cold junction, whereas the other contact, where heat is released, will warm up and is called the hot junction. One can use the cold junction to cool another body, providing that the heat generated at the hot junction can be removed from the semiconductor sufficiently quickly to reduce its conduction through the semiconductor to the cold junction. Furthermore, there will always be the Joule heating (I^2R) of the whole semiconductor sample since the bulk will always have a finite resistance.

A simplified schematic diagram of a practical single-element thermoelectric cooling device is shown in Figure 5.45. It uses two semiconductors, one n-type and the other p-type, each with ohmic contacts. The current direction therefore has opposite thermoelectric effects. On one side, the semiconductors share the same metal electrode. Effectively, the structure is an n-type and a p-type semiconductor connected in series through a common metal electrode. Typically, either Bi_2Te_3, Bi_2Se_3, or Sb_2Te_3 is used as the semiconductor material with copper usually as the metal electrode.

The current flowing the n-type semiconductor to the common metal electrode causes heat absorption, which cools this junction and hence the metal. The same current then enters the p-type semiconductor and causes heat absorption at this junction, which cools the same metal electrode. Thus the common metal electrode is cooled at both ends. The other ends of the semiconductor are hot junctions. They are connected to a large heat sink to remove the heat and thus prevent heat conduction through the semiconductors toward the cold junctions. The other face of the common metal electrode is in contact, through a thin ceramic plate (electrical insulator but thermal conductor), with the body to be cooled. In commercial Peltier devices, many of these elements are connected in series, as illustrated in Figure 5.46, to increase the cooling efficiency.

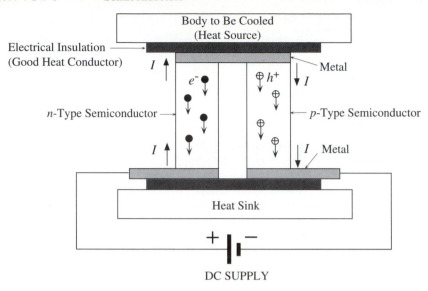

Figure 5.45   Cross section of a typical themoelectric cooler.

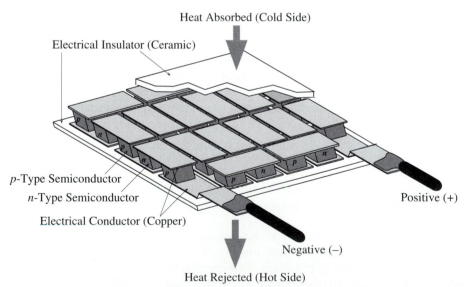

Figure 5.46   Typical structure of a commercial thermoelectric cooler.

Example 5.19 | **THE PELTIER COEFFICIENT**   Consider the motion of electrons across an ohmic contact between a metal and an n-type semiconductor and hence show that the rate of heat generation; Q', at the contact is approximately

$$Q' = \pm \Pi I$$

where Π, called the Peltier coefficient between the two materials, is given by

$$\Pi = \frac{1}{e}\left[(E_c - E_{Fn}) + \frac{3}{2}kT\right]$$

where $E_c - E_{Fn}$ is the energy separation of E_c from the Fermi level in the n-type semiconductor. The sign depends on the convention used for heat liberation or absorption.

Solution

We consider Figure 5.43a, which shows only the ohmic contact region between a metal and an n-type semiconductor when a current is passing through it. The majority of the applied voltage drops across the bulk of the semiconductor because the contact region, or the accumulation region, has an accumulation of electrons in the CB. The current is limited by the bulk resistance of the semiconductor. Thus, in the contact region we can take the Fermi level to be almost undisturbed and hence uniform, $E_{Fm} \approx E_{Fn}$. In the bulk of the metal, a conduction electron is at around E_{Fm} (same as E_{Fn}), whereas just at the end of the contact region in the semiconductor it is at E_c plus an average KE of $(3/2)kT$. The energy difference is the heat absorbed per electron going through the contact region. Since I/e is the rate at which electrons are flowing through the contact,

$$\text{Rate of energy absorption} = \left[\left(E_c + \frac{3}{2}kT\right) - E_{Fm}\right]\left(\frac{I}{e}\right)$$

or

$$Q' = \left[\frac{(E_c - E_{Fn}) + \frac{3}{2}kT}{e}\right]I = \Pi I$$

so that the Peltier coefficient is approximately given by the term in the square brackets. A more rigorous analysis gives Π as

$$\Pi = \frac{1}{e}[(E_c - E_{Fn}) + 2kT]$$

ADDITIONAL TOPICS

5.11   DIRECT AND INDIRECT BAND GAP SEMICONDUCTORS

E–k Diagrams   We know from quantum mechanics that when the electron is within a potential well of size L, its energy is quantized and given by

$$E_n = \frac{(\hbar k_n)^2}{2m_e}$$

where the wavevector k_n is essentially a quantum number determined by

$$k_n = \frac{n\pi}{L}$$

where $n = 1, 2, 3, \ldots$. The energy increases parabolically with the wavevector k_n.

We also know that the electron momentum is given by $\hbar k_n$. This description can be used to represent the behavior of electrons in a metal within which their average potential energy can be taken to be roughly zero. In other words, we take $V(x) = 0$ within the metal crystal and $V(x)$ to be large (e.g., $V(x) = V_o$) outside so that the electron is contained within the metal. This is the **nearly free electron model** of a metal that has been quite successful in interpreting many of the properties. Indeed, we were able to calculate the density of states $g(E)$ based on the three-dimensional potential well problem. It is quite obvious that this model is too simple since it does not take into account the actual variation of the electron potential energy in the crystal.

The potential energy of the electron depends on its location within the crystal and is periodic due to the regular arrangement of the atoms. How does a periodic potential energy affect the relationship between E and k? It will no longer simply be $E_n = (\hbar k_n)^2 / 2m_e$.

To find the energy of the electron in a crystal, we need to solve the Schrödinger equation for a periodic potential energy function in three dimensions. We first consider the hypothetical one-dimensional crystal shown in Figure 5.47. The electron potential energy functions for each atom add to give an overall potential energy function $V(x)$, which is clearly periodic in x with the periodicity of the crystal, a. Thus,

$$V(x) = V(x + a) = V(x + 2a) = \ldots$$

and so on. Our task is therefore to solve the Schrödinger equation

$$\frac{d^2\psi}{dx^2} + \frac{2m_e}{\hbar^2}[E + V(x)]\psi = 0 \qquad [5.52]$$

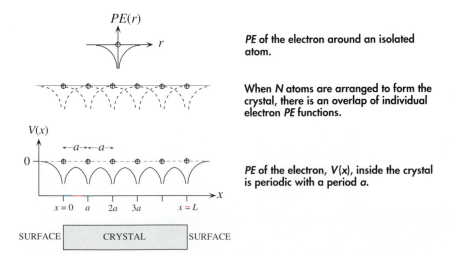

PE of the electron around an isolated atom.

When **N** atoms are arranged to form the crystal, there is an overlap of individual electron PE functions.

PE of the electron, V(x), inside the crystal is periodic with a period a.

Figure 5.47   The electron PE, V(x), inside the crystal is periodic with the same periodicity as that of the crystal, a.

Far away outside the crystal, by choice, V = 0 (the electron is free and PE = 0).

subject to the condition that the potential energy, $V(x)$, is periodic in a, that is,

$$V(x) = V(x + ma) \qquad m = 1, 2, 3, \ldots \qquad \text{[5.53]}$$

The solution of Equation 5.52 will give the electron wavefunction in the crystal and hence the electron energy. Since $V(x)$ is periodic, we should expect, by intuition at least, the solution $\psi(x)$ to be periodic. It turns out that the solutions to Equation 5.52, which are called **Bloch wavefunctions,** are of the form

$$\psi_k(x) = U_k(x) \exp(jkx) \qquad \text{[5.54]}$$

where $U_k(x)$ is a periodic function that depends on $V(x)$ and has the same periodicity, a, as $V(x)$. The term $\exp(jkx)$, of course, represents a traveling wave. We should remember that we have to multiply this by $\exp(-jEt/\hbar)$, where E is the energy, to get the overall wavefunction $\Psi(x, t)$. Thus the electron wavefunction in the crystal is a traveling wave that is modulated by $U_k(x)$.

There are many such Bloch wavefunction solutions to the one-dimensional crystal, each identified with a particular k value, say k_n, which acts as a kind of quantum number. Each $\psi_k(x)$ solution corresponds to a particular k_n and represents a state with an energy E_k. The dependence of the energy E_k on the wavevector k is what we call the E–k diagram. Figure 5.48 shows a typical E–k diagram for the hypothetical one-dimensional solid for k values in the range $-\pi/a$ to $+\pi/a$. Just as $\hbar k$ is the momentum of a free electron, $\hbar k$ for the Bloch electron is the momentum involved in its interaction with external fields, for example, those involved in the

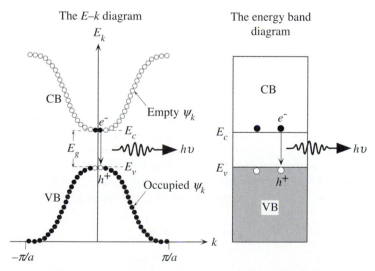

Figure 5.48   The E–k diagram of a direct bandgap semiconductor such as GaAs.

The E–k curve consists of many discrete points, each corresponding to a possible state, wavefunction $\psi_k(x)$, that is allowed to exist in the crystal. The points are so close that we normally draw the E–k relationship as a continuous curve. In the energy range E_v to E_c, there are no points ($\psi_k(x)$ solutions).

photon absorption process. Indeed, the rate of change of $\hbar k$ is the externally applied force, F_{ext}, on the electron such as that due to an electric field ($F_{ext} = e\mathcal{E}$). Thus, for the electron within the crystal,

$$\frac{d(\hbar k)}{dt} = F_{ext}$$

and consequently we call $\hbar k$ the **crystal momentum** of the electron.[3]

Inasmuch as the momentum of the electron in the x direction in the crystal is given by $\hbar k$, the E–k diagram is an **energy versus crystal momentum plot.** The states $\psi_k(x)$ in the lower E–k curve constitute the wavefunctions for the valence electrons and thus correspond to the states in the VB. Those in the upper E–k curve, on the other hand, correspond to the states in the conduction band (CB) since they have higher energies. All the valence electrons at absolute zero of temperature therefore fill the states, particular k_n values, in the lower E–k diagram.

It should be emphasized that an E–k curve consists of many discrete points, each corresponding to a possible state, wavefunction $\psi_k(x)$, that is allowed to exist in the crystal. The points are so close that we draw the E–k relationship as a continuous curve. It is clear from the E–k diagram that there is a range of energies, from E_v to E_c, for which there are no solutions to the Schrödinger equation and hence there are no $\psi_k(x)$ with energies in E_v to E_c. Furthermore, we also note that the E–k behavior is not a simple parabolic relationship except near the bottom of the CB and the top of the VB.

Above absolute zero of temperature, due to thermal excitation, however, some of the electrons from the top of the valence band will be excited to the bottom of the conduction band. According to the E–k diagram in Figure 5.48, when an electron and hole recombine, the electron simply drops from the bottom of the CB to the top of the VB without any change in its k value so that this transition is quite acceptable in terms of momentum conservation. We should recall that the momentum of the emitted photon is negligible compared with the momentum of the electron. The E–k diagram in Figure 5.48 is therefore for a **direct bandgap semiconductor.**

The simple E–k diagram sketched in Figure 5.48 is for the hypothetical one-dimensional crystal in which each atom simply bonds with two neighbors. In real crystals, we have a three-dimensional arrangement of atoms with $V(x, y, z)$ showing periodicity in more than one direction. The E–k curves are then not as simple as that in Figure 5.48 and often show unusual features. The E–k diagram for GaAs, which is shown in Figure 5.49a, as it turns out, has main features that are quite

---

[3] The actual momentum of the electron, however, is not $\hbar k$ because

$$\frac{d(\hbar k)}{dt} \neq F_{\text{external}} + F_{\text{internal}}$$

where $F_{\text{external}} + F_{\text{internal}}$ are all forces acting on the electron. The true momentum p_e satisfies

$$\frac{dp_e}{dt} = F_{\text{external}} + F_{\text{internal}}$$

However, as we are interested in interactions with external forces such as an applied field, we treat $\hbar k$ as if it were the momentum of the electron in the crystal and use the name **crystal momentum.**

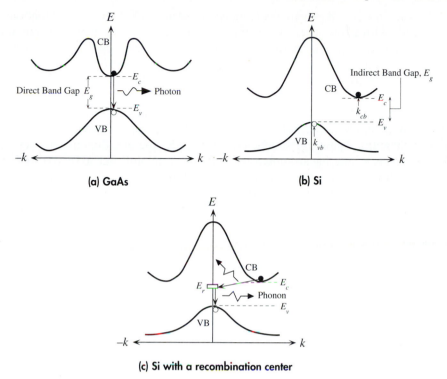

Figure 5.49
(a) In GaAs the minimum of the CB is directly above the maximum of the VB. GaAs is therefore a direct bandgap semiconductor.
(b) In Si, the minimum of the CB is displaced from the maximum of the VB and Si is an indirect bandgap semiconductor.
(c) Recombination of an electron and a hole in Si involves a recombination center.

similar to that sketched in Figure 5.48. GaAs is therefore a direct bandgap semiconductor in which electron–hole pairs can recombine directly and emit a photon. It is quite apparent that light-emitting devices use direct bandgap semiconductors to make use of direct recombination.

In the case of Si, the diamond crystal structure leads to an E–k diagram that has the essential features depicted in Figure 5.49b. We notice that the minimum of the CB is not directly above the maximum of the VB. An electron at the bottom of the CB therefore cannot recombine directly with a hole at the top of the VB because, for the electron to fall down to the top of the VB, its momentum must change from k_{cb} ro k_{vb}, which is not allowed by the law of conservation of momentum. Thus direct electron–hole recombination does not take place in Si and Ge. The recombination process in these elemental semiconductors occurs via a recombination center at an energy level E_r. The electron is captured by the defect at E_r, from where it can fall down into the top of the VB. The indirect recombination process is illustrated in Figure 5.49c. The energy of the electron is lost by the emission of phonons, that is, lattice vibrations. The E–k diagram in Figure 5.49b for Si is an example of an **indirect band gap semiconductor.**

In some indirect bandgap semiconductors such as GaP, the recombination of the electron with a hole at certain recombination centers results in photon emission. The E–k diagram is similar to that shown in Figure 5.49c except that the recombination centers at E_r are generated by the purposeful addition of nitrogen impurities to GaP. The electron transition from E_r to E_v involves photon emission.

Electron Motion and Drift   We can understand the response of a conduction band electron to an applied external force, for example, an applied field, by examining the E–k diagram. Again, for simplicity, we consider the one-dimensional crystal. The electron is wandering around the crystal quite randomly due to scattering from lattice vibrations. Thus the electron moves with a certain k value in the $+x$ direction, say k_+, as illustrated in the E–k diagram of Figure 5.50a. When it is scattered by a lattice vibration, its k value changes, perhaps to k_-, which is also shown in Figure 5.50a. This process of k changing randomly from one scattering to another scattering process continues all the time, so that over a long time the average value of k is zero, that is, average k_+ is the same as average k_-.

When an electric field is applied, say in the $-x$ direction, then the electron gains momentum in the $+x$ direction from the force of the field, $e\mathcal{E}_x$. With time, while the electron is not scattered, it moves up in the E–k diagram from k_{1+} to k_{2+} to k_{3+} and so on until a lattice vibration randomly scatters the electron to say k_{-1} (or to some other random k value) as shown in Figure 5.50b. Over a long time, the average of all k_+ is no longer equal to the average of all k_- and there is a net momentum in the $+x$ direction, which is tantamount to a drift in the same direction.

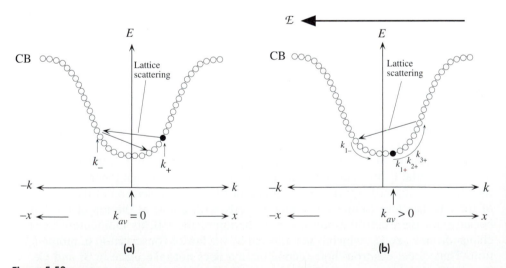

Figure 5.50
(a) In the absence of a field, over a long time, the average of all k values is zero; there is no net momentum in any one particular directon.
(b) In the presence of a field in the $-x$ direction, the electron accelerates in the $+x$ direction increasing its k value along x until it is scattered to a random k value. Over a long time, the average of all k values is along the $+x$ direction. Thus the electron drifts along $+x$.

Effective Mass   The usual definition of inertial mass of a particle in classical physics is based on

$$Force = Mass \times Acceleration$$

$$F = ma$$

When we treat the electron as a wave within the semiconductor crystal we have to determine whether we can still, in some way, use the convenient classical $F = ma$ relation to describe the motion of an electron under an applied force such as $e\mathcal{E}_x$ and if so what should be the apparent mass of the electron in the crystal?

We will evaluate the velocity and acceleration of the electron in the CB in response to an electric field, $\mathcal{E}_x$, along $-x$ that imposes an external force $F_{ext} = e\mathcal{E}_x$ in the $+x$ direction, as shown in Figure 5.50b. Our treatment will make use of the quantum mechanical $E-k$ diagram.

Since we are treating the electron as a wave, we have to evaluate the group velocity v_g, which, by definition, is $v_g = d\omega/dk$. We know that the time dependence of the wavefunction is $\exp(-jEt/\hbar)$ where the energy $E = \hbar\omega$ (ω is an "angular frequency" associated with the wave motion of the electron). Both E and ω depend on k. Thus, the group velocity is

$$v_g = \frac{1}{\hbar}\frac{dE}{dk} \qquad\qquad [5.55]$$

Thus the group velocity is determined by the **gradient** of the $E-k$ curve. In the presence of an electric field, the electron experiences a force $F_{ext} = e\mathcal{E}_x$ from which it gains energy and moves up in the $E-k$ diagram until, later on, it collides with a lattice vibration, as shown in Figure 5.50b. During a small time interval, δt, between collisions, the electron moves a distance, $v_g\,\delta t$, and hence gains energy, δE, which is

$$\delta E = F_{ext}v_g\,\delta t \qquad\qquad [5.56]$$

To find the acceleration of the electron and the effective mass, we somehow have to put this equation into a form that looks like $F_{ext} = m_e a$, where a is the acceleration. From Equation 5.56, the relationship between the external force and energy is

$$F_{ext} = \frac{1}{v_g}\frac{dE}{dt} = \hbar\frac{dk}{dt} \qquad\qquad [5.57]$$

where we used Equation 5.55 for v_g in Equation 5.56. Equation 5.57 is the reason for interpreting $\hbar k$ as the crystal momentum inasmuch as the rate of change of $\hbar k$ is the externally applied force.

The acceleration, a, is defined as dv_g/dt. We can use Equation 5.55,

$$a = \frac{dv_g}{dt} = \frac{d\left[\dfrac{1}{\hbar}\dfrac{dE}{dk}\right]}{dt} = \frac{1}{\hbar}\frac{d^2E}{dk^2}\frac{dk}{dt} \qquad\qquad [5.58]$$

From Equation 5.58, we can substitute for dk/dt in Equation 5.57, which is then a relationship between F_{ext} and a of the form

$$F_{ext} = \frac{\hbar^2}{\left[\dfrac{d^2E}{dk^2}\right]} a \qquad [5.59]$$

We know that the response of a free electron to the external force is $F_{ext} = m_e a$, where m_e is its mass in vacuum. Therefore it is quite clear from Equation 5.59 that the effective mass of the electron in the crystal is

Effective mass

$$m_e^* = \hbar^2\left[\frac{d^2E}{dk^2}\right]^{-1} \qquad [5.60]$$

Thus, the electron responds to an external force and moves as if its mass were given by Equation 5.60. The effective mass obviously depends on the E–k relationship, which in turn depends on the crystal symmetry and the nature of bonding between the atoms. Its value is different for electrons in the CB and for those in the VB, and moreover, it depends on the energy of the electron since it is related to the curvature of the E–k behavior (d^2E/dk^2). Further, it is clear from Equation 5.60 that the effective mass is a quantum mechanical quantity inasmuch as the E–k behavior is a direct consequence of the application of quantum mechanics (the Schrödinger equation) to the electron in the crystal.

It is interesting that, according to Equation 5.60, when the E–k curve is a downward concave as at the top of a band (e.g., Figure 5.48), the effective mass of an electron at these energies in a band is then negative. What does a negative effective mass mean? When the electron moves up on the E–k curve by gaining energy from the field, it actually decelerates, that is, moves more slowly. Its acceleration is therefore in the opposite direction to an electron at the bottom of the band. Electrons in the CB are at the bottom of a band so that their effective masses are positive quantities. At the top of a valence band, however, we have plenty of electrons. These electrons have negative effective masses and under the action of a field, they decelerate. Put differently, they accelerate in the opposite direction to the applied external force, F_{ext}. It turns out that we can describe the collective motion of these electrons near the top of a band by considering the motion of a few holes with positive masses.

It should be mentioned that Equation 5.60 defines the meaning of the effective mass in quantum mechanical terms. Its usefulness as a concept lies in the fact that we can measure it experimentally, for example, by cyclotron resonance experiments, and have actual values for it. This means we can simply replace m_e by m_e^* in equations that describe the effect of an external force on electron transport in semiconductors.

Holes   To understand the concept of a hole, we consider the E–k curve corresponding to energies in the VB, as shown in Figure 5.51a. If all the states are filled, then there are no empty states for the electrons to move into and consequently an electron cannot gain energy from the field. For each electron moving in the positive x direction with a momentum $\hbar k_+$, there is a corresponding electron with an equal and opposite momentum $\hbar k_-$ so that there is no net motion. For example, the

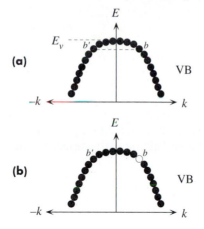

Figure 5.51
(a) In a full valence band, there is no net contribution to the current. There are equal numbers of electrons (e.g., at b and b') with opposite momenta.
(b) If there is an empty state (*hole*) at b at the top of the band, then the electron at b' contributes to the current.

electron at b is moving toward the right with k_{+b}, but its effect is canceled by that at b' moving toward the left with $k_{-b'}$. This cancellation of momenta by electron pairs applies to all the electrons since the valence band is assumed to be full. Thus, a full VB cannot contribute to the electric current.

Suppose that one of the states, labeled as b in Figure 5.51b, near the top of the valence band has a missing electron, or a hole, because the electron normally at b has been removed by some means of excitation to the conduction band. It is immediately obvious that the motion of the electron at b' toward the left, that is, $k_{-b'}$, is now *not* canceled, which means that this electron makes a net contribution to the current. We realize that the reason the presence of a hole makes conduction possible is the fact that the momenta of all the VB electrons are canceled except that at b'. It is also clear that in reaching this conclusion, we had to consider all the electrons in the valence band.

Let us maintain strict sign rules so that quantities such as the field ($\mathcal{E}_x$), group velocity (v_g), and acceleration (a) along the $+x$ direction are positive and those along the $-x$ direction are negative. If $\mathcal{E}_x$ is along $+x$ direction, then the acceleration of a *free* electron from *force*/mass is $[(-e)(\mathcal{E}_x)]/m_e$, which is negative and along $-x$ as we expect. Similarly, an electron at the bottom of the CB has a positive effective mass and an acceleration that is also negative. Our treatment of conduction in metals by electrons in Chapter 2 inherently assumed that electrons accelerated in the opposite direction to the applied field, that is, positive effective mass.

However, the electrons at the top of the VB have a negative effective mass, which we can write as $-|m_e^*|$. The acceleration a of the electron at b' contributing to the current is

$$a = \frac{-e\mathcal{E}_x}{-|m_e^*|} = \frac{+e\mathcal{E}_x}{+|m_e^*|}$$

which is positive, a along $\mathcal{E}_x$. This means that the acceleration of an electron with a negative effective mass at the top of a VB is equivalent to the acceleration of a positive charge $+e$ with an effective mass $|m_e^*|$. Put differently, we therefore can equivalently describe current conduction by the motion of the hole alone by assigning to it a positive charge and a positive effective mass.

Example 5.20 | **EFFECTIVE MASS**   Show that the effective mass of a free electron is the same as its mass in vacuum.

Solution

The expression for the energy of a free electron is

$$E = \frac{(\hbar k)^2}{2m_e}$$

The effective mass, by definition, is given by

$$m_e^* = \hbar^2 \left[\frac{d^2E}{dk^2} \right]^{-1}$$

Substituting $E = (\hbar k)^2/2m_e$ we get $m_e^* = m_e$. Since the energy of a conduction electron in a metal, within the nearly free electron model, will also have an energy $E = (\hbar k)^2/2m_e$, we can show that the effective mass of the electron in a metal is the same as the mass in vacuum.

Example 5.21 | **CURRENT DUE TO A MISSING ELECTRON IN THE VB**   First, let us consider a completely full valence band that contains, say, N electrons. $N/2$ of these are moving with momentum in the $+x$, and $N/2$ in the $-x$ direction. Suppose that the crystal is unit volume. An electron with charge $-e$ moving with a group velocity $\mathbf{v}_{gi}$ contributes to the current by an amount $-e\mathbf{v}_{gi}$. We can determine the current density $\mathbf{J}_N$ due to the motion of all the electrons (N of them) in the band,

$$\mathbf{J}_N = -e\sum_{i=1}^{N} \mathbf{v}_{gi} = 0$$

$\mathbf{J}_N$ is zero because for each value of $\mathbf{v}_{gi}$, there is a corresponding velocity equal in magnitude but opposite in direction (b and b' in Figure 5.51a). Our conclusion from this is that the contribution to the current density from a full valence band is nil, as we expect.

Suppose now that the jth electron is missing (b in Figure 5.51b), the net current density is due to $N - 1$ electrons in the band so that

$$\mathbf{J}_{N-1} = -e\sum_{i=1, i \neq j}^{N} \mathbf{v}_{gi} \qquad [5.61]$$

where the summation is for $i = 1$ to N and $i \neq j$ (jth electron is missing). We can write the sum as summation to N including the jth electron and minus the missing jth electron contribution,

$$\mathbf{J}_{N-1} = -e \sum_{i=1}^{N} \mathbf{v}_{gi} - (-e\mathbf{v}_{gj})$$

that is,

$$\mathbf{J}_{N-1} = +e\mathbf{v}_{gj} \qquad [5.62]$$

where we used $\mathbf{J}_N = 0$. We see that when there is a missing electron, there is a net current due to that empty state (jth). The current appears as the motion of a charge $+e$ with a velocity $\mathbf{v}_{gj}$ where $\mathbf{v}_{gj}$ is the group velocity of the missing electron. In other words, the current is due to the motion of a positive charge $+e$ at the site of the missing electron at k_j. One should note that Equation 5.61 describes the current by considering the motions of *all* the $N-1$ electrons, whereas Equation 5.62 describes the same current by simply considering the missing electron as if it were a positively charged particle ($+e$) moving with a velocity equal to that of the missing electron. Equation 5.62 is the convenient description universally adopted for a valence band containing missing electrons.

5.12 INDIRECT RECOMBINATION

We consider the recombination of minority carriers in an extrinsic indirect band gap semiconductor such as Si or Ge. As an example, we consider the recombination of electrons in a p-type semiconductor. In an indirect bandgap semiconductor, the recombination mechanism involves a recombination center, a third body that may be a crystal defect or an impurity, in the recombination process to satisfy the requirements of conservation of momentum. We can view the recombination process as follows. Recombination occurs when an electron is captured by the recombination center at the energy level E_r. As soon as the electron is captured, it will recombine with a hole because holes are abundant in a p-type semiconductor. In other words, since there are many majority carriers, the limitation on the rate of recombination is the actual capture of the minority carrier by the center. Thus, if τ_e is the electron recombination time, since the electrons will have to be captured by the centers, τ_e is given by

$$\tau_e = \frac{1}{S_r N_r v_{th}} \qquad [5.63]$$

where S_r is the capture (or recombination) cross section of the center, N_r is the concentration of centers, and v_{th} is the mean speed of the electron that you may take as its effective thermal velocity.

Equation 5.63 is valid under small injection conditions, that is, $p_{po} \gg n_p$. There is a more general treatment of indirect recombination called the Shockley–Read statistics of indirect recombination and generation, which is treated in more advanced semiconductor physics textbooks. That theory eventually arrives at the above expression for low-level injection conditions. We derived Equation 5.63 from a purely physical reasoning.

Gold is frequently added to silicon to aid recombination. It is found that the minority carrier recombination time is inversely proportional to the gold concentration, following Equation 5.63.

IMPORTANT TERMS

Acceptor atoms are dopants that have one less valency than the host atom. They therefore accept electrons from the VB and thereby create holes in the VB, which leads to $p > n$ and hence to a p-type semiconductor.

Average energy of an electron in the CB is $(3/2) kT$ as if the electrons were obeying Maxwell–Boltzmann statistics. This is only true for a nondegenerate semconductor.

Compensated semiconductor contains both donors and acceptors in the same crystal region that compensate for each other's effects. For example, if there are more donors than acceptors, $N_d > N_a$, then some of the electrons released by donors are captured by acceptors and the net effect is that $N_d - N_a$ number of electrons per unit volume are left in the CB.

Conduction band (CB) is a band of energies for the electron in a semiconductor where it can gain energy from an applied field and drift and thereby contribute to electrical conduction. The electron in the CB behaves as if it were a "free" particle with an effective mass, m_e^*.

Degenerate semiconductor has so many dopants that the electron concentration in the CB, or hole concentration in the VB, is comparable with the density of states in the band. Consequently, the Pauli exclusion principle is significant and Fermi–Dirac statistics must be used. The Fermi level is either in the CB for a n^+ type degenerate or in the VB for a p^+ type degenerate semiconductor. The superscript $+$ indicates a heavily doped semiconductor.

Diffusion is a random process by which particles move from high concentration regions to low concentration regions.

Donor atoms are dopants that have a valency one more than the host atom. They therefore donate electrons to the CB and thereby create electrons in the CB, that which leads to $n > p$ and hence to an n-type semiconductor.

Effective density of states (N_c) at the CB edge is a quantity that represents all the states in the CB per unit volume as if they were all at E_c. Similarly, N_v at the VB edge is a quantity that represents all the states in the VB per unit volume as if they were all at E_v.

Effective mass (m_e^*) of an electron is a quantum mechanical quantity that behaves like the inertial mass in classical mechanics, $F = ma$, in that it measures the object's inertial resistance to acceleration. It relates the acceleration a of an electron in a crystal to the applied external force F_{ext} by $F_{ext} = m_e^* a$. The external force is most commonly the force of an electric field, $e \, \mathcal{E}$, and excludes all internal forces within the crystal.

Einstein relation relates the diffusion coefficient D and the drift mobility μ of a given species of charge carriers through $(D/\mu) = (kT/e)$.

Electron affinity (χ) is the energy required to remove an electron from E_c to the vacuum level.

Energy of the electron in the crystal, whether in the CB or VB, depends on its momentum $\hbar k$ through the E–k behavior determined by the Schrödinger equation. E–k behavior is most conveniently represented graphically through E–k diagrams. For example, for an electron at the bottom of the CB, E increases as $(\hbar k)^2/m_e^*$ where $\hbar k$ is the momentum and m_e^* is the effective mass of the electron, which is determined from the E–k behavior.

Excess carrier concentration is the excess concentration *above* the thermal equilibrium value. Excess carriers are generated by an external excitation such as photogeneration.

Extrinsic semiconductor is a semiconductor that has been doped so that the concentration of one type of charge carrier far exceeds that of the other. Adding donor impurities releases electrons into the CB and n far exceeds p; thus, the semiconductor becomes n-type.

Fermi energy or level (E_F) may be defined in several equivalent ways. Fermi level is the energy level corresponding to the energy required to remove an electron from the semiconductor; there need not be any actual electrons at this energy level. The energy needed to remove an electron defines the work function Φ. We can define the Fermi level to be Φ below the vacuum level. E_F can also be defined as that energy value below which all states are full and above which all states are empty at absolute zero of temperature. E_F can also be defined through a difference. A difference in the Fermi energy, ΔE_F, in a system is the external electrical work done per electron either on the system or by the system just as electrical work done when a charge e moves through an electrostatic PE difference is $e\,\Delta V$. It can be viewed as a fundamental material property.

Intrinsic carrier concentration (n_i) is the electron concentration in the CB of an instrinsic semiconductor. The hole concentration in the VB is equal to the electron concentration.

Intrinsic semiconductor has an equal number of electrons and holes due to thermal generation across the bandgap, E_g. It corresponds to a pure semiconductor crystal in which there are no impurities or crystal defects.

Ionization energy is the energy required to ionize an atom, for example, to remove an electron.

Ionized impurity scattering limited mobility is the mobility of the electrons when their motion is limited by scattering from the ionized impurities in the semiconductor (e.g., donors and acceptors).

k is the wavevector of the electron's wavefunction. In a crystal the electron wavefunction, $\psi(x)$, is a *modulated travelling wave* of the form

$$\psi_k(x) = U_k(x)\,\exp(jkx)$$

where k is the wavevector and $U_k(x)$ is a periodic function that depends on the PE of interaction between the electron and the lattice atoms. k identifies all possible states, $\psi_k(x)$, that are allowed to exist in the crystal. $\hbar k$ is called the *crystal momentum* of the electron as its rate of change is the externally applied force to the electron, $d(\hbar k)/dt = F_{\text{external}}$.

Lattice scattering limited mobility is the mobility of the electrons when their motion is lim ited by scattering from thermal vibrations of the lattice atoms.

Majority carriers are electrons in an n-type and holes in a p-type semiconductor.

Mass action law in semiconductor science refers to the law $np = n_i^2$, which is valid under thermal equi-

librium conditions and in the absence of external biases and illumination.

Minority carrier diffusion length (L) is the mean distance a minority carrier diffuses before recombination, $L = \sqrt{D\tau}$, where D is the diffusion coefficient and τ is the minority carrier lifetime.

Minority carrier lifetime (τ) is the mean time for a minority carrier to disappear by recombination. $1/\tau$ is the mean probability per unit time that a minority carrier recombines with a majority carrier.

Minority carriers are electrons in a p-type and holes in an n-type semiconductor.

Nondegenerate semiconductor has electrons in the CB and holes in the VB that obey Boltzmann statistics. Put differently, the electron concentration, n, in the CB is much less than the effective density of states N_c and similarly $p \ll N_v$. It refers to a semiconductor that has not been heavily doped so that these conditions are maintained; typically, doping concentrations less than 10^{18} cm^{-3}.

Peltier effect is the phenomenon of heat absorption or liberation at the contact between two dissimilar materials as a result of a dc current passing through the junction. The rate of heat generation, Q', is proportional to the dc current, I, passing through the contact so that $Q' = +\Pi I$, where Π is called the Peltier coefficient and the sign depends on whether heat is absorbed or released.

Phonon is a quantum of energy associated with the virbrations of the atoms in the crystal, analogous to the photon. A phonon has an energy $\hbar\Omega$ where Ω is the frequency of the lattice vibration.

Photoconductivity is the change in the conductivity from dark to light, $\sigma_{\text{light}} - \sigma_{\text{dark}}$.

Photogeneration is the excitation of an electron into the CB by the absorption of a photon. If the photon is absorbed by an electron in the VB, then its excitation to the CB will generate an EHP.

Photoinjection is the photogeneration of carriers in the semiconductor by illumination. Photogeneration may be VB to CB excitation, in which case electrons and holes are generated in pairs.

Recombination of an electron–hole pair involves an electron in the CB falling down in energy into an empty state (hole) in the VB to occupy it. The result is the annihilation of an EHP. Recombination is direct when the electron falls directly down into an empty state in the VB as in GaAs. Recombination is indirect if the electron is first captured locally by a defect or an impurity, called a recombination center, and from there it falls down into an empty state (hole) in the VB as in Si and Ge.

Thermal equilibrium carrier concentrations are those electron and hole concentrations that are solely determined by the statistics of the carriers and the density of states in the band. Thermal equilibrium concentration obeys the mass action law, $np = n_i^2$.

Thermal velocity (v_{th}) of an electron in the CB is its mean (or effective) speed in the semiconductor as it moves around in the crystal. For a nondegenerate semiconductor, it can be obtained simply from $(1/2) m_e^* v_{th}^2 = (3/2) kT$.

Vacuum level is the energy level where the *PE* of the electron and the *KE* of the electron are both zero. It defines the energy level where the electron is just free from the solid.

Valence band (VB) is a band of energies for the electrons in bonds in a semiconductor. The valence band is made of all those states (wavefunctions) that constitute the bonding between the atoms in the crystal. At absolute zero of temperature, the VB is full of all the bonding electrons of the atoms. When an electron is excited to the CB, this leaves behind an empty state, which is called a hole. It carries positive charge and behaves as if it were a "free" positively charged entity with an effective mass of m_h^*. It moves around the VB by having a neighboring electron tunnel into the unoccupied state.

Work function (Φ) is the energy required to remove an electron from the solid to the vacuum level.

QUESTIONS AND PROBLEMS

5.1   Band gap and photodetection

a. Determine the maximum value of the energy gap that a semiconductor, used as a photoconductor, can have if it is to be sensitive to yellow light (600 nm).

b. A photodetector whose area is 5×10^{-2} cm^{-2} is irradiated with yellow light whose intensity is 2 mW cm^{-2}. Assuming that each photon generates one electron–hole pair, calculate the number of pairs generated per second.

c. From the known energy gap of the semiconductor GaAs ($E_g = 1.42$ eV), calculate the primary wavelength of photons emitted from this crystal as a result of electron–hole recombination.

d. Is the above wavelength visible?

e. Will a silicon photodetector be sensitive to the radiation from a GaAs laser? Why?

5.2   Minimum conductivity

a. Consider the conductivity of a semiconductor, $\sigma = en\mu_e + ep\mu_h$. Will doping always increase the conductivity?

b. Show that the minimum conductivity for Si is obtained when it is *p*-type doped such that the hole concentration is

$$p_m = n_i \sqrt{\frac{\mu_e}{\mu_h}}$$

and the corresponding minimum conductivity (maximum resistivity) is

$$\sigma_{min} = 2en_i \sqrt{\mu_e \mu_h}$$

c. Calculate p_m and σ_{min} for Si and compare with intrinsic values.

5.3 Compensation doping in Si

a. A Si wafer has been doped *n*-type with 10^{17} As atoms cm^{-3}.

1. Calculate the conductivity of the sample at 27 °C.
2. Where is the Fermi level in this sample at 27 °C with respect to the Fermi level (E_{Fi}) in intrinsic Si?
3. Calculate the conductivity of the sample at 127 °C.

b. The above *n*-type Si sample is further doped with 9×10^{16} boron atoms (*p*-type dopant) per centimeter cubed.

1. Calculate the conductivity of the sample at 27 °C.
2. Where is the Fermi level in this sample with respect to the Fermi level in the sample in (*a*) at 27 °C? Is this an *n*-type or *p*-type Si?

5.4 Temperature dependence of conductivity An *n*-type Si sample has been doped with 10^{15} phosphorus atoms cm^{-3}. The donor energy level for P in Si is 0.045 eV below the conduction band edge energy.

a. Calculate the room temperature conductivity of the sample.

b. Estimate the temperature above which the sample behaves as if intrinsic.

c. Estimate to within 20% the lowest temperature above which all the donors are ionized.

d. Sketch schematically the dependence of the electron concentration in the conduction band on the temperature as $\log(n)$ versus $1/T$, and mark the various important regions and critical temperatures. For each region draw an energy band diagram that clearly shows from where the electrons are excited into the conduction band.

e. Sketch schematically the dependence of the conductivity on the temperature as $\log(\sigma)$ versus $1/T$ and mark the various critical temperatures and other relevant information.

5.5 GaAs Ga has a valency of III and As has V. When Ga and As atoms are brought together to form the GaAs crystal, as depicted in Figure 5.52, the 3 valence electrons in each Ga and the 5 valence electrons in each As are

all shared to form four covalent bonds per atom. In the GaAs crystal with some 10^{23} or so equal numbers of Ga and As atoms, we have an average of four valence electrons per atom, whether Ga or As, so we would expect the bonding to be similar to that in the Si crystal: four bonds per atom. The crystal structure, however, is not that of diamond but rather that of zinc blende (Chapter 1).

a.   What is the average number of valence electrons per atom for a pair of Ga and As atoms and in the GaAs crystal?

b.   What will happen if Se or Te, from group VI, are substituted for an As atom in the GaAs crystal?

c.   What will happen if Zn or Cd, from Group II, are substituted for a Ga atom in the GaAs crystal?

d.   What will happen if Si, from Group IV, is substituted for an As atom in the GaAs crystal?

e.   What will happen if Si, from Group IV, is substituted for a Ga atom in the GaAs crystal? What do you think **amphoteric dopant** means?

f.   Based on the above discussion, what do you think the crystal structures of the III–V compound semiconductors AlAs, GaP, InAs, InP, and InSb will be?

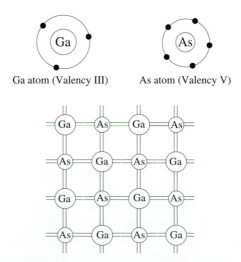

Ga atom (Valency III)   As atom (Valency V)

Figure 5.52   The GaAs crystal structure in two dimensions.

Average number of valence electrons per atom is four. Each Ga atom covalently bonds with four neighboring As atoms and vice versa.

5.6 **Doped GaAs** Consider the GaAs crystal at 300 K.

   a. Calculate the intrinsic conductivity and resistivity?

   b. In a sample containing only 10^{15} cm^{-3} ionized donors, where is the Fermi level? What is the conductivity of the sample?

   c. In a sample containing 10^{15} cm^{-3} ionized donors and 9×10^{14} cm^{-3} ionized acceptors, what is the free hole concentration?

5.7 **Degenerate semiconductor** Consider the general exponential expression for the concentration of electrons in the CB,

$$n = N_c \exp\left[-\frac{(E_c - E_F)}{kT}\right]$$

and the mass action law, $np = n_i^2$. What happens when the doping level is such that n approaches N_c and exceeds it? Can you still use the above expressions for n and p?

Consider an n-type Si that has been heavily doped and the electron concentration in the CB is 10^{20} cm^{-3}. Where is the Fermi level? Can you use $np = n_i^2$ to find the hole concentration? What is its resistivity? How does this compare with a typical metal? What use is such a semiconductor?

5.8 **Photoconductivity and speed** Consider two p-type Si samples both doped with 10^{15} B atoms cm^{-3}. Both have identical dimensions of length L (1 mm), width W (1 mm), and depth (thickness) D (0.1 mm). One sample, labeled A, has an electron lifetime of 1 μs whereas the other, labeled B, has an electron lifetime of 5 μs.

   a. At time $t = 0$, a laser light of wavelength 750 nm is switched on to illuminate the surface ($L \times W$) of both the samples. The incident laser light intensity on both samples is 10 mW cm^{-2}. At time $t = 50$ μs, the laser is switched off. Sketch the time evolution of the minority carrier concentration for both samples on the same axes.

   b. What is the photocurrent (current due to illumination alone) if each sample is connected to a 1 V battery.

*5.9 **Photoconductive gain** Consider the photoconductor shown in Figure 5.28 with ohmic contacts. The photoconductivity is given by Example 5.11,

$$\Delta\sigma = eG_{ph}\tau(\mu_e + \mu_h) \qquad \text{[5.64]}$$

where

$$G_{ph} = \frac{\eta I \lambda}{hcD}$$

is the charge carrier photogeneration rate (per second per unit volume) and τ is the minority carrier lifetime (hole lifetime in an n-type semiconductor).

If the contacts are ohmic, not limiting the current, then the photocurrent density is simply

$$J_{photo} = \Delta\sigma\frac{V}{L} = \Delta\sigma\mathcal{E} \qquad [5.65]$$

The drift velocity of the electrons in the photoconductor is $\mu_e\mathcal{E}$, so their transit time (time to cross the semiconductor) is

$$t_e = L/(\mu_e\mathcal{E}) \qquad [5.66]$$

a. Using Equation 5.64 to 5.66 show that the photocurrent is

$$I_{ph} = eG_{ph}\left(\frac{\tau}{t_e}\right)\left(1 + \frac{\mu_h}{\mu_e}\right)AL \qquad [5.67]$$

where $A = W \times L$ is the contact area.

b. The rate at which charge is being collected at the contacts is defined by the photocurrent current I_{ph}, whereas the rate of charge generation is $eG_{ph}(AL)$. The ratio is called the photoconductive gain. Show that

$$G_{ph} = \frac{\text{Rate of charge collection}}{\text{Rate of charge generation}} = \frac{\tau}{t_e}\left(1 + \frac{\mu_h}{\mu_e}\right) \qquad [5.68]$$

c. An n-type Si photoconductor has a length $L = 100$ μm and a cross-sectional area $A = 10^{-4}$ mm^2. For n-Si containing some 10^{17} donors cm^{-3}, a typical hole lifetime is 1 μs. If the applied bias to the photoconductor is 10 V, what is the photoconductive gain?

d. What are the transit times, t_e and t_h, of an electron and a hole across L?

e. It should be apparent that as electrons are much faster than holes, a photogenerated electron leaves the photoconductor very quickly. This leaves behind a drifting hole and therefore a positive charge in the semiconductor. Secondary (i.e., additional) electrons then flow into the photoconductor to maintain neutrality in the sample and the current continues to flow. These events will continue until the hole has disappeared by recombination, which takes on average a time τ. Thus, more charges flow through the contact per unit time than charges actually photogenerated per unit time. What will happen if the contacts are not ohmic, that is, they are Schottky junctions?

f. What can you say about the product $\Delta\sigma$ and the speed of response, which is proportional to $1/\tau$?

*5.10 **Compound semiconductor devices** Silicon and germanium crystalline semiconductors are what are called elemental group IV semiconductors. It is possible to have compound semiconductors from atoms in groups III and V. For example, GaAs is a compound semiconductor that has Ga from group III and As from group V, so that in the crystalline structure we have an "effective" or "mean" valency of IV per atom and the solid behaves like a semiconductor. Similarly GaSb (gallium antimonide) would be a III–V type semiconductor. Provided we have a stoichiometric compound, the semiconductor will be ideally intrinsic. If, however, there is an excess of Sb atoms in the solid GaSb, then we will have nonstoichiometry and the semi-conductor will be extrinsic. In this case, excess Sb atoms will act as donors in the GaSb structure. There are many useful compound semiconductors, the most important of which is GaAs. Some can be doped both n- and p-type, but many are one type only. For example, ZnO is a II–VI compound semiconductor with a direct bandgap of 3.2 eV, but unfortunately, due to the presence of excess Zn, it is naturally n-type and cannot be doped to p-type.

a. GaSb (gallium antimonide) is an interesting direct bandgap semiconductor with an energy bandgap, $E_g = 0.67$ eV, almost equal to that of germanium. It can be used as an LED (light-emitting diode) or laser diode material. What would be the wavelength of emission from a GaSb LED? Will this be visible?

b. Calculate the intrinsic conductivity of GaSb at 300 K taking $N_c = 2.3 \times 10^{19}$ cm^{-3}, $N_v = 6.1 \times 10^{19}$ cm^{-3}, $\mu_e = 5000$ cm^2 V^{-1} s^{-1}, and $\mu_h = 1000$ cm^2 V^{-1} s^{-1}. Compare with the intrinsic conductivity of Ge.

c. Excess Sb atoms will make gallium antimonide nonstoichiometric, that is, GaSb$_{1+\delta}$, which will result in an extrinsic semiconductor. Given that the density of GaSb is 5.4 g cm^{-3}, calculate δ (excess Sb) that will result in GaSb having a conductivity of 100 Ω^{-1} cm^{-1}. Will this be an n- or p-type semiconductor? You may assume that the drift mobilities are relatively unaffected by the doping.

5.11   Absorption coefficient Absorption coefficient, α, is proprotional to the occupied density of states, $g_{vb}(E)$ in the VB, *and* empty density of stages, $g_{cb}(E + h\nu)$ in the CB, at $h\nu$ above E. There are additional rules that involve momentum conservation that we will not consider in this intuitive analysis. Referring to the schematic density of states, $g(E)$ versus E, diagram of crystalline and amorphous Si in Figure 5.53, predict in a general way the qualitative behavior of the absorption coefficient, α, with photon energy.

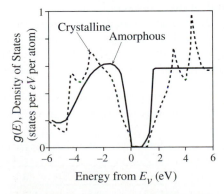

Figure 5.53   Density of states versus energy, $g(E)$ versus E, for crystalline and amorphous silicon

5.12   Excess minority carrier concentration Consider an n-type semiconductor and weak injection conditions. Assume that the minority carrier recombination time, τ_h, is constant (independent of injection—hence the weak injection assumption). The rate of change of the instantaneous hole concentration, $\partial p_n / \partial t$, due to recombination is given by

$$\frac{\partial p_n}{\partial t} = -\frac{p_n}{\tau_h} \qquad [5.69]$$

The net rate of increase (change) in p_n is the sum of the total generation rate G and the rate of change due to recombination, that is,

$$\frac{dp_n}{dt} = G - \frac{p_n}{\tau_h} \qquad [5.70]$$

By separating the generation term G into thermal generation G_o and photogeneration G_{ph} and considering the dark condition as one possible solution, show that

$$\frac{d\Delta p_n}{dt} = G_{ph} - \frac{\Delta p_{no}}{\tau_h} \qquad [5.71]$$

How does your derivation compare with Equation 5.25?

5.13   Schottky junction

a.   Consider a Schottky junction diode between Au and n-Si, doped with 10^{16} donors cm^{-3}. The cross-sectional area is 1 mm^2. Given the work function of Au as 5.1 eV, what is the theoretical barrier height, Φ_B, from the metal to the semiconductor?

b.   Given that the experimental barrier height Φ_B is about 0.8 eV, what is the reverse saturation current and the current when there is a forward bias of 0.3 V across the diode?

5.14   Schottky and ohmic contacts Consider an n-type Si sample doped with 10^{16} donors cm^{-3}. The length L is 100 μm; the cross-sectional area A is 10 μm $\times$ 10 μm. The two ends of the sample are labeled as B and C. The electron affinity (χ) of Si is 4.01 eV and the work functions, Φ, of four potential metals for contacts at B and C are listed in Table 5.3.

a.   Ideally, which metals will result in a Schottky contact?

b.   Ideally, which metals will result in an Ohmic contact?

Table 5.3   Work functions in eV

Cs	Li	Al	Au
1.8	2.5	4.25	5.1

c.   Sketch the I–V characteristics when both B and C are ohmic contacts. What is the relationship between I and V?

d.   Sketch the I–V characteristics when B is ohmic and C is a Schottky junction. What is the relationship between I and V?

e.   Sketch the I–V characteristics when both B and C are Schottky contacts. What is the relationship between I and V?

5.15   Peltier effect and electrical contacts Consider the Schottky junction and the ohmic contact shown in Figures 5.38 and 5.42 between a metal and n-type semiconductor.

a.   Is the Peltier effect similer in both contacts?

b.   Is the sign in $Q' = \pm \Pi I$ the same for both contacts?

c.   Which junction would you choose for a thermoelectric cooler? Give reasons.

*5.16   Peltier coolers and figure of merit (FOM) Consider the thermoelectric effect shown in Figure 5.44 in which a semiconductor has two contacts at its ends and is conducting an electric current I. We assume that the cold junction is at a temperature T_c and the hot junction is at T_h and that there is a temperature difference of $\Delta T = T_h - T_c$ between the two ends of the semiconductor. The current I flowing through the cold junction absorbs Peltier heat at a rate Q'_P, given by

$$Q'_P = \Pi I \qquad \text{[5.72]}$$

where Π is the Peltier coefficient for the junction between the metal and semiconductor, which we label as A. The current I flowing through the semiconductor generates heat due to the Joule heating of the semiconductor. The rate of Joule heat generated through the bulk of the semiconductor is

$$Q'_J = \left(\frac{L}{\sigma A}\right) I^2 \qquad \text{[5.73]}$$

We assume that half of this heat flows to the cold junction.

   In addition there is heat flow from the hot to the cold junction through the semiconductor, given by the thermal conduction equation

$$Q'_{TC} = \left(\frac{A\kappa}{L}\right)\Delta T \qquad \text{[5.74]}$$

   The net rate of heat absorption (cooling rate) at the cold junction is then

$$Q'_{\text{net cool}} = Q'_P - \frac{1}{2}Q'_J - Q'_{TC} \qquad \text{[5.75]}$$

By substituting from Equations 5.72 to 5.74 into Equation 5.75, obtain the net cooling rate in terms of the current I. Then by differentiating $Q'_{\text{net cool}}$

with respect to current, show that maximum cooling is obtained when the current is

$$I_m = \left(\frac{A}{L}\right)\Pi\sigma \qquad\qquad [5.76]$$

and the maximum cooling is

$$Q'_{max\ cool} = \frac{A}{L}\left[\frac{1}{2}\Pi^2\sigma - \kappa\Delta T\right] \qquad\qquad [5.77]$$

Under steady state operating conditions, the temperature difference, ΔT, reaches a steady-state value and the net cooling rate at the junction is then zero (ΔT is constant). From Equation 5.77 show that the maximum temperature difference achievable is

$$\Delta T_{max} = \frac{1}{2}\frac{\Pi^2\sigma}{\kappa} \qquad\qquad [5.78]$$

The quantity $\Pi^2\sigma/\kappa$ is defined as the figure of merit for the semiconductor as it determines the maximum ΔT achievable. The same expression also applies to metals, though we will not derive it here.

Use Table 5.4 to determine the FOM for various materials listed therein and discuss the significance of your calculations. Would you recommend a thermoelectric cooler based on a metal-to-metal junction?

*5.17 **Seebeck coefficient of semiconductors and thermal drift in semiconductor device** Consider an n-type semiconductor that has a temperature gradient across it. The right end is hot and the left end is cold, as depicted in Figure 5.54. There are more energetic electrons in the hot region than in the cold region. Consequently, electron diffusion occurs from hot to cold regions, which immediately exposes negatively charged donors in the hot region and therefore builds up an internal field and a built-in voltage, as shown in the figure. Eventually an equilibrium is reached when the diffusion of electrons is balanced by their drift driven by the built-in field. The net current must be zero. The Seebeck coefficient (or thermoelectric power) S measures this effect in terms of the voltage developed as a result of an applied temperature gradient as

$$S = \frac{dV}{dT} \qquad\qquad [5.79]$$

Table 5.4

Material	Π V	ρ Ωm	κ W m^{-1} K^{-1}	FOM
n-Bi$_2$Te$_3$	6.0×10^{-2}	10^{-5}	1.70	
p-Bi$_2$Te$_3$	7.0×10^{-2}	10^{-5}	1.45	
Cu	5.5×10^{-4}	1.7×10^{-8}	390	
W	3.3×10^{-4}	5.5×10^{-8}	167	

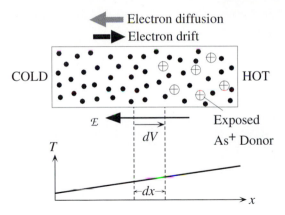

Figure 5.54 In the presence of a temperature gradient, there is an internal field and a voltage difference.

The Seebeck coefficient is defined as dV/dT, potential difference per unit temperature difference.

a. How is the Seebeck effect in a p-type semiconductor different than that for an n-type semiconductor when both are placed in the same temperature gradient in Figure 5.54?

b. Given that for an n-type semiconductor,

$$S_n = \frac{k}{e}\left[2 + \frac{(E_c - E_F)}{kT}\right] \qquad [5.80]$$

What are typical magnitudes for S_n in Si doped with 10^{14} and 10^{16} donors cm^{-3}? What is the significance of S_n at semiconductor device level?

c. Consider a pn junction Si device that has the p-side doped with 10^{18} acceptors cm^{-3} and the n-side doped 10^{14} donors cm^{-3}. Suppose that this pn junction forms the input stage of an op amp with a large gain, say 100. What will be the output signal if a small thermal fluctuation gives rise to a 1 °C temperature difference across the pn junction?

chapter
6

Semiconductor Devices

Most diodes are essentially *pn* junctions fabricated by forming a contact between a *p*-type and an *n*-type semiconductor. The junction possesses rectifying properties in that a current in one direction can flow quite easily whereas in the other direction it is limited by a leakage current that is generally very small. A transistor is a three-terminal solid-state device in which a current flowing between two electrodes is controlled by the voltage between the third and one of the other terminals. Transistors are capable of providing current and voltage gains, thereby enabling weak signals to be amplified. Transistors can also be used as switches just like electromagnetic relays. Indeed, the whole microcomputer industry is based on transistor switches. The majority of the transistors in microelectronics are essentially two types: **bipolar junction transistors** (BJTs) and **field effect transistors** (FETs). The appreciation of the underlying principles of the *pn* junction is essential to understanding the operation of the bipolar transistor. Field effect transistors operate on a totally different principle than bipolars. Their characteristics arise from the effect of the applied field on a conducting channel between two terminals. Integrated circuits (ICs) integrate literally millions of devices within a tiny piece of a single crystal of silicon called the **chip,** typically 1–10 mm in lateral size and 100–200 microns in thickness. The integrated devices and components in a chip are interconnected to implement not only sophisticated circuits but even subsystems and systems. The development of the Si chip revolutionized the electronics industry and led to a totally new philosophy and art of design and to the implementation of circuits, subsystems, and systems that could not even be envisioned by the use of discrete components. What are the basic principles of integrated circuit fabrication? The reader will find that there is an intimate connection between the properties of a device and the fabrication process. There are limits to achievable device performances.

6.1   IDEAL *pn* JUNCTION

6.1.1   No Applied Bias: Open Circuit

Consider what happens when one side of a sample of Si is doped *n*-type and the other *p*-type, as shown in Figure 6.1a. We assume that there is an abrupt discontinuity between the *p* and *n* regions, which we call the **metallurgical junction** and label as M in Figure 6.1a, where the fixed (immobile) ionized donors and the free electrons (in the conduction band, CB) in the *n*-region and fixed ionized acceptors and holes (in the valence band, VB) in the *p*-region are also shown.

Due to the hole concentration gradient from the *p*-side, where $p = p_{po}$, to the *n*-side, where $p = p_{no}$, holes diffuse toward the right. Similarly the electron con-

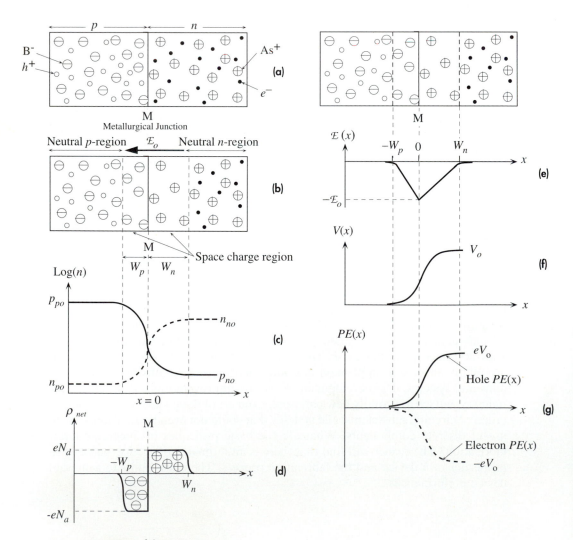

Figure 6.1   Properties of the *pn* junction.

centration gradient drives the electrons by diffusion toward the left. Holes and electrons diffusing toward each other meet and recombine around the junction region. The junction region consequently becomes depleted of free carriers in comparison with the bulk *p*- and *n*-regions far away from the junction. Note that we must, under equilibrium conditions (e.g., no applied bias or photoexcitation), have $pn = n_i^2$ everywhere. Electrons leaving the *n*-side near the junction M leave behind exposed positively charged donor ions, say As^+, of concentration N_d. Similarly, holes leaving the *p*-region near M expose negatively charged acceptor ions, say B^-, of concentration N_a. There is therefore a **space charge layer** (SCL) around M. Figure 6.1b shows the depletion region, or the space charge layer, around M, whereas Figure 6.1c illustrates the hole and electron concentration profiles in which the vertical concentration scale is logarithmic. The depletion region is also called the transition region.

It is clear that there is an internal electric field $\mathcal{E}_o$ from positive ions to negative ions, that is, $-x$ direction, that tries to drift the holes back into the *p*-region and electrons back into the *n*-region. This field drives the holes in the opposite direction to their diffusion. As shown in Figure 6.1b, $\mathcal{E}_o$ imposes a drift force on holes in the $-x$ direction, whereas the hole diffusion flux is in the $+x$ direction. A similar situation also applies for electrons with the electric field attempting to drift the electrons against diffusion from *n* to the *p*-region. It is apparent that as more and more holes diffuse toward the right, and electrons toward the left, the internal field around M will increase until eventually an "equilibrium" is reached when the rate of holes diffusing toward the right is just balanced by holes drifting back to the left, driven by the field $\mathcal{E}_o$. The electron diffusion and drift fluxes will also be balanced in equilibrium.

For uniformly doped *p*- and *n*-regions, the net space charge density, $\rho_{net}(x)$, across the semiconductor will be as shown in Figure 6.1d. (Why are the edges rounded?) The net space charge density, ρ_{net}, is negative and equal to $-eN_a$ in the SCL from $x = -W_p$ to $x = 0$ (where we take M to be) and then positive and equal to $+eN_d$ from $x = 0$ to W_n. The total charge on the left-hand side must be equal to that on the right-hand side for overall charge neutrality, so that

$$N_a W_p = N_d W_n \qquad \text{[6.1]}$$

In Figure 6.1, we arbitrarily assumed that the donor concentration is less than the acceptor concentration, $N_d < N_a$. From Equation 6.1 this implies that $W_n > W_p$, that is, the depletion region penetrates the *n*-side, the lightly doped side, more than the *p*-side, the heavily doped side. Indeed, if $N_a \gg N_d$, then the the depletion region is almost entirely on the *n*-side. We generally indicate heavily doped regions with the plus sign as a superscript, that is, p^+.

The electric field, $\mathcal{E}(x)$, and the net space charge density, $\rho_{net}(x)$, at a point are related in electrostatics[1] by

$$\frac{d\mathcal{E}}{dx} = \frac{\rho_{net}(x)}{\varepsilon}$$

[1] This is called **Gauss's law in point** form and comes from Gauss law in electrostatics. The integration of the electric field *E* over a closed surface, *S*, is related to the total charge enclosed

$$\int E \, dS = \frac{Q_{enclosed}}{\varepsilon}$$

where $E = qV$, where q is the charge of the carrier. Considering electrons $(q = -e)$, we see from Figure 6.1g that $E = 0$ on the *p*-side far away from M where $n = n_{po}$, and $E = -eV_o$ on the *n*-side away from M where $n = n_{no}$. Thus

$$\frac{n_{po}}{n_{no}} = \exp\left(-\frac{eV_o}{kT}\right) \qquad \text{[6.5a]}$$

This shows that V_o depends on n_{no} and n_{po} and hence on N_d and N_a. The corresponding equation for hole concentrations is clearly

$$\frac{p_{no}}{p_{po}} = \exp\left(-\frac{eV_o}{kT}\right) \qquad \text{[6.5b]}$$

Thus, rearranging Equations 6.5a and b we obtain

$$V_o = \frac{kT}{e} \ln\left(\frac{n_{no}}{n_{po}}\right) \qquad \text{and} \qquad V_o = \frac{kT}{e} \ln\left(\frac{p_{po}}{p_{no}}\right)$$

We can now write p_{po} and p_{no} in terms of the dopant concentrations inasmuch as $p_{po} = N_a$ and,

$$p_{no} = \frac{n_i^2}{n_{no}} = \frac{n_i^2}{N_d}$$

so that V_o becomes

$$V_o = \frac{kT}{e} \ln\left(\frac{N_a N_d}{n_i^2}\right) \qquad \text{[6.6]} \qquad \textit{Built-in voltage}$$

Clearly V_o has been conveniently related to the dopant and material properties via N_a, N_d, and n_i^2. The built-in voltage (V_o) is the voltage across a *pn* junction, going from *p*- to *n*-type semiconductor, in an open circuit. It is *not* the voltage across the diode, which is made up of V_o as well as the contact potentials at the metal-to-semiconductor junctions at the electrodes. If we add V_o and the contact potentials at the electroded ends, we will find zero.

Once we know the built-in potential from Equation 6.6, we can then calculate the width of the depletion region from Equation 6.4, namely

$$W_o = \left[\frac{2\varepsilon(N_a + N_d)V_o}{eN_a N_d}\right]^{1/2} \qquad \text{[6.7]} \qquad \textit{Depletion width}$$

Notice that the depletion width $W_o \propto V_o^{1/2}$, which results in the capacitance of the depletion region being voltage dependent, as we will see below.

THE BUILT-IN POTENTIALS FOR *Ge*, *Si*, AND *GaAs* *pn* JUNCTIONS   A *pn* junction diode has a concentration of 10^{17} acceptor atoms cm^{-3} on the *p*-side and a concentration of 10^{16} donor atoms cm^{-3} on the *n*-side. What will be the built-in potential for the semiconductor materials Ge, Si, and GaAs?   **Example 6.1**

Solution

The built-in potential is given by Equation 6.6, which requires the knowledge of the intrinsic concentration for each semiconductor. From Chapter 5 we can tabulate the following at 300 K:

Semiconductor	$E_g(eV)$	$n_i(cm^{-3})$	$V_o(V)$
Ge	0.7	2.40×10^{13}	0.37
Si	1.1	1.45×10^{10}	0.76
GaAs	1.4	1.79×10^6	1.22

Using

$$V_o = \left(\frac{kT}{e}\right) \ln\left(\frac{N_d N_a}{n_i^2}\right)$$

for Si with $N_d = 10^{17}$ cm^{-3} and $N_a = 10^{16}$ cm^{-3}, $kT/e = 0.0259$ at 300 K, and n_i from above, we obtain

$$V_o = (0.0259 \text{ V}) \ln\left[\frac{(10^{17})(10^{16})}{(1.45 \times 10^{10})^2}\right] = 0.756 \text{ V}$$

The results for all the three semiconductors are summarized in the last column of the above table.

Example 6.2 | **THE p^+n JUNCTION**   Consider a p^+n junction, which has a heavily doped p-side relative to the n-side, that is, $N_a \gg N_d$. Since the amount charge, Q, on both sides of the metallurgical junction must be the same (so that the junction is overall neutral)

$$Q = eN_a W_p = eN_d W_n$$

it is clear that the depletion region essentially extends into the n-side. According to Equation 6.7, when $N_d \ll N_a$, the width is

$$W_o = \left[\frac{2\varepsilon V_o}{eN_d}\right]^{1/2}$$

What is the depletion width for a pn junction Si diode that has been doped with 10^{18} acceptor atoms cm^{-3} on the p-side and 10^{16} donor atoms cm^{-3} on the n-side?

Solution

To apply the above equation, we need the built-in potential, which is

$$V_o = \left(\frac{kT}{e}\right) \ln\left(\frac{N_d N_a}{n_i^2}\right)$$

$$V_o = (0.0259 \text{ V}) \ln\left[\frac{(10^{18})(10^{16})}{(1.45 \times 10^{10})^2}\right] = 0.816 \text{ V}$$

Then with $N_d = 10^{16}$ cm^{-3}, that is, 10^{22} m^{-3}, $V_o = 0.816$ V, and $\varepsilon_r = 11.9$ in the equation for W_o

$$W_o = \left[\frac{2\varepsilon V_o}{eN_d} \right]^{1/2} = \left[\frac{2(11.9)(8.85 \times 10^{-12})(0.816)}{(1.6 \times 10^{-19})(10^{22})} \right]^{1/2}$$

$$= 3.3 \times 10^{-7} \text{ m} \quad \text{or} \quad 0.33 \ \mu\text{m}.$$

Nearly all of this region (99% of it) is on the *n*-side.

6.1.2   Forward Bias

Consider what happens when a battery is connected across a *pn* junction so that the positive terminal of the battery is attached to the *p*-side and the negative terminal to the *n*-side. Suppose that the applied voltage is *V*. It is apparent that the negative polarity of the supply will reduce the potential barrier V_o by *V*, as shown in Figure 6.2a. The reason for this is that the bulk regions outside the depletion width have high conductivities due to plenty of majority carriers in the bulk, in comparison with the depletion region in which there are mainly immobile ions. Thus, the applied voltage drops mostly across the depletion width, *W*. Consequently, *V* directly opposes V_o and the potential barrier against diffusion is reduced to $(V_o - V)$, as depicted in Figure 6.2b. This has drastic consequences because the probability that a hole will surmount this potential barrier and diffuse to the right now becomes proportional to $\exp[-e(V_o - V)/kT]$. In other words, the applied voltage effectively reduces the built-in potential and hence the built-in field, which acts against

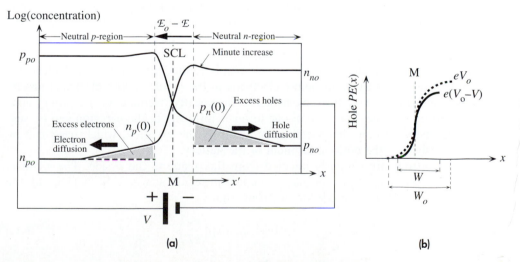

Figure 6.2   Forward-biased *pn* junction and the injection of minority carriers.
(a) Carrier concentration profiles across the device under forward bias.
(b) The hole potential energy with and without an applied bias. *W* is the width of the SCL with forward bias.

diffusion. Consequently many holes can now diffuse across the depletion region and enter the n-side. This results in the **injection of excess minority carriers,** holes, into the n-region. Similarly, excess electrons can now diffuse toward the p-side and enter this region and thereby become injected minority carriers.

The hole concentration

$$p_n(0) = p_n(x' = 0)$$

just outside the depletion region at $x' = 0$ (x' is measured from W_n) is due to the excess of holes diffusing as a result of the reduction in the built-in potential barrier. This concentration, $p_n(0)$, is determined by the probability of surmounting the new potential energy barrier $e(V_o - V)$,

$$p_n(0) = p_{po} \exp\left[\frac{-e(V_o - V)}{kT}\right] \tag{6.8}$$

This follows directly from the Boltzmann equation, by virtue of the hole potential energy rising by $e(V_o - V)$ from $x = -W_p$ to $x = W_n$, as indicated in Figure 6.2b, and at the same time the hole concentration falling from p_{po} to $p_n(0)$. By dividing Equation 6.8 by Equation 6.5b, we obtain the effect of the applied voltage directly, which shows how the voltage V determines the amount of excess holes diffusing and arriving at the n-region. Equation 6.8 divided by Equation 6.5b is

Law of the junction

$$p_n(0) = p_{no} \exp\left(\frac{eV}{kT}\right) \tag{6.9}$$

which is called the **law of the junction.** Equation 6.9 is an important equation that we will use again in dealing with pn junction devices. It describes the effect of the applied voltage V on the injected minority carrier concentration just outside the depletion region, $p_n(0)$. Obviously, with no applied voltage, $V = 0$ and $p_n(0) = p_{no}$, which is exactly what we expect.

Injected holes diffuse in the n-region and eventually recombine with electrons in this region as there are many electrons in the n-side. Those electrons lost by recombination are readily replenished by the negative terminal of the battery connected to this side. The current due to holes diffusing in the n-region can be sustained because more holes can be supplied by the p-region, which itself can be replenished by the positive terminal of the battery.

Electrons are similarly injected from the n-side to the p-side. The electron concentration $n_p(0)$ just outside the depletion region at $x = -W_p$ is given by the equivalent of Equation 6.9 for electrons, that is,

Law of the junction

$$n_p(0) = n_{po} \exp\left(\frac{eV}{kT}\right) \tag{6.10}$$

In the p-region, the injected electrons diffuse toward the positive terminal looking to be collected. As they diffuse they recombine with some of the many holes in this region. Those holes lost by recombination can be readily replenished by the positive terminal of the battery connected to this side. The current due to the diffusion of electrons in the *p*-side can be maintained by the supply of electrons from the *n*-side, which itself can be replenished by the negative terminal of the battery. It is apparent that an electric current can be maintained through a *pn* junction under forward bias, and that the current flow, surprisingly, seems to be due to the **diffusion of minority carriers.** There is, however, some drift of majority carriers as well.

If the lengths of the *p*- and *n*-regions are longer than the minority carrier diffusion lengths, then we will be justified to expect the hole concentration $p_n(x')$ on the *n*-side to fall exponentially toward the thermal equilibrium value, p_{no}, that is,

$$\Delta p_n(x') = \Delta p_n(0) \exp\left(-\frac{x'}{L_h}\right) \qquad \text{[6.11]}$$

where

$$\Delta p_n(x') = p_n(x') - p_{no}$$

is the excess carrier distribution and L_h is the **hole diffusion length,** defined by $L_h = \sqrt{D_h \tau_h}$ in which τ_h is the mean hole recombination lifetime (minority carrier lifetime) in the *n*-region. We base Equation 6.11 on our experience with minority carrier injection in Chapter 5.[3]

The hole diffusion current density $J_{D,\text{hole}}$ is therefore

$$J_{D,\text{hole}} = -eD_h \frac{dp_n(x')}{dx'} = -eD_h \frac{d\Delta p_n(x')}{dx'}$$

that is,

$$J_{D,\text{hole}} = \left(\frac{eD_h}{L_h}\right) \Delta p_n(0) \exp\left(-\frac{x'}{L_h}\right)$$

Although the above equation shows that the hole diffusion current depends on location, the total current at any location, however, is the sum of hole and electron contributions, which is independent of *x*, as indicated in Figure 6.3. The decrease in the minority carrier diffusion current with *x'* is made up by the increase in the current due to the drift of the majority carriers, as schematically shown in Figure 6.3. The field in the neutral region is not totally zero but a small value, just sufficient to drift the huge number of majority carriers there.

At $x' = 0$, just outside the depletion region, the hole diffusion current is

$$J_{D,\text{hole}} = \left(\frac{eD_h}{L_h}\right) \Delta p_n(0)$$

We can now use the law of the junction to substitute for $\Delta p_n(0)$ in terms of the applied voltage *V*. Writing

$$\Delta p_n(0) = p_n(0) - p_{no} = p_{no}\left[\exp\left(\frac{eV}{kT}\right) - 1\right]$$

---

[3] This is simply the solution of the continuity equation in the absence of an electric field, which is discussed in Chapter 5.

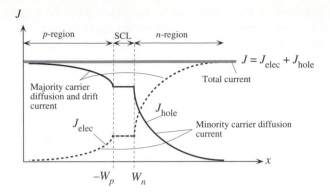

Figure 6.3   The total current anywhere in the device is constant. Just outside the depletion region, it is due to the diffusion of minority carriers.

and substituting in $J_{D,\,\text{hole}}$, we get

$$J_{D,\,\text{hole}} = \left(\frac{eD_h p_{no}}{L_h}\right)\left[\exp\left(\frac{eV}{kT}\right) - 1\right]$$

Thermal equilibrium hole concentration p_{no} is related to the donor concentration by

$$p_{no} = \frac{n_i^2}{n_{no}} = \frac{n_i^2}{N_d}$$

Thus,

$$J_{D,\,\text{hole}} = \left(\frac{eD_h n_i^2}{L_h N_d}\right)\left[\exp\left(\frac{eV}{kT}\right) - 1\right]$$

There is a similar expression for the electron diffusion current density $J_{D,\,\text{elec}}$ in the p-region. We will assume (quite reasonably) that the electron and hole currents do not change across the depletion region because, in general, the width of this region is narrow (reality is not quite like the schematic sketches in Figures 6.2 and 6.3). The electron current at $x = -W_p$ is the same as that at $x = W_n$. The total current density is then simply given by $J_{D,\,\text{hole}} + J_{D,\,\text{elec}}$, that is,

$$J = \left(\frac{eD_h}{L_h N_d} + \frac{eD_e}{L_e N_a}\right)n_i^2\left[\exp\left(\frac{eV}{kT}\right) - 1\right]$$

or

Ideal diode equation

$$J = J_{so}\left[\exp\left(\frac{eV}{kT}\right) - 1\right] \qquad [6.12]$$

This is the familiar diode equation with

$$J_{so} = \left[\left(\frac{eD_h}{L_h N_d}\right) + \left(\frac{eD_e}{L_e N_a}\right)\right]n_i^2$$

It is frequently called the **Shockley equation.** The constant J_{so} depends not only on the doping, N_d and N_a, but also on the material via n_i, D_h, D_e, L_h, and L_e. It is known as the **reverse saturation current density,** as explained below. Writing

$$n_i^2 = (N_c N_v) \exp\left(-\frac{eV_g}{kT}\right)$$

where $V_g = E_g/e$ is the band gap energy expressed in volts, we can write Equation 6.12 as

$$J = \left(\frac{eD_h}{L_h N_d} + \frac{eD_e}{L_e N_a}\right)\left[(N_c N_v)\exp\left(-\frac{eV_g}{kT}\right)\right]\left[\exp\left(\frac{eV}{kT}\right)-1\right]$$

that is,

$$J = J_1 \exp\left(-\frac{eV_g}{kT}\right)\left[\exp\left(\frac{eV}{kT}\right)-1\right]$$

or

$$J = J_1 \exp\left[\frac{e(V - V_g)}{kT}\right] \qquad \text{for} \qquad \frac{eV}{kT} \gg 1 \qquad \text{[6.13]}$$

where

$$J_1 = \left(\frac{eD_h}{L_h N_d} + \frac{eD_e}{L_e N_a}\right)(N_c N_v)$$

is a new constant.

The significance of Equation 6.13 is that it reflects the dependence of *I–V* characteristics on the band gap (via V_g), as displayed in Figure 6.4 for the three important semiconductors, Ge, Si, and GaAs. Notice that the voltage across the *pn* junction for an appreciable current of say ~0.1 mA is about 0.2 V for Ge, 0.6 V for Si, and 0.9 V for GaAs.

The diode equation, Equation 6.12, was derived by assuming that the lengths of the *p* and *n* regions outside the depletion region are long in comparison with the diffusion lengths L_h and L_e. Suppose that ℓ_p is the length of the *p*-side outside the

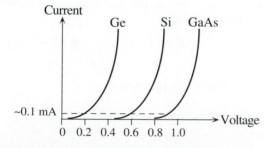

Figure 6.4   Schematic sketch of the *I–V* characteristics of Ge, Si, and GaAs *pn* junctions.

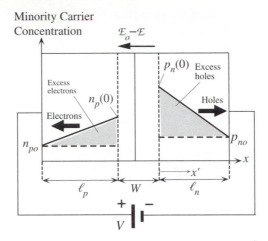

Figure 6.5   Minority carrier injection and diffusion in a short diode.

depletion region and ℓ_n is that of the n-side outside the depletion region. If ℓ_p and ℓ_n are shorter than the diffusion lengths L_e and L_h respectively, then we have what is called a **short diode** and consequently the minority carrier distribution profiles fall almost linearly with distance from the depletion region, as depicted in Figure 6.5. This can be readily proved by solving the continuity equation, but an intuitive explanation makes it clear. At $x' = 0$, the minority carrier concentration is determined by the law of the junction, whereas at the battery terminal there can be no excess carriers as the battery will simply collect these. Since the length of the neutral region is shorter than the diffusion length, there are practically no holes lost by recombination, and therefore the hole flow is expected to be uniform across ℓ_n. This can be so only if the driving force for diffusion, the concentration gradient, is linear.

The excess minority carrier gradient is

$$\frac{d\,\Delta p_n(x')}{dx'} = -\frac{[p_n(0) - p_{no}]}{\ell_n}$$

The current density, $J_{D,\text{hole}}$, due to the injection and diffusion of holes in the n-region as a result of forward bias is

$$J_{D,\text{hole}} = -eD_h\frac{d\,\Delta p_n(x')}{dx'} = eD_h\frac{[p_n(0) - p_{no}]}{\ell_n}$$

We can now use the law of the junction

$$p_n(0) = p_{no}\exp\!\left(\frac{eV}{kT}\right)$$

for $p_n(0)$ in the above equation and also obtain a similar equation for electrons diffusing in the p-region and then sum the two for the total current J,

$$J = \left(\frac{eD_h}{\ell_h N_d} + \frac{eD_e}{\ell_e N_a}\right) n_i^2 \left[\exp\left(\frac{eV}{kT}\right) - 1\right] \qquad \text{[6.14]} \quad \textit{Short diode}$$

It is clear that this expression is identical to that of a long diode, that is, Equation 6.12, if in the latter we replace the diffusion lengths, L_h and L_e, by the lengths ℓ_n and ℓ_p of the *n*- and *p*-regions outside the SCL.

So far we have assumed that, under a forward bias, the external current supplies the minority carriers diffusing and recombining in the neutral regions. However, some of the minority carriers will recombine in the depletion region. The external current therefore must also supply the carriers lost in the recombination process in the space charge layer. The recombination rate is a maximum near the middle of the depletion layer where the electron and hole concentrations are comparable, as apparent in Figure 6.2a. The potential around the center of the SCL is $(1/2)(V_o - V)$, as shown in Figure 6.2b, so that the carrier concentrations around the center of the SCL depend on $\exp[-e(V_o - V)/2kT]$ or on $\exp(eV/2kT)$ rather than on $\exp(eV/kT)$. This is understandable as only half of the applied voltage appears across half of the depletion layer. The rate at which electrons are recombining is then directly proportional to electron concentration in the middle of the SCL, $\exp(eV/2kT)$, and inversely proportional to the mean recombination time. From a quantitative analysis, the expression for the recombination current can be generally written as[4]

$$J_{\text{recom}} \approx J_{ro} \left[\exp\left(\frac{eV}{2kT}\right) - 1\right] \qquad \text{[6.15]}$$

where J_{ro} is a constant approximately given by

$$J_{ro} = \frac{en_i W}{2\tau_r}$$

where W is the width of the depletion region and τ_r is the mean minority carrier recombination time in W.

Equation 6.15 is the current that supplies the carriers that recombine in the depletion region. The total current into the diode will supply carriers for minority carrier diffusion in the neutral regions and recombination in the space charge layer so that it will be the sum of Equations 6.14 and 6.15.

$$J = J_{so} \exp\left(\frac{eV}{kT}\right) + J_{ro} \exp\left(\frac{eV}{2kT}\right); \qquad \left(V > \frac{kT}{e}\right) \qquad \text{[6.16]}$$

The above expression is often lumped into a single exponential as

$$J = J_o \exp\left(\frac{eV}{\eta kT}\right); \qquad \left(V > \frac{kT}{e}\right) \qquad \text{[6.17]} \quad \textit{The diode equation}$$

where J_o is a new constant and η is an **ideality factor,** which is 1 when the current is due to minority carrier diffusion in the neutral regions and 2 when it is due to

---

| [4] This is generally proved in advanced texts, involving what is called Shockley–Read–Hall statistics.

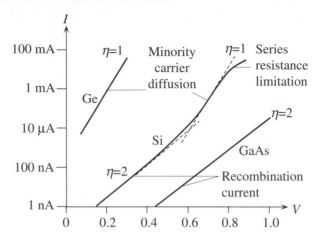

Figure 6.6   Schematic sketch of typical *I–V* characteristics of Ge, Si, and GaAs *pn* junctions as log(*I*) versus *V*.
The slope indicates $e/(\eta kT)$.

recombination in the space charge layer. Figure 6.6 shows typical expected *I–V* characteristics of *pn* junction Ge, Si, and GaAs diodes. At the highest currents, invariably, the bulk resistances of the neutral regions limit the current (why?). For Ge diodes, typically $\eta = 1$ and the overall *I–V* characteristics are due to minority carrier diffusion. In the case of GaAs, $\eta \approx 2$ and the current is limited by recombination in the space charge layer. For Si, typically, η changes from 2 to 1 as the current increases, indicating that both processes play an important role. In the case of heavily doped Si diodes, heavy doping leads to short minority carrier recombination times and the current is controlled by recombination in the space charge layer so that the $\eta = 2$ region extends all the way to the onset of bulk resistance limitation.

6.1.3   Reverse Bias

When a *pn* junction is reverse biased, as shown in Figure 6.7a, the applied voltage, as before, drops mainly across the depletion region, which becomes wider. The negative terminal will attract holes in the *p*-region to move away from the depletion region, which results in more exposed negative acceptor ions and thus a wider space charge layer. Similarly, the positive terminal will attract electrons away from the depletion region, which exposes more positively charged donors. The depletion width on the *n*-side also widens. The movement of electrons in the *n*-region toward the positive battery terminal cannot be sustained because there is no electron supply to this *n*-side. The *p*-side cannot supply electrons to the *n*-side because it has almost none. Furthermore the electrons from the battery cannot enter the *p*-side and transit across to the *n*-side because, before they get a chance, they will be recombined with holes. It is quite apparent that there can be no sustainable current in a reverse-biased *pn* junction.

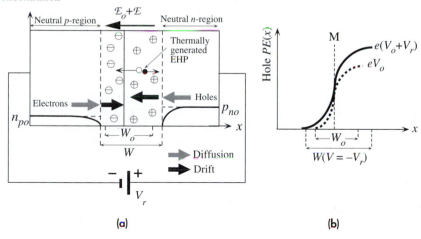

Figure 6.7   Reverse-biased *pn* junction

(a) Minority carrier profiles and the origin of the reverse current.

(b) Hole *PE* across the junction under reverse bias.

The applied voltage increases the built-in potential barrier, as depicted in Figure 6.7b, which impedes the diffusion of holes to the *n*-side and electrons to the *p*-side. The electric field in the depletion region is larger than the built-in internal field $\mathcal{E}_o$. The holes on the *n*-side near the depletion region get extracted by the field and are driven across the depletion width over to the *p*-side. The same number of holes can now exit the *p*-region to the battery, giving a small current. Similarly, the small concentration of electrons near the depletion region on the *p*-side get driven across the depletion width onto the *n*-side giving a small current. Examining Figure 6.7a, it is clear that this small reverse current is sustained by the diffusion of holes in the *n*-side from the bulk to the depletion layer, and similarly by electron diffusion in the *p*-side from the bulk to the depletion layer.

We can work out the reverse current density by evaluating the diffusion current densities, $J_{D,\text{hole}}$ and $J_{D,\text{elec}}$, as we did in the forward bias case. If we take a reasonably large reverse bias voltage, much greater than kT/e (25 mV at 20 °C), then the minority carrier concentrations just outside the depletion layer would be nearly zero, $p_n(0) = n_p(0) = 0$. This follows directly from the law of the junction with a reverse bias. The reverse current is then independent of the applied voltage and is called the **reverse saturation current.** Equation 6.12 with a negative voltage gives a diode current of $-AJ_{so}$, which is the reverse saturation current, hence the name given to J_{so}. The value of J_{so} depends only on the material via n_i, μ_h, μ_e, the dopant concentrations, and so on, but not on the voltage. Furthermore, as J_{so} depends on n_i^2, it is strongly a temperature dependent.

The thermal generation of electron–hole pairs (EHPs) in the space charge region can also contribute to the observed reverse current since the internal field in this layer will separate the electron and hole and drift them toward the neutral regions, as shown in Figure 6.7a. This drift will result in an external current in addition to the reverse current due to the diffusion of minority carriers. The theoretical evaluation of space charge layer generation current involves an in-depth knowledge of the charge carrier generation processes via recombination centers, which is discussed in advanced texts. Suppose that τ_g is the mean time to generate an electron–hole pair by virtue of the thermal vibrations of the lattice (τ_g is also called the thermal generation lifetime). We note that given τ_g, the rate of thermal generation per unit volume must be n_i/τ_g because it takes on average τ_g seconds to create n_i number of EHPs per unit volume. Furthermore, since WA(where $A = 1$) is the volume of the depletion region, the rate of EHP, or charge carrier, generation is Wn_i/τ_g. Therefore, the reverse current component due to thermal generation of electron–hole pairs within the depletion region should be given by

$$J_{gen} = \frac{eWn_i}{\tau_g}$$ [6.18]

The reverse bias widens the width W of the depletion layer and hence increases J_{gen}. The total reverse current density, J_{rev}, is the sum of the diffusion and generation components, that is,

$$J_{rev} = \left(\frac{eD_h}{\ell_h N_d} + \frac{eD_e}{\ell_e N_q} \right) n_i^2 + \frac{eWn_i}{\tau_g}$$ [6.19]

There is one more contribution that may be important in certain devices, that is, the surface leakage current due to charge carriers flowing on the surface, or along some interface, and bypassing the depletion layer. This component of the reverse current generally increases with the reverse bias. Figure 6.8 illustrates schematically the reverse I–V characteristics of an ideal pn junction and that of a common Si diode. The term **reverse leakage current** describes the total reverse current with all its possible components.

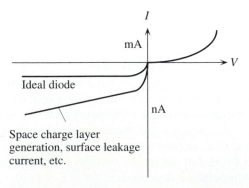

Figure 6.8   Reverse I–V characteristics of a pn junction (the positive and negative current axes have different scales)

FORWARD AND REVERSE-BIASED *Si* DIODE   An abrupt Si p^+n junction diode has a cross-sectional area of 1 mm^2, an acceptor concentration of 5×10^{18} boron atoms cm^{-3} on the *p*-side, and a donor concentration of 10^{16} arsenic atoms cm^{-3} on the *n*-side. The lifetime of holes in the *n*-region is 417 ns, whereas that of electrons in the *p*-region is 5 ns due to a greater concentration of impurities (recombination centers) on that side. Mean thermal generation lifetime (τ_g) is about 1 ms. The lengths of the *p*- and *n*-regions are 5 and 100 microns respectively. | **Example 6.3**

a.   Calculate the minority diffusion lengths and determine what type of a diode this is.

b.   What is the built-in potential across the junction?

c.   What is the current when there is a forward bias of 0.6 V across the diode at 27 °C? Assume that the current is by minority carrier diffusion.

d.   Estimate the forward current at 100 °C when the voltage across the diode remains at 0.6 V. Assume that the temperature dependence of n_i dominates over those of D, L, and μ.

e.   What is the reverse current when the diode is reverse biased by a voltage $V_r = 5$ V?

Solution

The general expression for the diffusion length is $L = \sqrt{D\tau}$ where D is the diffusion coefficient and τ is the carrier lifetime. D is related to the carrier mobility, μ, via the Einstein relationship, $D/\mu = kT/e$. We therefore need to know μ to calculate D and hence L. Electrons diffuse in the *p*-region and holes in the *n*-region, so we need μ_e in the presence of N_a acceptors and μ_h in the presence of N_d donors. From the drift mobility, μ versus dopant concentration in Figure 5.19, we have the following:

With $N_a = 5 \times 10^{18}$ cm^{-3}, $\mu_e = 120$ cm^2 V^{-1} s^{-1}
With $N_d = 10^{16}$ cm^{-3}, $\mu_h \approx 440$ cm^2 V^{-1} s^{-1}

Thus

$$D_e = \frac{kT\mu_e}{e} \approx (0.0259 \text{ V})(120 \text{ cm}^2 \text{ V}^{-1} \text{ s}^{-1}) = 3.10 \text{ cm}^2 \text{ s}^{-1}$$

$$D_h = \frac{kT\mu_h}{e} \approx (0.0259 \text{ V})(440 \text{ cm}^2 \text{ V}^{-1} \text{ s}^{-1}) = 11.39 \text{ cm}^2 \text{ s}^{-1}$$

Diffusion lengths are

$$L_e = \sqrt{D_e\tau_e} = \sqrt{[(3.10 \text{ cm}^2 \text{ s}^{-1})(5 \times 10^{-9}\text{s})]}$$

$$= 1.2 \times 10^{-4} \text{ cm} \quad \text{or} \quad 1.2 \text{ } \mu\text{m} < 5 \text{ } \mu\text{m}$$

$$L_h = \sqrt{D_h\tau_h} = \sqrt{[(11.39 \text{ cm}^2 \text{ s}^{-1})(417 \times 10^{-9}\text{s})]}$$

$$= 21.8 \times 10^{-4} \text{ cm} \quad \text{or} \quad 21.8 \text{ } \mu\text{m} < 100 \text{ } \mu\text{m}$$

We therefore have a long diode. The built-in potential is

$$V_o = \left(\frac{kT}{e}\right) \ln\left(\frac{N_d N_a}{n_i^2}\right) = (0.0259 \text{ V}) \ln\left[\frac{(5 \times 10^{18} \times 10^{16})}{(1.45 \times 10^{10})^2}\right] = 0.856 \text{ V}$$

To calculate the forward current when $V = 0.6$ V, we need to evaluate both the diffusion and recombination components to the current. It is likely that the diffusion component will exceed the recombination component at this forward bias (this can be easily verified). Assuming that the forward current is due to minority carrier diffusion in neutral regions,

$$I = I_{so}\left[\exp\left(\frac{eV}{kT}\right) - 1\right] \approx I_{so}\exp\left(\frac{eV}{kT}\right) \qquad \text{for } V \gg \frac{kT}{e} \qquad (= 0.0259 \text{ V})$$

where

$$I_{so} = AJ_{so} = Aen_i^2\left[\left(\frac{D_h}{L_h N_d}\right) + \left(\frac{D_e}{L_e N_a}\right)\right] \approx \frac{Aen_i^2 D_h}{L_h N_d}$$

as $N_a \gg N_d$. In other words, the current is mainly due to the diffusion of holes in the n-region. Thus,

$$I_{so} = \frac{(0.01 \text{ cm}^2)(1.6 \times 10^{-10} \text{ C})(1.45 \times 10^{10} \text{ cm}^{-3})^2(11.39 \text{ cm}^2 \text{ s}^{-1})}{(21.8 \times 10^{-4} \text{ cm})(10^{16} \text{ cm}^{-3})}$$

$$= 1.76 \times 10^{-13} \text{ A} \qquad \text{or} \qquad 0.176 \text{ pA}.$$

Then the diode current is

$$I \approx I_{so}\exp\left(\frac{eV}{kT}\right) = (1.76 \times 10^{-13} \text{ A})\exp\left[\frac{(0.6 \text{ V})}{(0.0259 \text{ V})}\right]$$

$$= 2.0 \times 10^{-3} \text{ A} \qquad \text{or} \qquad 2.0 \text{ mA}.$$

We note that when a forward bias of 0.6 V is applied, the built-in potential is reduced from 0.856 V to 0.256 V, which encourages minority carrier injection, that is, diffusion of holes from p- to n-side and electrons from n- to p-side. To find the current at 100 °C, first we assume that $I_{so} \propto n_i^2$. Then at $T = 273 + 100 = 373$ K, $n_i \approx 1.2 \times 10^{12}$ cm^{-3} (from n_i versus $1/T$ graph in Figure 5.16) so that

$$I_{so}(373 \text{ K}) \approx I_{so}(300 \text{ K})\left[\frac{n_i(373 \text{ K})}{n_i(300 \text{ K})}\right]^2$$

$$\approx (1.76 \times 10^{-13})\left[\frac{1.2 \times 10^{12}}{1.45 \times 10^{10}}\right]^2 = 1.21 \times 10^{-9} \qquad \text{or} \qquad 1.21 \text{ nA}.$$

At 100 °C, the forward current with 0.6 V across the diode is

$$I = I_{so}\exp\left(\frac{eV}{kT}\right) = (1.21 \times 10^{-9} \text{ A})\exp\left[\frac{(0.6 \text{ V})(300 \text{ K})}{(0.0259 \text{ V})(373 \text{ K})}\right] = 0.15 \text{ A}.$$

When a reverse bias of V_r is applied, the potential difference across the depletion region becomes $V_o + V_r$ and the width, W, of the depletion region is

$$W = \left[\frac{2\varepsilon(V_o + V_r)}{eN_d}\right]^{1/2} = \left[\frac{2(11.9)(8.85 \times 10^{-12})(0.856 + 5)}{(1.6 \times 10^{-19})(10^{22})}\right]^{1/2}$$

$$= 0.88 \times 10^{-6} \text{ m} \qquad \text{or} \qquad 0.88 \text{ } \mu\text{m}$$

The thermal generation current with $V_r = 5$ V is

$$I_{gen} = \frac{eAWn_i}{\tau_g} = \frac{(1.6 \times 10^{-19} \text{ C})(0.01 \text{ cm}^2)(0.88 \times 10^{-4} \text{ cm})(1.45 \times 10^{10} \text{ cm}^{-3})}{(10^{-6} \text{ s})}$$

$$= 2.0 \times 10^{-9} \text{ A} \qquad \text{or} \qquad 2 \text{ nA}$$

This thermal generation current is much greater than the reverse saturation current I_{so} (0.176 pA). The reverse current is therefore dominated by I_{gen} and it is 2 nA.

6.2   *pn* JUNCTION BAND DIAGRAM

6.2.1   Open Circuit

Figure 6.9a shows the energy band diagrams for a *p*-type and an *n*-type semiconductor of the same material (same E_g) when the semiconductors are isolated from each other. In the *p*-type material the Fermi level E_{Fp} is Φ_p below the vacuum level and is close to E_v. In the *n*-type material the Fermi level E_{Fn} is Φ_n below the vacuum level and is close to E_c. The separation $E_c - E_{Fn}$ determines the electron concentration, n_{no}, in the *n*-type and $E_{Fp} - E_v$ determines the hole concentration, p_{po}, in the *p*-type semiconductor under thermal equilibrium conditions.

An important property of the Fermi energy, E_F, is that in a system in equilibrium, the Fermi level must be spatially continuous. A difference in Fermi levels, ΔE_F, is equivalent to electrical work eV, which is either done on the system or extracted from the system. When the two semiconductors are brought together, as in Figure 6.9b, the Fermi level must be uniform through the two materials and the junction at M, which marks the position of the metallurgical junction. Far away from M, in the bulk of the *n*-type semiconductor, we should still have an *n*-type semiconductor and $E_c - E_{Fn}$ should be the same as before. Similarly, $E_{Fp} - E_v$ far away from M inside the *p*-type material should also be the same as before. These features are sketched in Figure 6.9b keeping E_{Fp} and E_{Fn} the same through the whole system and, of course, keeping the bandgap, $E_c - E_v$, the same. Clearly, to draw the energy band diagram, we have to bend the bands, E_c and E_v, around the junction at M because E_c on the *n*-side is close to E_{Fn} whereas on the *p*-side it is far away from E_{Fp}. How do bands bend and what does it mean?

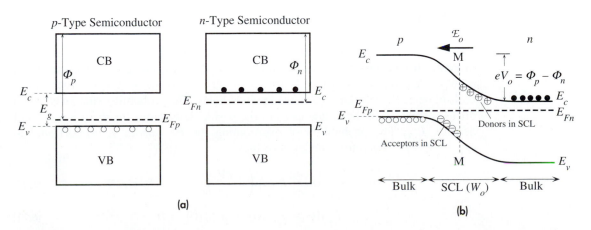

(a)

(b)

Figure 6.9

(a) Two isolated *p*- and *n*-type semiconductors (same material).
(b) A *pn* junction band diagram when the two semiconductors are in contact. The Fermi level must be uniform in equilibrium. The metallurgical junction is at M. The region around M contains the space charge layer (SCL). On the *n*-side of M, SCL has the exposed positively charged donors, whereas on the *p*-side it has the exposed negatively charged acceptors.

As soon as the two semiconductors are brought together to form the junction, electrons diffuse from the *n*-side to the *p*-side and as they do so they deplete the *n*-side near the junction. Thus E_c must move away from E_{Fn} toward M, which is exactly what is sketched in Figure 6.9b. Holes diffuse from the *p*-side to the *n*-side and the loss of holes in the *p*-type material near the junction means that E_v moves away from E_{Fp} toward M, which is also in the figure.

Furthermore, as electrons and holes diffuse toward each other, most of them recombine and disappear around M, which leads to the formation of a depletion region or the space charge layer, as we saw in Figure 6.1. The electrostatic potential energy (*PE*) of the electron decreases from 0 inside the *p*-region to $-eV_o$ inside the *n*-region, as shown in Figure 6.1g. The total energy of the electron must therefore decrease going from the *p*- to the *n*-region by an amount eV_o. In other words, the electron in the *n*-side at E_c must overcome a *PE* barrier to go over to E_c in the *p*-side. This *PE* barrier is eV_o, where V_o is the built-in potential that we evaluated in Section 6.1. Band bending around M therefore accounts not only for the variation of electron and hole concentrations in this region but also for the effect of the built-in potential (and hence the built-in field as the two are related).

In Figure 6.9b we have also schematically sketched in the positive donor (at E_d) and the negative acceptor (at E_a) charges in the SCL around M to emphasize that there are exposed charges near M. These charges are, of course, immobile and, generally, they are not shown in band diagrams. It should be noted that in the SCL region, marked as W_o, the Fermi level is close to neither E_c nor E_v, compared with the bulk semiconductor regions. This means that both *n* and *p* in this zone are much less than their bulk values n_{no} and p_{po}. The metallurgical junction zone has been depleted of carriers compared with the bulk. Any applied voltage must therefore drop across the SCL.

6.2.2   Forward and Reverse Bias

The energy band diagram of the *pn* junction under open circuit conditions is shown in Figure 6.10a. There is no net current, so the diffusion current of electrons from *n*- to *p*-side is balanced by the electron drift current from *p*- to *n*-side driven by the built-in field, $\mathcal{E}_o$. Similar arguments apply to holes. The probability that an electron diffuses from E_c in the *n*-side to E_c in the *p*-side determines the diffusion current density, J_{diff}. The probability of overcoming the *PE* barrier is proportional to $\exp(-eV_o/kT)$. Therefore, under zero bias,

$$J_{diff}(0) = B \exp\left(-\frac{eV_o}{kT}\right) \qquad [6.20]$$

$$J_{net}(0) = J_{diff}(0) + J_{drift}(0) = 0 \qquad [6.21]$$

where B is a proportionality constant and $J_{drift}(0)$ is the current due to the drift of electrons by $\mathcal{E}_o$. Clearly $J_{drift}(0) = -J_{diff}(0)$, that is, drift is in the opposite direction to diffusion.

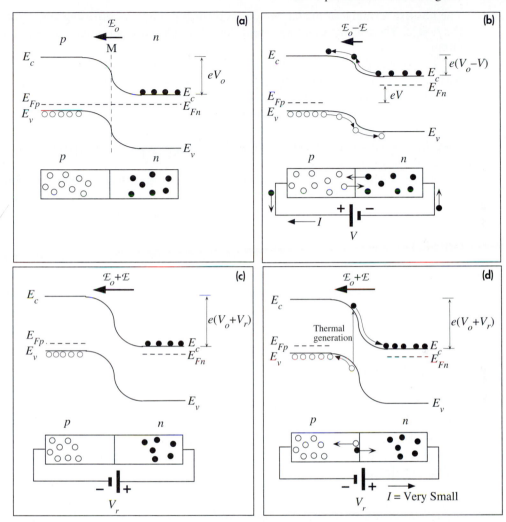

Figure 6.10 Energy band diagrams for a *pn* junction (a) open circuit, (b) forward bias, (c) reverse bias conditions, (d) thermal generation of electron–hole pairs in the depletion region results in a small reverse current.

When the *pn* junction is forward biased, the majority of the applied voltage drops across the depletion region so that the applied voltage is in opposition to the built-in potential, V_o. Figure 6.10b shows the effect of forward bias, which is to reduce the *PE* barrier from eV_o to $e(V_o - V)$. The electrons at E_c in the *n*-side can now readily overcome the *PE* barrier and diffuse to the *p*-side. The diffusing electrons from the *n*-side can be replenished easily by the negative terminal of the battery connected to this side. Similarly holes can now diffuse from *p*- to *n*-side.

The positive terminal of the battery can replenish those holes diffusing away from the p-side. There is therefore a current flow through the junction and around the circuit.

The probability that an electron at E_c in the n-side overcomes the new PE barrier and diffuses to E_c in the p-side is now proportional to $\exp[-e(V_o - V)/kT]$. The latter increases enormously even for small forward voltages. The new diffusion current due to electrons diffusing from n- to p-side is

$$J_{diff}(V) = B \exp\left[-\frac{e(V_o - V)}{kT}\right]$$

There is still a drift current due to electrons being drifted by the new field, $\mathcal{E}_o - \mathcal{E}$ ($\mathcal{E}$ is the applied field), in SCL. This drift current now has the value $J_{drift}(V)$. The net current is the diode current under forward bias

$$J = J_{diff}(V) - J_{drift}(V)$$

J_{drift} is difficult to evaluate. As a first approximation we can assume that although $\mathcal{E}_o$ has decreased to $\mathcal{E}_o - \mathcal{E}$, there is, however, an increase in the electron concentration in the SCL due to diffusion so that we can take J_{drift} to remain the same as $J_{drift}(0)$. Thus

$$J \approx J_{diff}(V) - J_{drift}(0) = B \exp\left[-\frac{e(V_o - V)}{kT}\right] - B \exp\left(-\frac{eV_o}{kT}\right)$$

Factoring leads to

$$J \approx B \exp\left(-\frac{eV_o}{kT}\right)\left[\exp\left(\frac{eV}{kT}\right) - 1\right]$$

We should also add to this the hole contribution, which has a similar form with a different constant B. The diode current–voltage relationship then becomes the familiar diode equation,

$$J = J_o\left[\exp\left(\frac{eV}{kT}\right) - 1\right]$$

where J_o is a temperature-dependent constant.[5]

When a reverse bias, $V = -V_r$, is applied to the pn junction, the voltage again drops across the SCL. In this case, however, V_r adds to the built-in potential V_o so that the PE barrier becomes $e(V_o + V_r)$, as shown in Figure 6.10c. The field in the SCL at M increases to $\mathcal{E}_o + \mathcal{E}$, where $\mathcal{E}$ is the applied field.

The diffusion current due to electrons diffusing from E_c in the n-side to E_c in the p-side is now almost negligible because it is proportional to $\exp[-e(V_o + V_r)/kT]$, which rapidly becomes very small with V_r. There is, however, a small reverse current arising from the drift component. When an electron–hole pair (EHP) is thermally generated in the SCL, as shown in Figure 6.10d, the field here separates the pair. The electron falls down the PE hill, down E_c, to the n-side to be collected by the battery. Similarly the hole falls down

---

| [5] The derivation is similar to that for the Schottky diode, but there were more assumptions here.

its own *PE* hill (energy increases downward for holes) to make it to the *p*-side. The process of falling down a *PE* hill is the same process as being driven by a field, in this case by $\mathcal{E}_o + \mathcal{E}$. Under reverse bias conditions, there is therefore a small reverse current that is *independent* of the bias but dependent on the rate of thermal generation of EHPs in the SCL.

THE BUILT-IN VOLTAGE V_o FROM THE ENERGY BAND DIAGRAM   The energy band treatment allows a simple way to calculate V_o. When the junction is formed in Figure 6.9 from a to b, E_{Fp} and E_{Fn} must shift and line up. Using the energy band diagrams in this figure and semiconductor equations for n and p, derive an expression for the built-in voltage V_o in terms of the material and doping properties N_d, N_a, and n_i.

Example 6.4

Solution

The shift in E_{Fp} and E_{Fn} to line up is clearly $\Phi_p - \Phi_n$, the work function difference. Thus the *PE* barrier eV_o is $\Phi_p - \Phi_n$. From Figure 6.9, we have

$$eV_o = \Phi_p - \Phi_n = (E_c - E_{Fp}) - (E_c - E_{Fn})$$

But on the *p*- and *n*-sides, the electron concentrations in thermal equilibrium are given by

$$n_{po} = N_c \exp\left[-\frac{(E_c - E_{Fp})}{kT}\right]$$

$$n_{no} = N_c \exp\left[-\frac{(E_c - E_{Fn})}{kT}\right]$$

From these equations, we can now substitute for $(E_c - E_{Fp})$ and $(E_c - E_{Fn})$ in eV_o. The N_c cancel and we obtain

$$eV_o = kT \ln\left(\frac{n_{no}}{n_{po}}\right)$$

Since $n_{po} = n_i^2/N_a$ and $n_{no} = N_d$, we readily obtain the built-in potential, V_o,

$$V_o = \left(\frac{kT}{e}\right) \ln\left[\frac{(N_a N_d)}{n_i^2}\right]$$

6.3   DEPLETION LAYER CAPACITANCE OF THE *pn* JUNCTION

It is apparent that the depletion region of a *pn* junction has positive and negative charges separated over a distance W similar to a parallel plate capacitor. The stored charge in the depletion region, however, unlike the case of a parallel plate capacitor, does not depend linearly on the voltage. It is useful to define an incremental capacitance that relates the incremental charge stored to an incremental voltage change across the *pn* junction.

The width of the depletion region is given by

$$W = \left[\frac{2\varepsilon(N_a + N_d)(V_o - V)}{eN_a N_d}\right]^{1/2}$$

[6.22]

where, for forward bias, V is positive, which reduces V_o, and, for reverse bias, V is negative, so V_o is increased. We are interested in obtaining the capacitance of the depletion region under dynamic conditions, that is, when V is a function of time. When the applied voltage V changes by dV, to $V + dV$, then W also changes via Equation 6.22, and as a result, the amount of charge in the depletion region becomes $Q + dQ$, as shown in Figure 6.11a. The **depletion layer capacitance** C_{dep} is defined by

$$C_{dep} = \left|\frac{dQ}{dV}\right|$$

[6.23]

where the amount of charge (on any one side of the depletion layer) is

$$|Q| = eN_d W_n A = eN_a W_p A$$

and $W = W_n + W_p$. We can therefore substitute for W in Equation 6.22 in terms of Q and then differentiate it to obtain dQ/dV. The final result for the depletion capacitance is

$$C_{dep} = \frac{\varepsilon A}{W} = \frac{A}{(V_o - V)^{1/2}}\left[\frac{e\varepsilon(N_a N_d)}{2(N_a + N_d)}\right]^{1/2}$$

[6.24]

We should note that C_{dep} is given by the same expression as that for the parallel plate capacitor, $\varepsilon A/W$, but with W being voltage dependent by virtue of Equation 6.22. The $C_{dep} - V$ behavior is sketched in Figure 6.11b. Notice that C_{dep}

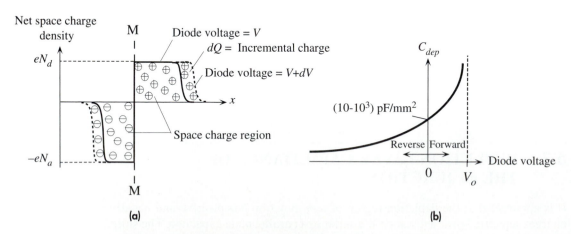

(a)   **(b)**

Figure 6.11   The depletion region behaves like a capacitor.
(a) The charge in the depletion region depends on the applied voltage just as in a capacitor.
(b) The incremental capacitance of the depletion region increases with forward bias and decreases with reverse bias. Its value is typically in the range of picofarads per mm² of device area.

decreases with increasing reverse bias, which is expected since the separation of the charges increases via $W \propto (V_o - V)^{1/2}$. The capacitance C_{dep} is present under both forward and reverse bias conditions.

The voltage dependence of the depletion capacitance is utilized in **varactor diodes** (varicaps), which are employed as voltage-dependent capacitors in tuning circuits. A varactor diode is reverse biased to prevent conduction, and its depletion capacitance is varied by the magnitude of the reverse bias.

6.4 DIFFUSION (STORAGE) CAPACITANCE AND THE SMALL-SIGNAL MODEL

The diffusion or storage capacitance arises under forward bias only. As shown in Figure 6.2a, when the p^+n junction is forward biased, we have stored a positive charge on the n-side by the continuous injection and diffusion of minority carriers. Similarly, a negative charge has been stored on the p^+-side by electron injection, but the magnitude of this negative charge is small for the p^+n junction. When the applied voltage is increased from V to $V + dV$, as shown in Figure 6.12, then $p_n(0)$ changes from $p_n(0)$ to $p_n'(0)$. If dQ is the additional minority carrier charge injected into the n-side, as a result of a small increase, dV, in V, then the incremental **storage** or **diffusion capacitance** C_{diff} is defined as $C_{diff} = dQ/dV$. At voltage V, the injected positive charge, Q, on the n-side is disappearing by recombination at a rate Q/τ_h. The diode current, I, is therefore Q/τ_h, from which

$$Q = \tau_h I = \tau_h I_o \left[\exp\left(\frac{eV}{kT}\right) - 1 \right] \qquad \text{[6.25]}$$

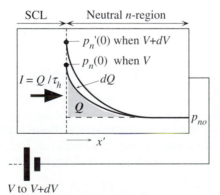

V to $V+dV$

Figure 6.12 Consider the injection of holes into the n-side during forward bias. Storage or diffusion capacitance arises because when the diode voltage increases from V to $V + dV$, more minority carriers are injected and more minority carrier charge is stored in the n-region.

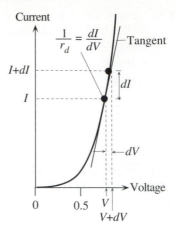

Figure 6.13   The incremental resistance of the diode is defined as dV/dI, which is the inverse of the tangent at I.

Thus,

Diffusion capacitance

$$C_{diff} = \frac{dQ}{dV} = \frac{\tau_h e I}{kT} = \frac{\tau_h I(\text{mA})}{25}$$ **[6.26]**

where we used $e/kT \approx 1/0.025$ at room temperature. Generally the value of the diffusion capacitance, typically in the nanofarads range, far exceeds that of the depletion layer capacitance.

Suppose that the voltage V across the diode is increased by an infinitesimally small amount dV, as shown in an exaggerated way in Figure 6.13. This gives rise to a small increase dI in the diode current. We define the **incremental resistance,** r_d, of the diode as dV/dI, so that

Incremental resistance

$$r_d = \frac{dV}{dI} = \frac{kT}{eI} = \frac{25}{I(\text{mA})}$$ **[6.27]**

The incremental resistance is therefore the inverse of the slope of the I–V characteristics at a point and hence depends on the current I. It relates the changes in the diode current and voltage arising from the **diode action** alone, by which we mean the modulation of the rate of minority carrier diffusion by the diode voltage. We could have equivalently defined an incremental conductance by

$$g_d = \frac{dI}{dV} = \frac{1}{r_d}$$

From Equations 6.26 and 6.27 we have

$$r_d C_{diff} = \tau_h$$ **[6.28]**

The incremental resistance, r_d, and diffusion capacitance, C_{diff}, of a diode determine its response to small ac signals under forward bias conditions. By *small*

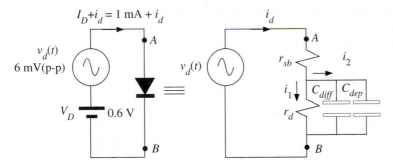

Figure 6.14   The small signal ac equivalent circuit of a forward-biased *pn* junction diode.

we mean voltages smaller than the thermal voltage kT/e or 25 mV at room temperature. The diode in Figure 6.14 has a dc voltage V_D (e.g., 0.6 V) across it and hence a dc current I_D (e.g., 1 mA) flowing throught it. There is, however, an ac source, v_d, in series with the dc supply that provides a small ac signal (e.g., 6 mV peak-to-peak, p-p) across the diode. What is the response of the circuit to the ac signal, v_d, alone? In other words, what is the ac current i_d in the circuit due to v_d?

The ac signal modulates the total voltage, V, across the diode about its dc value, V_D. The resulting current has two components. The first component, i_1, is that which results from the diode action alone. For small signals, as in the present example, can take $v_d = \delta V$ and $i_1 = \delta I$ and from Equation 6.27, which represents the diode action, $\delta V = r_d \, \delta I$ or $v_d = r_d \, i_1$. At $I_D = 1$ mA, $r_d = 25 \, \Omega$, and given $v_d = 6$ mV p-p, $i_1 = 0.24$ mA p-p. The second current component, i_2, charges and discharges the capacitances C_{dep} and C_{diff}. This current is proportional to the rate of change of v_d, just as it would be in an ordinary capacitor,

$$i_2 = (C_{diff} + C_{dep}) \frac{dv_d}{dt}$$

For sinusoidal signals, of course, the above equation is equivalent to using an impedance

$$z_d = \frac{1}{[j\omega(C_{diff} + C_{dep})]}$$

across which v_d appears. Thus, as far as the ac source is concerned, the diode looks like a resistance, r_d, in parallel with the capacitances C_{dep} and C_{diff}, as shown in Figure 6.14. The finite resistance of the neutral semiconductor regions gives rise to some inevitable voltage drop so that not all the applied voltage appears across the junction. As shown in Figure 6.14, we use a series resistance r_{sb} to take into account the series bulk resistance of the neutral regions. Typically $r_{sb} < 1 \, \Omega$ and can often be neglected in comparison with r_d. To finish our example, we will neglect C_{dep} and take $f = 10$ kHz($\omega = 62832$ rad/s), and $\tau_h = 1 \, \mu$s. From Equation 6.28, $C_{diff} = 40$ nF and hence $z_d = (1/j) 398 \, \Omega$. Thus i_2 has a p-p value, v_d/z_d, which is 0.015 mA. Since this is a capacitive current, it is 90° out of phase with i_1. At this frequency, the ac current is determined essentially by r_d. At higher frequencies, z_d will be smaller and eventually z_d will shunt r_d, when the diode will be behaving almost as a capacitor. The equivalent circuit for small ac signals shown in Figure 6.14 is called the small-signal equivalent circuit model of the diode under forward bias. When the diode is reverse biased, r_d is very large, C_{diff} disappears, and the diode exhibits only C_{dep}. For small-signal ac models, we do not include the dc supplies, which only provide the necessary dc bias. The ac current simply passes through the dc supply as if it had an internal resistance of zero.

Example 6.5 | **INCREMENTAL RESISTANCE AND CAPACITANCE**   An abrupt Si p^+n junction diode of cross-sectional area (A) 1 mm^2 with an acceptor concentration of 5×10^{18} boron atoms cm^{-3} on the p-side and a donor concentration of 10^{16} arsenic atoms cm^{-3} on the n-side is forward biased to carry a current of 5 mA. The lifetime of holes in the n-region is 417 ns, whereas that of electrons in the p-region is 5 ns. What are the small-signal ac resistance, incremental storage, and depletion capacitances of the diode?

Solution

This is the same diode we considered in Example 6.3 for which the built-in potential was 0.856 V and $I_{so} = 0.176$ pA. The current through the diode is 5 mA when

$$I = I_{so}\exp\left(\frac{eV}{kT}\right) \quad \text{or} \quad V = \left(\frac{kT}{e}\right)\ln\left(\frac{I}{I_{so}}\right) = (0.0259)\ln\left(\frac{5 \times 10^{-3}}{0.176 \times 10^{-12}}\right) = 0.623 \text{ V}$$

The incremental diode resistance is given by

$$r_d = \frac{25}{I(\text{mA})} = \frac{25}{5} = 5 \ \Omega$$

The depletion capacitance per unit area with $N_a \gg N_d$ is

$$C_{dep} = A\left[\frac{e\varepsilon(N_aN_d)}{2(N_a + N_d)(V_o - V)}\right]^{1/2} \approx A\left[\frac{e\varepsilon N_d}{2(V_o - V)}\right]^{1/2}$$

At $V = 0.623$ V, with $V_o = 0.856$ V, $N_d = 10^{22}$ m^{-3}, $\varepsilon_r = 11.9$, and $A = 10^{-6}$ m^2, the above gives

$$C_{dep} = 10^{-6}\left[\frac{(1.6 \times 10^{-19})(11.9)(8.85 \times 10^{-12})(10^{22})}{2(0.856 - 0.623)}\right]^{1/2}$$

$$= 6.01 \times 10^{-10} \text{ F} \quad \text{or} \quad 601 \text{ pF}$$

The incremental diffusion capacitance C_{diff} due to holes injected and stored in the n-region is

$$C_{diff} = \frac{\tau_h I(\text{mA})}{25} = \frac{(417 \times 10^{-9})(5)}{25} = 8.3 \times 10^{-8} \text{ F} \quad \text{or} \quad 83 \text{ nF}$$

Clearly the diffusion capacitance (83 nF) that arises during forward bias completely overwhelms the depletion capacitance (601 pF).

We note that there is also a diffusion capacitance due to electrons injected and stored in the p-region. However, electron lifetime in the p-region is very short (here 5 ns) so that the value of this capacitance is much smaller than that due to holes in the n-region. In calculating the diffusion capacitance, we normally consider the minority carriers that have the longest recombination lifetime, here τ_h. These are the carriers that take a long time to disappear by recombination when the bias is suddenly switched off.

6.5   REVERSE BREAKDOWN: AVALANCHE AND ZENER BREAKDOWN

The reverse voltage across a *pn* junction cannot be increased without limit. Eventually the *pn* junction breaks down either by the Avalanche or Zener breakdown mechanisms, which lead to large reverse currents, as shown in Figure 6.15. In the $V = -V_{br}$ region, the reverse current increases dramatically with the reverse bias. If unlimited, the large reverse current will increase the power dissipated, which in turn raises the temperature of the device, which leads to a further increase in the reverse current and so on. If the temperature does not burn out the device, for example, by melting the contacts, then the breakdown is recoverable. If the current is limited by an external resistance to value within the power dissipation specifications, then there is no reason why the device cannot operate under breakdown conditions.

6.5.1   Avalanche Breakdown

As the reverse bias increases, the field in the SCL can become so large that an electron drifting in this region can gain sufficient kinetic energy to impact on a Si atom and ionize it, or break a Si–Si bond. The phenomenon by which a drifting electron gains sufficient energy from the field to ionize a host crystal atom by bombardment is termed **impact ionization.** The accelerated electron must gain at least an energy equal to E_g as impact ionization breaks a Si–Si bond, which is tantamount to exciting an electron from the valence band to the conduction band. Thus an additional electron–hole pair is created by this process.

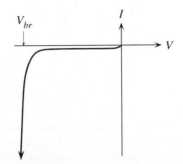

Figure 6.15   Reverse *I–V* characteristics of a *pn* junction.

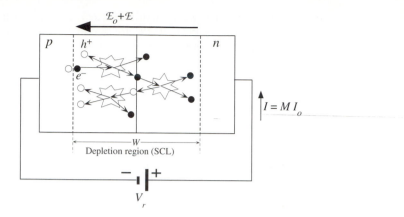

Figure 6.16   Avalanche breakdown by impact ionization

Consider what happens when a thermally generated electron just inside the SCL in the p-side is accelerated by the field. The electron accelerates and gains sufficient energy to collide with a host Si atom and release an EHP by impact ionization, as depicted in Figure 6.16. It will lose at least E_g amount of energy, but it can accelerate and head for another ionizing collision further along the depletion region until it reaches the neutral n-region. The EHPs generated by impact ionization themselves can now be accelerated by the field and will themselves give rise to further EHPs by ionizing collisions and so on, leading to an **avalanche effect.** One initial carrier can thus create many carriers in the SCL through an avalanche of impact ionization.

If the reverse current in the SCL in the absence of impact ionization is I_o, then due to the avalanche of ionizing collisions in the SCL, the reverse current becomes MI_o where M is the multiplication. It is the net number of carriers generated by the avalanche effect per carrier in the SCL. Impact ionization depends strongly on the electric field. Small increases in the reverse bias can lead to dramatic increases in the multiplication process. Typically

$$M = \frac{1}{1 - \left(\dfrac{V_r}{V_{br}}\right)^n}$$ [6.29]

where V_r is the reverse bias, V_{br} is the breakdown voltage, and n is an index in the range three to five. It is clear that the reverse current, MI_o, increases sharply with V_r near V_{br}, as depicted in Figure 6.15. Indeed, the voltage across a diode under reverse breakdown remains around V_{br} for very large current variations (several orders of magnitude). If the reverse current under breakdown is limited by an appropriate external resistor, R, as shown in Figure 6.17, to prevent destructive power dissipation in the diode, then the voltage across the diode remains approximately at V_{br}. Thus, as long as $V_r > V_{br}$, the diode clamps the voltage between A and B to approximately V_{br}. The reverse current in the circuit is then $(V_r - V_{br})/R$.

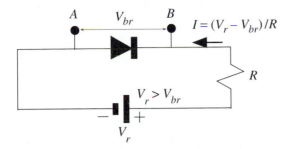

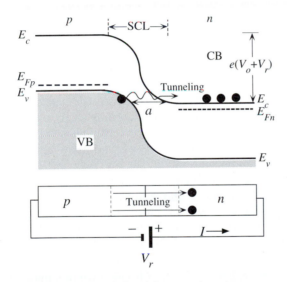

Figure 6.17   If the reverse breakdown current when $V_r > V_{br}$ is limited by an external resistance, R, to prevent destructive power dissipation, then the diode can be used to clamp the voltage between A and B to remain approximately V_{br}.

Figure 6.18   Zener breakdown involves electrons tunneling from the VB of p-side to the CB of n-side when the reverse bias reduces E_c to line up with E_v.

Since the electric field in the SCL depends on the width of the depletion region W, which in turn depends on the doping parameters, V_{br} also depends on the doping, as discussed in Example 6.6.

6.5.2   Zener Breakdown

Heavily doped pn junctions have narrow depletion widths, which lead to large electric fields within this region. When a reverse bias is applied to a pn junction, the energy band diagram of the n-side can be viewed as being lowered with respect to the p-side, as depicted in Figure 6.18. For a sufficient reverse bias (typically less

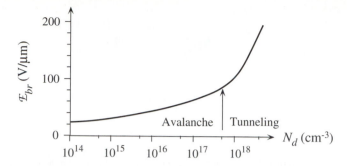

Figure 6.19   The breakdown field $\mathcal{E}_{br}$ in the depletion layer for the onset of reverse breakdown versus doping concentration N_d in the lightly doped region in a one-sided (p^+n or pn^+) abrupt pn junction.

Avalanche and tunneling mechanisms are separated by the arrow.

| SOURCE: Data extracted from M. Sze and G. Gibbons, *Solid State Electronics*, 9, no. 831 (1966).

than 10 V), E_c on the n-side may be lowered to be below E_v on the p-side. This means that electrons at the top of the VB in the p-side are now at the same energy level as the empty states in the CB in the n-side. As the separation between the VB and CB narrows, shown as a ($<W$), the electrons easily tunnel from the VB in p-side to the CB in n-side, which leads to a current. This process is called the **Zener effect.** As there are many electrons in the VB and many empty states in the CB, the tunneling current can be substantial. The reverse voltage, V_r, which starts the tunneling current and hence the Zener breakdown, is clearly that which lowers E_c on n-side to be below E_v on p-side and thereby gives a separation that encourages tunneling. In nonquantum mechanical terms, one may intuitively view the Zener effect as the strong electric field in the depletion region ripping some of those electrons in the Si–Si bonds and thereby releasing them for conduction.

Figure 6.19 shows the dependence of the breakdown field, $\mathcal{E}_{br}$, in the depletion region for the onset of avalanche or Zener breakdown in a one-sided (p^+n or pn^+) abrupt junction on the dopant concentration, N_d, in the lightly doped side. At high fields, the tunneling becomes the dominant reverse breakdown mechanism.

Example 6.6 | **AVALANCHE BREAKDOWN**   Consider a uniformly doped abrupt p^+n junction ($N_a \gg N_d$) reverse biased by $V = -V_r$.

a.   What is the relationship between the depletion width W and the potential difference $(V_o + V_r)$ across W?

b.   If avalanche breakdown occurs when the maximum field in the depletion region, $\mathcal{E}_o$, reaches the breakdown field $\mathcal{E}_{br}$, show that the breakdown voltage $V_{br}(\gg V_o)$ is then given by

$$V_{br} = \frac{\varepsilon \mathcal{E}_{br}^2}{2eN_d}$$

c.   An abrupt Si p^+n junction has boron doping of 10^{19} cm^{-3} on the p-side and phosphorus doping of 10^{16} cm^{-3} on the n-side. The dependence of avalanche breakdown field on the impurity concentration is shown in Figure 6.19.
1.   What is the reverse breakdown voltage of this Si diode?
2.   Calculate the reverse breakdown voltage when the phosphorus doping is increased to 10^{17} cm^{-3}.

Solution

One can assume that all the applied reverse bias drops across the depletion layer so that the new voltage across W is now $V_o + V_r$. We have to integrate $dE/dx = \rho_{net}/\varepsilon$ as before across W to find the maximum field. The most important fact to remember here is that the pn junction equations relating W, E_o, V_o, N_o, N_d, and so on remain the same but with V_o replaced with $V_o + V_r$ since the applied reverse bias of V_r increases V_o to $V_o + V_r$. Then from Equation 6.4,

$$W^2 = \frac{2\varepsilon(V_o + V_r)[N_a^{-1} + N_d^{-1}]}{e} \approx \frac{2\varepsilon(V_o + V_r)}{eN_d}$$

since $N_a \gg N_d$. The maximum field that corresponds to the breakdown field E_{br} is given by

$$E_o = -\frac{2(V_o + V_r)}{W}$$

Thus, from these two equations we can eliminate W and obtain $V_{br} = V_r$ as

$$V_{br} = \frac{\varepsilon E_{br}^2}{2eN_d}$$

Given $N_d \gg N_a$ we have a p^+n junction with $N_d = 10^{16}$ cm^{-3}. The depletion region extends into the n-region, so the maximum field actually occurs in the n-region. Here the breakdown field E_{br} depends on the doping level as given in the graph of critical field at breakdown, E_{br} versus doping concentration, N_d, in Figure 6.19. Taking $E_{br} \approx$ 40 V/μm or 4.0×10^5 V cm^{-1} at $N_d = 10^{16}$ cm^{-3} and using the above equation for V_{br}, we get $V_{br} = 52$ V.
    When $N_d = 10^{17}$ cm^{-3}, E_{br} from the graph is about 6×10^5 V cm^{-1}, which leads to $V_{br} = 11.8$ V.

6.6   BIPOLAR TRANSISTOR (BJT)

6.6.1   Common Base (CB) dc Characteristics

As an example, we will consider the *pnp* bipolar junction transistor (BJT) whose basic structure is shown in Figure 6.20a. The *pnp* transistor has three differently doped semiconductor regions. These regions of different doping occur within the same single crystal by the variation of acceptor and donor concentrations resulting from the fabrication process. The most heavily doped p-region (p^+) is called the **emitter.** In contact with this region is the lightly doped n-region, which is called the **base.** The next region is the p-type doped **collector.** The base region has the most narrow width for reasons discussed below. Although the three regions in

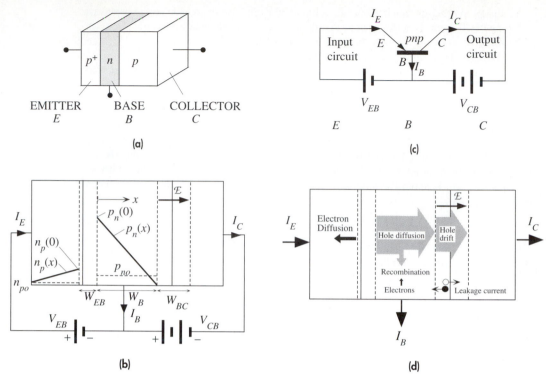

Figure 6.20

(a) A schematic illustration of the *pnp* bipolar transistor with three differently doped regions.
(b) The *pnp* bipolar operated under normal and active conditions.
(c) The CB configuration with input and output circuits identified.
(d) The illustration of various current components under normal and active conditions.

Figure 6.20a have identical cross-sectional areas, in practice, due to the fabrication process, the cross-sectional area increases from the emitter to the collector and the collector region has an extended width. For simplicity, we will assume that the cross-sectional area is uniform, as in Figure 6.20a.

The *pnp* BJT connected as shown in Figure 6.20b is said to be operating under normal and active conditions, which means that the base–emitter (BE) junction is forward biased and the base–collector (BC) junction is reverse biased. The circuit in Figure 6.20b, in which the base is common to both the collector and emitter bias voltages, is known as the common base (CB) configuration.[6] Figure 6.20c shows the CB transistor circuit with the BJT represented by its circuit symbol. The arrow identifies the emitter junction and points in the direction of current flow when the EB junction is forward biased. Figure 6.20c also identifies the emitter circuit, where V_{EB} is connected, as the input circuit. The collector circuit, where V_{CB} is connected, is the output circuit.

| [6]CB should not be confused with conduction band abbreviation.

The base–emitter junction is simply called the **emitter junction** and the base–collector junction is called the **collector junction.** As the emitter is heavily doped, the base–emitter depletion region, W_{EB}, extends almost entirely into the base. Generally, the base and collector regions have comparable doping, so the base–collector depletion region, W_{BC}, extends to both sides. The width of the neutral base region outside the depletion regions is labeled as W_B. All these parameters are shown and defined in Figure 6.20b.

We should note that all the applied voltages drop across the depletion widths. The applied collector–base voltage V_{CB} reverse biases the BC junction and hence increases the field in the depletion region at the collector junction.

Since the EB junction is forward biased, minority carriers are then injected into the emitter and base exactly as they are in the forward-biased diode. Holes are injected into the base and electrons into the emitter, as depicted in Figure 6.20d. Hole injection into the base, however, far exceeds the electron injection into the emitter because the emitter is heavily doped. We can then assume that the emitter current is entirely due to holes injected from the emitter into the base. We therefore neglect the contribution to I_E arising from the injection of electrons from the base into the emitter.

Injected holes into the base must diffuse toward the collector junction because there is a hole concentration gradient in the base. Hole concentration, $p_n(W_B)$, just outside the depletion region at the collector junction is zero because the increased field sweeps all the holes here across the junction into the collector (collector junction is reverse biased).

The hole concentration, $p_n(0)$, in the base just outside the emitter junction depletion region is given by the law of the junction. Measuring x from this point (Figure 6.20b),

$$p_n(0) = p_{no} \exp\left(\frac{eV_{EB}}{kT}\right) \tag{6.30}$$

whereas at the collector end, $x = W_B$, $p_n(W_B) = 0$.

If no holes are lost by recombination in the base, then all the injected holes diffuse to the collector junction. There is no field in the base to drift the holes. Their motion is by diffusion. When they reach the collector junction, they are quickly swept across into the collector by the internal field $\mathcal{E}$ in W_{BC}. It is apparent that all the injected holes from the emitter become collected by the collector. The collector current is then the same as the emitter current. The only difference is that the emitter current flows across a smaller voltage difference, V_{EB}, whereas the collector current flows through a larger voltage difference, V_{CB}. This means a net gain in power from the emitter circuit to the collector circuit.

Since the current in the base is by diffusion, to evaluate the emitter and collector currents we must know the hole concentration gradient at $x = 0$ and $x = W_B$ and therefore we must know the hole concentration profile, $p_n(x)$, across the base.[7] In the first instance, we can approximate the $p_n(x)$ profile in the base as

---

[7] The actual concentration profile can be calculated by solving the steady-state continuity equation, which can be found in more advanced texts.

a straight line from $p_n(0)$ to $p_n(W_B) = 0$, as shown in Figure 6.20b. This is only true in the absence of any recombination in the base as in the short diode case. The emitter current is then

$$I_E = -eAD_h\left(\frac{dp_n}{dx}\right)_{x=0} = eAD_h\frac{p_n(0)}{W_B} \qquad \text{[6.31]}$$

We can substitute for $p_n(0)$ from Equation 6.30 to obtain

Emitter current

$$I_E = \frac{eAD_hp_{no}}{W_B}\exp\left(\frac{eV_{EB}}{kT}\right) \qquad \text{[6.32]}$$

It is apparent that I_E is determined by V_{EB}, the forward bias applied across the EB junction, and the base width, W_B. In the absence of recombination, the collector current is the same as the emitter current, $I_C = I_E$. The control of the collector current, I_C, in the output (collector) circuit by V_{EB} in the input (emitter) circuit is what constitutes the **transistor action.**

A small number of the diffusing holes in the narrow base inevitably become lost by recombination with the large number of electrons present in this region, as depicted in Figure 6.20d. Thus the collector current in general is less than the emitter current. The ratio of the collector current to the emitter current is defined as the **CB current gain** (or current transfer ratio), α, of the transistor,

Definition of CB current gain

$$\alpha = \frac{I_C}{I_E} \qquad \text{[6.33]}$$

Inasmuch as some recombination is inevitable in the base, α is always less than unity. Typically, α is in the range 0.99–0.999. It is not difficult to evaluate α in terms of the base width, diffusion coefficient, and minority carrier lifetime by a simple semiquantitative argument.

If τ_h is the hole (minority carrier) lifetime in the base, then $1/\tau_h$ is the probability per unit time that a hole will recombine and disappear. We also know that in time t, a particle diffuses a distance x, given by $x = \sqrt{2Dt}$, where D is the diffusion coefficient. The time, τ_t, it takes for a hole to diffuse across W_B is then given by

$$\tau_t = \frac{W_B^2}{2D_h}$$

Base minority carrier transit time

This diffusion time is called the **transit time** of the minority carriers across the base. The probability of recombination in time τ_t is then τ_t/τ_h. The probability of not recombining and therefore diffusing across is $(1 - \tau_t/\tau_h)$. Since I_E represents the holes entering the base per unit time, $I_E(1 - \tau_t/\tau_h)$ represents the number of holes leaving the base per unit time (without recombining), which is the collector current I_C. The current transfer ratio is then

$$\alpha = 1 - \frac{\tau_t}{\tau_h} \qquad \text{[6.34]}$$

CB current gain

As electrons are lost by recombination, the base must be replenished with electrons, which are supplied by the external battery in the form of a small base current, I_B (Figure 6.20d). The base current therefore depends on the number of holes lost in the base. The number of holes entering the base per unit time is represented by I_E and the number recombining per unit time is then represented by $I_E(\tau_t/\tau_h)$, which is the base current I_B,

$$I_B = \frac{\tau_t}{\tau_h} I_E \qquad\qquad \text{[6.35]} \qquad \textit{Base current}$$

The ratio of the collector current to the base current is defined as the **current gain** β of the transistor.[8] By using Equations 6.33, 6.34, and 6.35, we can relate β to α:

$$\beta = \frac{I_C}{I_B} = \frac{\alpha}{1-\alpha} \approx \frac{\tau_h}{\tau_t} \qquad\qquad \text{[6.36]}$$

The base–collector junction in Figure 6.20b is reverse biased, which leads to a leakage current into the collector terminal even in the absence of an emitter current. This leakage current is due to thermally generated electron–hole pairs in the depletion region, W_{BC}, being drifted by the internal field, as schematically illustrated in Figure 6.20d. Suppose that we open circuit the emitter ($I_E = 0$). Then the collector current is simply the leakage current, denoted by I_{CBO}. The base current is then $-I_{CBO}$ (flowing out from the base terminal). In the presence of an emitter current, I_E, we have

$$I_C = \alpha I_E + I_{CBO} \qquad\qquad \text{[6.37]} \qquad \textit{Active region collector current}$$

$$I_B = (1-\alpha)I_E - I_{CBO} \qquad\qquad \text{[6.38]} \qquad \textit{Active region base current}$$

The above equations give the collector and base currents in terms of the input current I_E, which in turn depends on V_{EB}. They only hold when the collector junction is reverse biased and the emitter junction is forward biased, which is defined as the **active region** of the BJT. It should be emphasized that what constitutes the transistor action is the control of I_E, and hence I_C, by V_{EB}.

The dc characteristics of the CB-connected BJT as in Figure 6.20b are normally represented by plotting the collector current, I_C, as a function of V_{CB} for various fixed values of the emitter current. A typical example of such dc characteristics for a *pnp* transistor is illustrated in Figure 6.21. The following characteristics are apparent. The collector current when $I_E = 0$ is the CB junction leakage current, I_{CBO}, typically a fraction of a microampere. As long as the collector is negatively biased with respect to the base, the CB junction is reverse biased and the collector current is given by $I_C = \alpha I_E + I_{CBO}$, which is close to the emitter current when $I_E \gg I_{CBO}$. When the polarity of V_{CB} is changed, the CB junction becomes forward biased. The collector junction is then like a forward-biased diode and the collector current is the difference between the forward-biased CB junction current and the forward-biased EB junction current. As they are in opposite directions, they subtract.

We note that I_C increases slightly with V_{CB} even when I_E is constant. In our treatment I_C did not directly depend on V_{CB}, which simply reverse biased the

---

[8] β is a useful parameter when the transistor is used in what is called the common–emitter (CE) configuration, in which the input current is made to flow into the base.

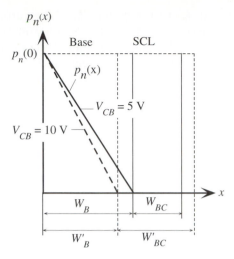

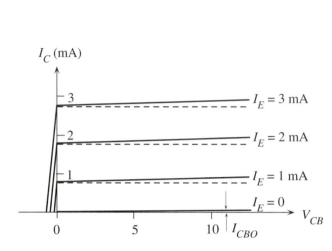

Figure 6.21   DC I–V characteristics of the *pnp* bipolar transistor (exaggerated to highlight various effects)

Figure 6.22   The Early effect
When the BC reverse bias increases, the depletion width W_{BC}, increases to W'_{BC}, which reduces the base width W_B to W'_B. As $p_n(0)$ is constant (constant V_{EB}), the minority carrier concentration gradient becomes steeper and the collector current, I_C, increases.

collector junction to collect the diffusing holes. In our discussions we assumed that the base width W_B does not depend on the applied voltages. This is only approximately true. Suppose that we increase the reverse bias V_{CB} (for example, from 5 to 10 V). Then the base–collector depletion width W_{BC} also increases, as schematically depicted in Figure 6.22. Consequently the base width, W_B, gets slightly narrower, which leads to a slightly shorter base transit time, τ_t. The current gain

$$\alpha = 1 - \frac{\tau_t}{\tau_h}$$

is then slightly larger, which leads to a small increase in I_C. The modulation of the base width W_B by V_{CB} is not very strong, which means that the slopes of the $I_C - V_{CB}$ lines at a fixed I_E are very small in Figure 6.21. The base width modulation by V_{CB} is called the **Early effect.**

6.6.2   Common Base Amplifier

According to Equation 6.32, the emitter current depends exponentially on V_{EB},

$$I_E = I_{EO} \exp\left(\frac{eV_{EB}}{kT}\right) \qquad [6.39]$$

where I_{EO} is a constant. It is therefore apparent that small changes in V_{EB} lead to large changes in I_E. Since $I_C \approx I_E$, we see that small variations in V_{EB} cause large

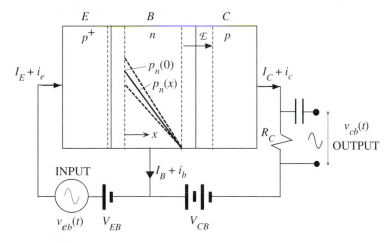

Figure 6.23   A *pnp* transistor operated in the active region in the common base amplifier configuration

The applied (input) signal v_{eb} modulates the dc voltage across the EB junction and hence modulates the injected hole concentration up and down about the dc value $p_n(0)$. The solid line shows $p_n(x)$ when only the dc bias V_{EB} is present. The dashed lines show how $p_n(x)$ is modulated up and down by the signal v_{eb} superimposed on V_{EB}.

changes in I_C in the collector circuit. This can be fruitfully used to obtain voltage amplification, as shown in Figure 6.23. The dc voltage V_{EB} forward biases the EB junction, which means it provides a dc current I_E. The input signal is the ac voltage v_{eb} applied in series with the dc bias voltage V_{EB} to the EB junction. The applied signal v_{eb} modulates the total voltage across the EB junction and hence, by virtue of Equation 6.30, modulates the injected hole concentration, $p_n(0)$, up and down about the dc value determined by V_{EB}, as depicted in Figure 6.30. This variation in $p_n(0)$ alters the concentration gradient and therefore gives rise to a change in I_E, and hence a nearly identical change in I_C. The change in the collector current can be converted to voltage change by using a resistor R_C in the collector circuit, as shown in Figure 6.23. The output is taken as an ac voltage across R_C.

For simplicity we will assume that changes, δV_{EB} and δI_E, in the dc values of V_{EB} and I_E are small, which means that δV_{EB} and δI_E can be related by differentiating Equation 6.39. We are hence tacitly assuming an operation under small signals. Further, we will take the changes to represent the ac signal magnitudes, $v_{eb} = \delta V_{EB}$, $i_e = \delta I_E$, $i_c \approx \delta I_C \approx \delta I_E \approx i_e$, and $v_{cb} = \delta V_{CB}$.

The output signal voltage, v_{cb}, across R_C corresponds to the change in V_{CB},

$$v_{cb} = \delta V_{CB} = R_C \, \delta I_C = R_C \, \delta I_E$$

The variation in the emitter current, δI_E, depends on the variation δV_{EB} in V_{EB}, which can be determined by differentiating Equation 6.39,

$$\frac{\delta I_E}{\delta V_{EB}} = \frac{e}{kT} I_E$$

By definition, δV_{EB} is the input signal, v_{eb}. The change δI_E in I_E is the input signal current (i_e) flowing into the emitter as a result of δV_{EB}. Therefore the quantity $\delta V_{EB}/\delta I_E$ represents an input resistance, r_e, seen by the source v_{eb}:

CB input resistance

$$r_e = \frac{\delta V_{EB}}{\delta I_E} = \frac{kT}{eI_E} = \frac{25}{I_E(\text{mA})}$$ [6.40]

The output signal is then

$$v_{cb} = R_C \, \delta I_E = R_C \frac{v_{eb}}{r_e}$$

so that the voltage amplification is

CB voltage gain

$$A_V = \frac{v_{cb}}{v_{eb}} = \frac{R_C}{r_e}$$ [6.41]

To obtain a voltage gain we obviously need $R_C > r_e$, which is invariably the case by the appropriate choice of I_E, hence r_e, and R_C. For example, when the BJT is biased so that I_E is 10 mA, r_e is 2.5 Ω and if R_C is chosen to be 50 Ω then the gain is 20.

Example 6.7 | **A pnp TRANSISTOR**   Consider a *pnp* Si BJT that has the following properties. The emitter region mean acceptor doping is 10^{19} cm^{-3}, the base region mean donor doping is 2×10^{16} cm^{-3}, and the collector region mean acceptor doping is 1×10^{16} cm^{-3}. The transistor emitter and base widths are 2 microns each. The effective cross-sectional area of the device is 0.02 mm^2. The hole lifetime in the base is approximately 250 ns.

a.   Calculate the CB current transfer ratio α.

b.   Calculate the current gain β.

c.   Suppose that the dc bias voltage V_{EB} is such that the emitter dc current is 2.5 mA. What is V_{EB} and the small signal input resistance r_e of the transistor in the CB configuration?

d.   What should R_C be to obtain a gain of 10?

e.   What is the base current?

f.   What frequency range of the input voltages can the CB amplifier amplify?

Solution

In the base, $N_d = 2 \times 10^{16}$ cm^{-3}, so from the dependence of hole drift mobility on the doping concentration (Figure 5.19), we obtain the hole drift mobility $\mu_h = 410$ cm^2 V^{-1} s^{-1} (minority carriers in the base). From the Einstein relationship we can easily find the diffusion coefficient of holes,

$$D_h = \left[\frac{kT}{e}\right]\mu_h = (0.0259)(410) = 10.6 \text{ cm}^2 \text{ s}^{-1}$$

The minority carrier transit time across the base is

$$\tau_t = \frac{W_B^2}{2D_h} = \frac{(2 \times 10^{-4} \text{ cm})^2}{2(10.6 \text{ cm}^2 \text{ s}^{-1})} = 1.89 \times 10^{-9} \text{ s} \quad \text{or} \quad 1.89 \text{ ns}$$

The CB current gain is

$$\alpha = 1 - \frac{\tau_t}{\tau_h} = 1 - \frac{1.89 \times 10^{-9}\,\text{s}}{250 \times 10^{-9}\,\text{s}} = 0.99245$$

The current gain β of the transistor is

$$\beta = \frac{\alpha}{(1 - \alpha)} = \left[\frac{0.99245}{1 - 0.99245}\right] = 131.5$$

The emitter current is

$$I_E = I_{E0} \exp\left(\frac{eV_{EB}}{kT}\right)$$

where

$$I_{E0} = \frac{eAD_h p_{no}}{W_B} = \frac{eAD_h n_i^2}{N_d W_B} = \frac{(1.6 \times 10^{-19})(0.02 \times 10^{-2})(10.6)(1.45 \times 10^{10})^2}{(2 \times 10^{16})(2 \times 10^{-4})}$$

$$= 1.79 \times 10^{-14}\,\text{A}$$

Thus

$$V_{EB} = \left(\frac{kT}{e}\right) \ln\left(\frac{I_E}{I_{E0}}\right) = (0.0259) \ln\left(\frac{2.5 \times 10^{-3}}{1.79 \times 10^{-14}}\right) = 0.663\,\text{V}$$

The input resistance for small ac signals is

$$r_e = \frac{kT}{eI_E} = \frac{25}{I_E(\text{mA})} = \frac{25}{2.5} = 10\,\Omega$$

The voltage gain A_V is R_C/r_e, which must be 10. Since r_e is 10 Ω, R_C must be 100 Ω.
    The base current is

$$I_B = I_E - I_C = I_E(1 - \alpha) = 2.5 \times 10^{-3}(1 - 0.99245)$$

$$= 1.89 \times 10^{-5}\,\text{A} \quad \text{or} \quad 18.9\,\mu\text{A}$$

The transit time of minority carriers across the base is τ_t. If the input signal changes before the minority carriers have diffused across the base, then the collector current cannot respond to the changes in the input. Thus, if the frequency of the input signal is greater than $1/\tau_t$, the minority carriers will not have time to transit the base and the collector current will remain unmodulated by the input signal. One can set the upper frequency limit at around $1/\tau_t$, or 530 MHz.

EMITTER INJECTION EFFICIENCY γ | **Example 6.8**

a.   Consider a *pnp* transistor with the parameters as defined in Figure 6.20. Show that the injection efficiency of the emitter, defined as

$$\gamma = \frac{\text{Emitter current due to minority carriers injected into the base}}{\text{Total emitter current}}$$

is given by

$$\gamma = \frac{1}{1 + \dfrac{N_d W_B \mu_{e(\text{emitter})}}{N_a W_E \mu_{h(\text{base})}}}$$

b. How would you modify the CB current gain α to include the emitter injection efficiency?

c. Calculate the emitter injection efficiency for the *pnp* transistor in Example 6.7, which has an acceptor doping of 10^{19} cm^{-3} in the emitter, donor doping of 2×10^{16} cm^{-3} in the base, emitter and base widths of 2 μm and a minority carrier lifetime of 250 ns in the base. What are its modified α and β?

Solution

When the BE junction is forward biased, holes are injected into the base, giving an emitter current $I_{E(\text{hole})}$, and electrons are injected into the emitter, giving an emitter current $I_{E(\text{electron})}$. The total emitter current is therefore

$$I_E = I_{E(\text{hole})} + I_{E(\text{electron})}$$

Only the holes injected into the base are useful in giving a collector current because only they can reach the collector. Injection efficiency is defined as

$$\gamma = \frac{I_{E(\text{hole})}}{I_{E(\text{hole})} + I_{E(\text{electron})}} = \frac{1}{1 + \dfrac{I_{E(\text{electron})}}{I_{E(\text{hole})}}}$$

But

$$I_{E(\text{hole})} = \frac{eAD_{h(\text{base})}n_i^2}{N_d W_B} \exp\left(\frac{V_{EB}}{kT}\right)$$

$$I_{E(\text{electron})} = \frac{eAD_{e(\text{emitter})}n_i^2}{N_a W_E} \exp\left(\frac{V_{EB}}{kT}\right)$$

When we substitute into the definition of γ and use $D = \mu kT/e$, we obtain

Emitter injection efficiency

$$\gamma = \frac{1}{1 + \dfrac{N_d W_B \mu_{e(\text{emitter})}}{N_a W_E \mu_{h(\text{base})}}}$$

The hole component of the emitter current is given as γI_E. Of this, a fraction $(1 - \tau_t/\tau_h)$ will give a collector current. Thus, the emitter-to-collector current transfer ratio α, taking into account the emitter injection efficiency, is

Emitter-to-collector current transfer ratio

$$\alpha = \gamma\left(1 - \frac{\tau_t}{\tau_h}\right)$$

In the emitter, $N_{a(\text{emitter})} = 10^{19}$ cm^{-3}, $\mu_{e(\text{emitter})} = 110$ cm^2 V^{-1} s^{-1}, and in the base, $N_{d(\text{base})} = 2 \times 10^{16}$ cm^{-3}, $\mu_{h(\text{base})} = 410$ cm^2 V^{-1} s^{-1}. The emitter injection efficiency is

$$\gamma = \frac{1}{1 + \dfrac{(2 \times 10^{16})(2)(110)}{(10^{19})(2)(410)}} = 0.99946$$

The transit time $\tau_t = W_B^2/2D_h = 1.89 \times 10^{-9}$ s (as before), so the overall α is

$$\alpha = 0.99946\left(1 - \frac{1.89 \times 10^{-9}}{250 \times 10^{-9}}\right) = 0.99191$$

and the overall β is

$$\beta = \frac{\alpha}{(1 - \alpha)} = 122.6$$

In this particular example the α and hence β of the BJT are primarily determined by minority carrier recombination in the base because if we assume full injection, $\gamma = 1$, $\alpha = 0.99246$, and $\beta = 131.6$. In many transistors, however, τ_t/τ_h is very small and γ is lower, so γ controls the overall current gain.

6.6.3   Common Emitter (CE) dc Characteristics

The *pnp* bipolar transistor when connected in the common emitter (CE) configuration has the emitter common to both the input and output circuits, as shown in Figure 6.24. The advantage of the CE configuration is that the input current is the current flowing between the ac source and the base, which is the base current, I_B. This current is much smaller than the emitter current by about a factor of β. In the CE configuration, the dc voltage V_{CE} must be greater than V_{BE} to reverse bias the collector junction and collect the diffusing holes in the base.

The dc characteristics of the BJT in the CE configuration are normally given as I_C versus V_{CE} for various values of fixed base currents, I_B, as shown in Figure 6.25. The characteristics can be readily understood by Equations 6.37 and 6.38. We should note that, in practice, we are essentially adjusting V_{BE} to keep I_B constant because, by Equation 6.38,

$$I_B = (1 - \alpha)I_E - I_{CBO}$$

and I_E depends on V_{BE} via Equation 6.39.

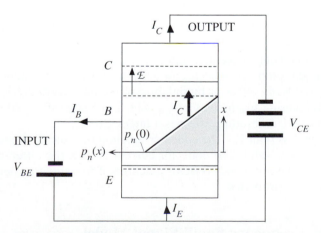

Figure 6.24   A *pnp* transistor operated in the active region in the common emitter configuration

The dc voltage across the BE junction, V_{BE}, controls the current I_E and hence I_B and I_C. The input current is the current that flows between V_{BE} and the base, which is I_B. The output current is the current flowing between V_{CE} and the collector, which is I_C.

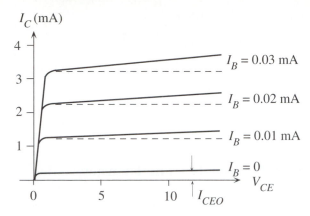

Figure 6.25   DC *I–V* characteristics of the *pnp* bipolar transistor in the CE configuration (exaggerated to highlight various effects)

Increasing I_B requires increasing V_{BE}, which increases I_C. Using Equations 6.37 and 6.38, we can obtain I_C in terms of I_B alone,

$$I_C = \beta I_B + \frac{1}{(1 - \alpha)} I_{CBO}$$

or

Active region collector current

$$I_C = \beta I_B + I_{CEO} \qquad\qquad [6.42]$$

where

$$I_{CEO} = \frac{I_{CBO}}{(1 - \alpha)} \approx \beta I_{CBO}$$

is the leakage current into the collector when the base is open circuited. This is much larger in the CE circuit than in the CB configuration.

Even when I_B is kept constant, I_C still exhibits a small increase with V_{CE}, which, according to Equation 6.42, indicates an increase in the current gain β with V_{CE}. This is due to the Early effect or modulation of the base width by V_{CB}, shown in Figure 6.22. Increasing V_{CE} increases V_{CB}, which increases W_{BC}, reduces W_B, and hence shortens τ_t. The resulting effect is a larger $\beta(\approx \tau_h/\tau_t)$.

When V_{CE} is less than V_{BE}, the collector junction becomes forward biased and Equation 6.42 is not valid. The collector current is then the difference between forward currents of emitter and collector junctions. The transistor operating in this region is said to be **saturated.**

6.6.4   Low-Frequency Small-Signal Model

The *pnp* bipolar transistor in the CE amplifier configuration is shown in Figure 6.26. The input circuit has a dc bias V_{BE} to forward bias the BE junction and the output circuit has a dc voltage, V_{CE} (larger than V_{BE}), to reverse bias the BC

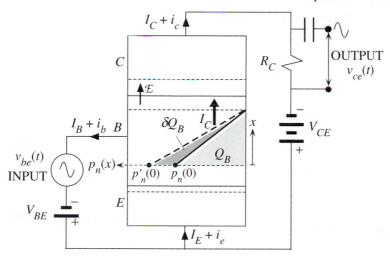

Figure 6.26   A *pnp* transistor operated in the active region in the common emitter amplifier configuration

The applied signal v_{be} modulates the dc voltage across the BE junction and hence modulates the injected hole concentration up and down about the dc value $p_n(0)$. The solid line shows $p_n(x)$ when only the dc bias V_{BE} is present. The dashed lines show how $p_n(x)$ is modulated up by a positive small signal v_{be} superimposed on V_{BE}.

junction. An input signal in the form of a small ac signal, v_{be}, is applied in series with the bias voltage, V_{BE}, and modulates the voltage across the BE junction about its dc value, V_{BE}. The varying voltage across the BE modulates $p_n(0)$ up and down about its dc value, which leads to a varying emitter current and hence to an almost identically varying collector current in the output circuit. The variation in the collector current is converted to an output voltage signal by the collector resistance R_C.

Since the BE junction is forward biased, the relationship between I_E and V_{BE} in Equation 6.39, of course, remains valid. We can differentiate Equation 6.39,

$$I_E = I_{EO} \exp\left(\frac{eV_{BE}}{kT}\right)$$

to relate small variations in I_E and V_{BE} as in the presence of small signals superimposed on dc values. For small signals, we have $v_{be} = \delta V_{BE}$, $i_b = \delta I_B$, $i_e = \delta I_E$, $i_c = \delta I_C$. Then, from Equation 6.42 we immediately see that $\delta I_C = \beta \delta I_B$, so that $i_c = \beta i_b$. Since $\alpha \approx 1$, $i_e \approx i_c$.

What is the advantage of the CE circuit over the CB configuration? First, the input current is the base current, which is about a factor of β smaller than the emitter current. The ac input resistance of the CE circuit is therefore a factor of β higher than that of the CB circuit. This means that the amplifier does not load the ac source; the input resistance of the amplifier is much greater than the internal (or output) resistance of the ac source. The small-signal input resistance r_{be} is

$$r_{be} = \frac{v_{be}}{i_b} = \frac{\delta V_{BE}}{\delta I_B} \approx \beta \frac{\delta V_{BE}}{\delta I_E} = \frac{\beta kT}{eI_E} \approx \frac{\beta \, 25}{I_C(\text{mA})} \qquad \text{[6.43]}$$

where we differentiated Equation 6.39. It is clear that for a collector current of 1 mA and a typical β of 100, the small-signal input resistance, r_e, in the CB amplifier is 25 Ω, whereas in the CE amplifier, the input resistance, r_{be}, is 2,500 Ω.

What is the CE voltage gain? The output signal, v_{ce} develops across R_C,

$$v_{ce} = \delta V_{CE} = R_C \, \delta I_C = R_C i_c$$

The voltage amplification is

$$A_V = \frac{v_{ce}}{v_{be}} = \frac{R_C i_c}{r_{be} i_b} = \frac{R_C \beta}{r_{be}} \approx \frac{R_C I_C(\text{mA})}{25} \qquad \text{[6.44]}$$

which is the same as that in the CB configuration. However, in the CE configuration the output to input current ratio, $i_c/i_b = \beta$, whereas this is almost unity in the CB configuration. Consequently, the CE configuration provides a greater power amplification, which is the second advantage of the CE circuit.

The input signal, v_{be}, gives rise to an output current, i_c. This input voltage to output current conversion is defined into a parameter called the **mutual conductance,** or **transconductance, g_m.**

Transconductance

$$g_m = \frac{i_c}{v_{be}} \approx \frac{\delta I_E}{\delta V_{BE}} = \frac{I_E(\text{mA})}{25} = \frac{1}{r_e} \qquad \text{[6.45]}$$

The voltage amplification of the CE amplifier is then

Voltage gain

$$A_V = g_m R_C \qquad \text{[6.46]}$$

We generally find it convenient to use a small-signal equivalent circuit for the low-frequency behavior of a BJT in the CE configuration. Between the base and emitter, the applied ac source voltage, v_s, sees only an input resistance of r_{be}, as shown in Figure 6.27. To underline the importance of the input resistance, the output (or the internal) resistance of the ac source is also shown. In the output circuit there is a voltage-controlled current source, i_c, which generates a current of $g_m v_{be}$. The current, i_c, passes through the load (or collector) resistance R_C across

AC source   Small signal equivalent circuit   Load

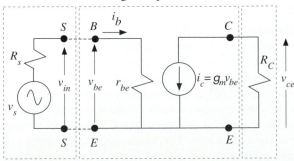

Figure 6.27   Low-frequency small-signal simplified equivalent circuit of the bipolar transistor in the CE configuration with a load resistor R_C in the collector circuit

which the voltage signal develops. As we are only interested in ac signals, the batteries are taken as a short circuit path for the ac current, which means that the internal resistances of the batteries are taken as zero. This model, of course, is only valid under normal and active operating conditions and small signals about dc values.

The bipolar transistor general dc current equation

$$I_C = \beta I_B$$

where $\beta \approx \tau_h/\tau_t$ is a material dependent constant, implies that the ac small-signal collector current is

$$\delta I_C = \beta \delta I_B \qquad \text{or} \qquad i_c = \beta i_b$$

Thus, the CE dc and ac small-signal current gains are the same. This is a reasonable approximation in the low frequency range, typically at frequencies below $1/\tau_h$. It is useful to have a relationship between β, g_m, and r_{be}. Using Equations 6.43 and 6.44 we have

$$\beta = g_m r_{be} \qquad\qquad \text{[6.47]} \quad \begin{array}{l}\beta \text{ at low}\\ \text{frequencies}\end{array}$$

In transistor data books, the dc current gain I_C/I_B, is denoted as h_{FE}, whereas the small-signal ac current gain, i_c/i_b, is denoted as h_{fe}. Except at high frequencies, $h_{fe} \approx h_{FE}$.

CE LOW FREQUENCY SMALL-SIGNAL EQUIVALENT CIRCUIT Consider a BJT with a β of 100. | **Example 6.9**
It is used as a CE amplifier in which the collector current is 2.5 mA and R_C is 1 kΩ. If the ac source has an rms voltage of 1 mV and an output resistance of 50 Ω, what is the rms output voltage? What are the input and output power and the overall power amplification?

Solution

As the collector current is 2.5 mA, the input resistance and the transconductance are

$$r_{be} = \frac{\beta\,25}{I_C(\text{mA})} = \frac{(100)(25)}{2.5} = 1000 \ \Omega$$

$$g_m = \frac{I_C(\text{mA})}{25} = \frac{2.5}{25} = 0.1 \ \text{A/V}$$

The voltage gain of the BJT small-signal equivalent circuit is

$$A_V = \frac{v_{ce}}{v_{be}} = g_m R_C = (0.1)(1000) = 100$$

When the ac source is connected to the B and E terminals (Figure 6.27), the input resistance r_{be} of the BJT loads the ac source so that v_{be} across BE is

$$v_{be} = v_s \frac{r_{be}}{(r_{be} + r_s)} = (1 \ \text{mV}) \frac{1000 \ \Omega}{(1000 \ \Omega + 50 \ \Omega)} = 0.952 \ \text{mV}$$

The output voltage is therefore

$$v_{ce} = A_V v_{be} = 100(0.952 \ \text{mV}) = 95.2 \ \text{mV}$$

The loading effect makes the output less than 100 mV. To reduce the loading of the ac source, we need to increase r_{be}, that is, reduce the collector current, but that also reduces the gain. So to keep the gain the same, we need to reduce I_C and increase R_C. However, R_C cannot be increased indefinitely because there is a dc voltage drop across R_C, given by $I_C R_C$, which must be less than the supply voltage and the maximum variation in the output signal.

The power amplification of the CE BJT itself is

$$A_P = \frac{i_c v_{ce}}{i_b v_{be}} = \beta A_V$$

$$A_P = (100)(100) = 10{,}000$$

The input power into the *BE* terminals is

$$P_{in} = v_{be} i_b = \frac{v_{be}^2}{r_{be}} = \frac{(0.952 \times 10^{-3}\,\text{V})^2}{1000\,\Omega} = 9.06 \times 10^{-10}\,\text{W} \qquad \text{or} \qquad 0.906\,\text{nW}$$

The output power is

$$P_{out} = P_{in} A_P = (9.06 \times 10^{-10})(10{,}000) = 9.06 \times 10^{-6}\,\text{W} \qquad \text{or} \qquad 9.06\,\mu\text{W}$$

6.7   JUNCTION FIELD EFFECT TRANSISTOR (JFET)

6.7.1   General Principles

The basic structure of the junction field effect transistor (JFET) with an *n*-type channel (*n*-channel) is depicted in Figure 6.28a. An *n*-type semiconductor slab is provided with contacts at its ends to pass current through it. These terminals are called **source** (*S*) and **drain** (*D*). Two of the opposite faces of the *n*-type semiconductor are heavily *p*-type doped to some small depth so that an *n*-type channel is formed between the source and drain terminals, as shown in Figure 6.28a. The two p^+ regions are normally electrically connected and are called the **gate** (*G*). As the gate is heavily doped, the depletion layers extend almost entirely into the *n*-channel, as shown in the figure. For simplicity we will assume that the two gate regions are identical (both p^+ type) and that the doping in the *n*-type semiconductor is uniform. We will define the *n*-channel to be the region of conducting *n*-type material contained between the two depletion layers.

The basic and idealized symmetric structure in Figure 6.28a is useful in explaining the principle of operation as discussed below but does not truly represent the structure of a typical practical device. A simplified schematic sketch of the cross section of a more practical device (as, for example, fabricated by the planar technology) is shown in Figure 6.28b where it is apparent that the two gate regions do not have identical doping and that, except for one of the gates, all contacts are on one surface.

We first consider the behavior of the JFET with the gate and source shorted ($V_{GS} = 0$), as shown in Figure 6.29a. The resistance between *S* and *D* is essentially

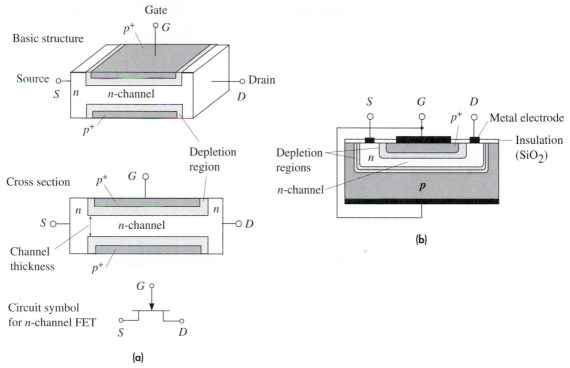

Figure 6.28

(a) The basic structure of the junction field effect transistor (JFET) with an *n*-channel. The two p^+ regions are electrically connected and form the gate.

(b) A simplified sketch of the cross section of a more practical *n*-channel JFET.

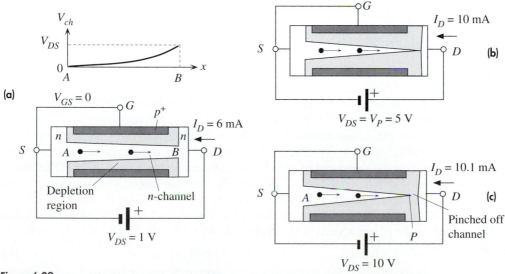

Figure 6.29

(a) The gate and source are shorted ($V_{GS} = 0$) and V_{DS} is small.

(b) V_{DS} has increased to a value that allows the two depletion layers to just touch, when $V_{DS} = V_P(5 \text{ V})$ and the p^+n junction voltage at the drain end, $V_{GD} = -V_{DS} = -V_P = -5$ V.

(c) V_{DS} is large ($V_{DS} > V_P$) so a short length of the channel is pinched off.

451

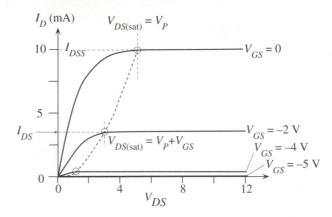

Figure 6.30   Typical I_D versus V_{DS} characteristics of a JFET for various fixed gate voltages V_{GS}

the resistance of the conducting n-channel between A and B, R_{AB}. When a positive voltage is applied to D with respect to $S(V_{DS} > 0)$, then a current flows from D to S, which is called the **drain current,** I_D. There is a voltage drop along the channel, between A and B, as indicated in Figure 6.29a. The voltage in the n-channel is zero at A and V_{DS} at B. As the voltage along the n-channel is positive, the p^+n junctions between the gates and the n-channel become progressively more reverse biased from A and to B. Consequently the depletion layers extend more into the channel and thereby decrease the thickness of the conducting channel from A to B.

Increasing V_{DS} increases the widths of the depletion layers, which penetrate more into the channel and hence result in more channel narrowing towards the drain. The resistance of the n-channel, R_{AB}, therefore increases with V_{DS}. The drain current therefore does not increase linearly with V_{DS} but falls below it because

$$I_D = \frac{V_{DS}}{R_{AB}}$$

and R_{AB} increases with V_{DS}. Thus I_D versus V_{DS} exhibits a sublinear behavior, as shown in the $V_{DS} < 5$ V region in Figure 6.30.

As V_{DS} increases further, the depletion layers extend more into the channel and eventually, when $V_{DS} = V_P(= 5V)$, the two depletion layers around B meet at point P at the drain end of the channel, as depicted in Figure 6.29b. The channel is then said to be "pinched off" by the two depletion layers. The voltage V_P is called the **pinch-off voltage.** It is equal to the magnitude of reverse bias needed across the p^+n junctions to make them just touch at the drain end. Since the actual bias voltage across the p^+n junctions at the drain end (B) is V_{GD}, the pinch-off occurs whenever

Pinch-off condition

$$V_{GD} = -V_P \qquad \qquad \textbf{[6.48]}$$

In the present case, gate to source is shorted, $V_{GS} = 0$, so that $V_{GD} = -V_{DS}$ and pinch-off occurs when $V_{DS} = V_P(5$ V$)$. The drain current from pinch-off onwards, as shown in Figure 6.30, does not increase significantly with V_{DS} for reasons given below. Beyond $V_{DS} = V_P$, there is a short pinched-off channel of length ℓ_{po}.

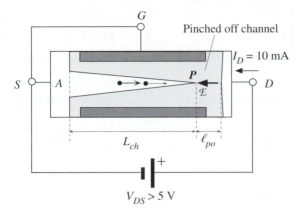

Figure 6.31   The pinched-off channel and conduction for $V_{DS} > V_P(5\text{ V})$

The pinched-off channel is a reverse-biased depletion region that separates the drain from the n- channel, as depicted in Figure 6.31. There is a very strong electric field, $\mathcal{E}$, in this pinched-off region in the D to S direction, from positive donors in the depletion layer toward the electrons in the n-channel. Electrons in the n-channel drift toward P, and when they arrive at P, they are swept across the pinched-off channel by $\mathcal{E}$. This process is similar to minority carriers in the base of a BJT reaching the collector junction depletion region, where the internal field there sweeps them across the depletion layer into the collector. Consequently the drain current is actually determined by the resistance of the conducting n-channel over L_{ch} from A to P in Figure 6.31 and not by the pinched-off channel.

As V_{DS} increases, most of the additional voltage simply drops across ℓ_{po} as this region is depleted of carriers and hence highly resistive. Point P, where the depletion layers first meet, moves slightly toward A, thereby slightly reducing the channel length, L_{ch}. Point P must still be at a potential V_P because it is this potential that just makes the depletion layers touch. Thus the voltage drop across L_{ch} remains as V_P. Beyond pinch-off then

$$I_D = \frac{V_P}{R_{AP}} \qquad (V_{DS} > V_P)$$

Since R_{AP} is determined by L_{ch}, which decreases slightly with V_{DS}, I_D increases slightly with V_{DS}. In many cases, I_D is conveniently taken to be saturated at a value I_{DSS} for $V_{DS} > V_P$. Typical I_D versus V_{DS} behavior is shown in Figure 6.30.

We now consider what happens when a negative voltage, say $V_{GS} = -2$ V, is applied to the gate with respect to the source, as shown in Figure 6.32a with $V_{DS} = 0$. The p^+n junctions are now reverse biased from the start, the channel is narrower, and the channel resistance is now larger than in the $V_{GS} = 0$ case. The drain current that flows when a small V_{DS} is applied, as in Figure 6.32b, is now smaller than in the $V_{GS} = 0$ case as apparent in Figure 6.30. The p^+n junctions are now progressively more reverse biased from V_{GS} at the source end to $V_{GD} = V_{GS} - V_{DS}$ at the drain end. We therefore need a smaller $V_{DS}(3$ V) to pinch

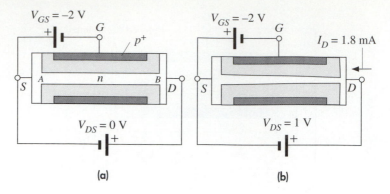

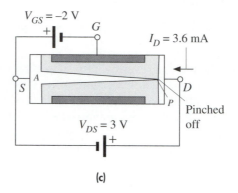

Figure 6.32

(a) The JFET with a negative V_{GS} voltage has a narrower n-channel at the start.
(b) Compared to the $V_{GS} = 0$ case, the same V_{DS} gives less I_D as the channel is narrower.
(c) The channel is pinched off at $V_{DS} = 3$ V sooner than the $V_{GS} = 0$ case, where it was $V_{DS} = 5$ V.

off the channel, as shown in Figure 6.32c. When $V_{DS} = 3$ V, the G to D voltage, V_{GD}, across the p^+n junctions at the drain end is -5 V, which is $-V_P$, so the channel becomes pinched-off. Beyond pinch-off, I_D is nearly saturated just as in the $V_{GS} = 0$ case, but its magnitude is obviously smaller as the thickness of the channel at A is smaller; compare Figures 6.29 and 6.32. In the presence of V_{GS}, the pinch-off occurs at $V_{DS} = V_{DS(sat)}$, and from Equation 6.48.

Pinch-off condition

$$V_{DS(sat)} = V_P + V_{GS} \qquad \text{[6.49]}$$

where V_{GS} is a negative voltage (reducing V_P). Beyond pinch-off when $V_{DS} > V_{DS(sat)}$, the point P where the channel is *just pinched* still remains at potential $V_{DS(sat)}$, given by Equation 6.49.

For $V_{DS} > V_{DS(sat)}$, I_D becomes nearly saturated at a value denoted as I_{DS}, which is indicated in Figure 6.30. When G and S are shorted ($V_{GS} = 0$), I_{DS} is called I_{DSS} (which stands for I_{DS} with shorted gate to source). Beyond pinch-off, with negative V_{GS}, I_{DS} is

$$I_D \approx I_{DS} \approx \frac{V_{DS(sat)}}{R_{AP}(V_{GS})} = \frac{V_P + V_{GS}}{R_{AP}(V_{GS})} \qquad V_{DS} > V_{DS(sat)} \qquad \text{[6.50]}$$

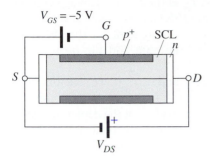

Figure 6.33   When $V_{GS} = -5$ V, the depletion layers close the whole channel from the start, at $V_{DS} = 0$.

As V_{DS} is increased, there is a very small drain current, which is the small reverse leakage current due to thermal generation of carriers in the depletion layers.

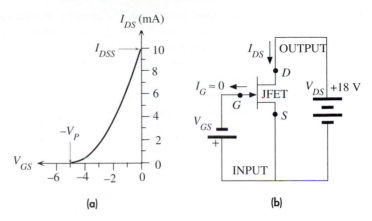

Figure 6.34
(a) Typical I_{DS} versus V_{GS} characteristics of a JFET.
(b) The dc circuit where V_{GS} in the gate–source circuit (input) controls the drain current I_{DS} in the drain–source (output) circuit in which V_{DS} is kept constant and large $(V_{DS} > V_P)$.

where $R_{AP}(V_{GS})$ is the effective resistance of the conducting n-channel from A to P (Figure 6.32b), which depends on the channel thickness and hence on V_{GS}. The resistance increases with more negative gate voltage as this increases the reverse bias across the p^+n junctions, which leads to the narrowing of the channel. For example, when $V_{GS} = -4$ V, the channel thickness at A becomes narrower than in the case with $V_{GS} = -2$ V, thereby increasing the resistance, R_{AP}, of the conducting channel and therefore decreasing I_{DS}. Further, there is also a reduction in the drain current by virtue of $V_{DS(sat)}$ decreasing with negative V_{GS}, as apparent in Equation 6.50. Figure 6.30 shows the effect of the gate voltage on the I_D versus V_{DS} behavior. The two effects, that from $V_{DS(sat)}$ and that from $R_{AP}(V_{GS})$ in Equation 6.50, lead to I_{DS} almost decreasing parabolically with $-V_{GS}$.

When the gate voltage is such that $V_{GS} = -V_P (= -5$ V) with the source and drain shorted $(V_{DS} = 0)$, then the two depletion layers touch over the entire channel length and the whole channel is closed, as illustrated in Figure 6.33. The channel is said to be off. The only drain current that flows when a V_{DS} is applied is due to the thermally generated carriers in the depletion layers. This current is very small.

Figure 6.30 summarizes the full I_D versus V_{DS} characteristics of the n-channel JFET at various gate voltages, V_{GS}. It is apparent that I_{DS} is relatively independent of V_{DS} and that is controlled by the gate voltage, V_{GS}, as expected by Equation 6.50. This is analogous to the BJT in which the collector current, I_C, is controlled by the base-emitter bias voltage, V_{BE}. Figure 6.34a shows the dependence of I_{DS} on the gate voltage V_{GS}. The transistor action is the control of the drain current, I_{DS}, in the drain-source (output) circuit by the voltage, V_{GS}, in the gate–source (input) circuit), as shown in Figure 6.34b. This control is only possible if $V_{DS} > V_{DS(sat)}$. When $V_{GS} = -V_p$, the drain current is nearly zero because the channel has been totally pinched off. This gate–source voltage is denoted by $V_{GS(off)}$ as the drain current has been switched off. Furthermore, we should note that as V_{GS} reverse biases the p^+n junction, the current into the gate, I_G, is the reverse leakage current of these junctions. It is usually very small. In some JFETs, I_G is as low as a fraction of a nanoampere. We should also note that the circuit symbol for the JFET, as shown in Figure 6.28a, has an arrow to identify the gate and the pn junction direction.

Is there a convenient relationship between I_{DS} and V_{GS}? If we calculate the effective resistance, R_{AP}, of the n-channel between A and P, we can obtain its dependence on the channel thickness, and thus on the widths of the depletion layers and hence on V_{GS}. We can then find I_{DS} from Equation 6.50. It turns out that a simple parabolic dependence seems to represent the data reasonably well,

*Beyond
pinch-off*

$$I_{DS} = I_{DSS}\left[1-\left(\frac{V_{GS}}{V_{GS(off)}}\right)\right]^2 \qquad \text{[6.51]}$$

where $V_{GS(off)}$ is defined as $-V_P$, that is, that gate–source voltage that just pinches off the channel. The pinchoff voltage, V_P, here is a positive quantity because it was introduced through $V_{DS(sat)}$. $V_{GS(off)}$ however is negative, $-V_P$. We should note two important facts about the JFET. Its name originates from the effect that modulating the electric field in the reverse-biased depletion layers (by changing V_{GS}) varies the depletion layer penetration into the channel and hence the resistance of the channel. The transistor action hence can be thought of as being based on a field effect. Since there is a p^+n junction between the gate and the channel, the name has become JFET. This junction in reverse bias provides the isolation between the gate and channel.

Secondly, the region beyond pinch-off, where Equations 6.50 and 6.51 hold, is commonly called the **current saturation region,** as well as **constant current region** and **pentode region.** The term **saturation** should not be confused with similar terms used for saturation effects in bipolar transistors. A saturated BJT cannot be used as an amplifier, but JFETs are invariably used as amplifiers in the saturated current region.

6.7.2   JFET Amplifier

The transistor action in the JFET is the control of I_{DS} by V_{GS}, as shown in Figure 6.34. The input circuit is therefore the gate–source circuit containing V_{GS} and the output circuit is the drain–source circuit in which the drain current, I_{DS}, flows. The JFET is almost never used with the pn junction between the gate and channel forward biased ($V_{GS} > 0$) as this would lead to a very large gate current and near shorting of the gate to source voltage. With V_{GS} limited to negative voltages, the maximum current in the output circuit can only be I_{DSS}, as shown in Figure 6.34a. The maximum input voltage, V_{GS}, should therefore give an I_{DS} less than I_{DSS}.

Figure 6.35a shows a simplified illustration of a typical JFET voltage amplifier. As the source is common to both the input and output circuits, this is called a **common source** (CS) **amplifier.** The input signal is the ac source, v_{gs}, connected in series with a negative dc bias voltage, V_{GG}, of -1.5 V in the GS circuit. First we

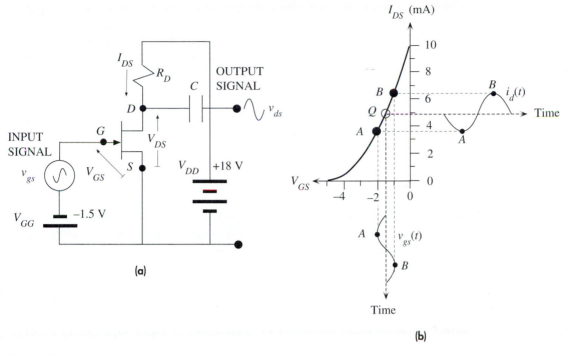

Figure 6.35
(a) Common source (CS) ac amplifier using a JFET.
(b) Explanation of how I_D is modulated by the signal, v_{gs}, in series with the dc bias voltage V_{GG}.

will find out what happens when there is no ac signal in the circuit ($v_{gs} = 0$). The dc supply (-1.5 V) in the input provides a negative dc voltage to the gate and therefore gives a dc current I_{DS} in the output circuit (less than I_{DSS}). Figure 6.35b shows that when $V_{GS} = -1.5$ V, point Q on the I_{DS} versus V_{GS} characteristics gives $I_{DS} = 4.9$ mA. Point Q, which determines the dc operation, is called the **quiescent point.**

The ac source, v_{gs}, is connected in series with the nagative dc bias voltage. It therefore modulates V_{GS} up and down about -1.5 V with time, as shown in Figure 6.35b. Suppose that v_{gs} varies sinusoidally between -0.5 V and $+0.5$ V. Then, as shown in Figure 6.35b when v_{gs} is -0.5 V (point A) $V_{GS} = -2.0$ V and the drain current is given by point A on the I_{DS}–V_{GS} curve and is about 3.6 mA. When v_{gs} is $+0.5$ V (point B) then $V_{GS} = -1.0$ V and drain current is given by point B on the I_{DS}–V_{GS} curve and is about 6.4 mA. The input variation from -0.5 V to $+0.5$ V has thus been converted to a drain current variation from 3.6 mA to 6.4 mA as indicated in Figure 6.35b. We could have just as easily calculated the drain current from Equation 6.51. Table 6.1 summarizes what happens to the drain current as the ac input voltage is varied about zero.

The change in the drain current with respect to its dc value is the output signal current denoted as i_d. Thus at A,

$$i_d = 3.6 - 4.9 = -1.3 \text{ mA}$$

Table 6.1   Voltage and current in the common source amplifier of Figure 6.35a

v_{gs} (V)	V_{GS} (V)	I_{DS} (mA)	i_d (mA)	$V_{DS} = V_{DD} - I_{DS}R_D$	v_{ds} (V)	Voltage Gain	Comment
0	-1.5	4.9	0	8.2	0		dc conditions, point Q
-0.5	-2.0	3.6	-1.3	10.8	$+2.6$	-5.2	Point A
$+0.5$	-1.0	6.4	$+1.5$	5.2	-3.0	-6	Point B

and at B,

$$i_d = 6.4 - 4.9 = 1.5 \text{ mA}$$

The variation in the output current is not quite symmetric as that in the input signal, v_{gs}, because the I_{DS}–V_{GS} relationship, Equation 6.51, is not linear.

The drain current variations in the DS circuit are converted to voltage variations by the resistance R_D. The voltage across DS is

$$V_{DS} = V_{DD} - I_{DS}R_D \qquad \text{[6.52]}$$

where V_{DD} is the bias battery voltage in the DS circuit. Thus, variations in I_{DS} result in variations in V_{DS} that are in the opposite direction or 180° out of phase. The ac output voltage between the D and S is tapped out through a capacitor C, as shown in the figure. The capacitor C simply blocks the dc. Suppose that $R_D = 2000\ \Omega$ and $V_{DD} = 18$ V, then using Equation 6.52 we can calculate the dc value of V_{DS} and also the minimum and maximum values of V_{DS}, as shown in Table 6.1.

It is apparent that as v_{gs} varies from -0.5 V, at A, to $+0.5$ V, at B, V_{DS} varies from 8.2 V to 5.2 V, respectively. The change in V_{DS} with respect to dc is what constitutes the output signal, v_{ds}, as only the ac is tapped out. From Equation 6.52, the change in V_{DS} is related to the change in I_{DS} by

$$v_{ds} = -R_D i_d \qquad \text{[6.53]}$$

Thus the output, v_{ds}, changes from -3.0 V to 2.6 V. The peak-to-peak voltage amplification is

$$A_{V(\text{pk-pk})} = \frac{\Delta V_{DS}}{\Delta V_{GS}} = \frac{v_{ds(\text{pk-pk})}}{v_{gs(\text{pk-pk})}} = \frac{-3 \text{ V} - (2.6 \text{ V})}{0.5 \text{ V} - (-0.5 \text{ V})} = -5.6$$

The negative sign represents the fact that the output and input voltages are out of phase by 180°. This can be seen from Table 6.1 that even though the ac input signal, v_{gs}, is symmetric about zero, ± 0.5 V, the ac output signal, v_{ds}, is not symmetric, which is due to the I_{DS} versus V_{GS} curve being nonlinear, and thus varies between -3.0 V and 2.6 V. If we were to calculate the voltage amplification for the most negative input signal, we would find -5.2, whereas for the most positive input signal it would be -6. The peak-to-peak voltage amplification, which was -5.6, represents a mean gain taking both negative and positive input signals into account.

The amplification can of course be increased by increasing R_D, but we must maintain V_{DS} at all times above $V_{DS(sat)}$ (beyond pinch-off) to ensure that the drain current, I_{DS}, in the output circuit is only controlled by V_{GS} in the input circuit.

When the signals are small about dc values, we can use differentials to represent small signals. For example, $v_{gs} = \delta V_{GS}$, $i_d = \delta I_{DS}$, $v_{ds} = \delta V_{DS}$, and so on. The variation δI_{DS} due to δV_{GS} about the dc value may be used to define a **mutual transconductance** g_m (sometimes denoted as g_{fs}) for the JFET,

$$g_m = \frac{dI_{DS}}{dV_{GS}} \approx \frac{\delta I_{DS}}{\delta V_{GS}} = \frac{i_d}{v_{gs}}$$

Definition of JFET transconductance

This transconductance can be found by differentiating Equation 6.51,

$$g_m = \frac{dI_{DS}}{dV_{GS}} = -\frac{2I_{DSS}}{V_{GS(off)}}\left[1 - \left(\frac{V_{GS}}{V_{GS(off)}}\right)\right]$$

$$g_m = -\frac{2[I_{DSS}I_{DS}]^{1/2}}{V_{GS(off)}} \qquad \text{[6.54]}$$

JFET transconductance

The output signal current is

$$i_d = g_m v_{gs}$$

so that using Equation 6.53, the small-signal voltage amplification is

$$A_V = \frac{v_{ds}}{v_{gs}} = \frac{-R_D(g_m v_{gs})}{v_{gs}} = -g_m R_D \qquad \text{[6.55]}$$

Small-signal voltage gain

Equation 6.55 is only valid under small-signal conditions in which the variations about the dc values are small compared with the dc values themselves. The negative sign indicates that v_{ds} and v_{gs} are 180° out of phase.

THE JFET AMPLIFIER   Consider the *n*-channel JFET common source amplifier shown in Figure 6.35a. The JFET has an I_{DSS} of 10 mA and a pinch-off voltage, V_P, of 5 V as in Figure 6.35b. Suppose that the gate dc bias voltage supply $V_{GG} = -1.5$ V, the drain circuit supply $V_{DD} = 18$ V, and $R_D = 2000\ \Omega$. What is the voltage amplification for small signals? How does this compare with the peak-to-peak amplification, -5.6, found for an input signal that had a peak-to-peak value of 1 V?

Example 6.10

Solution

We first calculate the operating conditions at the bias point with no ac signals. This corresponds to point Q in Figure 6.35b. The dc bias voltage V_{GS} across the gate to source is -1.5 V. The resulting dc drain current I_{DS} can be calculated from Equation 6.51 with $V_{GS(off)} = -V_P = -5$ V:

$$I_{DS} = I_{DSS}\left[1 - \left(\frac{V_{GS}}{V_{GS(off)}}\right)\right]^2 = (10\ \text{mA})\left[1 - \left(\frac{-1.5}{-5}\right)\right]^2 = 4.9\ \text{mA}$$

The transconductance at this dc current (at Q) is given by

$$g_m = -\frac{2[I_{DSS}I_{DS}]^{1/2}}{V_{GS(off)}} = -\frac{2[(10 \times 10^{-3})(4.9 \times 10^{-3})]^{1/2}}{-5} = 2.8 \times 10^{-3}\ \text{A/V}$$

The voltage amplification of small signals about point Q is

$$A_V = -g_m R_D = -(2.8 \times 10^{-3})(2000) = -5.6.$$

This turns out to be the same as the peak-to-peak voltage amplification we calculated in Table 6.1. When the input ac signal, v_{gs}, varies between -0.5 and $+0.5$ V, as in Table 6.1, the output signal is not symmetric. It varies between -3 V and 2.8 V, so the voltage gain depends on the input signal. The amplifier is then said to exhibit nonlinearity.

6.8   METAL-OXIDE-SEMICONDUCTOR FIELD EFFECT TRANSISTOR (MOSFET)

6.8.1   Field Effect and Inversion

The metal-oxide-semiconductor field effect transistor is based on the effect of a field penetrating into a semiconductor. Its operation can be understood by first considering a parallel plate capacitor with metal electrodes and vacuum as insulation in between, as shown in Figure 6.36a. When a voltage V is applied between the plates, charges $+Q$ and $-Q$ (where $Q = CV$) appear on the plates and there is an electric field given by $\mathcal{E} = V/L$. The origins of these charges are the conduction electrons for $-Q$ and exposed positively charged metal ions for $+Q$. Metallic bonding is based on all the valence electrons forming a sea of conduction electrons and permeating the space between metal ions that are fixed at crystal lattice sites. As the electrons are mobile they are readily displaced by the field. Thus in the lower plate $\mathcal{E}$ displaces some of the conduction electrons to the surface to form $-Q$. In the top plate $\mathcal{E}$ displaces some electrons from the surface into the bulk to expose positively charged metal ions to form $+Q$.

Suppose that the plate area is 1 cm^2 and spacing is 0.1 μm and that we apply 2 V across it. The capacitance C is 8.85 nF and the magnitude of charge Q on each plate is 1.77×10^{-8} C, which corresponds to 1.1×10^{11} electrons. A typical metal such as copper has something like 1.9×10^{15} atoms per cm^2 on the surface. Thus, there will be that number of positive metal ions and electrons on the surface (assuming one conduction electron per atom). The charges $+Q$ and $-Q$ can therefore be generated by the electrons and metal ions at the surface alone. For example, if one in every 1.7×10^4 electrons on the surface moves one atomic spacing (~ 0.3 nm) into the bulk, then the surface will have a charge of $+Q$ due to exposed positive metal ions. It is clear that, for all practical purposes, the electric field does not penetrate into the metal and terminates at the metal surface.

The same is not true when one of the electrodes is a semiconductor, as shown in Figure 6.36b where the structure now is a metal-insulator-semiconductor. Suppose that we replace the lower metal with a p-type semiconductor with an acceptor concentration of 10^{15} cm^{-3}. The number of acceptor atoms on the surface[9] is 1×10^{10} cm^{-2}. We may assume that at room temperature all the acceptors are ionized and thus negatively charged. It is immediately apparent that we do not have a sufficient number of negative acceptors at the surface to generate the charge $-Q$.

| [9] Surface concentration of atoms (atoms per unit area) can be found from $n_{\text{surf}} \approx (n_{\text{bulk}})^{2/3}$.

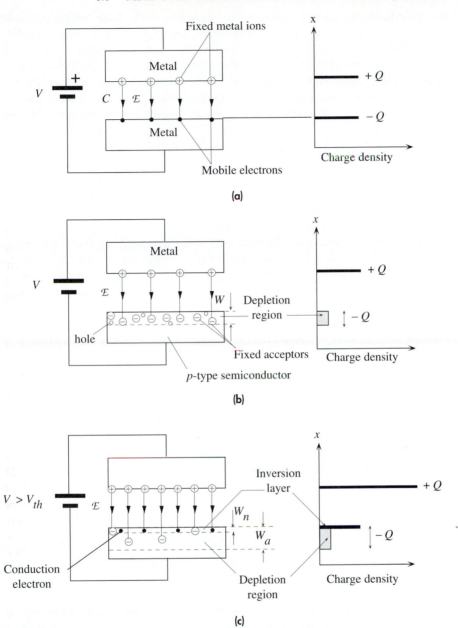

Figure 6.36 The field effect

(a) In a metal-air-metal capacitor, all the charges reside on the surface.

(b) Illustration of field penetration into a p-type semiconductor.

(c) As the field increases, eventually when $V > V_{th}$, an inversion layer is created near the surface in which there are conduction electrons.

We must therefore also expose negative acceptors in the bulk, which means that the field must penetrate into the semiconductor. Holes in the surface region of the semiconductor become repelled toward the bulk and thereby expose more negative acceptors. We can estimate the width W into which the field penetrates since the total negative charge exposed, $eAWN_a$, must be Q, We find that W is of the order of 1 μm, which is something like 4,000 atomic layers. Our conclusion is that the field penetrates into a semiconductor by an amount that depends on the doping concentration.

The penetrating field into the semiconductor drifts away most of the holes in this region and thereby exposes negatively charged acceptors to make up the charge $-Q$. The region into which the field penetrates has lost holes and is therefore depleted of its equilibrium concentration of holes. We refer to this region as a **depletion layer.** As long as $p > n$ even though $p \ll N_a$, this still has p-type characteristics as holes are in the majority.

If the voltage increases further, $-Q$ also increases, as the field becomes stronger and penetrates more into the semiconductor but eventually it becomes more difficult to make up the charge $-Q$ by simply extending the depletion layer width W into the bulk. It becomes possible (and more favorable) to attract conduction electrons into the depletion layer and form a thin electron layer of width W_n near the surface. The charge $-Q$ is now made up of the fixed negative charge of acceptors in W_a and of conduction electrons in W_n, as shown in Figure 6.36c. Further increases in the voltage do not change the width W_a of the depletion layer but simply increase the electron concentration in W_n. Where do these electrons come from as the semiconductor is doped p-type? Some are attracted into the depletion layer from the bulk, where they were minority carriers. But most are thermally generated by the breaking of Si–Si bonds (i.e., across the bandgap) in the depleted layer. Thermal generation in the depletion layer generates electron–hole pairs that become separated by the field. The holes are then drifted by the field into the bulk and the electrons toward the surface. Recombination of the thermally generated electrons and holes with other carriers is greatly reduced because the depletion layer has so few carriers. Since the electron concentration in the electron layer exceeds the hole concentration and this layer is within a normally p-type semiconductor, we call this an **inversion layer.**

It is now apparent that increasing the field in the metal-insulator-semiconductor device first creates a depletion layer and then an inversion layer at the surface when the voltage exceeds some threshold value, V_{th}. This is the basic principle of the field effect device. As long as $V > V_{th}$, any increase in the field and hence $-Q$ leads to more electrons in the inversion layer, whereas the width of the depletion layer W_a and hence the quantity of fixed negative charge remain constant. The insulator between the metal and the semiconductor, that is, vacuum in Figure 6.36, is typically SiO_2 in many devices.

6.8.2   Enhancement MOSFET

Figure 6.37 shows the basic structure of an enhancement n-channel MOSFET device (NMOSFET). A metal-insulator-semiconductor structure is formed between a p-type Si substrate and an aluminum electrode, which is called the gate

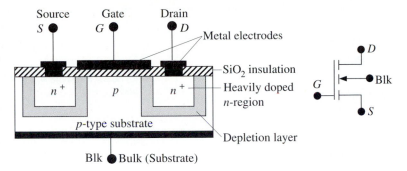

Figure 6.37   The basic structure of the enhancement MOSFET and its circuit symbol

(G). The insulator is the SiO_2 oxide grown during fabrication. There are two n^+ doped regions at the ends of the MOS that form the source (S) and drain (D). A metal contact is also made to the p-type Si substrate (or the bulk), which in many devices is connected to the source terminal as shown in Figure 6.37. Further, many MOSFETs have a degenerately doped polycrystalline Si material as the gate that serves the same function as the metal electrode.

With no voltage applied to the gate, S to D is an n^+pn^+ structure that is always reverse biased whatever the polarity of the source to drain voltage. However, if the substrate (bulk) is connected to the source, a negative V_{DS} will forward bias the n^+p junction between the drain and the substrate. As the n-channel MOSFET device is not normally used with a negative V_{DS}, we will not consider this polarity.

When a positive voltage less than V_{th} is applied to the gate, $V_{GS} < V_{th}$, as shown in Figure 6.38a, the p-type semiconductor under the gate develops a depletion layer as a result of the expulsion of holes into the bulk, just as in Figure 6.36b. Since S and D are isolated by a low conductivity p-doped region that has a depletion layer from S to D, no current can flow for any positive V_{DS}.

With $V_{DS} = 0$, as soon as V_{GS} is increased beyond the threshold voltage, V_{th}, an n-channel inversion layer is formed within the depletion layer under the gate and immediately below the surface, as shown in Figure 6.38b. This n-channel links the two n^+ regions of source and drain. We then have a continuous n-type material with electrons as mobile carriers between the source and drain. When a small V_{DS} is applied, a drain current, I_D, flows that is limited by the resitance of the n-channel R_{n-ch}

$$I_D = \frac{V_{DS}}{R_{n-ch}} \qquad\qquad \text{[6.56]}$$

Thus, I_D initially increases with V_{DS} almost linearly, as shown in Figure 6.38b.

The voltage variation along the channel is from zero at A (source end) to V_{DS} at B (drain end). The gate to n-channel voltage is then V_{GS} at A and $V_{GD} = V_{GS} - V_{DS}$ at B. Thus point A depends only on V_{GS} and remains undisturbed by V_{DS}. As V_{DS} increases, the voltage at B (V_{GD}) deceases and thereby causes less inversion. This means that the channel gets narrower from A to B and its resistance, R_{n-ch}, increases with V_{DS}. I_D versus V_{DS} then falls increasingly below the $I_D \propto V_{DS}$ line. Eventually when the gate to n-channel voltage at B decreases to just below V_{th}, the

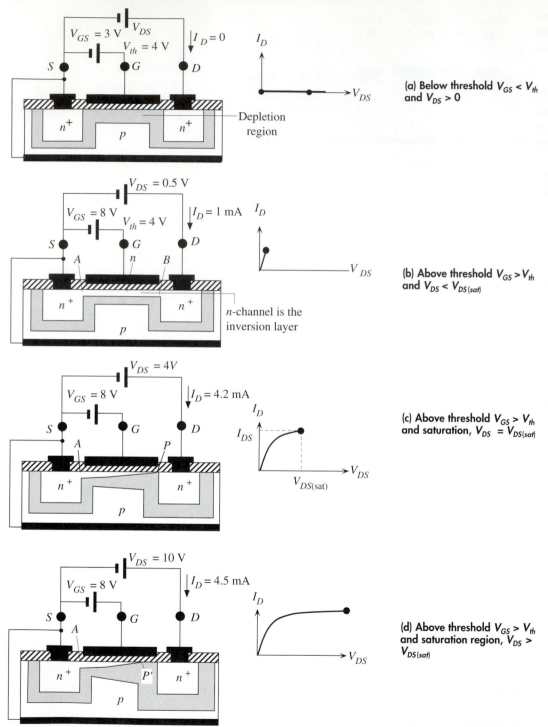

Figure 6.38 The MOSFET I_D versus V_{DS} characteristics

inversion layer at B disappears and a depletion layer is exposed, as illustrated in Figure 6.38c. The n-channel becomes pinched off at this point, P. This occurs when $V_{DS} = V_{DS(sat)}$, satisfying

$$V_{GD} = V_{GS} - V_{DS(sat)} = V_{th} \qquad \text{[6.57]}$$

It is apparent that the whole process of the narrowing of the n-channel and its eventual pinch-off is similar to the operation of the n-channel JFET. When the drifting electrons in the n-channel reach P, the large electric field within the very narrow depletion layer at P sweeps the electrons across into the n^+ drain. The current is limited by the supply of electrons from the n-channel to the depletion layer at P, which means that it is limited by the effective resistance of the n-channel between A and P.

When V_{DS} exceeds $V_{DS(sat)}$, the additional V_{DS} drops mainly across the highly resistive depletion layer at P, which extends slightly to P' toward A, as shown in Figure 6.38d. At P', the gate to channel voltage must still be just V_{th} as this is the voltage required to just pinch off the channel and just eliminate inversion. The widening of the depletion layer (from B to P') at the drain end with V_{DS}, however, is small compared with the channel length, AB. The resistance of the channel from A to P' does not change significantly with increasing V_{DS}, which means that the drain current is then nearly saturated at I_{DS},

$$I_D \approx I_{DS} \approx \frac{V_{DS(sat)}}{R_{AP'(n-ch)}} \qquad V_{DS} > V_{DS(sat)} \qquad \text{[6.58]}$$

As $V_{DS(sat)}$ depends on V_{GS}, so does I_{DS}. The overall I_{DS} versus V_{DS} characteristics for various fixed gate voltages, V_{GS}, of a typical enhancement MOSFET is shown in Figure 6.39a. It can be seen that there is only a slight increase in I_{DS} with V_{DS} beyond $V_{DS(sat)}$. The I_{DS} versus V_{GS} when $V_{DS} > V_{DS(sat)}$ characteristics are shown in

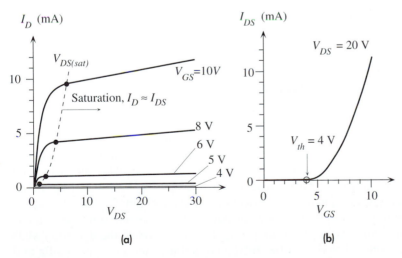

Figure 6.39   (a) Typical I_D versus V_{DS} characteristics of an enhancement MOSFET ($V_{th} = 4$ V) for various fixed gate voltages V_{GS}. (b) Dependence of I_{DS} on V_{GS} at a given $V_{DS}(> V_{DS(sat)})$.

Figure 6.39b. It is apparent that as long as $V_{DS} > V_{DS(sat)}$, the saturated drain current, I_{DS}, in the source–drain (or output) circuit is almost totally controlled by the gate voltage V_{GS} in the source–gate (or input) circuit. This is what constitutes the MOSFET action. Variations in V_{GS} then lead to variations in the drain current, I_{DS} (just as in the JFET), which forms the basis of the MOSFET amplifier. The term *enhancement* refers to the fact that a gate voltage exceeding V_{th} is required to enchance a conducting channel between the source and drain. This contrasts with the JFET where the gate voltage depletes the channel and decreases the drain current.

The experimental relationship between I_{DS} and V_{GS} (when $V_{DS} > V_{DS(sat)}$) has been found to be best described by a parabolic equation similar to that for the JFET, except that now V_{GS} enhances the channel when $V_{GS} > V_{th}$ so that I_{DS} exists only when $V_{GS} > V_{th}$,

Enhancement NMOSFET

$$I_{DS} = K(V_{GS} - V_{th})^2 \qquad\qquad [6.59]$$

where K is a constant. For an ideal MOSFET, it can be expressed as

$$K = \frac{Z\mu_e\varepsilon}{2Lt_{ox}}$$

where μ_e is the electron drift mobility in the channel, L and Z are the length and width of the gate controlling the channel, and ε and t_{ox} are the permittivity ($\varepsilon_r\varepsilon_o$) and thickness of the oxide insulation under the gate. According to Equation 6.59, I_{DS} is independent of V_{DS}. The shallow slopes of the I_D versus V_{DS} lines beyond $V_{DS(sat)}$ in Figure 6.39a can be accounted for by writing Equation 6.59 as

Enhancement NMOSFET

$$I_{DS} = K(V_{GS} - V_{th})^2(1 - \lambda V_{DS}) \qquad\qquad [6.60]$$

where λ is a constant that is typically 0.01 V^{-1}. If we extend the I_{DS} versus V_{DS} lines, they intersect the $-V_{DS}$ axis at $1/\lambda$, which is called the **Early voltage.** It should be apparent that I_{DSS}, which is I_{DS} with the gate and source shorted ($V_{GS} = 0$), is zero and is not a useful quantity in describing the behavior of the enhancement MOSFET.

6.8.3   Depletion MOSFET

The basic difference between the depletion and enhancement MOSFETs is that the depletion device already has a conducting channel established between the source and the drain by the diffusion of dopants during fabrication. The function of the gate voltage is to modulate the channel conductance by the field effect. Figure 6.40a shows the basic structure of a depletion *n*-channel MOSFET. The *n*-channel is diffused in during fabrication and has a doping level comparable with the *p*-type bulk. There is therefore a depletion layer extending into the channel from the bulk along its length from A to B. The operation of the depletion MOSFET is almost identical to that of the JFET and is illustrated in Figure 6.40a to d.

The presence of a conducting channel means that there is conduction even when the gate is shorted to the source ($V_{GS} = 0$). For small V_{DS}, I_D increases with

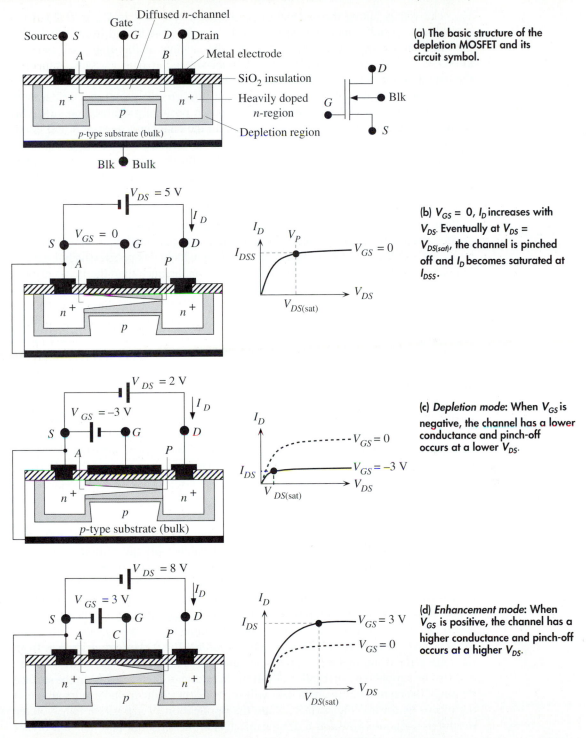

(a) The basic structure of the depletion MOSFET and its circuit symbol.

(b) $V_{GS} = 0$, I_D increases with V_{DS}. Eventually at $V_{DS} = V_{DS(sat)}$, the channel is pinched off and I_D becomes saturated at I_{DSS}.

(c) *Depletion mode*: When V_{GS} is negative, the channel has a lower conductance and pinch-off occurs at a lower V_{DS}.

(d) *Enhancement mode*: When V_{GS} is positive, the channel has a higher conductance and pinch-off occurs at a higher V_{DS}.

Figure 6.40   The depletion MOSFET I_D versus V_{DS} characteristics

V_{DS}. The voltage along the channel increases from zero at A to V_{DS} at B. Thus the n-channel to p-bulk junction is progressively more reverse biased from A to B. The depletion layer therefore extends more into the channel at the drain end. Further, although the gate voltage is zero, the channel voltage at B is V_{DS} with respect to the gate. Put differently, the gate appears to have a voltage $-V_{DS}$ with respect to the channel at B. The resulting field (directed toward the gate) penetrates into the channel, expels electrons, and hence exposes positively charged donors. This creates a depletion layer whose penetration into the channel is zero at A and maximum at B. As V_{DS} is increased, both depletion layers penetrate more into the channel and eventually, when $V_{DS} = V_P(5 \text{ V})$, they pinch off the channel at P, as shown in Figure 6.40b. Beyond pinch-off, I_D is nearly saturated at I_{DS} just as in the JFET. The physical process that leads to the saturation of I_D is the same as that in the JFET.

When a negative bias is applied to the gate, for example, $V_{GS} = -3$ V, as shown in Figure 6.40c, then the field from the gate penetrates into the channel and thereby creates a depletion layer even at A. The depletion layer penetration at A is determined by $V_{GS}(= -3 \text{ V})$ and at B by $V_{GD}(= V_{GS} - V_{DS})$, which gets progressively larger as V_{DS} increases. The channel therefore has a lower conductance than the $V_{GS} = 0$ case. Further, as V_{DS} increases, the depletion layer from the channel–bulk junction also penetrates into the channel and this penetration is largest at the drain end. Thus as V_{DS} increases, the channel is narrowed by both depletion layers and is pinched off at P when $V_{DS} = V_{DS(sat)}$. The drain current is smaller than the $V_{GS} = 0$ case because the gate voltage has already narrowed the channel. Since the application of a negative voltage to the gate always depletes the n-channel, this is called the **depletion mode of operation.**

The drain current, beyond pinch-off, when $V_{DS} > V_{DS(sat)}$, is almost saturated at I_{DS} and is essentially controlled by V_{GS}. As V_{GS} becomes more negative, the field from the gate penetrates further into the channel and depletes it even more. Not surprisingly, it turns out that I_{DS} and V_{DS} can be related by the same expression as that for the n-channel JFET and frequently written in the form

Depletion
NMOSFET

$$I_{DS} = I_{DSS}\left[1 - \left(\frac{V_{GS}}{V_{GS(off)}}\right)\right]^2 \tag{6.61}$$

where I_{DSS} is the saturated drain current when $V_{GS} = 0$ (Figure 6.40b) and $V_{GS(off)} = -V_P$ is the gate–source voltage that pinches off the channel.

Can the depletion MOSFET operate in the enhancement mode where it will enhance the conductance of the channel? If we apply a positive voltage to the gate, as depicted in Figure 6.40d, then the field from the gate will again penetrate the semiconductor, but now it needs additional negative charges in the channel. Additional electrons must then accumulate in the channel near the surface. Where do these electrons come from? There are two sources. First, they are drifted in from the source and drain regions, which are n^+ and hence have plenty of electrons. They can also be supplied by thermal generation (breaking of Si–Si bonds) in the surface region of the channel as the field here will separate the electron–hole pairs and drift the electrons toward the surface. Thus the application of a positive gate voltage results in a channel that has a greater electron concentration and hence a higher

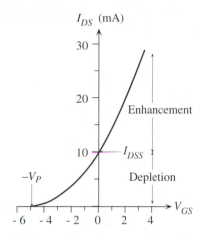

Figure 6.41   Typical I_{DS} versus V_{GS} characteristics of a depletion n-channel MOSFET

conductance. The drain current therefore increases with positive V_{GS}. Figure 6.41 shows how I_{DS} is controlled by V_{GS} and identifies the depletion and enhancement modes of operation.

We can illustrate the effect of a positive V_{GS} on $V_{DS(sat)}$ with an example. When V_{GS} is zero, $V_{DS} = 5$ V was sufficient to cause pinch-off, as in Figure 6.40b. Suppose that now $V_{GS} = 3$ V, but $V_{DS}(= 5$ V$)$ is the same. The voltage along the channel is zero at A and 5 V at B. The gate voltage with respect to the channel is 3 V at A but $3 - 5$ or -2 V at B. The field from the gate cannot start forming a depletion layer until the gate voltage with respect to the channel is negative, which will be at some point, marked as C in Figure 6.40d, along the channel. Thus the channel between A and C does not have a depletion layer penetrating from the surface, which is another factor that increases the conductance of the channel. The gate to channel voltage at B is only -2 V, whereas previously it was -5 V. Thus the depletion layer from the surface does not penetrate as much as before and the two depletion layers therefore cannot pinch off the channel. We have to increase V_{DS} to 8 V to achieve the pinch-off as in Figure 6.40d.

When a gate voltage, V_{GS}, is applied to a MOSFET, whether enhancement or depletion type, the gate current, I_G, that flows is almost totally negligible because the oxide insulation has a very high resistance. The leakage current through the oxide insulation is of the order of 10^{-14}–10^{-11} A or even smaller. In fact, MOSFETs have a lower gate current than JFETs. If, on the other hand, an ac voltage is applied to the gate, then there is an ac current that flows between the gate and the source. This ac current is due to the capacitance, C_{gs}, between the gate and source, which includes the oxide.

It should be mentioned that all the above discussions for the n-channel MOSFET (enhancement and depletion) can be equally applied to the corresponding p-channel MOSFET (PMOSFET) with the appropriate reversal of carrier types and voltages. It may be of interest to mention that although the BJT is a device based on the **diffusion of minority carriers,** FETs, whether JFET or MOSFET, operate on the principle of **majority carrier drift.**

6.8.4   Threshold Voltage

The threshold voltage is an important parameter in MOSFET devices. Its control in device fabrication is therefore essential. Figure 6.42a shows an idealized MOS structure where all the electric field lines from the metal pass through the oxide and penetrate the p-type semiconductor. The charge $-Q$ is made up of fixed negative acceptors in a surface region of W_a and of conduction electrons in the inversion layer at the surface, as shown in the figure. The voltage drop across the MOS

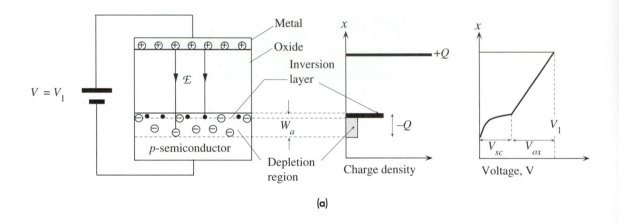

(a)

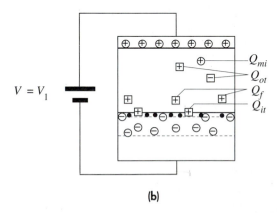

(b)

Figure 6.42

(a) The threshold voltage and the ideal MOS structure

(b) In practice, there are several charges in the oxide and at the oxide-semiconductor interface that affect the threshold voltage: Q_{mi} = Mobile ionic charge (e.g., Na^+), Q_{ot} = Trapped oxide charge, Q_f = Fixed oxide charge, and Q_{it} = Charge trapped at the interface.

structure, however, is not uniform. As the field penetrates the semiconductor, there is a voltage drop, V_{sc}, across the field penetration region of the semiconductor by virtue of $\mathcal{E} = -dV/dx$, as shown in the figure. The field terminates on both electrons in the invesion layer and acceptors in W_a so that within the semiconductor $\mathcal{E}$ is not uniform and therefore the voltage drop is not constant. But the field in the oxide is uniform, as we assumed there were no charges inside the oxide. The voltage drop across the oxide is constant and is V_{ox}, as shown in the figure. As the applied voltage is V_1, we must have $V_{sc} + V_{ox} = V_1$. The actual voltage drop V_{sc} across the semiconductor determines the condition for inversion. We can show this as follows. If the acceptor doping concentration is 10^{16} cm^{-3}, then the Fermi level E_F in the bulk of the p-type semiconductor must be 0.347 eV below E_{Fi} in intrinsic Si. To make the surface n-type we need to shift E_F at the surface to go just above E_{Fi}. Thus we need to shift E_F from bulk to surface by at least 0.347 eV. We have to bend the energy band by 0.347 eV at the surface. Since the voltage drop across the semiconductor is V_{sc} and the corresponding electrostatic PE change is eV_{sc}, this must be 0.347 eV   or $V_{sc} = 0.347$ V. The gate voltage for the start of inversion will then be $V_{ox} + 0.347$ V. By inversion, however, we generally infer that the electron concentration at the surface is comparable to the hole concentration in the bulk. This means that we actually have to shift E_F above E_{Fi} by another 0.347 eV so that the gate threshold voltage, V_{th}, must be $V_{ox} + 0.694$ V.

In practice there are a number of other important effects that must be considered in evaluating the threshold voltage. Invariably there are charges both within the oxide and at the oxide–semiconductor interface that alter the field penetration into the semiconductor and hence the threshold voltage needed at the gate to cause inversion. Some of these are depicted in Figure 6.42b and can be qualitatively summarized as follows.

There may be some mobile ions within the SiO_2, such as alkaline ions (Na^+, K^+), which are denoted as Q_{mi} in Figure 6.42b. These may be introduced unintentionally, for example, during cleaning and etching processes in the fabrication. In addition there may be various trapped (immobile) charges within the oxide, Q_{ot}, due to structural defects, for example, an interstitial Si^+. Frequently these oxide trapped charges are created as a result of radiation damage (irradiation by x-rays or other high energy beams). They can be reduced by annealing the device.

A significant number of fixed positive charges (Q_f) exist in the oxide region close to the interface. They are believed to originate from the nonstoichiometry of the oxide near the oxide–semiconductor interface. They are generally attributed to positively charged Si^+ ions. During the oxidation process, a Si atom is removed from the Si surface to react with the oxygen diffusing in through the oxide. When the oxidation process is stopped suddenly, there are unfulfilled Si ions in this region. Q_f depends on the crystal orientation and on the oxidation and annealing processes. The semiconductor to oxide interface itself is a sudden change in the structure from crystalline Si to amorphous oxide. The semiconductor surface itself will have various defects, as discussed in Chapter 1. There is some inevitable mismatch between the two structures at the interface, and consequently there are broken bonds, dangling bonds, point defects such as vacancies and Si^+, and other defects at this interface that trap charges (e.g., holes). All these interface charges are represented as Q_{it} in Figure 6.42b. Q_{it} depends not only on the crystal orientation but also on the chemical composition of the interface. Both Q_f and Q_{it} overall represent a positive charge that effectively reduces the gate voltage needed for inversion. They are smaller for the (100) surface than the (111) surface so that (100) is the preferred surface for the Si MOS device.

In addition to various charges in the oxide and at the interface shown in Figure 6.42b, there will also be a voltage difference, denoted as V_{FB}, between the semiconductor surface and the metal surface, even in the absence of an applied voltage. V_{FB} arises from the work function difference between the metal and the p-type semiconductor, as discussed in Chapter 4. The metal work function is generally smaller than the semiconductor work function, which means that the semiconductor surface will have an accumulation of electrons and the metal surface will have positive charges (exposed metal ions). The gate voltage needed for inversion will therefore also depend on V_{FB}. Since V_{FB} is normally positive and Q_f and Q_{it} are also positive, there may already be an inversion layer formed at the semiconductor surface even without a positive gate voltage. The fabrication of an enhancement MOSFET then requires special fabrication procedures, such as ion implantation, to obtain a positive and predictable V_{th}.

The simplest way to control the threshold gate voltage is to provide a separate electrode to the bulk of an enhancement MOSFET, as shown in Figure 6.37, and to apply a bias voltage to the bulk with respect to the source to obtain the desired V_{th} between the gate and source. This technique has the disadvantage of requiring an additional bias supply for the bulk and also adjusting the bulk to source voltage almost individually for each MOSFET.

6.8.5   Ion Implanted MOS Transistors and Poly-Si Gates

The most accurate method of controlling the threshold voltage is by ion implantation, as the number of ions that are implanted into a device and their location can be closely controlled. Furthermore, ion implantation can also provide a self-alignment of the edges of the gate electrode with the source and drain regions. In the case of an n-channel enhancement MOSFET, it is generally desirable to keep the p-type doping in the bulk low to avoid small V_{DS} for reverse breakdown between the drain and the bulk (see Figure 6.37). Consequently, the surface, in practice, already has an inversion layer (without any gate voltage) due to various fixed positive charges residing in the oxide and at the interface, as shown in Figure 6.42b (positive Q_f and Q_{it} and V_{FB}). It then becomes necessary to implant the surface region under the gate with boron acceptors to remove the electrons and restore this region to a p-type behavior.

The ion implantation process is carried out in vacuum where the required impurity ions are generated and then accelerated toward the device. The energy of the arriving ions and hence their penetration into the device can be readily controlled. Typically, the device is implanted with B acceptors under the gate oxide, as shown in Figure 6.43. The distribution of implanted acceptors as a function of distance into the device from the surface of the oxide is also shown in the figure. The position of the peak depends on the energy of the ions and hence on the accelerating voltage. The peak of the concentration of implanted acceptors is

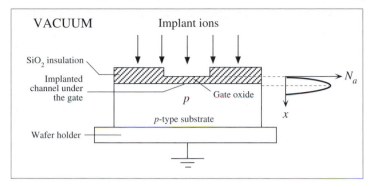

Figure 6.43   Schematic illustration of ion implantation for the control of V_{th}.

made to occur just below the surface of the semiconductor. Since ion implantation involves the impact of energetic ions with the crystal structure, it results in the inevitable generation of various defects within the implanted region. The defects are almost totally eliminated by annealing the device at an elevated temperature. Annealing also broadens the acceptor implanted region as a result of increased diffusion of implanted acceptors.

Ion implantation also has the advantage of providing self-alignment of the drain and source with the edges of the gate electrode. In a MOS transistor, it is important that the gate electrode extends all the way from the source to the drain regions so that the channel formed under the gate can link the two regions; otherwise, an incomplete channel will be formed. To avoid the possibility of forming an incomplete channel, it is necessary to allow for some overlap, as shown in Figure 6.44a, between the gate and source and drain regions because of various tolerances and variations involved in the fabrication of a MOSFET by conventional masking and diffusional techniques. The overlap, however, results in additional capacitances between the gate and source and the gate and drain and adversely

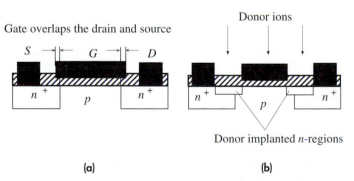

Figure 6.44   (a) There is an overlap of the gate electrode with the source and drain regions and hence additional capacitance between the gate and drain. (b) n^+-type ion implantation extends the drain and source to line up with the gate.

affects the high frequency (or transient) response of the device. It is therefore desirable to align the edges of the gate electrode with the source and drain regions. Suppose that the gate electrode is made narrower so that it does not extend all the way between the source and drain regions, as shown in Figure 6.44b. If the device is now ion implanted with donors, then donor ions passing through the thin oxide will extend the n^+ regions up to the edges of the gate and thereby align the drain and source with the edges of the gate. The thick metal gate is practically impervious to the arriving donor ions.

Another method of controlling V_{th} is to use silicon instead of Al for the gate electrode. This technique is called **silicon gate technology.** Typically, the silicon for the gate is vacuum deposited (e.g., by chemical vapor deposition using silane gas) onto the oxide, as shown in Figure 6.45. As the oxide is noncrystalline, the Si gate is polycrystalline (rather than a single crystal) and is therefore called a poly-Si gate. Normally it is heavily doped to ensure that it has sufficiently low resistivity to avoid RC time constant limitations in charging and discharging the gate capacitance during transient or ac operations. The advantage of the poly-Si gate is that its work function depends on the doping (type and concentration) and can be controlled so that V_{FB} and hence V_{th} can also be controlled. There are also additional advantages in using the poly-Si gate. For example, it can be raised to high temperatures (Al melts at 660 °C). It can be used as a mask over the gate region of the semiconductor during the formation of the source and drain regions. If ion implantation is used to deposit donors into the semiconductor, then the n^+ source and drain regions are self-aligned with the poly-Si gate, as shown in Figure 6.45.

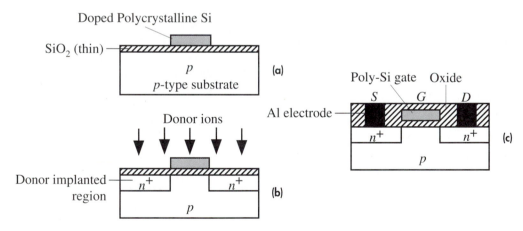

Figure 6.45   The poly-Si gate technology. (Poly-Si is deposited onto the oxide and the areas outside the gate dimesions are etched away. (b) The poly-Si gate acts as a mask during ion implantion of donors to form the n⁺ source and drain regions., (b) A simplified schematic sketch of the final poly-Si MOS transistor.

THE ENHANCEMENT *NMOSFET*   A particular enhancement NMOS has a gate with a width (Z) of 50 μm, length (L) of 10 μm, and SiO$_2$ thickness of 450 Å. The relative permittivity of SiO$_2$ is 3.9. The *p*-type bulk is doped with 10^{16} acceptors cm^{-3}. Its threshold voltage is 4 V. Estimate the drain current when $V_{GS} = 8$ V and $V_{DS} = 20$ V, given $\lambda = 0.01$. What is the effect of increasing the oxide thickness? | **Example 6.11**

Solution

Since $V_{DS} > V_{th}$, we can assume that the drain current is saturated and we can use the I_{DS} versus V_{GS} relationship in Equation 6.59,

$$I_{DS} = K(V_{GS} - V_{th})^2(1 + \lambda V_{DS})$$

where

$$K = \frac{Z\mu_e\varepsilon}{2Lt_{ox}}$$

The electron mobility when $N_a = 10^{16}$ cm^{-3} is 1300 cm^2 V^{-1} s^{-1} (Chapter 5). Thus

$$K = \frac{Z\mu_e\varepsilon_r\varepsilon_0}{2Lt_{ox}} = \frac{(50 \times 10^{-6})(1300 \times 10^{-4})(3.9 \times 8.85 \times 10^{-12})}{2(10 \times 10^{-6})(450 \times 10^{-10})} = 0.00025$$

When $V_{GS} = 8$ V and $V_{DS} = 20$ V, with $\lambda = 0.01$, we have

$$I_{DS} = 0.00025(8 - 4)^2[1 + (0.01)(20)] = 0.0048 \text{ A} \qquad \text{or} \qquad 4.8 \text{ mA.}$$

Increasing the oxide thickness decreases K and hence I_{DS} becomes less sensitive to V_{GS}.

6.9   SEMICONDUCTOR DEVICE FABRICATION[10]

6.9.1   Discrete Devices and Integrated Circuits

The replacement of vacuum tubes by semiconductor devices in the mid 1950s was a major development in electronic engineering. It led to miniaturized and highly reliable circuits. The next major and revolutionary development, which led to a totally new philosophy of electronic design, was the mass production[11] of very sophisticated circuits, called **integrated circuits (ICs),** in single Si crystal chips,

---

[10] Semiconductor device fabrication processes since the late 1960s have gone though rapid changes and advances. The fabrication of an integrated circuit today can easily involve more than a hundred individual processes (steps) and, further, there are usually several alternatives for each individual step. In a general text it is impossible to go into the details of each process. An overall view of some of the processes typically used in fabrication is sufficient to appreciate the complexity of this field.

[11] The mass production of ICs and the resulting incredible cost effectiveness in producing exceedingly reliable complicated circuits is somewhat reminiscent of the mass production of the Model T by Ford, which meant affordable cars for almost everyone.

within the same single-crystal volume, typically 1 mm or so in lateral dimension and 100–200 μm in thickness. By the late 60s, it became possible to integrate a variety of very complex circuits, with nearly all their interconnections and active and passive components, into a single chip. Today, the IC technology can integrate not only an individual circuit but also subsystems and even systems into a single chip. It is not unusual to find an IC with hundreds of thousands of components incorporated into it. It has become possible to design circuits and implement functions that were virtually impossible (and some could not even be envisioned) with discrete devices. The major breakthrough in semiconductor fabrication that lead to today's IC technology was the development of the **planar process** (by Fairchild Semiconductor Co., in the early 60s), which used the diffusion of impurities into a semiconductor wafer through unmasked areas on the surface. All doping and interconnection processes were carried out on one plane, one surface of the wafer. By diffusing various impurities into a Si wafer at selected areas, it was possible to achieve any desired doping. This allowed the simultaneous fabrication of thousands of devices at various steps and led to batch processing. The enormous advantages of mass production and miniaturization cannot be overemphasized. The devices are fabricated within a single wafer, a thin slice of a Si crystal. During fabrication many wafers (perhaps 20–30) are processed together. Each wafer yields up to a few thousand integrated circuits, so one fabrication, perhaps taking a few days, yields several thousand or so ICs, as depicted in Figure 6.46 where typical dimensions are also shown to emphasize the resulting miniaturization. Each IC chip itself has many thousands of components that have been integrated. The design and fabrication of these high component–density integrated circuits is what is generally meant by **microelectronics.** There are, of course, many electronic designs that still need discrete devices. The fabrication of a discrete device requires the same basic steps as those used in fabricating complicated ICs, the latter generally having many more repeated steps and a more complicated photolithographic step, as discussed below.

Monolithic (Greek: *mono–*single, *lithos–*stone) **IC technology** integrates all the necessary active and passive components of a circuit within *the same single Si crystal volume*, the chip, typically within several microns of the crystal. In contrast, in the **film IC technology,** various components and their interconnections are deposited onto an insulating substrate such as a ceramic. The majority of ICs are monolithic due to the economic advantages of this technology for many applications. Generally, ICs are classified as **bipolar transistor** or **MOS** depending on the type of transistor integrated into the structure. They are also classified according to their applications as **analog** or **digital** ICs.

6.9.2   Monolithic IC Fabrication: Planar Process

Bipolar Transistor Integration   Typically, all integrated components within an IC, whether active (e.g., bipolar transistors) or passive (e.g., resistors), are contained within a thin layer at the surface of a slice of a single Si crystal called the **wafer.** The wafer acts simply as a mechanical support for all the components and is called

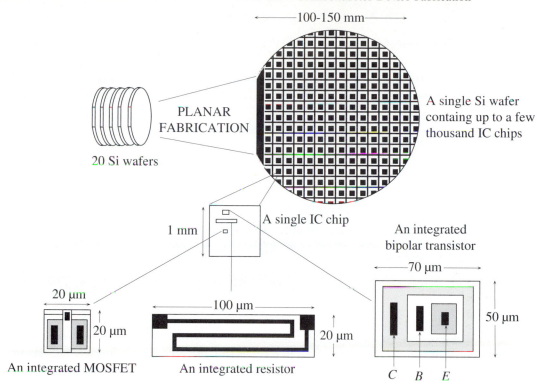

Figure 6.46 Mass production of IC chips by the planar technology leads to miniaturized, extremely reliable, and cost-efficient integrated circuits (the flat edges on the wafers identify the crystal orientations for the fabrication process).

a **substrate.** The final structure of a typical integrated *npn* bipolar transistor within a Si chip is shown in Figure 6.47. Inside the crystal, the collector region completely surrounds the base, which itself surrounds the emitter. The reasons for this structure will become apparent below and one should not worry about the shapes of the boundaries between the doped regions at this point. The base and the emitter are therefore contained within an **n-type island** that forms the collector. The surface has an oxide layer and metal electrodes passing through it to make contacts with the emitter, base, and collector regions. The n^+ buried layer is a highly conducting layer that runs laterally along the collector region and is in contact with it. It provides a low resistance path for the collector current. The n^+-region between the metal electrode and the lightly doped n-type collector ensures a good ohmic contact to be made to the collector. Otherwise, the contact between the Al and the lightly doped collector will be Schottky type and will alter the BJT characteristics. This BJT must be isolated from other components within the crystal. The whole n-type collector region is surrounded by a p^+-type material and the buried layer by the p-type wafer below it. If the p-regions are kept at the most negative potential, which is frequently the ground potential, then the pn junctions around the collector will be reverse biased and the collector will be effectively isolated from other

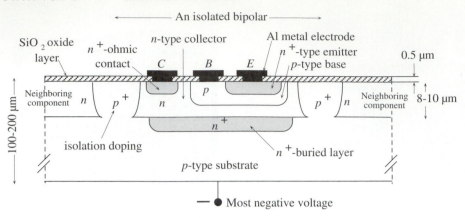

Figure 6.47   Typical structure of an integrated *npn* bipolar transistor that is isolated from other components by the reverse-biased *pn* junction between the collector and the surrounding *p*-type semiconductor

components. Isolation, in this case, is therefore achieved by fabricating each component within an *n*-type island in a negatively biased *p*-type material. The islands are therefore separated from each other by reverse-biased *pn* junctions.

Fabrication Process: Overview   To illustrate the basic steps involved in the monolithic IC fabrication process, we will consider the fabrication of two neighboring integrated *npn* BJTs within an IC chip. There will be many other components, up to a few hundred thousand, within the chip that are fabricated simultaneously with these bipolars. Typical steps in integrating two BJTs in an IC and the structures of the devices at various steps are shown in Figure 6.48a through 6.48t and will be described one at a time below. The final interconnection enables a Darlington pair to be manufactured, as shown in Figure 6.48u. In the planar technology, the fabrication process takes place plane by plane at one surface of the wafer so that the deepest doped region tends to be fabricated first.

The first process involves fabricating the n^+ buried layers of all the *npn* bipolars. The process starts with a wafer, a thin slice of *p*-type single Si crystal, which is cut from a cylindrical single Si crystal grown usually by the Czochralski technique (Chapter 1). The surface of the wafer is oxidized in an oxidation furnace to form a thin layer of SiO_2 as in Figure 6.48a. Doping of the Si crystal requires either the diffusion or implantation of dopant atoms into the crystal at selective areas. The success of the IC technology is due to the fact that silicon dioxide, SiO_2, grown on the surface of a Si crystal is an excellent barrier against diffusion. It can be easily etched to open windows to allow the diffusion of dopants through selective areas. The SiO_2 layer adheres to the Si crystal without any cracks and pores and forms a continuous layer with an amorphous structure. In addition to acting as a diffusion mask, it also passivates the crystal surface. Without the passivation, the surface will

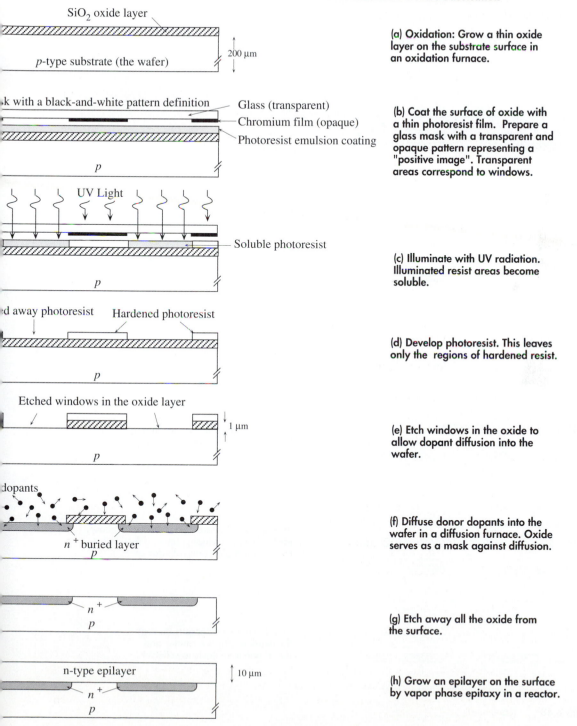

SiO$_2$ oxide layer

p-type substrate (the wafer)

200 μm

(a) Oxidation: Grow a thin oxide layer on the substrate surface in an oxidation furnace.

k with a black-and-white pattern definition
Glass (transparent)
Chromium film (opaque)
Photoresist emulsion coating

p

(b) Coat the surface of oxide with a thin photoresist film. Prepare a glass mask with a transparent and opaque pattern representing a "positive image". Transparent areas correspond to windows.

UV Light

Soluble photoresist

p

(c) Illuminate with UV radiation. Illuminated resist areas become soluble.

d away photoresist   Hardened photoresist

p

(d) Develop photoresist. This leaves only the regions of hardened resist.

Etched windows in the oxide layer

1 μm

p

(e) Etch windows in the oxide to allow dopant diffusion into the wafer.

dopants

n^+ buried layer
p

(f) Diffuse donor dopants into the wafer in a diffusion furnace. Oxide serves as a mask against diffusion.

n^+
p

(g) Etch away all the oxide from the surface.

n-type epilayer

10 μm

n^+
p

(h) Grow an epilayer on the surface by vapor phase epitaxy in a reactor.

Figure 6.48   Typical steps in integrating two bipolar transistors in an IC.

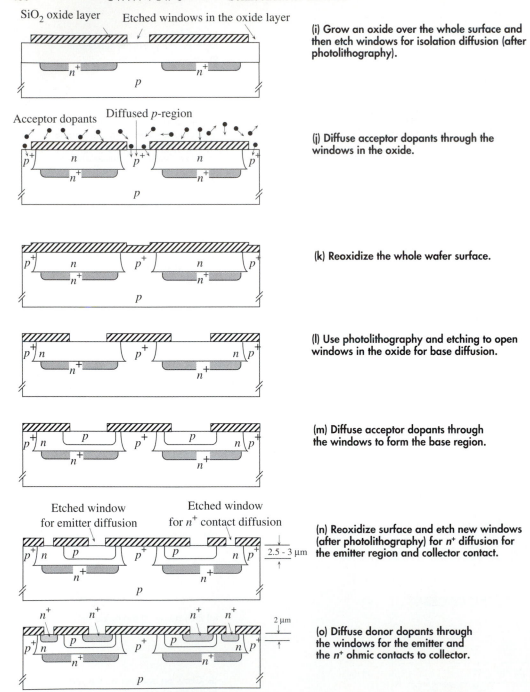

SiO₂ oxide layer   Etched windows in the oxide layer

(i) Grow an oxide over the whole surface and then etch windows for isolation diffusion (after photolithography).

Acceptor dopants   Diffused p-region

(j) Diffuse acceptor dopants through the windows in the oxide.

(k) Reoxidize the whole wafer surface.

(l) Use photolithography and etching to open windows in the oxide for base diffusion.

(m) Diffuse acceptor dopants through the windows to form the base region.

Etched window for emitter diffusion   Etched window for n^+ contact diffusion

(n) Reoxidize surface and etch new windows (after photolithography) for n^+ diffusion for the emitter region and collector contact.

2.5 - 3 μm

2 μm

(o) Diffuse donor dopants through the windows for the emitter and the n^+ ohmic contacts to collector.

Figure 6.48   (continued)

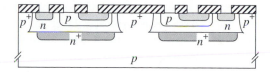

(p) Reoxidze surface, and etch new windows (after photolithography) for contacts.

Al metallization

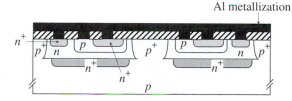

(q) Metallization involves depositing Al over the surface. Al forms electrodes to the exposed semiconductor regions defined by windows in step (p).

Al interconnection   Al electrodes (contacts)

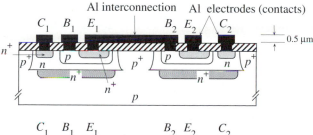

0.5 μm

(r) Using a suitable mask, etch away unnecessary Al to define contacts and interconnections.

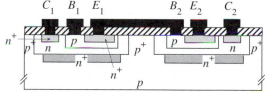

(s) A simplified and idealized representation of the structure in step (r) without the curved boundary effects of diffusion.

n^+ emitter p-base n^+ contact to collector

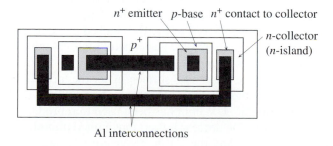

n-collector (n-island)

Al interconnections

(t) The top view for a Darlington pair — two connected integrated npn bipolars.

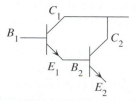

(u) The Darlington pair.

Figure 6.48   (concluded)

have sufficient conductance to leak a current over the surface in device applications. Further, SiO_2 is an excellent insulator and serves as a dielectric layer in MOS devices. The oxidation process can occur in various ambients. **Dry oxidation** involves the oxidation of Si in a pure oxygen gas atmosphere, whereas **wet oxidation** takes place in a steam environment. Wet oxidation is faster than dry oxidation and is used more frequently. The SiO_2 is simply referred to as **oxide** in semiconductor fabrication.

As the oxide is an excellent barrier against atomic diffusion, we have to open diffusion windows in the oxide at areas where there are to be n^+ buried layers. A process called **lithography** provides a means of defining windows in the oxide. Lithography literally means "writing on stone" (obviously "stone" here implies the wafer surface). Lithography is a process by which a pattern is defined on the surface of the wafer, which identifies the crystal regions to be doped. Equivalently, it defines regions of the oxide where windows are needed to access the crystal for the doping process. The surface of the oxidized wafer is coated with a thin **photoresist** (or simply **resist**) emulsion as in Figure 6.48b. We will consider a **positive resist,** which becomes chemically soluble when exposed to ultraviolet (UV) exposure (a negative resist becomes insoluble when exposed to UV light). A black-and-white photomask is fabricated in which the white (transparent) areas correspond to regions where diffusion windows are needed in the oxide. The photomask is a positive image of the diffusion window pattern needed in the oxide. Typically the photomask is a transparent glass with a chromium film deposited on it that defines the opaque pattern, as shown in Figure 6.48b. The wafer surface is then illuminated by UV light through the photomask, as depicted in Figure 6.48c. The photoresist regions under the transparent areas of the photomask become soluble by the UV light (UV photons break molecular bonds in the resist). Subsequently, the photoresist is developed in a suitable chemical solution in which the soluble regions are chemically washed away and only the insoluble regions remain, as in Figure 6.48d. The photoresist mask that remains on the oxide is cured (hardened) to resist the etchant to be used for removing the oxide. Windows are then etched in the oxide by using an etchant (e.g., HF acid), as in Figure 6.48e. The etchant removes the oxide outside the photoresist coated areas. It does not, however, attack the Si crystal surface, so overetching does not damage the crystal surface. After the formation of diffusion windows in the oxide, the photoresist mask on the oxide is stripped off by chemically dissolving it. In this example, UV light has been used to reproduce the diffusion window pattern in the resist. The process is therefore named **photolithography** ("using light to write on stone") and the UV light-sensitive resist is called **photoresist.** It is also possible to form a resist mask on the oxide by writing with an electron beam instead of UV light, in which case the process becomes **electron-beam lithography.**

Once the Si crystal surface has been accessed through the diffusion windows in the oxide, donor dopants are diffused into the crystal. The diffusion process occurs at a high temperature (typically 900–1150 °C) in the presence of an appropriate dopant gas (phosphorus gas). The surface regions where the diffused donors exceed the bulk acceptor concentration become n-type by compensation doping. The temperature and gas pressure are such that heavy n-type doping is achieved to form the n^+ buried layers. The diffusion process is based on the random motions of atoms from high to low concentration regions, which means that there is some inevitable lateral spread in the doped *n*-region underneath the windows, as shown in Figure 6.48f.

The n^+ buried layers only serve as high conductance paths for the collector current in each BJT. Their doping levels and crystal defect content are not critical. On the other hand, the subsequent layers of diffused regions forming the base and the emitter must be free of crystal defects and impurities and those doping levels must by closely controlled. The wafers cut and prepared from the bulk-grown crystals are generally not suitable for direct use in device fabrication as the crystal surface may not be totally free of flaws from the wafer preparation process and the wafer invariably has oxygen impurities. This problem is overcome by growing a nearly perfect crystal of controlled doping in the form of a thin layer (typically $8-10$ μm) on top of this wafer. The wafer surface therefore acts as a foundation (a seed crystal) on which a new crystal is grown. The wafer is a mechanical support for the new crystal to be grown and is therefore called a **substrate.** All the devices are fabricated within this thin layer of near perfect Si crystal. The thin layer of near perfect Si crystal grown on the Si wafer is called an **epilayer,** which stands for an epitaxial layer (Greek: *epis*–layered, *taxis*–ordered).

The oxide that was used in the n^+ buried layer diffusion is removed by etching, as in Figure 6.48g, and then a thin lightly doped *n*-type epilayer is grown on the surface, as shown in Figure 6.48h. The epilayer is frequently deposited by **chemical vapor deposition** (CVD). All other diffused and doped regions occur within the epilayer. We must, however, isolate individual components from each other. This is achieved by embedding each component in an *n*-type island and with *p*-type material between the islands so that each component is isolated by reverse-biased *pn* junctions. The whole crystal surface is reoxidized and photolithography is used to define the next set of window patterns on the oxide. Windows are then etched in the oxide, around each collector region, as in Figure 6.48i. Acceptor impurities are then diffused all the way to the substrate to form p^+-regions around each collector, which results in isolated *n*-type islands, as shown in Figure 6.48j. Each isolated *n*-region forms an individual collector.

The next step is the diffusion of *p*-type dopants for the base. The whole crystal surface is reoxidized, as in Figure 6.48k. The new oxide grows on the surface of the already-present oxide as well as the surface of the Si crystal. The growth rate, however, is faster on the Si crystal surface than on the already-present oxide surface. For simplicity and clarity, we will assume a flat oxide surface. Photolithography is used once again to pattern a photoresist mask on the oxide and hence define the new diffusion windows. Diffusion windows are then etched where base regions are to be formed, as in Figure 6.48l. Boron gas is then diffused though the windows to form a lightly doped *p*-type base, as in Figure 6.48m. After the base diffusion, the whole surface is reoxidized. Once again, using photolithography, new diffusion windows are etched in the oxide where the n^+ emitter regions and n^+ ohmic contacts to the collector are to be formed, as in Figure 6.48n. Either by diffusion or by ion implantation, donor dopants are introduced to form the n^+ emitter regions and the n^+ contact regions of the collectors, as in Figure 6.48o. Frequently, low temperature ion implantation is preferred to high temperature diffusion for the n^+ final doping. Ion implantation provides a much closer control of the doping process.

The bipolars fabricated at the end of step (o) need contacts and interconnections. The whole surface is reoxidized. Once again, using photolithography, the oxide is masked by a photoresist pattern. Windows are etched at places where contacts are needed to the BJT, as in Figure 6.48p. Aluminum is then coated over the whole surface of the wafer, as in Figure 6.48q. Contacts are made to the necessary semiconductor regions through the oxide windows. The aluminum covering the whole surface area is then etched selectively, using an appropriate mask, to remove all the unwanted Al but leaving the metal where interconnections are needed and where external contact pads are required, as in Figure 6.48r. The final structure of the two bipolars in Figure 6.48r is frequently idealized as in Figure 4.48s without the curved diffusion boundaries. The top view in Figure 6.48t shows the metal interconnections for a Darlington pair. After the formation of interconnections, each chip on the wafer is electrically tested and those that fail are marked and later discarded. The yields, anything between 20–90%, are highly variable depending on the complexity of the circuit and the number of pretrials.

Following the testing step, the wafer is diced, scribed with a diamond stylus, to individual IC chips. Each working chip is then mounted, bonded to leads, and encapsulated in a package that is suitable for its application.

Dopant Diffusion Profiles and Ion Implantation   The base and emitter regions of the *npn* BJT in Figure 6.48 are formed by the diffusion of acceptor and donor dopants respectively. The epilayer (the collector), due to its method of deposition (e.g., CVD), has a relatively uniform donor concentration. However, atomic diffusion involves atomic motions from high to low concentration regions. The dopant concentration during the diffusion process therefore decreases from the surface into the bulk. The exact profile depends on the diffusion process for the dopants (e.g., diffusion coefficient of the dopants), temperature, and time. At the end of a diffusion process when the temperature is dropped, there is a "frozen" dopant concentration profile. Figure 6.49 depicts the acceptor and donor dopant concentration profiles in the epilayer after the base and emitter diffusions have been completed. All doping effects occur as a consequence of compensation doping. At around 3 μm from the surface, the *p*-type dopant concentration, introduced by base diffusion, is the same as the *n*-type dopant concentration in the epilayer. This point is the metallurgical junction, C, between the base and collector. Between 2 to 3 μm, the acceptor concentration exceeds the donor concentration so that this region is *p*-type and forms the base. The region from the surface to the 2 μm point, E, has excess donor concentration over the acceptor concentration, and this layer forms the emitter. It is apparent that the base width is controlled by the intersections of the dopant concentration profiles. The impurity concentration profile is a sensitive function of temperature and time, which means that the base width is difficult to control. Indeed, high temperatures experienced during emitter diffusion, even though for a short duration, will affect the acceptor concentration profile formed during base diffusion.

Diffusional doping results in concentration profiles that are difficult to control reproducibly at the micron level. In addition, atomic diffusion is a random process, so there is an inevitable lateral spread during diffusion, as apparent in Fig-

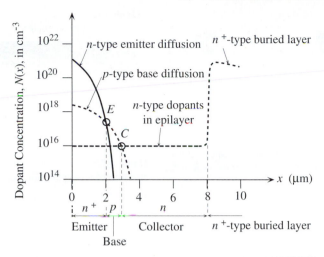

Figure 6.49 Typical dopant concentration profiles across an *npn* bipolar fabricated by diffusion processes.

ure 6.48m. Both these problems are easily overcome by ion implantation. Ion implantation involves bombarding the wafer surface in vacuum by dopant ions that are accelerated through large voltage differences. The dopant ions impinge the surface at almost normal incidence and penetrate the sample through windows in the oxide mask, as depicted in Figure 6.50 for ion implantation of the emitter region of an *npn* BJT. The penetration distance of dopants depends on their energy and hence the voltage difference that accelerates them. A typical implanted donor concentration profile is qualitatively shown in Figure 6.50. There is a donor concentration peak that depends on the accelerating voltage. The lateral spread underneath a window is virtually absent, which leads to an excellent alignment with any mask feature. Since ion implantation is carried out at low temperatures, it does not

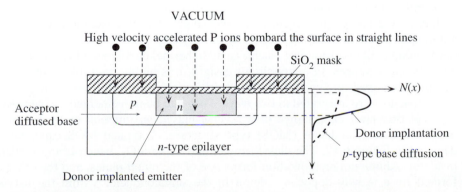

Figure 6.50 Ion implantation is a highly controlled form of doping that exhibits virtually no lateral spreading underneath the window.

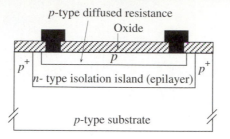

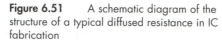

Figure 6.51   A schematic diagram of the structure of a typical diffused resistance in IC fabrication

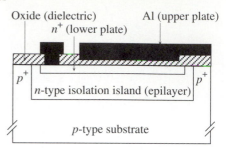

Figure 6.52   A schematic diagram of the structure of a typical MOS capacitor

affect dopant concentration profiles in other regions of the device. Its main drawback is the generation of crystal flaws and defects in the implanted regions. The incoming dopant ions lose their energy through collisions with the Si crystal structure, which leads to the dislodging of host atoms. Crystal defects are almost totally eliminated by annealing the device at an elevated temperature that is well below typical diffusion temperatures. Although ion implantation is costly, due to its many advantages, it is now replacing diffusional doping in many fabrication technologies.

Other Components Integration   The example in Figure 6.48 illustrated the integration of an *npn* BJT. Other components such as FET devices, resistors, and capacitors are integrated by using similar fabrication processes. The only circuit component that cannot be integrated is the inductor. It is possible, however, to obtain an inductance by using active devices (bipolar transistors) and capacitors.

Figure 6.51 depicts one type of integrated resistor, which is called a **diffused resistance.** The resistor is essentially the lightly doped base region of an *npn* transistor that is made sufficiently thin and long to provide the necessary resistance. In the case of large resistance values, typically greater than 10 kΩ, the thin strip is frequently folded to obtain a long length (Figure 6.46) and conserve area. As in the BJT case, the resistor is embedded in an *n*-type island for isolation from other components. Diffused resistors generally occupy chip areas greater than BJTs and have a number of disadvantages, such as a high thermal coefficient of resistivity and parasitic capacitances. In many circuit designs, it is highly desirable to minimize the number of resistors or replace them with active devices.

Capacitors are integrated in two ways. They can be implemented by the use of the depletion layer capacitance of a reverse-biased *pn* junction or by having a metal-oxide-semiconductor (MOS) type structure. A typical MOS capacitor is depicted in Figure 6.52. The dielectric medium is the SiO_2, which is an excellent insulator. Aluminum metallization forms one of the metal plates, and the other is formed by a heavily doped n^+-region in the semiconductor within the *n*-type isolation island. The capacitance depends on the oxide thickness, with typical practical values less than 1 nF mm^{-2}. It is apparent that the integration of a

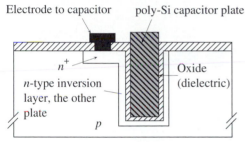

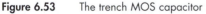

Figure 6.53   The trench MOS capacitor

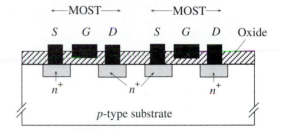

Figure 6.54   Same channel MOS transistors are self-isolated from each other.

capacitor can easily consume prime chip area. When large capacitance values are needed in a small chip area, there are a number of other methods for integrating capacitances. Using a different dielectric medium than SiO_2 with a larger relative permittivity will increase the capacitance. Further, one can use the vertical dimension of the chip to obtain large areas for the capacitor plates. This is illustrated in the trench capacitor shown in Figure 6.53. A deep trench is etched in the p-type substrate and the surface is oxidized. The trench is then filled with poly-Si (heavily doped and hence highly conducting polycrystalline Si) by CVD. The poly-Si in the trench acts as one of the plates of the capacitor, and the oxide on the trench surface forms the dielectric medium. The other plate is the conducting thin n-type inversion layer that is formed in the p-type Si around the trench surface during fabrication. This n-type inversion layer is accessed through an n^+ contact. Trench capacitors are sometimes used in dynamic random access memories (DRAMs).

The integration of MOS transistors is less complicated than bipolar ones and leads to smaller areas. Figure 6.54 shows two typical integrated neighboring n-channel MOS transistors. First, it is apparent that the two devices are self-isolated from each other when the p-type bulk is the most negative voltage. Then n^+ source and drain of each MOST is surrounded by the p-type bulk. The n-channel that forms between the source and drain of each MOST is also isolated by the p-type bulk. Further, there are fewer process steps. Consequently, it becomes highly desirable to design circuits with MOS instead of bipolar transistors. When the circuit contains mixed channel MOS transistors, it becomes necessary to isolate the n-channel MOS. Figure 6.55 depicts the structure of a pair of complementary MOS (CMOS) transistors. The substrate is n-type. The p-type islands, formed by diffusion, contain the n-channel MOSs. Isolation is achieved by reverse biasing the pn junction between the p-island and n-substrate. The control of the threshold gate voltage and the alignment of the gate with the source and drain in MOSs are achieved by using heavily doped polycrystalline silicon as the gate and ion implantation to form the source and drain, as illustrated in Figure 6.45.

Isolation   The isolation of integrated bipolars in Figure 6.48 was achieved by embedding each BJT in an n-type island in the p-type substrate that is connected to the most negative voltage, as in Figure 6.47. Thus, isolation involves surrounding

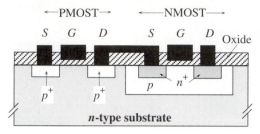

Figure 6.55   CMOS transistors

NMOST is embedded in an isolated *p*-island in an *n*-type bulk.

each BJT with a reverse-biased *pn* junction and is called **pn junction isolation.** This has drawbacks. There is a capacitive load at each collector. There are also capacitive feedback paths between different BJT collectors. Both of these limit the high frequency response of the IC. Another disadvantage is the inefficient use of the chip area for the isolation process. The isolation channel diffusion in Figure 6.48j must provide sufficient spacing between the *n*-type islands. Since it is diffused though the whole epilayer thickness ($8-10$ μm), it has considerable lateral spreading. There are a number of different isolation processes that lead to a better isolation of each component. One isolation process, developed by Fairchild Semiconductor and called **isoplanar oxide isolation,** involves surrounding each component with an oxide wall. The basic principle is illustrated in Figure 6.56a and b. Silicon nitride, Si_3N_4, is a good barrier against oxidation and is used as a mask for the selective oxidation of the silicon surface. The whole surface of the wafer is coated with a Si_3N_4 film using a CVD process. Then deep trenches are etched through the Si_3N_4 and about halfway through the epilayer, as in Figure 6.56a. During oxidation, the oxide that is formed grows both outward from the Si surface and inward into the crystal at about the same rate. Eventually the oxide reaches the *p*-type substrate and hence surrounds the region to be isolated with an oxide wall, as shown in Figure 6.56b. Figure 6.56c shows the structure of an *npn* isoplanar bipolar transistor. Its top view prior to metallization is shown in Figure 6.56d. There is an oxide wall surrounding the whole device. An oxide wall also isolates the n^+ emitter from the n^+ collector contact region, both of which are formed simultaneously. The dimensions of an isoplanar bipolar transistor are about half the size of a *pn* junction isolated bipolar transistor. Even though the collector still has a *pn* junction isolation from the substrate, the *pn* junction capacitance is now much reduced because there are no p^+ regions isolating the collectors as in Figure 6.47. It is also possible to isolate each component by totally surrounding its region with an oxide. This type of total dielectric isolation requires special fabrication steps, which are discussed in advanced textbooks.

Lithography   Lithography is a process by which a pattern is defined on the surface of the wafer that identifies the crystal regions to be doped. Photolithography is illustrated in Figure 6.48b–e. First, a high-contrast pattern that has transparent

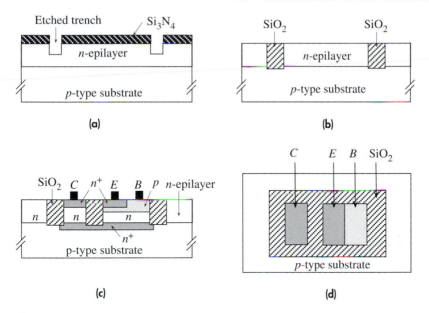

Figure 6.56   Dielectric isolation and the isoplanar bipolar transistor
(a) A local window is etched through the Si_3N_4 barrier and about halfway through the epilayer to surround the region to be isolated.
(b) Oxidation results in the formation of an oxide wall all the way to the substrate. There is now an isolated n-type epilayer surrounded by an oxide wall.
(c) Typical cross-sectional structure of an isoplanar bipolar transistor where three oxide walls are used for isolation.
(d) The top view before the metallization step.

and opaque features is generated on a mask. Transparency to UV light is obtained by making the mask from a quartz plate and using a chromium film pattern to generate the required opaque pattern. Then the image of this pattern (or the negative of this image) is transferred to the wafer surface by shining UV light through it, as depicted in Figure 6.48c. The lithographic step is repeated many times during the fabrication process. The smaller are the diffusion windows, the greater are the number of devices per unit area. What are the minimum dimensions of various features in the pattern? What is the minimum line width in the pattern that can be put on the wafer?

If the mask is in close proximity to the wafer surface, as in Figure 6.48b, then the mask features are the same size as the pattern features generated on the wafer surface. This puts tight constraints on mask fabrication because the pattern features on the mask then have to be controlled to close tolerances, less than a micron. Furthermore, the mask has to be the same size as the wafer surface (10–15 cm in diameter) and the mask must have an identical pattern for each chip area. The photolithographic technology therefore prefers to use pattern projection in which the image of the mask pattern is projected with optical reduction onto one chip area, as illustrated in Figure 6.57. The mask is N (typically 5) times larger than the

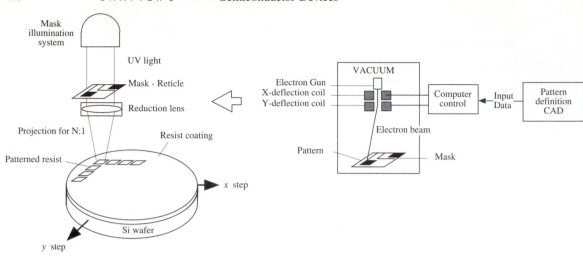

Figure 6.57   A UV illumination system is used to project the mask pattern onto the wafer surface. The projection demagnifies the mask image so that the pattern on the resist surface is *N* times smaller. The wafer is stepped and the exposure is repeated to pattern the neighboring chip on the wafer. The mask is generated (patterned) using an election beam whose scan is computer controlled.

actual pattern to be generated per chip. The mask is called a **reticle**[12] and represents the pattern required for one chip (one cell). An expensive UV illumination and sophisticated lens system project the reticle image onto a small wafer surface area that defines the chip. Following this exposure, the wafer is stepped in the *x* and *y* directions and the projection is repeated to expose the neighboring chip area, as shown in the Figure 6.57. The system using this optical projection and wafer stepping to print the recticle pattern onto the wafer surface is called a **reduction stepper and repeater,** or simply a **reduction stepper.** There are obvious advantages to the reduction stepper system. First, the mask is *N* times larger than the device, so tolerance requirements on the mask pattern features are not critical. For example, a tolerance of 0.2 micron at the device level requires a tolerance of one micron in mask fabrication. Moreover, flaws in the mask pattern are demagnified when projected onto the wafer surface so they become insignificantly small and do not cause faults. A mask defect that extends about one micron in size in the mask becomes 0.2 micron at the device level and within the device tolerance levels.

The fabrication of masks or reticles for the photolithographic process itself requires a lithographic type of process. We will consider one method of making a mask, a reticle, for the reduction stepper. The lithographic process in this case uses an electron beam to write the pattern onto the mask because the electron beam can be electronically controlled to close tolerances (typically ~50 nm). Figure 6.57

---

[12] Reticle in optics refers to a grid of fine wires placed in the focus of the objective lens of a telescope. Here it refers to the patterned mask placed in the projection system.

illustrates a typical electron beam lithography for the generation of the mask. The required pattern on a mask is first generated in software by using computer-aided design (CAD). This computer-generated description of the mask pattern is then coded to control an electron beam. The beam scans the mask surface and is on or off depending on the coded pattern. The mask is typically a quartz plate with a uniform thin chromium coating. It is covered with a uniform resist coating that is sensitive to the electron beam exposure. The resist becomes soluble in those areas where it has been exposed to the electron beam. After the mask has been scanned, it is developed, in other words, exposed areas are etched away. This leaves a mask that has a chromium pattern on a quartz plate.

Lithography is one of the most important steps in microelectronics fabrication. It determines the minimum line width in the fabrication process and hence the ultimate device density (number of devices per square area of the chip). Present lithographic research attempts to develop lithographic processes that will push the minimum line width to around 0.1 micron (the so-called point one technology). There are essentially three contenders. First is a refinement of the present UV system by various modifications and also by the reduction of UV wavelength. Second is the use of x-rays since x-rays have shorter wavelengths. Third is the electron beam lithography. The reader is referred to more advanced texts on these subjects.

SCALING IN MICROELECTRONICS (VLSI)   The number of devices integrated per unit area of   **Example 6.12**
Si wafer over the last two decades has been increasing to implement more and more functions within a single chip. The device density ultimately depends on the size of a single device. We consider scaling down a single MOST to integrate more devices per unit area. Scaling can be done to achieve either constant field scaling or to achieve constant voltage. Consider scaling to maintain the fields constant within the device. Linear device dimensions are scaled by a factor of $1/\kappa$, where $\kappa > 1$, and the doping is scaled by a factor of κ. Predict the effects on the device performance.

Solution

As the fields in the device are maintained constant, all voltages, including the threshold voltage, scale down by $1/\kappa$. Greater doping increases the conductivity by κ, but the resistance remains unchanged. The drain current scales down by $1/\kappa$. We can also show this by using

$$I_{DS(sat)} \propto \left(\frac{Z}{L}\right)\left(\frac{1}{t_{ox}}\right)(\text{Voltage})^2 \propto \frac{1}{\kappa}$$

The gate capacitance is now smaller so the time constant, τ, associated with the charging and discharging of this capacitance is also smaller by $1/\kappa$. Thus the switching times are shorter by $1/\kappa$ and we have faster devices. The power dissipation is smaller but so is the delivered output power. The number of logic operations per second per unit area depends on $1/\tau LZ$ so that this shoots up by κ^3, an excellent result. The main problem with this type of scaling is that the voltage levels scale down and lose compatibility with standard voltage levels used elsewhere in the system. Table 6.2 summarizes some of the features of this type of constant field scaling.

Table 6.2   Constant field MOST scaling and typical effects on device performance

Parameter	Scaling Factor	Comment
Device dimensions, L, Z, t_{ox} (L = channel length, Z = gate width, t_{ox} = oxide thickness)	$1/\kappa$	Dimensions scaled down but Z/L (gate width to length) is the same.
Doping	κ	Increase doping to keep resistance constant (see below).
Voltage = Field × Length	$1/\kappa$	Constant field.
Threshold voltage, $V_{th} \propto \mathcal{E}_{th} \times t_{ox}$	$1/\kappa$	Ideally, for inversion we need a field $\mathcal{E}_{th}$ from the gate into the channel.
Resistivity	$1/\kappa$	Reduced.
Resistance = $\rho L/A$	1	Remains constant.
Current = Voltage/Resistance	$1/\kappa$	Smaller current.
$I_{DS(sat)} \propto (Z/L)(1/t_{ox})(\text{Voltage})^2$	$1/\kappa$	Smaller current.
Gate area = $L \times Z$	$1/\kappa^2$	Reduced significantly. Advantage.
Gate capacitance, $C = \varepsilon ZL/t_{ox}$	$1/\kappa$	Reduced.
$\tau = RC$, time constants (time delays)	$1/\kappa$	Faster device. Advantage.
Power per device, IV	$1/\kappa^2$	Smaller power dissipation. Advantage.
Number of logic operations per second per unit area = $1/\tau A$	κ^3	A large increase. Excellent.
Output power = Current × Output voltage = $I \times (\mathcal{E} \times L)$	$1/\kappa^2$	Poor output power performance.

ADDITIONAL TOPICS

6.10   HIGH-FREQUENCY SMALL-SIGNAL BJT MODEL

We consider the *pnp* bipolar transistor depicted in Figure 6.24. Under normal dc operating conditions, the minority carrier (hole) concentration profile and therefore the injected minority carrier (hole) charge in the base will be constant, as shown in Figure 6.24. The base current simply replaces those majority carriers (electrons) continuously consumed in recombination. If Q_B is the total amount of injected minority carrier charge in the base due to the dc voltage V_{BE}, which is shown in Figure 6.26, then the dc base current I_B is simply Q_B/τ_h, as $1/\tau_h$ is the mean rate of recombination,

DC base current

$$I_B = \frac{Q_B}{\tau_h}$$

Suppose that we now increase V_{BE} by δV_{BE}, which leads to an increase in $p_n(0)$ to $p_n'(0)$, which is shown as the dashed line in Figure 6.26. With more minority carriers injected, there is now an additional stored charge in the base, as indicated by the gray shaded area and labeled as δQ_B in Figure 6.26. Since the stored charge, Q_B, in the base depends on V_{BE}, there is a capacitive effect appearing between the BE terminals. We represent the capacitive effect across the base–emitter terminals by defining a small-signal diffusion[13] (or storage) capacitance, C_{diff}, by

$$C_{diff} = \frac{\delta Q_B}{\delta V_{BE}}$$

Base diffusion capacitance

The stored charge, neglecting recombination, as shown in Figure 6.26, is

$$Q_B = \frac{1}{2} e A W_B p_n(0) = \frac{1}{2} e A W_B p_{no} \exp\left(\frac{e V_{BE}}{kT}\right)$$

Differentiating this with respect to V_{BE} we obtain

$$C_{diff} = \frac{e Q_B}{kT} = \frac{e \tau_h I_B}{kT} = \frac{\tau_h e I_C}{\beta kT} = \frac{\tau_h}{r_{be}}$$

where we have used Equation 6.44. This is the capacitance that has to be charged and discharged as the signal v_{be} modulates V_{BE}. Its value is generally greater than the capacitance of the BE depletion region. For example, typically, for a transistor with $\beta = 100$ and a minority carrier lifetime of $1~\mu s$, when $I_C = 1$ mA, $r_{be} = 2.5~k\Omega$ and

$$C_{diff} = \frac{\tau_h}{r_{be}} = \frac{10^{-6}}{2,500} = 0.4~\text{nF}$$

There is also the capacitance of the depletion region, C_{dep}, between the base and the emitter. Its value for an abrupt junction was derived during the treatment of the *pn* junction and depends on the width of the *BE* depletion region, W_{BE},

$$C_{dep} = \frac{\varepsilon A}{W_{BE}}$$

where W_{BE} decreases with increasing V_{BE}. The total small-signal capacitance across BE is therefore

$$C_{be} = C_{diff} + C_{dep}$$

The small-signal equivalent circuit of the BJT must therefore also include the capacitance C_{be}, as shown in Figure 6.58a. As the base–collector junction is reverse biased, there is, in addition, a depletion region capacitance between the base and the collector terminals, which is shown as C_{bc} in Figure 6.58a. It is apparent that C_{bc} provides a feedback path for the output current into the input, which, generally, deteriorates the gain of the BJT amplifier.

---

[13] The stored charge in the base is due to the diffusion of injected minority carriers in this region.

AC source   Small signal equivalent circuit   Load

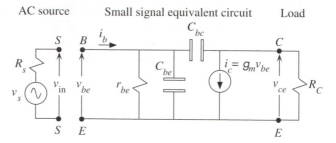

(a) The small signal model including storage and depletion region
capacitances.

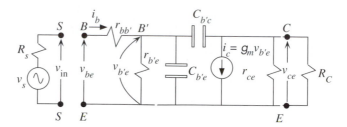

(b) The high frequer / small signal model including storage and
depletion regior ,pacitances and base spread and collector-emitter
resistances.

Figure ʿ ᴐ   Simplified small-signal HF model of the bipolar
transisiʊr in the CE configuration

The structure and geometry of nearly all bipolar transistors is such that recombination invariably occurs over a region rather than at one specific location. Figure 6.59 shows a simplified schematic sketch of a typical *pnp* BJT fabricated by the commonly used planar technology. It is apparent that the device is not symmetric. The surface area of the collector junction is larger than that of the emitter junction. We define the active base region as the volume of base that contains the majority of the emitter to collector hole current, which is indicated (only roughly) in the figure. The base current has to supply electrons to the closest and farthest points of recombination, which means that it flows over the whole active region of the base, though its magnitude gets smaller farther away from the base terminal, B. Since the base material has a finite resistivity, there is therefore a distributed voltage drop along the active region. In other words, the voltage across the base–emitter, V_{BE}, gets smaller as we move away from B. Thus the minority carrier injection and hence I_E get less father away from B. Suppose that point B' is some mean point in the active base region, as roughly shown in Figure 6.59, that represents an **effective** or a **true base point** such that $V_{B'E}$ is the effective base–emitter voltage that controls the emitter current. Thus, by definition,

$$I_E = I_{EO} \exp\left(\frac{eV_{B'E}}{kT}\right)$$

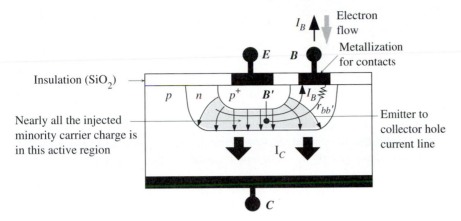

Figure 6.59   A simplified representation of the structure of a typical *pnp* bipolar transistor (for example, as fabricated by diffusion processes)

The base current flows through a finite semiconductor region to replenish the electrons lost by recombination. It has to supply electrons to the closest and also to the farthest point in the base. Point B′ represents a mean point in the active base region that acts as an effective base point.

Then we can calculate (or obtain by measurement) the effective resistance, say $r_{bb'}$, of the base region, which gives rise to the voltage drop $V_{BE} - V_{B'E}$:

$$V_{BE} - V_{B'E} = I_B r_{bb'}$$

where $r_{bb'}$ is called the **base spread resistance.** It takes into account not only the distributed voltage drop in the active base region due to the flow of I_B but also some of the base material outside the active base to the base terminal, B. Figure 6.59 shows that $r_{bb'}$ is placed between the base terminal, B, and the true base point, B', and has the base current, I_B, flowing through it.

As B' represents the true base point, the capacitances C_{be} and C_{bc} involving the active base are now between B' and E, and B' and C so that they become $C_{b'e}$ and $C_{b'c}$. The modified small-signal equivalent curcuit is shown in Figure 6.58b. The model, in addition, has a resistance r_{ce} placed between the CE terminals due to the following effect. Suppose that V_{BE} is kept constant and V_{CE} is increased by an amount δV_{CE}. This increases the reverse bias V_{CB} by δV_{CB}. Consequently the base width, W_B, gets narrower, as shown in Figure 6.22 (due to the Early effect), which leads to a steeper minority carrier concentration gradient and hence to a greater collector current, say by an amount δI_C. Since δV_{CE} leads to δI_C, the two output parameters are related just as they would be in a resistance. We define a small signal collector–emitter resistance r_{ce} by

$$r_{ce} = \frac{\delta V_{CE}}{\delta I_C} \approx \frac{v_{ce}}{i_c}$$

The modulation of the base width W_B by V_{CE} is not very strong and hence δI_C is generally very small. Consequently r_{ce} is quite large, typically greater than 50 kΩ. It is represented as a resistance across the CE terminals of the small-signal equivalent circuit as shown in Figure 6.58b.

There are other effects, usually of secondary nature, in the small-signal equivalent circuit that are beyond the scope of this book. Their inclusion does not significantly affect the predictions of the model. Small-signal equivalent circuits, such as that in Figure 6.24b, are most useful in obtaining the frequency response.

For example, suppose that an ac source, v_s, is connected across the base–emitter (in series with a dc bias, which is not shown in small-signal equivalent circuits). We adjust v_s so that the input current, i_b, is always constant. What is the current gain, β, at different frequencies? The impedance between B' and E, $z_{b'e}$, is given by $r_{b'e}$ and $C_{b'e}$ in parallel. At low frequencies, we can neglect the impedance of $C_{b'e}$ so that there is only $r_{b'e}$. The low-frequency gain, β, is then

Low frequency β

$$\beta = \frac{i_c}{i_b} = \frac{g_m v_{b'e}}{\dfrac{v_{b'e}}{r_{b'e}}} = g_m r_{b'e} = \beta_o$$

where by definition $\beta_o = g_m r_{b'e}$ and represents the low-frequency current gain as derived previously in Equation 6.47.

At high frequencies, $z_{b'e}$ becomes shunted by the small impedance of $C_{b'e}$ so that

$$z_{b'e} = \frac{1}{j\omega C_{b'e}}$$

The magnitude of the current amplification at high frequencies is

High frequency β

$$|\beta| = \frac{i_c}{|i_b|} = \frac{g_m v_{b'e}}{\left|\dfrac{v_{b'e}}{z_{b'e}}\right|} = \frac{g_m}{\omega C_{b'e}}$$

We see that the current gain decreases with the frequency in the high-frequency range. At high frequencies, $C_{b'e}$ shunts $B'E$ and thereby reduces $v_{b'e}$, which results in a smaller output current, $i_c = g_m v_{b'e}$. The overall current amplification is shown in Figure 6.60, where the current gain exhibits a cutoff fre-

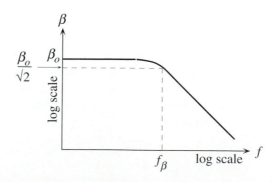

Figure 6.60   The frequency dependence of the current gain β shown on a log–log plot. The cutoff frequency is when $\beta = \beta_o/\sqrt{2}$.

quency at $f = f_\beta$, where its magnitude has fallen by a factor of $\sqrt{2}$. This occurs when the impedance of $C_{b'e}$ is equal to $r_{b'e}$

$$f_\beta = \frac{1}{2\pi r_{b'e} C_{b'e}}$$ [6.62]   *β cutoff frequency*

6.11   *pn* JUNCTION–GENERATED SHOT NOISE

When a *pn* junction is forward biased, minority carriers become injected and diffuse across the neutral regions. The injection of the minority carriers across the SCL is a statistical process. For example, electrons are not injected from *n*- to *p*-side at well-defined intervals but arrive at random intervals. Sometimes more and sometimes less electrons are injected into the *p*-side even though, over a long time, the average injection rate is constant. This leads to fluctuations in the diode current about the dc value, as depicted in Figure 6.61. We can appreciate this by noting that although the mean electron energy in the *CB* in the *n*-side is $(3/2)\,kT$ above E_c, some electrons have less and some more than $(3/2)\,kT$. In fact, their energy distribution can be described by Boltzmann statistics. Further, some will be moving towards the *p*-side and some away. At any instant, only those electrons that have an energy above the potential energy barrier $e(V_o - V)$ and happen to be moving toward the *p*-side become injected into the *p*-side. Similar arguments, of course, also apply to holes. Since the injection and arrival of minority carriers in the neutral regions are reminiscent to random shots, the resulting fluctuations in the diode current are known as **shot noise.** What is the rms (root mean square) value of the current fluctuations, or the noise current, in the forward-biased *pn* junction?

The fluctuations in carrier injection and hence in the diode current mean that there are fluctuations in the stored charges in the neutral regions. Suppose that at one instant the excess current over the average diode current is ΔI. The corresponding amount, ΔQ_h, of charge stored in the neutral region is $\tau_h \, \Delta I$. The stored energy is $(1/2)(\Delta Q_h)^2/C_{diff}$ or $(1/2)(\tau_h \, \Delta I)^2/C_{diff}$, where C_{diff} is the diffussion capaci-

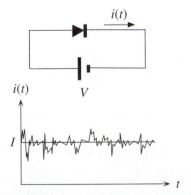

Figure 6.61   The diode current exhibits shot noise.

tance that dominates over the depletion layer capacitance. The mean value of this stored energy can only be about $(1/2)\,kT$ as the fluctuations are due to the Boltzmann energy distribution of electrons in the CB and holes in the VB. Thus

$$\frac{1}{2}\frac{\overline{(\tau_h\,\Delta I)^2}}{C_{diff}} = \frac{1}{2}kT$$

The mean square current fluctuation is then

$$\overline{(\Delta I)^2} = \frac{kTC_{diff}}{\tau_h^2}$$

From Section 6.1, we can use

$$C_{diff} = \frac{\tau_h eI}{kT}$$

to obtain

$$\overline{(\Delta I)^2} = \frac{eI}{\tau_h}$$

The quantity $1/\tau_h$ represents the rate of recombination. The changes in the stored charge can easily follow the current fluctuations if the frequency of the fluctuations is less than $1/\tau_h$ (or, put differently, if the fluctuations are slower than the rate of recombination). Thus the frequency bandwidth, B, of the current fluctuations is about $1/\tau_h$. We define the shot noise current, i_{sn}, as the rms value of the current fluctuation; then

$$i_{sn} = \sqrt{\overline{(\Delta I)^2}} = \sqrt{eIB}$$

This result is based on an intuitive and a semiquantitative argument. A more rigorous approach arrives at the same result but with a factor 2 under the root sign

Shot noise

$$i_{sn} = \sqrt{2eIB}$$

Example 6.13 | **SHOT NOISE**   A forward-biased *pn* junction is carrying a current of 1 mA. Given that the minority carrier recombination time is 1 μs, what is the shot noise current?

Solution

The frequency bandwidth B can be taken to be at most $1/\tau_h$, or 1 MHz. Then the shot noise is

$$i_{sn} = [2eIB]^{1/2} = [2 \times 1.6 \times 10^{-19} \times 10^{-3} \times 10^6]^{1/2}$$

$$= 1.79 \times 10^{-8}\ \text{A} \qquad \text{or} \qquad 17.9\ \text{nA}$$

Under normal and active operating conditions, a BJT has the base–emitter forward biased so there is shot noise in the emitter and hence the collector current.

IMPORTANT TERMS

Accumulation occurs when an applied voltage to the gate (or metal electrode) of a MOS device causes the semiconductor under the oxide to have a greater number of majority carriers than the equilibrium value. Majority carriers have been accumulated at the surface of the semiconductor under the oxide.

Active device is a device that exhibits gain (current or voltage, or both) and has a directional electronic function. Transistors are active devices, whereas resistors, capacitors, and inductors are passive devices.

Avalanche breakdown is the enormous increase in the reverse current in a *pn* junction when the applied reverse field is sufficiently high to cause the generation of electron–hole pairs by impact ionization in the space charge layer.

Base spread resistance ($r_{bb'}$) is an effective resistance that represents the voltage drop from the external base terminal (B) to the actual base point (B'). It is the voltage $V_{B'E}$ that controls the collector current.

Base width modulation (the Early effect) is the modulation of the base width by the voltage appearing across the base–collector junction. An increase in the base to collector voltage increases the collector junction depletion layer width, which results in the narrowing of the base width.

Bipolar junction transistor (BJT) is a transistor whose normal operation is based on the injection of carriers from the emitter into the base region, where they become minority carriers, and their subsequent diffusion to the collector, where they give rise to a collector current. The voltage between the base and the emitter controls the collector current.

Built-in voltage (V_o) is the voltage across a *pn* junction, going from *p*- to *n*-type semiconductor, in an open circuit.

Channel is the conducting strip between the source and drain regions of a MOSFET.

Chip is a piece (or a volume) of a semiconductor crystal that contains many integrated active and passive components to implement a circuit.

Collector junction is the metallurgical junction between the base and the collector of a bipolar transistor.

Critical electric field is the field in the space charge (or depletion) region at reverse breakdown (avalanche or Zener).

Depletion layer (or space charge layer, SCL) is a region around the metallurgical junction where recombination of electrons and holes has depleted this region of its large number of equilibrium majority carriers.

Depletion MOSFET is a MOSFET device that already has a conducting channel between the source and the drain in the absence of a gate voltage. In the depletion mode of operation, the applied gate voltage reduces the conductance of the channel and hence reduces the drain current. In the enhancement mode of operation, the gate voltage polarity enhances the conductance of the source to drain channel and increases the drain current.

Depletion (space charge) layer capacitance is the incremental capacitance (dQ/dV) due to the change in the exposed dopant charges in the depletion layer as a result of the change in the voltage across the *pn* junction.

Diffusion is the flow of particles of a given species from high to low concentration regions by virtue of their random thermal motions.

Diffusion (storage) capacitance is the *pn* junction capacitance due to the injuction and storage of minority carriers in the neutral regions when a forward bias is applied.

Emitter junction is the metallurgical junction between the emitter and the base.

Enhancement MOSFET is a MOSFET device that needs a gate to source voltage above the threshold voltage to form a conducting channel between the source and the drain. In the absence of a gate voltage, there is no conduction between the source and drain. In its usual mode of operation, the gate voltage enhances the conductance of the source to drain inversion layer and increases the drain current.

Field effect transistor (FET) is a transistor whose normal operation is based on controlling the conductance of a channel between two electrodes by the application of an external field. The effect of the applied field is to control the current flow. The current is due to majority carrier drift from the source to the drain and is controlled by the voltage applied to the gate.

Forward bias is the application of an external voltage to a *pn* junction such that the positive terminal is connected to the *p*-side and the negative to the *n*-side. The applied voltage reduces the built-in potential.

Impact ionization is the process by which a high electric field accelerates a free charge carrier (electron in the CB), which then impacts with a Si–Si bond to generate a free electron–hole pair. The impact excites an electron from E_v to E_c.

Incremental conductance (g_d) is the change in the forward current per unit change in the voltage across the diode, dI/dV. Incremental resistance is the inverse of the incremental conductance.

Integrated circuit (IC) is a chip of a semiconductor crystal in which many active and passive components have been miniaturized and integrated together to form a sophisticated circuit.

Inversion occurs when an applied voltage to the gate (or metal electrode) of a MOS device causes the semiconductor under the oxide to develop a conducting layer (or a channel) at the surface of the semiconductor. The conducting layer has opposite polarity carriers to the bulk semiconductor and hence is termed an inversion layer.

Ion implantation is a process that is used to bombard a sample in vacuum with ions of a given species of atom. First the dopant atoms are ionized in vaccum and then accelerated by applying voltage differences to impinge on a sample to be doped. The sample is grounded to neutralize the implanted ions.

Law of the junction relates the injected minority carrier concentration just outside the depletion layer to the applied voltage. For holes in the *n*-side, it is

$$p_n(0) = p_{no} \exp\left(\frac{eV}{kT}\right)$$

where $p_n(0)$ is the hole concentration just outside the depletion layer.

Lithography is a process by which a pattern is defined on the surface of the wafer that identi- fies the crystal regions to be doped or metallic interconnections.

Long diode is a *pn* junction with neutral regions longer than the minority carrier diffusion lengths.

Mask is a high-contrast pattern on a transparent plate that identifies the processing pattern (e.g., diffusion regions) needed on the wafer surface. For example, a typical mask may be a quartz plate with a patterned chromium film on it.

Metallurgical junction is where there is an effective junction between the *p*-type and *n*-type doped regions in the crystal. It is where the donor and acceptor concentrations are equal or where there is a transition from *n*- to *p*-type doping.

Metal-oxide-semiconductor transistor (MOST) is a field effect transistor in which the conductance between the source and drain is controlled by the voltage supplied to the gate electrode, which is insulated from the channel by an oxide layer.

Minority carrier injection is the flow of electrons into the *p*-side and holes into the *n*-side of a *pn* junction when a voltage is applied to reduce the built-in voltage across the junction.

MOS is short for a metal-insulator-semiconductor structure in which the insulator is typically silicon oxide. It can also be a different type of dielectric; for example, it can be the nitride Si_3N_4.

NMOS is an enhancement type *n*-channel MOS-FET.

Ohmic contact is a contact that can supply charge carriers to a semiconductor at a rate determined by charge transport through the semiconductor and not by the contact properties itself. Thus, the current is limited by the conductivity of the semiconductor and not the contact.

Passive device or component is a device that exhibits no gain and no directional function. Resistors, capacitors, and inductors are passive components.

Photolithography uses light in the lithographic process to define a pattern on the surface of the wafer.

Pinch-off voltage is the gate to source voltage needed to just pinch off the conducting channel between the source and drain with no source to drain voltage applied. It is also the source to drain voltage that just pinches off the channel when the gate and source are shorted. Beyond pinch-off, the drain current is almost constant and controlled by V_{GS}.

PMOS is an enhancement type *p*-channel MOS-FET.

Poly-Si gate is short for a polycrystalline and highly doped Si gate.

Recombination current flows under forward bias to replenish the carriers recombining in the space charge (depletion) layer. Typically, it is described by $I = I_{ro}[\exp(eV/2kT) - 1]$.

Resist is an irradiation-sensitive chemical emulsion that is coated as a thin layer onto the surface of a wafer. Its chemical solubility (or etchability) depends on irradiation (UV light, x-rays, or electron beam). If the resist is sensitive to light (UV light), it is called a photoresist.

Reticle is a patterned mask that is placed in a UV light projection system to project the image of the pattern onto the wafer surface for the lithographic process.

Reverse bias is the application of an external voltage to a *pn* junction such that the positive terminal is connected to the *n*-side and the negative to the *p*-side. The applied voltage increases the built-in potential.

Reverse saturation current is the reverse current that would flow in a reverse-biased ideal *pn* junction obeying the Shockley equation.

Scaling describes shrinking the dimensions of devices to integrate more devices per unit area and to increase the speed of operation.

Shockley diode equation relates the diode current to the diode voltage through $I = I_o[\exp(eV/kT) - 1]$. It is based on the injection and diffusion of injected minority carriers by the application of a forward bias.

Short diode is a *pn* junction in which the neutral regions are shorter than the minority carrier diffusion lengths.

Shot noise is the fluctuations in the current passing through a *pn* junction (or a Schottky diode) as a consequence of the fluctuations in the number of carriers injected across the potential energy barrier. In general, it refers to fluctuations in the current through any device due to fluctuations in the emitted (or injected) carriers.

Small-signal equivalent circuit of a transistor replaces the transistor with an equivalent circuit that consists of resistances, capacitances, and dependent sources (current or voltage). The equivalent circuit represents the transistor behavior under small-signal ac conditions. The batteries are replaced with short circuits (or their internal resistances). Small signals imply small variations about dc values.

Stepper is short for stepper and repeater, which is an optical-based system that projects the pattern of a reticle (mask) onto a small area on the surface of the wafer. This projected image forms the processing pattern for one chip. The wafer is stepped in the *x* and *y* directions to repeat the imaging of other chips.

Substrate is a single mechanical support that carries active and passive devices. For example, in integrated circuit technology, typically, many integrated circuits are fabricated on a single silicon crystal wafer that serves as the substrate.

Thermal generation current is the current that flows in a reverse-biased *pn* junction as a result of the thermal generation of electron–hole pairs in the depletion layer that become separated and swept across by the built-in field.

Threshold voltage is the gate voltage needed to establish a conducting channel between the source and drain of an enhancement MOST.

Transistor is a three-terminal solid-state device in which a current flowing between two electrodes is controlled by the voltage between the third and one of the other terminals or by a current flowing into the third terminal.

Ultraviolet (UV) radiation is an electromagnetic wave with wavelength typically in the range 4 nm to 400 nm.

Very large scale integration (VLSI) is the integration of $10^6 – 10^7$ devices into a single Si chip to implement very complex circuits (for example, microprocessor chips, memories, etc.).

X-rays are electromagnetic radiation of wavelength typically in the range 10 pm – 1 nm. X-ray photons have high energies.

Zener breakdown is the enormous increase in the reverse current in a *pn* junction when the applied voltage is sufficient to cause the tunneling of electrons from the valence band in the *p*-side to the conduction band in the *n*-side. Zener breakdown occurs in *pn* junctions that are heavily doped on both sides so that the depletion layer width is narrow.

QUESTIONS AND PROBLEMS

6.1 **The built-in voltage**   There is a rigorous derivation of the built-in voltage across a pn junction. Inasmuch as in equilibrium there is no net current through the pn junction, drift of electrons due to the built-in field, $\mathcal{E}(x)$, must be just balanced by their diffusion due to the concentration gradient, dn/dx. We can thus set the total electron and hole current densities (drift + diffusion) through the depletion region to zero. Considering holes alone,

$$J_{\text{hole}}(x) = ep(x)\mu_h\mathcal{E}(x) - eD_h\frac{dp}{dx} = 0$$

By using the definition of the electric field $\mathcal{E}(x)$ in terms of $V(x)$ and the Einstein relation between D_h and μ_h, show that the above equation reduces to

$$-ep\ dV - kT\ dp = 0$$

By integrating this equation show that

$$V_o = \frac{kT}{e}\ln\!\left(\frac{p_{po}}{p_{no}}\right)$$

6.2 **The pn junction**   Consider an abrupt Si pn^+ junction that has 10^{15} acceptors cm^{-3} on the p-side and 10^{19} donors on the n-side. The minority carrier recombination times are $\tau_e = 490$ ns for electrons in the p-side and $\tau_h = 2.5$ ns for holes in the n-side. The cross-sectional area is 1 mm^2. Assuming a long diode, calculate the current, I, through the diode at room temperature when the voltage, V, across it is 0.6 V. What are V/I and the incremental resistance (r_d) of the diode and why are they different?

*6.3 **The Si pn junction**   Consider a long pn junction diode with an acceptor doping, N_a, of 10^{18} cm^{-3} on the p-side and donor concentration of N_d on the n-side. The diode is forward biased and has a voltage of 0.6 V across it. The diode cross-sectional area is 1 mm^2. The minority carrier recombination time, τ, depends on the dopant concentration, $N_{\text{dopant}}(\text{cm}^{-3})$, through the following approximate relation

$$\tau = \frac{5 \times 10^{-7}}{(1 + 2 \times 10^{-17}N_{\text{dopant}})}$$

a.   Suppose that $N_d = 10^{15}$ cm^{-3}. Then the depletion layer extends essentially into the n-side and we have to consider minority carrier recombination time, τ_h, in this region. Calculate the diffusion and recombination contributions to the total diode current. What is your conclusion?

b.   Suppose that $N_d = N_a$. Then W extends equally to both sides and, further, $\tau_e = \tau_h$. Calculate the diffusion and recombination contributions to the diode current. What is your conclusion?

Table 6.3   Capacitance at various values of reverse bias (V_r)

V_r (V)	1	2	3	5	10	15	20
C (pF)	46.0	37.1	32.0	26.0	19.0	15.9	14

6.4   Junction capacitance of a pn junction   The capacitance (C) of a reverse-biased abrupt Si p^+n junction has been measured as a function of the reverse bias voltage, V_r, as listed in Table 6.3. The pn junction cross-sectional area is 500 μm $\times$ 500 μm. By plotting $1/C^2$ versus V_r, obtain the built-in potential, V_o, and the donor concentration, N_d, in the n-region. What is N_a?

6.5   Temperature dependence of diode properties

a.   Consider the reverse current in a pn junction. Show that

$$\frac{\delta I_{rev}}{I_{rev}} \approx \left(\frac{E_g}{\eta kT}\right)\frac{\delta T}{T}$$

where $\eta = 2$ for Si and GaAs, in which thermal generation in the depletion layer dominates the reverse current, and $\eta = 1$ for Ge, in which the reverse current is due to minority carrier diffusion to the depletion layer. It is assumed that $E_g \gg kT$ at room temperature. Order the semiconductors Ge, Si, and GaAs according to the sensitivity of reverse current to temperature.

b.   Consider a forward-biased pn junction carrying a *constant* current I. Show that the change in the voltage across the pn junction per unit change in the temperature is given by

$$\frac{dV}{dT} = -\frac{V_g - V}{T}$$

where $V_g = E_g/e$ is the energy gap expressed in volts. Calculate typical values for dV/dT for Ge, Si, and GaAs assuming that, typically, $V = 0.2$ V for Ge, 0.6 V for Si, and 0.9 V for GaAs. What is your conclusion? Can one assume that, typically, $dV/dT \approx -2$ mV °C^{-1} for these diodes?

6.6   Avalanche breakdown   Consider a Si p^+n junction diode that is required to have an avalanche breakdown voltage of 25 V. Given the breakdown field $\mathcal{E}_{br}$ in Figure 6.19, what should be the donor doping concentration?

6.7   Design of a pn junction diode   Design an abrupt Si pn^+ junction that has a reverse breakdown voltage of 100 V and provides a current of 10 mA when the voltage across it is 0.6 V. Assume that, if N_{dopant} is in cm^{-3}, the minority carrier recombination time is given by

$$\tau = \frac{5 \times 10^{-7}}{(1 + 2 \times 10^{-17}N_{dopant})}$$

Mention any assumptions made.

*6.8 Minority carrier profiles (the hyperbolic functions)   Consider a *pnp* BJT under normal operating conditions in which the EB junction is forward biased and the BC junction is reverse biased. The field in the neutral base region outside the depletion layers can be assumed to be negligibly small. The continuity equation for holes, $p_n(x)$, in the *n*-type base region is

$$D_h \frac{d^2 p_n}{dx^2} - \frac{p_n - p_{no}}{\tau_h} = 0 \qquad \text{[6.63]}$$

where $p_n(x)$ is the hole concentration at x from just outside the depletion region and p_{no} and τ_h are the equilibrium hole concentration and hole recombination lifetime in the base.

a.   What are the boundary conditions at $x = 0$ and $x = W_B$, just outside the collector region depletion layer? (Consider the law of the junction.)

b.   Show that the following expression for $p_n(x)$ is a solution of the continuity equation

$$p_n(x) = p_{n0} \left[\exp\left(\frac{eV}{kT}\right) - 1 \right] \frac{\left[\sinh\left(\dfrac{W_B - x}{L_h}\right) \right]}{\sinh\left(\dfrac{W_B}{L_h}\right)} + p_{no} \left[1 - \frac{\sinh\left(\dfrac{x}{L_h}\right)}{\sinh\left(\dfrac{W_B}{L_h}\right)} \right] \qquad \text{[6.64]}$$

where $V = V_{EB}$ and $L_h = \sqrt{D_h \tau_h}$.

c.   Show that Equation 6.64 satisfies the boundary conditions.

*6.9 The *pnp* bipolar transistor   Consider a *pnp* transistor in common base configuration and under normal operating conditions. The emitter–base junction is forward biased and the base–collector junction is reverse biased. The emitter, base, and collector dopant concentrations are $N_{d(E)}$, $N_{d(B)}$, and $N_{a(C)}$ respectively where $N_{a(E)} \gg N_{d(B)} \geq N_{a(C)}$. For simplicity, assume uniform doping in all the regions. The base and emitter widths are W_B and W_E, respectively, both much shorter than the minority carrier diffusion lengths, L_h and L_e. The minority carrier lifetime in the base is the hole recombination time, τ_h. The minority carrier mobility in the base and emitter are denoted by μ_h and μ_e respectively.

The minority carrier concentration profile in the base can be represented by Equation 6.64.

a.   Assuming that the emitter injection efficiency is unity show that

$$1. \quad I_E \approx \frac{e A D_h n_i^2 \coth\left(\dfrac{W_B}{L_h}\right)}{L_h N_{d(B)}} \exp\left(\frac{e V_{EB}}{kT}\right)$$

2. $I_C \approx \dfrac{eAD_h n_i^2 \; \text{cosech}\left(\dfrac{W_B}{L_h}\right)}{L_h N_{d(B)}} \; \exp\left(\dfrac{eV_{EB}}{kT}\right)$

3. $\alpha \approx \text{sech}\left(\dfrac{W_B}{L_h}\right)$

4. $\beta \approx \dfrac{\tau_h}{\tau_t}$ where $\tau_t = \dfrac{W_B^2}{2D_h}$ is the base transit time.

b. Consider the total emitter current, I_E, through the EB junction, which has diffusion and recombination components as expressed below

$$I_E = I_{E(so)} \exp\left(\frac{eV_{EB}}{kT}\right) + I_{E(ro)} \exp\left(\frac{eV_{EB}}{2kT}\right)$$

Only the hole component of the diffusion current (first term) can contribute to the collector current. Show that when $N_{a(E)} \gg N_{d(B)}$, the emitter injection efficiency, γ, is given by

$$\gamma \approx \left[1 + \frac{I_{E(ro)}}{I_{E(so)}} \exp\left(-\frac{eV_{EB}}{2kT}\right)\right]^{-1}$$

How does $\gamma < 1$ modify the expressions derived in part (*a*)? What is your conclusion (consider small and large emitter currents, or $V_{EB} = 0.4$ and 0.7 V)?

6.10 **A high voltage *npn* transistor**   Consider an idealized high voltage *npn* BJT with the properties listed in Table 6.4. The base region has a relatively uniform doping. The emitter and collector donor concentrations are mean values. The cross-sectional area is 0.025 cm². The transistor is biased to operate in the normal active mode. The base–emitter voltage is 0.6 V and the base–collector voltage is 100 V.

a. Calculate the reverse breakdown voltage across the base–collector junction.

b. Calculate the depletion layer width extending from the collector into the base. What is the width, W_B, of the neutral base region?

c. Calculate α and hence β for this transistor. Comment on how these should depend on V_{CB} (remember part (*b*)).

Table 6.4   Properties of an *npn* BJT

Emitter Width	Emitter Doping	Hole Lifetime in Emitter	Base Width	Base Doping	Electron Lifetime in Base	Collector Doping
20 μm	2×10^{18} cm^{-3}	24 ns	20 μm	2×10^{15} cm^{-3}	480 ns	2×10^{16} cm^{-3}

  d.   What are the emitter, collector, and base currents?

  e.   What is the emitter injection efficiency? Does it affect α and β?

  f.   What is the collector current when $V_{CB} = 110$ V but $V_{EB} = 0.6$ V? What is the incremental collector output resistance defined as $\Delta V_{CB}/\Delta I_C$?

6.11   **The JFET pinch-off voltage**   Consider the symmetric n-channel JFET shown in Figure 6.62. The width of each depletion region extending into the n-channel is W. The thickness, or depth, of the channel, defined between the two metallurgical junctions, is $2a$. Assuming an abrupt pn junction and $V_{DS} = 0$, show that when the gate to source votage is $-V_p$ the channel is pinched off where

$$V_P = \frac{a^2 e N_d}{2\varepsilon} - V_o$$

where V_o is the built-in potential between p^+n junction and N_d is the donor concentration of the channel.

  Calculate the pinch-off voltage of a JFET that has an acceptor concentration of 10^{19} cm^{-3} in the p^+ gate, a channel donor doping of 10^{16} cm^{-3}, and a channel thickness (depth), $2a$, of 2 μm.

6.12   **The JFET**   Consider an n-channel JFET that has a symmetric p^+n gate–channel structure as shown in Figures 6.28a and 6.62. Let L be the gate length, Z the gate width, and $2a$ the channel thickness. The pinch-off voltage is given by

$$V_P = \frac{a^2 e N_d}{2\varepsilon} - V_o$$

where V_o is the built-in potential between the p^+ gate and the n-channel and N_d is the channel donor concentration. The drain saturation current, I_{DSS}, is the drain current when $V_{GS} = 0$. This occurs when $V_{DS} = V_{DS(sat)} = V_P$ (Figure 6.30) so that $I_{DSS} = V_P G_{ch}$, where G_{ch} is the conductance of the channel between the source and the pinched-off point (Figure 6.31). Taking into account the shape of the channel at pinch-off, if G_{ch} is about 1/3 of the conductance of the free or unmodulated (rectangular) channel, show that

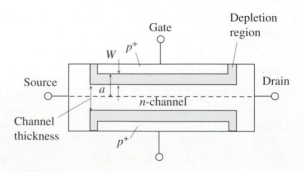

Figure 6.62   A symmetric JFET

$$I_{DSS} = V_P \left[\frac{1}{3} \frac{(e\mu_e N_d)(2a)Z}{L} \right]$$

A particular n-channel JFET with a symmetric p^+n gate–channel structure has a pinch-off voltage of 3.9 V and an I_{DSS} of 5.5 mA. If the gate and channel dopant concentrations are $N_a = 10^{19}$ cm^{-3} and $N_d = 10^{15}$ cm^{-3} respectively, find the channel thickness $2a$ and Z/L. If $L = 10$ μm, what is Z? What is the gate–source capacitance when the JFET has no voltage supplies connected to it?

6.13 The JFET amplifier   Consider an n-channel JFET that has a pinch-off voltage (V_P) of 5 V and $I_{DSS} = 10$ mA. It is used in a common source configuration as in Figure 6.35a in which the gate to source bias voltage (V_{GS}) is -1.5 V.

a.   If a small signal voltage gain of 10 is needed, what should be the drain resistance (R_D) and a reasonable drain voltage supply (V_{DD}) given that the maximum rated V_{DS} is 25 V?

b.   If an ac signal of 3 V peak-to-peak is applied to the gate in series with the dc bias voltage, what will be the ac output voltage peak-to-peak? What is the voltage gain for positive and negative input signals? What is your conclusion?

6.14 The Enhancement NMOSFET amplifier   Consider an n-channel enhancement NMOS that has a gate width (Z) of 150 μm, channel length (L) of 10 μm, and oxide thickness (t_{ox}) of 500 Å. The substrate acceptor doping (N_a) is 10^{16} cm^{-3} and the threshold voltage (V_{th}) is 2 V.

a.   Calculate the drain current when $V_{GS} = 5$ V and $V_{DS} = 5$ V and assuming $\lambda = 0.01$.

b.   What is the small-signal voltage gain if the NMOSFET is connected as a common source amplifier, as shown in Figure 6.63, with a drain

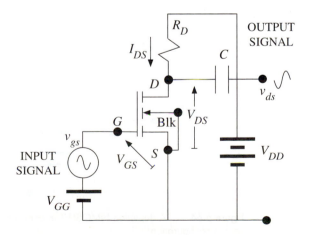

Figure 6.63   NMOSFET amplifier

are not conducted from one plate of the capacitor to the other through the dielectric. Dielectric materials often serve to insulate current-carrying conductors or conductors at different voltages. Why can we not simply use air as insulation between high voltage conductors? When the electric field inside an insulator exceeds a critical field called the **dielectric strength,** the medium suffers dielectric breakdown and a large discharge current flows through the dielectric. Some 40 percent of utility generator failures are linked to insulation failures in the generator. Dielectric breakdown is probably one of the oldest electrical engineering problems and that which has been most widely studied and never fully explained.

7.1   MATTER POLARIZATION AND RELATIVE PERMITTIVITY

7.1.1   Relative Permittivity: Definition

We first consider a parallel plate capacitor with vacuum as the dielectric medium between the plates, as shown in Figure 7.1a. The plates are connected to a constant voltage supply V. Let Q_o be the charge on the plates. This charge can be easily measured. The capacitance C_o of the parallel plate capacitor in free space, as in Figure 7.1a, is defined by

$$C_o = \frac{Q_o}{V} \qquad \text{[7.1]}$$

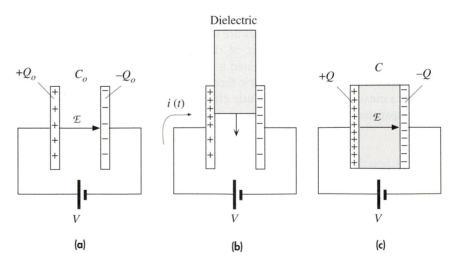

Figure 7.1

(a) Parallel plate capacitor with free space between the plates.

(b) As a slab of insulating material is inserted between the plates, there is an external current flow indicating that more charge is stored on the plates.

(c) The capacitance has been increased due to the insertion of a medium between the plates.

The electric field, directed from high to low potential, is defined by the gradient of the potential, $\mathcal{E} = dV/dx$. Thus, the electric field $\mathcal{E}$ between the plates is just V/d where d is the separation of the plates.

Consider now what happens when a dielectric slab (a slab of any nonconducting material) is inserted into this parallel plate capacitor, as shown in Figure 7.1b and c with V kept the same. During the insertion of the dielectric slab, there is an external current flow that indicates that there is additional charge being stored on the plates. The charge on the electrodes increases from Q_o to Q. We can easily measure the extra charge $Q - Q_o$ flowing from the battery to the plates by integrating the observed current in the circuit during the process of insertion, as shown in Figure 7.1b. Because there is now a greater amount of charge stored on the plates, the capacitance of the system in Figure 7.1c is larger than that in Figure 7.1a by the ratio Q to Q_o. The relative permittivity (or the dielectric constant) ε_r is defined to reflect this increase in the capacitance or the charge storage ability by virtue of having a dielectric medium. If C is the capacitance with the dielectric medium as in Figure 7.1c, then by definition

$$\varepsilon_r = \frac{Q}{Q_o} = \frac{C}{C_o} \qquad [7.2]$$

Definition of relative permittivity

The increase in the stored charge is due to the polarization of the dielectric by the applied field, as explained below. It is important to remember that when the dielectric medium is inserted, the electric field remains unchanged, provided that the insulator fills the whole space between the plates as indicated in Figure 7.1c. The voltage V remains the same and therefore so does the gradient V/d, which means that $\mathcal{E}$ remains constant.

7.1.2   Dipole Moment and Electronic Polarization

An electrical dipole moment is simply a separation between a negative and positive charge of equal magnitude, Q, as shown in Figure 7.2. If **a** is the vector from the negative to the positive charge, the electric dipole moment is defined as a vector by[1]

$$\mathbf{p} = Q\mathbf{a} \qquad [7.3]$$

Definition of dipole moment

The region that contains the $+Q$ and $-Q$ charges has zero net charge. Unless the two charge centers coincide, this region will nonetheless, by virtue of the definition in Equation 7.3, contain a dipole moment.

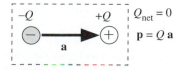

Figure 7.2   The definition of electric dipole moment.

---

[1] The definition of a dipole moment and the properties of dipoles were covered in Chapter 1 in the explanation of secondary bonding.

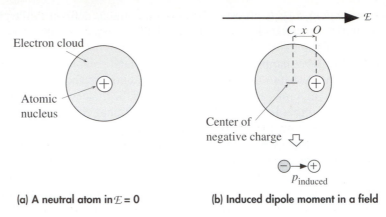

(a) A neutral atom in $\mathcal{E} = 0$ (b) Induced dipole moment in a field

Figure 7.3   The origin of electronic polarization.

The net charge within a neutral atom is zero. Furthermore, on average, the center of negative charge of the electrons coincides with the positive nuclear charge, which means that the atom has no net dipole moment, as indicated in Figure 7.3a. However, when this atom is placed in an external electric field, it will develop an induced dipole moment. The electrons, being much lighter than the positive nucleus, become easily displaced by the field, which results in the separation of the negative charge center from the positive charge center, as shown in Figure 7.3b. This separation of negative and positive charges and the resulting induced dipole moment are termed **polarization.** An atom is said to be **polarized** if it possesses an effective dipole moment, that is, if there is a separation between the centers of negative and positive charge distributions.

The induced dipole moment depends on the electric field causing it. We define a quantity called the **polarizability,** α, to relate the induced dipole moment p_{induced} to the field $\mathcal{E}$ causing it,

Definition of polarizability

$$p_{\text{induced}} = \alpha \mathcal{E} \tag{7.4}$$

where α is a coefficient called the polarizability of the atom. It depends on the polarization mechanism. Since the polarization of a neutral atom involves the displacement of electrons, α is called **electronic polarization** and denoted as α_e. Inasmuch as the electrons in an atom are not rigidly fixed, all atoms possess a certain amount of electronic polarizability.

In the absence of an electric field, the center of mass of the orbital motions of the electrons coincides with the positively charged nucleus and the electronic dipole moment is zero. Suppose that the atom has Z number of electrons orbiting the nucleus and all the electrons are contained within a certain sphere of radius. When an electric field $\mathcal{E}$ is applied, the light electrons become displaced in the opposite direction to $\mathcal{E}$ so that their center of mass, C, is shifted by some distance x with respect to the nucleus, O, which we take to be the origin, as shown in Figure 7.3b. As the electrons are "pushed" away by the applied field, the coulombic attraction between the electrons and nuclear charge "pulls in" the electrons. The force on the electrons, due to $\mathcal{E}$, trying to separate them away from the nuclear charge is $Ze\mathcal{E}$. The restoring force, F_r, which is the coulombic attractive force between the electrons and the nucleus, can be taken to be proportional to the displacement x, provided that the latter is small.[2] The restoring force F_r is obviously zero when C coincides with $O(x = 0)$. We can write

$$F_r = -\beta x$$

where β is a constant and the negative sign indicates that F_r is always directed toward the nucleus, O (Figure 7.3b). In equilibrium, the net force on the negative charge is zero or

$$Ze\mathcal{E} = \beta x$$

from which x is known. Therefore the **magnitude** of the induced electronic dipole moment p_e is given by

$$p_e = (Ze)x = \left(\frac{Z^2e^2}{\beta}\right)\mathcal{E} \qquad \text{[7.5]}$$

Electronic polarization

As expected, p_e is proportional to the applied field. The electronic dipole moment in Equation 7.5 is valid under static conditions, that is, when the electric field is a dc field. Suppose that we suddenly remove the applied electric field polarizing the atom. There is then only the restoring force, $-\beta x$, which always acts to pull the electrons toward the nucleus, O. The equation of motion of the negative charge center is then (from $F = ma$)

$$-\beta x = Zm_e\frac{d^2x}{dt^2}$$

Thus the displacement at any time is

$$x(t) = x_o\cos(\omega_o t)$$

where

$$\omega_o = \left[\frac{\beta}{Zm_e}\right]^{1/2} \qquad \text{[7.6]}$$

is the oscillation frequency of the center of mass of the electron cloud about the nucleus and x_o is the displacement before the removal of the field. After the removal of the field, the electronic charge cloud executes simple harmonic motion about the nucleus with a frequency determined by Equation 7.6. It is analogous to a mass on a spring being pulled and let go. The system then executes simple harmonic motion. The oscillations of course die out with time. In the atomic case, a sinusoidal displacement implies that the electronic charge cloud has an acceleration

$$\frac{d^2x}{dt^2} = -x_o\omega_o^2\cos(\omega_o t)$$

[2] It may be noticed that even if F_r is a complicated function of x, it can still be expanded in a series in terms of powers of x, that is, x, x^2, x^3, and so on, and for small x only the x term is significant, $F_r = -\beta x$.

It is well known from classical electromagnetism that an accelerating charge radiates electromagnetic energy just like a radio antenna. Consequently the oscillating charge cloud loses energy, and thus its amplitude of oscillation decreases. (Recall that the average energy is proportional to the square of the amplitude of the displacement.)

From the expression derived for p_e in Equation 7.5, we can find the electronic polarizability α_e from Equation 7.4,

Static electronic polarizability

$$\alpha_e = \frac{Ze^2}{m_e \omega_o^2} \tag{7.7}$$

7.1.3 Polarization Vector P

When a material is placed in an electric field, the atoms and the molecules of the material become polarized so that we have a distribution of dipole moments in the material. We can visualize this effect with the insertion of the dielectric slab into the parallel plate capacitor, as depicted in Figure 7.4a. The placement of the dielectric slab into an electric field polarizes the molecules in the material. The induced dipole moments all point in the direction of the field. Consider the polarized medium alone, as shown in Figure 7.4b. In the bulk of the material, the dipoles are aligned head to tail. Every positive charge has a negative charge next to it and vice versa. There is therefore no net charge within the bulk. But the positive charges of the dipoles appearing at the right-hand face are not canceled by negative charges

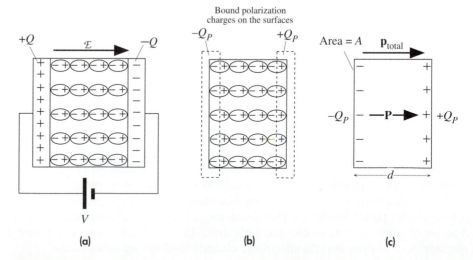

Figure 7.4
(a) When a dielectric is placed in an electric field, bound polarization charges appear on the opposite surfaces.
(b) The origin of these polarization charges is the polarization of the molecules of the medium.
(c) We can represent the whole dielectric in terms of its surface polarization charges $+Q_P$ and $-Q_P$.

of any dipoles at this face. There is therefore a surface charge, $+Q_P$, on the right-hand face that results from the polarization of the medium. Similarly, there is a negative charge, $-Q_P$, with the same magnitude appearing on the left-hand face due to the negative charges of the dipoles at this face. We see that charges $+Q_P$ and $-Q_P$ appear on the opposite surfaces of a material when it becomes polarized in an electric field, as shown in Figure 7.4c. These charges are **bound** and are a direct result of the polarization of the molecules. They are termed **surface polarization charges.** Figure 7.4c emphasizes this aspect of dielectric behavior in an electric field by showing the dielectric and its polarization charges only.

We represent the polarization of a medium by a quantity called **polarization, P**, which is defined as the dipole moment per unit volume,

$$\mathbf{P} = \frac{1}{\text{Volume}}[\mathbf{p}_1 + \mathbf{p}_2 + \ldots + \mathbf{p}_N] \qquad [7.8a]$$

Definition of polarization vector

where $\mathbf{p}_1$, $\mathbf{p}_2$, $\ldots$, $\mathbf{p}_N$ are the dipole moments induced at N molecules in the volume. If $\mathbf{p}_{av}$ is the average dipole moment per molecule, then an equivalent definition of $\mathbf{P}$ is

$$\mathbf{P} = N\mathbf{p}_{av} \qquad [7.8b]$$

where N is the number of molecules per unit volume.[3] There is an important relationship, given below, between $\mathbf{P}$ and the polarization charges Q_P on the surfaces of the dielectric. It should be emphasized for future discussions that if polarization arises from the effect of the applied field, as shown in Figure 7.4a, which is usually the case, $\mathbf{p}_{av}$ must be the *average dipole moment per atom in the direction of the applied field*. In that case we often also denote $\mathbf{p}_{av}$ as the induced average dipole moment per molecule, $p_{induced}$.

To calculate the polarization, $\mathbf{P}$, for the polarized dielectric in Figure 7.4b, we need to sum all the dipoles in the medium and divide by the volume, Ad, as in Equation 7.8a. However, the polarized medium can be simply represented as in Figure 7.4c in terms of surface charge $+Q_P$ and $-Q_P$, which are separated by the thickness distance d. We can view this arrangement as one big dipole moment p_{total} from $-Q_P$ to $+Q_P$. Thus

$$p_{total} = Q_P d$$

Since the polarization is defined as the total dipole moment per unit volume, the magnitude of $\mathbf{P}$ is

$$P = \frac{p_{total}}{\text{Volume}} = \frac{Q_P d}{Ad} = \frac{Q_P}{A}$$

But Q_P/A is the surface polarization charge density, σ_P, so that

$$P = \sigma_p \qquad [7.9]$$

Polarization and bound surface charge density

Polarization is a vector and Equation 7.9 only gives its magnitude. For the rectangular slab in Figure 7.4c, the direction of P is normal to the surface. For $+\sigma_p$

| [3] In this chapter we use n for the refractive index and N for the number of atoms or molecules per unit volume.

External field

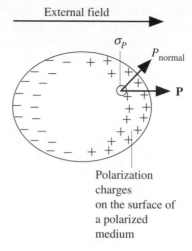

Polarization
charges
on the surface of
a polarized
medium

Figure 7.5   Polarization charge
density on the surface of a polarized
medium is related to the normal
component of the polarization vector.

(right face), it comes out from the surface and for $-\sigma_P$ (left face), it is directed into the surface. Although Equation 7.9 is derived for one specific geometry, the rectangular slab, it can be generalized as follows. *The charge per unit area appearing on the surface of a polarized medium is equal to the component of the polarization vector normal to this surface.* If P_{normal} is the component of **P** normal to the surface where the polarization charge density is σ_P, as shown in Figure 7.5, then,

$$P_{normal} = \sigma_p$$

The polarization, P, induced in a dielectric medium when it is placed in an electric field depends on the field itself. The induced dipole moment per molecule within the medium depends on the electrical field by virtue of Equation 7.4. To express the dependence of P on the field, $\mathcal{E}$, we define a quantity called the **electric susceptibility, χ_e,** by

Definition of electric susceptibility

$$P = \chi_e \varepsilon_o \mathcal{E} \qquad\qquad [7.10]$$

Equation 7.10 shows an *effect*, P, due to a *cause*, $\mathcal{E}$, and the quantity χ_e relates the effect to its cause. Put differently, χ_e acts as a proportionality constant. It may depend on the field itself, in which case the effect is nonlinearly related to the cause. Further, electronic polarizability is defined by

$$p_{induced} = \alpha_e \mathcal{E}$$

so that

$$P = N p_{induced} = N \alpha_e \mathcal{E}$$

where N is the number of molecules per unit volume. Then from Equation 7.10, χ_e and α_e are related by

$$\chi_e = \frac{1}{\varepsilon_o} N \alpha_e \qquad [7.11]$$

It is important to differentiate the difference between *free* and *polarization* (or *bound*) charges. The charges stored on the metal plates are free because they can move in the metal. For example both Q_o and Q, before and after the dielectric insertion in Figure 7.1, are free charges that arrive on the plates from the battery. The polarization charges $+Q_P$ and $-Q_P$, on the other hand, are bound to the molecules. They cannot move within the dielectric or on its surface.

The field $\mathcal{E}$ *before* the dielectric was inserted (Figure 7.1a) is given by

$$\mathcal{E} = \frac{V}{d} = \frac{Q_o}{C_o d} = \frac{Q_o}{\varepsilon_o A} = \frac{\sigma_o}{\varepsilon_o} \qquad [7.12]$$

where $\sigma_o = Q_o/A$ is the free surface charge density without any dielectric medium between the plates, as in Figure 7.1a.

After the insertion of the dielectric, this field remains the same V/d, but the free charges on the plates are different. The free surface charge on the plates is now Q. In addition there is a bound polarization charge on the dielectric surfaces next to the plates, as shown in Figure 7.4a. It is apparent that the flow of current during the insertion of the dielectric, Figure 7.1b, is due to the additional free charges $Q - Q_o$, needed on the capacitor plates to neutralize the opposite polarity polarization charges Q_P appearing on the dielectric surfaces. The total charge (see Figure 7.4a) due to that on the plate plus that appearing on the dielectric surface, $Q - Q_P$, must be the same as before, Q_o, so that the field, as given by Equation 7.12, does not change inside the dielectric, that is,

$$Q - Q_P = Q_o$$

or

$$Q = Q_o + Q_P$$

Dividing by A, defining $\sigma = Q/A$ as the free surface charge density on the plates with the dielectric inserted, and using Equation 7.12, we obtain

$$\sigma = \varepsilon_o \mathcal{E} + \sigma_P$$

Since $\sigma_P = P$ and $P = \chi_e \varepsilon_o \mathcal{E}$, Equations 7.9 and 7.10, we can eliminate σ_P to obtain

$$\sigma = \varepsilon_o (1 + \chi_e)\mathcal{E}$$

From the definition of the relative permittivity in Equation 7.2 we have

$$\varepsilon_r = \frac{Q}{Q_o} = \frac{\sigma}{\sigma_o}$$

so that substituting for σ and using Equation 7.12 we obtain

$$\varepsilon_r = 1 + \chi_e \qquad [7.13]$$

Relative permittivity and electric susceptibility

In terms of electronic polarization, from Equation 7.11, this is

*Relative
permittivity and
polarizability*

$$\varepsilon_r = 1 + \frac{N\alpha_e}{\varepsilon_o}$$

[7.14]

The significance of Equation 7.14 is that it relates the microscopic polarization mechanism that determines α_e to the macroscopic property, ε_r.

7.1.4   Local Field $\mathcal{E}_{\text{loc}}$ and Clausius–Mossotti Equation

Equation 7.14, which relates ε_r to electronic polarizability, α_e, is only approximate because it assumes that the field acting on an individual atom or molecule is the field $\mathcal{E}$, which is assumed to be uniform within the dielectric. In other words, the induced polarization, $p_{\text{induced}} \propto \mathcal{E}$. However, the induced polarization depends on the actual field experienced by the molecule. It is apparent from Figure 7.4a that there are polarized molecules within the dielectric with their negative and positive charges separated so that the field is not constant *on the atomic scale* as we move through the dielectric. This is depicted in Figure 7.6. The field experienced by an individual molecule is actually different than $\mathcal{E}$, which represents the average field in the dielectric. As soon as the dielectric becomes polarized, the field at some arbitrary point depends not only on the charges on the plates (Q) but also on the orientations of all the other dipoles around this point in the dielectric. When averaged over some distance, say a few thousand molecules, this field becomes $\mathcal{E}$, as shown in Figure 7.6.

The actual field experienced by a molecule in a dielectric is defined as the **local field** and denoted by $\mathcal{E}_{\text{loc}}$. It depends not only on the free charges on the plates but

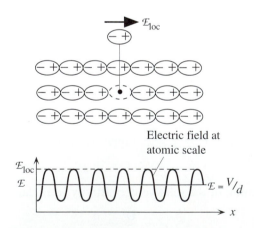

Figure 7.6   The electric field inside a polarized dielectric at the atomic scale is not uniform.

The local field is the actual field that acts on a molecule. It can be calculated by removing that molecule and evaluating the field at that point from the charges on the plates and the dipoles surrounding the point.

also on the arrangement of all the polarized molecules around this point. In evaluating $\mathcal{E}_{loc}$ we simply remove the molecule from this point and calculate the field at this point coming from all sources, including neighboring polarized molecules, as visualized in Figure 7.6. $\mathcal{E}_{loc}$ will depend on the amount of polarization the material has experienced. The greater the polarization, the greater is the local field because there are bigger dipoles around this point. $\mathcal{E}_{loc}$ depends on the arrangement of polarized molecules around the point of interest and hence depends on the crystal structure. In the simplest case of a material with a cubic crystal structure, or a liquid (no crystal structure), the local field, $\mathcal{E}_{loc}$, acting on a molecule increases with polarization as[4]

$$\mathcal{E}_{loc} = \mathcal{E} + \frac{1}{3\varepsilon_o}P$$

The induced polarization in the molecule now depends on this local field, $\mathcal{E}_{loc}$, rather than the average field $\mathcal{E}$. Thus

$$p_{induced} = \alpha_e \mathcal{E}_{loc}$$

The fundamental definition of electric susceptibility by the equation

$$P = \chi_e \varepsilon_o \mathcal{E}$$

is unchanged, which means that $\varepsilon_r = 1 + \chi_e$, Equation 7.13, remains intact. The polarization is defined by $P = Np_{induced}$, and $p_{induced}$ can be related to $\mathcal{E}_{loc}$ and hence to $\mathcal{E}$ and P. Then

$$P = (\varepsilon_r - 1)\varepsilon_o \mathcal{E}$$

can be used to eliminate $\mathcal{E}$ and P and obtain a relationship between ε_r and α_e. This is the Clausius–Mossotti equation,

$$\frac{\varepsilon_r - 1}{\varepsilon_r + 2} = \frac{N\alpha_e}{3\varepsilon_o} \qquad \text{[7.15]}$$

Clausius– Mossotti equation

This equation allows the calculation of the macroscopic property ε_r from microscopic polarization phenomena, namely α_e.

ELECTRONIC POLARIZABILITY OF A VAN DER WAALS SOLID   The electronic polarizability of | **Example 7.1**
the Ar atom is 1.7×10^{-40} F m². What is the static dielectric constant of solid Ar (below 84 K) if its density is 1.8 g cm⁻³?

Solution

To calculate ε_r we need the number of Ar atoms per unit volume, N, from the density, d. If $M_{at} = 39.95$ is the relative atomic mass of Ar and N_A is Avogadro's number, then we have

$$N = \frac{N_A d}{M_a} = \frac{(6.02 \times 10^{23} \text{ mol}^{-1})(1.8 \text{ g cm}^{-3})}{(39.95 \text{ g mol}^{-1})} = 2.71 \times 10^{22} \text{ cm}^{-3}$$

[4] This field is called the **Lorentz field** and the proof, though not difficult, is not necessary for the present introductory treatment of dielectrics.

with $N = 2.71 \times 10^{28}$ m^{-3} and $\alpha_e = 1.7 \times 10^{-40}$ F m^2, we have

$$\varepsilon_r = 1 + \frac{N\alpha_e}{\varepsilon_o} = 1 + \frac{(2.71 \times 10^{28})(1.7 \times 10^{-40})}{(8.85 \times 10^{-12})} = 1.52$$

If we use the Clausius–Mossotti equation, we get

$$\varepsilon_r = \frac{1 + \dfrac{2N\alpha_e}{3\varepsilon_o}}{1 - \dfrac{N\alpha_e}{3\varepsilon_o}} = 1.63$$

The two values are different by about 7%. The simple relationship in Equation 7.14 underestimates the relative permittivity.

7.2   ELECTRONIC POLARIZATION: COVALENT SOLIDS

When a field is applied to a solid substance, the constituent atoms or molecules become polarized, as we visualized in Figure 7.4a. The electron clouds within each atom become shifted by the field, and this gives rise to electronic polarization. This type of electronic polarization within an atom, however, is quite small compared with the polarization due to the valence electrons in the covalent bonds within the solid. For example, in crystalline silicon, there are electrons shared with neighboring Si atoms in covalent bonds, as shown in Figure 7.7a. These valence electrons form bonds (i.e., become shared) between the Si atoms because they are already loosely bound to their parent atoms. If this were not the case, the solid would be van der Waals solid with atoms held together by secondary bonds (e.g., solid Ar below 83.8 K). In the covalent solid, the valence electrons therefore are not rigidly tied to the ionic cores left in the Si atoms. Although intuitively we often view these valence electrons as living in covalent bonds between the ionic Si cores, they nonetheless belong to the whole crystal because they can tunnel from bond to bond and exchange places with each other. We refer to their wavefunctions as delocalized, that is, not localized to any particular Si atom. When an electric field is applied, the negative charge distribution associated with these valence electrons becomes readily shifted with respect to the positive charges of the ionic Si cores, as depicted in Figure 7.7b and the crystal exhibits polarization, or develops a polarization vector. One can appreciate the greater flexibility of electrons in covalent bonds compared with those in individual ionic cores by comparing the energy involved in freeing each. It takes perhaps 1–2 eV to break a covalent bond to free the valence electron, but it takes more than 10 eV to free an electron from an individual ionic Si core. Thus, the valence electrons in the bonds readily respond to an applied field and become displaced. This type of electronic polarization, due to the displacement of electrons in covalent bonds, is responsible for the large dielectric constants of covalent crystals. For example $\varepsilon_r = 11.9$ for the Si crystal and $\varepsilon_r = 16$ for the Ge crystal.

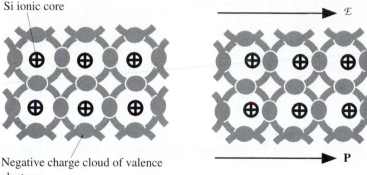

Si ionic core

Negative charge cloud of valence electrons

(a)   (b)

Figure 7.7
(a) Valence electrons in covalent bonds in the absence of an applied field.
(b) When an electric field is applied to a covalent solid, the valence electrons in the covalent bonds are shifted very easily with respect to the positive ionic cores. The whole solid becomes polarized due to the collective shift in the negative charge distribution of the valence electrons.

ELECTRONIC POLARIZABILITY OF COVALENT SOLIDS   Consider a pure Si crystal that has $\varepsilon_r = 11.9$. | **Example 7.2**

a.   What is the electronic polarizability due to valence electrons per Si atom (if one could portion the observed crystal polarization to individual atoms)?

b.   Suppose that a Si crystal sample is electroded on opposite faces. By how much is the local field greater than the applied field?

c.   What is the resonant frequency f_0 corresponding to ω_o?

From the density of the Si crystal, the number of Si atoms per unit volume, N, is given as 5×10^{28} m^{-3}.

Solution

a.   Given the number of Si atoms, we can apply the Clausius–Mossotti equation to find α_e

$$\alpha_e = \frac{3\varepsilon_o}{N}\frac{\varepsilon_r - 1}{\varepsilon_r + 2} = \frac{3(8.85 \times 10^{-12})}{(5 \times 10^{28})}\frac{11.9 - 1}{11.9 + 2} = 4.17 \times 10^{-40} \text{ F m}^2$$

This is larger, for example, than the electronic polarizability of an isolated Ar atom, which has more electrons.

b.   The local field is

$$\mathcal{E}_{\text{loc}} = \mathcal{E} + \frac{1}{3\varepsilon_o}P$$

But, by definition,

$$P = \chi_e \varepsilon_o \mathcal{E} = (\varepsilon_r - 1)\varepsilon_o \mathcal{E}$$

Substituting for P,

$$\mathcal{E}_{\text{loc}} = \mathcal{E} + \frac{1}{3}(\varepsilon_r - 1)\mathcal{E}$$

so that the local field with respect to the applied field is

$$\frac{\mathcal{E}_{\text{loc}}}{\mathcal{E}} = \frac{1}{3}(\varepsilon_r + 2) = 4.63$$

The local field is a factor of 4.63 greater than the applied field.

c.   Since polarization is due to valence electrons and there are 4 per Si atom, we can use Equation 7.7,

$$\omega_0 = \left[\frac{Ze^2}{m_e \alpha_e}\right]^{1/2} = \left[\frac{4(1.6 \times 10^{-19})^2}{(9.1 \times 10^{-31})(4.17 \times 10^{-40})}\right]^{1/2}$$

$$= 1.65 \times 10^{16} \text{ rad s}^{-1}$$

The resonant frequency f_o is

$$f_o = \frac{\omega_o}{2\pi} = 2.6 \times 10^{15} \text{ Hz}$$

This frequency is typically associated with electromagnetic waves of wavelength in the ultraviolet region.

7.3   POLARIZATION MECHANISMS

In addition to electronic polarization, we can identify a number of other polarization mechanisms that may also contribute to the relative permittivity.

7.3.1   Ionic Polarization

This type of polarization occurs in ionic crystals such as NaCl, KCl, LiBr, and so forth. The ionic crystal has distinctly identifiable ions, for example, Na^+ and Cl^-, located at well-defined lattice sites so that each pair of oppositely charged neighboring ions has a dipole moment. As an example, we consider the one-dimensional NaCl crystal depicted as a chain of alternating Na^+ and Cl^- ions in Figure 7.8a. In the absence of an applied field, the solid has no net polarization because the dipole moments of equal magnitude are lined up head to head and tail to tail so that the net dipole moment is zero. The dipole moment, p_+, in the positive x direction has the same magnitude as p_- in the negative x direction, so the net dipole moment

$$p_{\text{net}} = p_+ - p_- = 0$$

In the presence of a field $\mathcal{E}$ along the x direction, however, the Cl^- ions are pushed in the $-x$ direction and the Na^+ ions in the $+x$ direction about their equilibrium

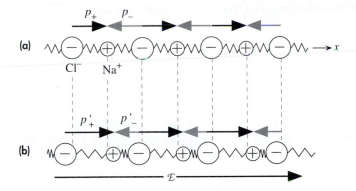

Figure 7.8
(a) A NaCl chain in the NaCl crystal without an applied field.
Average or net dipole moment per ion is zero.
(b) In the presence of an applied field, the ions become slightly
displaced, which leads to a net average dipole moment per ion.

positions. Consequently, the dipole moment p_+ in the $+x$ direction *increases* to p'_+ and the dipole moment p_- *decreases* to p'_-, as shown in Figure 7.8b. The net dipole moment is now no longer zero. The net dipole moment, or the average dipole moment, per ion pair is now $(p'_+ - p'_-)$, which depends on the electric field $\mathcal{E}$. Thus the induced average dipole moment per ion pair, p_{av}, depends on the field $\mathcal{E}$. The ionic polarizability α_i is defined in terms of the local field experienced by the ions,

$$p_{av} = \alpha_i \mathcal{E}_{loc} \qquad [7.16]$$

The larger the α_i, the greater the induced dipole moment. Generally, α_i is larger than the electronic polarizability α_e by a factor of 10 or more, which leads to ionic solids having large dielectric constants. The polarization P, exhibited by the ionic solid is therefore given by

$$P = N_i p_{av} = N_i \alpha_i \mathcal{E}_{loc}$$

where N_i is the number of ion pairs per unit volume. By relating the local field to $\mathcal{E}$ and using

$$P = (\varepsilon_r - 1)\varepsilon_o \mathcal{E}$$

we can again obtain the Clausius–Mossotti equation, but now due to ionic polarization,

$$\frac{\varepsilon_r - 1}{\varepsilon_r + 2} = \frac{1}{3\varepsilon_o} N_i \alpha_i \qquad [7.17]$$

Each ion also has a core of electrons that become displaced in the presence of an applied field with respect to their positive nuclei and therefore also contribute to the polarization of the solid. This electronic polarization simply adds to the ionic polarization. Its magnitude is invariably much smaller than the ionic contribution in these solids.

7.3.2   Orientational (Dipolar) Polarization

Certain molecules possess permanent dipole moments. For example, the HCl molecule shown in Figure 7.9a has a permanent dipole moment p_o from the Cl^- ion to the H^+ ion. In the liquid or gas phases, these molecules, in the absence of an electric field, are randomly oriented as a result of thermal agitation, as shown in Figure 7.9b. When an electric field $\mathcal{E}$ is applied, $\mathcal{E}$ tries to align the dipoles parallel to itself, as depicted in Figure 7.9c. The Cl^- and H^+ charges experience forces in opposite directions. But the nearly rigid bond between Cl^- and H^+ holds them together, which means that the molecule experiences a torque, τ, about its center of mass.[5] This torque acts to rotate the molecule to align p_o with $\mathcal{E}$. If all the molecules were to simply rotate and align with the field, the polarization of the solid would be

$$P = Np_o$$

where N is the number of molecules per unit volume. However, due to their thermal energy, the molecules move around randomly and collide with each other and with the walls of the container. These collisions destroy the dipole alignments. Thus the

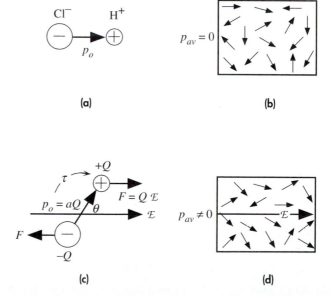

Figure 7.9

(a) A HCl molecule possesses a permanent dipole moment, p_o.
(b) In the absence of a field, thermal agitation of the molecules results in zero net average dipole moment per molecule.
(c) A dipole such as HCl placed in a field experiences a torque that tries to rotate it to align p_o with the field $\mathcal{E}$.
(d) In the presence of an applied field, the dipoles try to rotate to align with the field against thermal agitation. There is now a net average dipole moment per molecule along the field.

---

[5] The oppositely directed forces also slightly stretch the Cl^-–H^+ bond, but we neglect this effect.

thermal energy tries to randomize the orientations of the dipole moments. A snapshot of the dipoles in the material in the presence of a field can be pictured as in Figure 7.9d in which the dipoles have different orientations. There is, nonetheless, a net average dipole moment per molecule, p_{av}, that is finite and directed along the field. Thus the material exhibits net polarization, which leads to a dielectric constant that is determined by this **orientational polarization.**

To find the induced average dipole moment, p_{av} along $\mathcal{E}$, we need to know the average potential energy, E_{dip}, of a dipole placed in a field $\mathcal{E}$ and how this compares with the average thermal energy $(5/2)kT$ per molecule as in the present case of five degrees of freedom. E_{dip} represents the average external work done by the field in aligning the dipoles with the field. If $(5/2)kT$ is much greater than E_{dip}, then the average thermal energy of collisions will prevent any dipole alignment with the field. If, however, E_{dip} is much greater than $(5/2)kT$ then the thermal energy is insufficient to destroy the dipole alignments.

A dipole at an angle θ to the field experiences a torque τ that tries to rotate it, as shown in Figure 7.9c. Work done, dW, by the field in rotating the dipole by $d\theta$ is $\tau \, d\theta$ (as in $F \, dx$). This work, dW, represents a small change, dE, in the potential energy of the dipole. No work is done if the dipole is already aligned with $\mathcal{E}$, when $\theta = 0$, which corresponds to the minimum in PE. On the other hand, maximum work is done when the torque has to rotate the dipole from $\theta = 180°$ to $\theta = 0°$ (either clockwise or counterclockwise, it doesn't matter). The torque experienced by the dipole, according to Figure 7.9c is given by

$$\tau = (F \sin \theta)a \qquad \text{or} \qquad \mathcal{E}p_o \sin \theta$$

where

$$p_o = aQ$$

If we take $PE = 0$ when $\theta = 0$, then the maximum PE is when $\theta = 180°$, or

$$E_{max} = \int_0^\pi p_o \mathcal{E} \sin \theta \, d\theta = 2p_o \mathcal{E}$$

The average dipole potential energy, E_{dip}, is then $(1/2)E_{max}$ or $p_o\mathcal{E}$. For orientational polarization to be effective, this energy must be greater than the average thermal energy. The average dipole moment, p_{av}, along $\mathcal{E}$ is directly proportional to the magnitude of p_o itself and also proportional to the average dipole energy to average thermal energy ratio, that is,

$$p_{av} \propto p_o \frac{p_o \mathcal{E}}{\frac{5}{2}kT}$$

If we were to do the calculation properly using Boltzmann statistics for the distribution of dipole energies among the molecules, that is, the probability that the dipole has an energy E is proportional to $\exp(-E/kT)$, then we would find that when $p_o E < kT$ (generally the case),

$$p_{av} = \frac{1}{3} \frac{p_o^2 \mathcal{E}}{kT} \qquad \qquad \textbf{[7.18]}$$

It turns out that the intuitively derived expression above is roughly the same as Equation 7.18. Strictly, of course, we should use the local field acting on each molecule, in which case $\mathcal{E}$ is simply replaced by $\mathcal{E}_{loc}$. From Equation 7.18 we can define a dipolar orientational polarizability α_d per molecule by

$$\alpha_d = \frac{1}{3}\frac{p_o^2}{kT} \qquad [7.19]$$

It is apparent that, in contrast to the electronic and ionic polarization, dipolar orientational polarization is strongly temperature dependent. α_d decreases with temperature, which means that the relative permittivity ε_r also decreases with temperature. Dipolar orientational polarization is normally exhibited by polar liquids (e.g., water, alcohol, acetone, and various electrolytes) and polar gases (e.g., gaseous HCl and steam). It can also occur in solids if there are permanent dipoles within the solid structure, even if dipolar rotation involves a discrete jump of an ion from one site to another, such as in various glasses.

7.3.3   Interfacial Polarization

Interfacial polarization occurs whenever there is an accumulation of charge at an interface between two materials or between two regions within a material. The simplest example is interfacial polarization due to the accumulation of charges in the dielectric near one of the electrodes, as depicted in Figures 7.10a and b. Invariably materials, however perfect, contain crystal defects, impurities, and various mobile charge carriers such as electrons (e.g., from donor-type impurities),

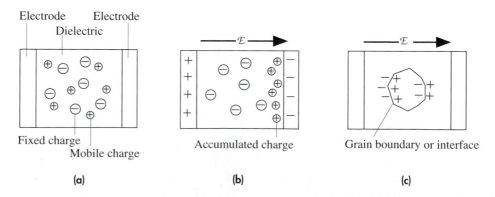

Figure 7.10

(a) A crystal with equal number of mobile positive ions and fixed negative ions. In the absence of a field, there is no net separation between all the positive charges and all the negative charges.
(b) In the presence of an applied field, the mobile positive ions migrate toward the negative electrode and accumulate there. There is now an overall separation between the negative charges and positive charges in the dielectric. The dielectric therefore exhibits interfacial polarization.
(c) Grain boundaries and interfaces between different materials frequently give rise to interfacial polarization.

holes, or ionized host or impurity ions. In the particular example in Figure 7.10a, the material has an equal number of positive ions and negative ions, but the positive ions are assumed to be far more mobile. For example, if present, the H^+ ion (which is a proton) and the Li^+ ion in ceramics and glasses are more mobile than negative ions in the structure because they are relatively small. Under the presence of an applied field, these positive ions migrate to the negative electrode. The positive ions, however, cannot leave the dielectric and enter the crystal structure of the metal electrode. They therefore simply pile up at the interface and give rise to a positive space charge near the electrode. These positive charges at the interface attract more electrons to the negative electrode. This additional charge on the electrode, of course, appears as an increase in the dielectric constant. The term **interfacial polarization** arises because the positive charges accumulating at the interface and the remainder of negative charges in the bulk together constitute dipole moments that appear in the polarization vector **P** (**P** sums all the dipoles within the material per unit volume).

Another typical interfacial polarization mechanism is the trapping of electrons or holes at defects at the crystal surface, at the interface between the crystal and the electrode. In this case we can view the positive charges in Figure 7.10a as holes and negative charges as immobile ionized acceptors. We assume that the contacts are blocking and do not allow electrons or holes to be injected, that is, exchanged between the electrodes and the dielectric. In the presence of a field, the holes drift to the negative electrode and become trapped in defects at the interface, as in Figure 7.10b.

Grain boundaries frequently lead to interfacial polarization as they can trap charges migrating under the influence of an applied field, as indicated in Figure 7.10c. Dipoles between the trapped charges increase the polarization vector. Interfaces also arise in heterogeneous dielectric materials, for example, when there is a dispersed phase within a continuous phase. The principle is then the same as schematically illustrated in Figure 7.10c.

7.3.4   Total Polarization

In the presence of electronic, ionic, and dipolar polarization mechanisms, the average induced dipole moment per molecule will be the sum of all the contributions in terms of the local field,

$$p_{av} = \alpha_e \mathcal{E}_{loc} + \alpha_i \mathcal{E}_{loc} + \alpha_d \mathcal{E}_{loc}$$

Each effect adds linearly to the net dipole moment per molecule, which is a fact verified by experiments. Interfacial polarization cannot be simply added to the above as $\alpha_{if}\mathcal{E}_{loc}$ because it occurs at interfaces and cannot be put into an average polarization per molecule in the bulk. Further, the fields are not well defined at the interfaces. The dielectric constant ε_r under electronic, ionic, and dipolar polarization is then given by

$$\frac{\varepsilon_r - 1}{\varepsilon_r + 2} = \frac{1}{3\varepsilon_o}[N_e\alpha_e + N_i\alpha_i + N_d\alpha_d]$$

[7.20]   *Clausius-Mossotti Equation*

Table 7.1   Typical examples of polarization mechanisms.

Example	Polarization	Static ε_r	Comment
Ar gas	Electronic	1.0005	Small N in gases: $\varepsilon_r \approx 1$
Ar liquid (T < 87.3 K)	Electronic	1.53	van der Waals bonding
Si crystal	Electronic polarization due to valence electrons	11.9	Covalent solid; bond polarization
NaCl crystal	Ionic	5.90	Ionic crystalline solid
CsCl crystal	Ionic	7.20	Ionic crystalline solid
Water	Orientational	80	Dipolar liquid
Nitromethane (27 °C)	Orientational	34	Dipolar liquid
PVC (polyvinyl chloride)	Orientational	7	Dipolar orientation partly hindered in the solid

Table 7.1 summarizes the various polarization mechanisms and the corresponding static (or very-low-frequency) dielectric constant. Typical examples where one mechanism dominates over others are also listed.

Example 7.3

DIPOLAR POLARIZATION OF WATER   Given the static dielectric constant of water as 80 and its density as 1 g cm^{-3}, calculate the permanent dipole moment, p_o, per water molecule, assuming that it is the orientational polarization of individual molecules that gives rise to the dielectric constant. If the permanent dipole moment, p_o, of the water molecule is 6.1×10^{-30} C m, can we treat the liquid as an independent collection of H_2O molecules?

Solution

To apply the Clausius–Mossotti equation, we need the number of H_2O molecules per unit volume. The molecular mass, M_{mol}, of H_2O is 18×10^{-3} kg mol^{-1} and its density d is 10^3 kg m^{-3}. The number of H_2O molecules per unit volume, N_d, is

$$N_d = \frac{N_A d}{M_{mol}} = \frac{(6.022 \times 10^{23} \text{ mol}^{-1})(10^3 \text{ kg m}^{-3})}{18 \times 10^{-3} \text{ kg mol}^{-1}} = 3.35 \times 10^{28} \text{ m}^{-3}$$

Then

$$\alpha_d = \frac{3\varepsilon_o}{N_d} \frac{\varepsilon_r - 1}{\varepsilon_r + 2} = \frac{3(8.85 \times 10^{-12})}{(3.35 \times 10^{28})} \frac{80 - 1}{80 + 2} = 7.65 \times 10^{-40} \text{ F m}^2$$

Using the expression for orientational polarization, we have

$$p_o = \sqrt{\alpha_d 3kT} = \sqrt{(7.65 \times 10^{-40})3(1.38 \times 10^{-23})(300)} = 3.1 \times 10^{-30} \text{ C m}$$

This is half the actual permanent dipole moment of H_2O. The reason for the difference is that the individual H_2O molecules are not totally free to rotate. In the liquid, H_2O molecules cluster together through hydrogen bonding, so rotation of individual molecules is limited by this bonding. The theory of orientational polarization above did not take into account any bonding between the dipolar molecules.

7.4   FREQUENCY DEPENDENCE: DIELECTRIC CONSTANT AND DIELECTRIC LOSS

The static dielectric constant is an effect of polarization under dc conditions. When the applied field, or the voltage across a parallel plate capacitor, is a sinusoidal signal, then the polarization of the medium under these ac conditions leads to an ac dielectric constant that is generally different than the static case. As an example we will consider orientational polarization involving dipolar molecules. The sinusoidally varying field changes magnitude and direction continuously and it tries to line up the dipoles one way and then the other way and so on. If the instantaneous induced dipole moment, p, per molecule can instantaneously follow the field variations, then at any instant

$$p = \alpha_d \mathcal{E} \qquad \qquad \textbf{[7.21]}$$

and the polarizability α_d has its expected maximum value from dc conditions, that is,

$$\alpha_d = \frac{p_o^2}{3kT} \qquad \qquad \textbf{[7.22]}$$

There are two factors opposing the immediate alignment of the dipoles with the field. First is that thermal agitation tries to randomize the dipole orientations. Collisions in the gas phase, random jolting from lattice vibrations in the liquid and solid phases, for example, aid the randomization of the dipole orientation. Second, the molecules rotate in a viscous medium by virtue of their interactions with neighbors, which is particularly strong in the liquid and solid states and means that the dipoles cannot respond instantaneously to the changes in the applied field. If the field changes too rapidly, then the dipoles cannot follow the field and, as a consequence, remain randomly oriented. At high frequencies, therefore, α_d will be zero as the field cannot induce a dipole moment. At low frequencies, of course, the dipoles can respond rapidly to follow the field and α_d has its maximum value. It is clear that α_d changes from its maximum value in Equation 7.22 to zero as the frequency of the field is increased. We need to find the behavior of α_d as a function of frequency, ω, so that we can determine the dielectric constant, ε_r, by the Clausius–Mossotti equation.

Suppose that after a prolonged application, corresponding to dc conditions, the field across the dipolar gaseous medium is suddenly decreased from $\mathcal{E}_o$ to $\mathcal{E}$ at a time we define as zero, as shown in Figure 7.11. The field $\mathcal{E}$ is smaller than $\mathcal{E}_o$ so that the induced dc dipole moment per molecule should be smaller and given by $\alpha_d(0) \, \mathcal{E}$ where $\alpha_d(0)$ is α_d at $\omega = 0$, dc conditions. Therefore, the induced dipole moment per molecule has to decrease, or *relax*, from $\alpha_d(0)\mathcal{E}_o$ to $\alpha_d(0)\mathcal{E}$. In a gas medium the molecules would be moving around randomly and their collisions with each other and the walls of the container randomize the induced dipole per molecule. Thus the decrease, or the **relaxation process,** in the induced dipole moment is achieved by random collisions. Assuming that τ is the average time, called the **relaxation time,** between molecular collisions, then this is the mean time

Then,

$$\varepsilon_r' = 1 + \frac{[\varepsilon_r(0) - 1]}{1 + (\omega\tau)^2}$$

$$\varepsilon_r'' = \frac{[\varepsilon_r(0) - 1]\omega\tau}{1 + (\omega\tau)^2}$$

which are called **Debye equations.** They reflect the behavior of ε_r' and ε_r'' as a function of frequency, shown in Figure 7.12. The imaginary part ε_r'' that represents the dielectric loss exhibits a peak at $\omega = 1/\tau$, which is called a **Debye loss peak.** Many gases and some liquids with dipolar molecules exhibit this type of behavior. In the case of solids, the peak is much broader due to a large spread of relaxation times.

Example 7.5 | **DIELECTRIC LOSS PER UNIT CAPACITANCE AND THE LOSS ANGLE, δ**   Obtain the dielectric loss per unit capacitance in a capacitor in terms of the loss tangent. Obtain the phase difference between the current through the capacitor and that through R_P. What is the significance of δ?

Solution

We consider the equivalent circuit in Figure 7.13. The power loss in the capacitor is due to R_P. If V is the rms value of the voltage across the capacitor, then the power dissipated per unit capacitance, W_{cap}, is

$$W_{cap} = \frac{V^2}{R_P} \times \frac{1}{C} = V^2 \frac{\omega\varepsilon_o\varepsilon_r'' A}{d} \times \frac{d}{\varepsilon_o\varepsilon_r'A} = V^2 \frac{\omega\varepsilon_r''}{\varepsilon_r'}$$

or

$$W_{cap} = V^2\omega \tan \delta$$

As $\tan \delta$ is frequency dependent and peaks at some frequency, so does the power dissipated per unit capacitance. A clear design objective would be to keep W_{cap} as small as possible. Further, for a given voltage, W_{cap} does not depend on the dielectric geometry. For a given voltage and capacitance, we therefore cannot reduce the power dissipation by simply changing the dimensions of the dielectric.

Consider the rms currents through R_P and C, I_{loss} to I_{cap} respectively, and their ratio,

$$\frac{I_{loss}}{I_{cap}} = \frac{V}{R_P} \times \frac{1}{\frac{1}{j\omega C}} = \frac{\omega\varepsilon_o\varepsilon_r'' A}{d} \times \frac{d}{j\omega\varepsilon_o\varepsilon_r'A} = -j \tan \delta$$

As expected, the two are 90° out of phase ($-j$) and the loss current (through R_P) is a factor, $\tan \delta$, of the capacitive current (through C). The ratio of I_{cap} and the total current, $I_{total} = I_{cap} + I_{loss}$, is

$$\frac{I_{cap}}{I_{total}} = \frac{I_{cap}}{I_{cap} + I_{loss}} = \frac{1}{1 + \frac{I_{loss}}{I_{cap}}} = \frac{1}{1 - j \tan \delta}$$

The phase angle between I_{cap} and I_{total} is determined by the negative of the phase of the denominator term, $[1 - j \tan \delta]$. Thus the phase angle between I_{cap} and I_{total} is δ, where I_{cap} leads I_{total} by δ. δ is also called the loss angle. When the loss angle is zero, I_{cap} and I_{total} are equal and there is no loss in the dielectric.

Table 7.2   Dielectric properties of three insulators

Material	ε_r'	$\tan\delta$	$\omega\tan\delta$	ε_r'	$\tan\delta$	$\omega\tan\delta$
		$f = 60$ Hz			$f = 1$ MHz	
Polycarbonate	3.17	9×10^{-4}	0.34	2.96	1×10^{-2}	6.2×10^4
Silicone rubber	3.7	2.25×10^{-2}	8.48	3.4	4×10^{-3}	2.5×10^4
Epoxy with mineral filler	5	4.7×10^{-2}	17.7	3.4	3×10^{-2}	18×10^4

DIELECTRIC LOSS PER UNIT CAPACITANCE   Consider the three dielectric materials listed in Table 7.2 with their dielectric constant ε_r' (usually simply stated as ε_r) and loss factors $\tan\delta$. At a given voltage, which dielectric will have the lowest power dissipation per unit capacitance at 60 Hz? Is this also true at 1 MHz? | **Example 7.6**

Solution

The power dissipated at a given voltage per unit capacitance depends only on $\omega\tan\delta$ so we do not need to use ε_r'. Calculating $\omega\tan\delta$ or $(2\pi f)\tan\delta$, we find the values listed in the table at 60 Hz and 1 MHz. At 60 Hz, polycarbonate has the lowest power dissipation per unit capacitance, but at 1 MHz it is silicone rubber.

DIELECTRIC LOSS AND FREQUENCY   Calculate the heat generated per second due to dielectric loss per cm^3 of cross-linked polyethylene, XLPE (typical power cable insulator), and alumina, Al_2O_3 (typical substrate in thin and thick film electronics), at 60 Hz and 1 MHz at a field of 100 kV cm^{-1}. Their properties are given in Table 7.3. What is your conclusion? | **Example 7.7**

Solution

The power dissipated per unit volume is

$$W_{vol} = (2\pi f)\mathcal{E}^2\varepsilon_o\varepsilon_r'\tan\delta$$

We can calculate W_{vol} by substituting the properties of individual dielectrics at the given frequency f. For example, for XLPE at 60 Hz,

$$W_{vol} = (2\pi\, 60\text{ Hz})(100 \times 10^3 \times 10^2\text{ V m}^{-1})^2(8.85 \times 10^{-12}\text{ F m}^{-1})(2.3)(3 \times 10^{-4})$$

$$= 230.3\text{ W m}^{-3}$$

Table 7.3   Dielectric loss per unit volume for two insulators (κ is the thermal conductivity)

Material	ε_r'	$\tan\delta$	Loss (W cm^{-3})	ε_r'	$\tan\delta$	Loss (W cm^{-3})	κ (W cm^{-1} K^{-1})
		$f = 60$ Hz			$f = 1$ MHz		
XLPE	2.3	3×10^{-4}	0.230 mW	2.3	4×10^{-4}	5.12 W	0.005
Alumina	8.5	1×10^{-3}	2.84 mW	8.5	1×10^{-3}	47.3 W	0.33

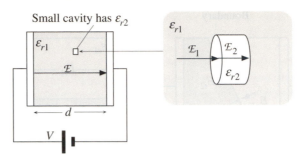

Figure 7.19   Field in the cavity is higher than the field in the solid.

which gives

$$\mathcal{E}_2 = 5\left(\frac{V}{d}\right)$$

Air insulation in a 100 micron (0.1 mm) thick cavity breaks down when $\mathcal{E}_2$ is typically 100 kV cm^{-1}. From $\mathcal{E}_2 = 5(V/d)$, a voltage of 20 kV will result in the breakdown of air in the cavity and hence a discharge current. This is called a **partial discharge** as only a partial breakdown of the insulation, that in the cavity, has occurred between the electrodes. Under an ac voltage, the discharge in the cavity can often be sustained by the capacitive current through the surrounding dielectric. Without this cavity, the dielectric would accept a greater voltage across it, which in this case is typically greater than 100 kV.

Example 7.8   **FIELD INSIDE A THIN DIELECTRIC WITHIN A SECOND DIELECTRIC**   When the dielectric fills the whole space between the plates of a capacitor, the net field within the dielectric is the same as before, $\mathcal{E} = V/d$. Explain what happens when a dielectric slab of thickness $t \ll d$ is inserted in the middle of the space between the plates, as shown in Figure 7.20. What is the field inside the dielectric?

Solution

The problem is illustrated in Figure 7.20 and has symmetry in that the field in air on either side of the dielectric is the same and $\mathcal{E}_1$. The boundary conditions give

$$\varepsilon_{r1}\mathcal{E}_1 = \varepsilon_{r2}\mathcal{E}_2$$

Further, the integral of the field from one plate to the other must be V because $dV/dx = \mathcal{E}$. Examining Figure 7.20, we see that the integration is

$$\mathcal{E}_1(d - t) + \mathcal{E}_2 t = V$$

We now have to eliminate $\mathcal{E}_1$ between the above two equations and obtain $\mathcal{E}_2$, which can be done by algebraic manipulation,

$$\mathcal{E}_2 = \frac{\varepsilon_{r1}}{\varepsilon_{r2} - \dfrac{t}{d}(\varepsilon_{r2} - \varepsilon_{r1})}\left(\frac{V}{d}\right) \tag{7.33}$$

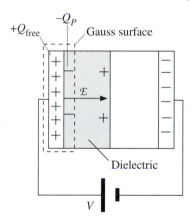

Figure 7.21   A convenient Gauss surface for calculating the field inside the dielectric is a very thin rectangular surface enclosing the surface of the dielectric.

The total charges enclosed are the free charges on the electrodes and the polarization charges on the surface of the dielectric.

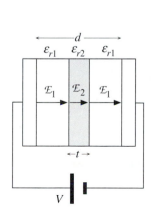

Figure 7.20   A thin slab of dielectric is placed in the middle of a parallel plate capacitor.

The field inside the thin slab is $\mathcal{E}_2$.

If $t \ll d$, then this approximates to

$$\mathcal{E}_2 = \frac{\varepsilon_{r1}}{\varepsilon_{r2}}\left(\frac{V}{d}\right) \qquad \text{and} \qquad \mathcal{E}_1 = \left(\frac{V}{d}\right) \qquad (t \ll d) \qquad \text{[7.34]}$$

Clearly $\mathcal{E}_1$ in the air space remains the same as the applied field V/d. Since $\varepsilon_{r1} = 1$ (air) and $\varepsilon_{r2} > 1$, $\mathcal{E}_2$ in the thin dielectric slab is smaller than the applied field V/d. On the other hand, if we have air space between two dielectric slabs, then the field in this air space will be greater than the field inside the two dielectric slabs. Indeed, if the applied voltage is sufficiently large, the field in the air gap can cause dielectric breakdown of this region.

GAUSS'S LAW WITHIN A DIELECTRIC AND FREE CHARGES   Gauss's law in Equation 7.29 contains the total charge Q_{total}, enclosed within the surface. Generally, these enclosed charges are free charges Q_{free}, due to the free carriers on the electrode, and bound charges, Q_P, due to polarization charges on the dielectric surface. Apply Gauss's law using a Gaussian rectangular surface enclosing the left electrode and the dielectric surface in Figure 7.21. Show that the electric field $\mathcal{E}$ in the dielectric can be expressed in terms of free charges only, Q_{free}, through

Example 7.9

$$\oint_{\text{surface}} \mathcal{E}_n \, dA = \frac{Q_{\text{free}}}{\varepsilon_o \varepsilon_r} \qquad \text{[7.35]}$$

Free charges and field in a dielectric

where ε_r is the relative permittivity of the dielectric medium.

Solution

We apply Gauss's law to a hypothetical rectangular surface enclosing the left electrode and the dielectric surface. The field $\mathcal{E}$ in the dielectric is normal and outwards at the Gauss surface in Figure 7.21. Thus $\mathcal{E}_n = \mathcal{E}$ in the left-hand side of Equation 7.29.

$$\varepsilon_o A\mathcal{E} = Q_{total} = Q_{free} - Q_P = Q_{free} - AP = Q_{free} - A\varepsilon_o(\varepsilon_r - 1)\mathcal{E}$$

where we have used $P = \varepsilon_o(\varepsilon_r - 1)\mathcal{E}$. Rearranging,

$$\varepsilon_o\varepsilon_r A\mathcal{E} = Q_{free}$$

Since $A\mathcal{E}$ is effectively the surface integral of $\mathcal{E}_n$, the above corresponds to writing Gauss's law in a dielectric in terms of free charges as

$$\oint_{surface} \mathcal{E}_n \, dA = \frac{Q_{free}}{\varepsilon_o\varepsilon_r}$$

The above equation assumes that polarization P and $\mathcal{E}$ are linearly related,

$$P = \varepsilon_o(\varepsilon_r - 1)\mathcal{E}$$

We note that if we only use free charges in Gauss's law, then we simply multiply ε_o by the dielectric constant of the medium. The above proof is by no means a rigorous derivation.

7.6   DIELECTRIC STRENGTH AND INSULATION BREAKDOWN

7.6.1   Dielectric Strength: Definition

A defining property of a dielectric medium is not only its ability to increase capacitance but also, and equally important, its insulating behavior or low conductivity so that the charges are not simply conducted from one plate of the capacitor to the other through the dielectric. Dielectric materials are widely used as insulating media between conductors at different voltages to prevent the ionization of air and hence current flashovers between conductors. The voltage across a dielectric material and hence the field within it cannot, however, be increased without limit. Eventually a voltage is reached that causes a substantial current to flow between the electrodes, which appears as a short between the electrodes and leads to what is called **dielectric breakdown.** In gaseous and many liquid dielectrics, the breakdown does not generally permanently damage the material. This means that if the voltage causing breakdown is removed, then the dielectric can again sustain voltages until the voltage is sufficiently high to cause breakdown again. In solid dielectrics the breakdown process invariably leads to the formation of a permanent conducting channel and hence to permanent damage. The **dielectric strength** $\mathcal{E}_{br}$ is the maximum field that can be applied to an insulating medium without causing

Table 7.4   Dielectric strength; typical values at room temperature and 1 atm

Dielectric Medium	Dielectric Strength	Comment
Atmosphere at 1 atm pressure	31.7 kV cm^{-1} at 60 Hz	1 cm gap. Breakdown by electron avalanche by impact ionization.
SF$_6$ gas	79.3 kV cm^{-1} at 60 Hz	Used in high voltage circuit breakers to avoid discharges.
Polybutene	>138 kV cm^{-1} at 60 Hz	Liquid dielectric used as oil filler and HV pipe cables.
Transformer oil	128 kV cm^{-1} at 60 Hz	—
Amorphous silicon dioxide (SiO$_2$) in MOS technology	10 MV cm^{-1} dc	Very thin oxide films without defects. Instrinsic breakdown limit.
Borosilicate glass	10 MV cm^{-1} duration of 10 μs 6 MV cm^{-1} duration of 30 s	Instrinsic breakdown. Thermal breakdown.
Polypropylene	295–314 kV cm^{-1}	Likely to be thermal breakdown or electrical treeing.

dielectric breakdown. Beyond $\mathcal{E}_{br}$, dielectric breakdown takes place. The dielectric strength of solids depends on a number of factors besides simply the molecular structure, such as the impurities in the material, microstructural defects (e.g., microvoids), sample geometry, nature of the electrodes, temperature, and ambient conditions (e.g., humidity), as well as the duration and frequency of the applied field. Dielectric strength is different under dc and ac conditions. There are also **aging effects** that slowly degrade the properties of the insulator and reduce the dielectric strength. For engineers involved in insulation, the dielectric strength of solids is therefore one of the most difficult parameters to interpret and use. For example, the breakdown field also depends on the thickness of the insulation because thicker insulators have more volume and hence a greater probability of containing a microstructural defect (e.g., a microcavity) that can initiate a dielectric breakdown. Table 7.4 shows some typical dielectric strengths for various dielectrics used in electrical insulation. Unpressurized gases have lower breakdown strengths than liquids and solids.

7.6.2   Dielectric Breakdown and Partial Discharges: Gases

Due to cosmic radiation, there are always a few free electrons in a gas. If the field is sufficiently large, then one of these electrons can be accelerated to sufficiently large kinetic energies to impact ionize a neutral gas molecule and produce an additional free electron and a positively charged gas ion. Both the first and liberated electrons are now available to accelerate in the field again and further impact ionize more neutral gas molecules, and so on. Thus, an avalanche of impact ionization processes creates many free electrons and positive gas ions in the gas, which give rise to a discharge current between the electrodes. The process is similar to avalanche breakdown in a reverse-biased *pn* junction. The breakdown in gases depends on the pressure. The concentration of gas molecules is greater at higher pressures. This means that the mean separation between molecules, and, hence, the mean free path of a free electron, is shorter. Shorter mean free paths inhibit the free electrons from accelerating to reach impact ionization energies unless the field is increased. Thus, generally, $\mathcal{E}_{br}$ increases with the gas pressure. The 60 Hz breakdown field for an air gap of 1 cm at room temperature and at atmospheric pressure is about 31.7 kV cm^{-1}. On the other hand, the gas sulfurhexafluoride, SF$_6$, has a dielectric strength of 79.3 kV cm^{-1} and an even higher strength when pressurized. SF$_6$ is therefore used instead of air in high voltage circuit breakers.

A **partial discharge** occurs when only a local region of the dielectric is exhibiting discharge so that the discharge does not directly connect the two electrodes. For example, for the cylindrical conductor carrying a high voltage above a grounded plate, as in Figure 7.22a, the electric field is greatest on the surface of the conductor facing the ground. This field initiates discharge locally in this region because the field is sufficiently high to give rise to an electron avalanche effect. Away from the conductor, however, the field is not sufficiently strong to continue the electron avalanche discharge. This type of local discharge in high field regions is termed **corona discharge.** Voids and cracks occurring within solid dielectrics and discontinuities at the dielectric–electrode interface can also lead to partial discharges as the field in these voids is higher than the average field in the dielectric, and, further, the dielectric strength in the gas (e.g., atmosphere) in the void is less than that of the continuous solid insulation. Figure 7.22b and c depict two examples of partial discharges occuring in voids, one inside the solid (perhaps an air or gas bubble introduced during the processing of the dielectric) and the other (perhaps in the form of a crack) at the solid–electrode interface. In practice, a variety of factors can lead to microvoids and microcavities inside solids as well as at interfaces. Partial discharges in these voids physically and chemically erode the surrounding dielectric region and lead to an overall deterioration of the dielectric strength. If uncontrolled, they can eventually give rise to a major breakdown.

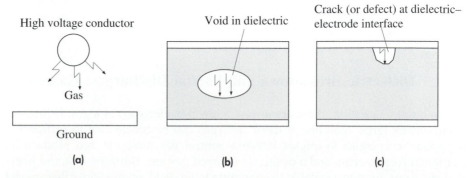

Figure 7.22
(a) The field is greatest on the surface of the cylindrical conductor facing the ground. If the voltage is sufficiently large, this field gives rise to a corona discharge.
(b) The field in a void within a solid can easily cause partial discharge.
(c) The field in the crack at the solid–metal interface can also lead to a partial discharge.

7.6.3   Dielectric Breakdown: Liquids

The processes that lead to the breakdown of insulation in liquids are not as clear as the electron avalanche effect in gases. In impure liquids with small conductive particles in suspension, it is believed that these impurities coalesce end to end to form a conducting bridge between the electrodes and thereby give rise to discharge. In some liquids, the discharge initiates as partial discharges in gas bubbles entrapped in the liquid. These partial discharges can locally raise the temperature and vaporize more of the liquid and hence increase the size of the bubble. The eventual discharge can be a series of partial discharges in entrapped gas bubbles. Moisture absorption and absorption of gases from the ambient generally deteriorate the dielectric strength. Oxidation of certain liquids, such as oils, with time produces more acidic and hence higher conductivity inclusions or regions that eventually give discharge. In some liquids, the discharge involves the emission of a large number of electrons from the electrode into the liquid due to field emission at high fields. This is a discharge process by electrode injection.

7.6.4   Dielectric Breakdown: Solids

There are various major mechanisms that can lead to dielectric breakdown in solids. The most likely mechanism depends on the dielectric material's condition and sometimes on extrinsic factors such as the ambient conditions, moisture absorption being a typical example.

Intrinsic Breakdown or Electronic Breakdown   The most common type of electronic breakdown is an **electron avalanche breakdown.** A free electron in the conduction band (CB) of a dielectric in the presence of a large field can be accelerated to sufficiently large energies to collide with and ionize a host atom of the solid. The electron gains an energy $e\mathcal{E}_{br}\ell$ when it moves a distance ℓ under an applied field $\mathcal{E}_{br}$. If this energy is greater than the bandgap energy E_g, then the electron, as a result of a collision with the lattice vibrations, can excite an electron from the valence band to the conduction band, that is, break a bond. Both the primary and the released electron can further impact ionize other host atoms and thereby generate an electron avalanche effect that leads to a substantial current. The initial conduction electrons for the avalanche are either present in the CB or are injected from the metal into the CB as a result of field-assisted thermal emission from the Fermi energy in the metal to the CB in the dielectric. Taking typical values, $E_g \approx 5$ eV and ℓ to be of the order of the mean free path for lattice scattering, say ~50 nm, one finds $\mathcal{E}_{br} \sim 1$ MV cm^{-1}. Obviously, $\mathcal{E}_{br}$ depends on the choice of ℓ, but its order of magnitude indicates voltages that are quite large. This type of breakdown represents an upper theoretical limit that is probably approached by only certain dielectrics—those that have practically no defects. Usually, microstructural defects lead to a lower dielectric strength than the limit indicated by intrinsic breakdown. Silicon dioxide (SiO_2) films with practically no structural defects in present MOS (metal-oxide-semiconductor) capacitors (as in the gates of MOSFETs) probably exhibit an intrinsic breakdown.

If dielectric breakdown does not occur by an electron avalanche effect (perhaps due to short mean free paths in the insulator), then another insulation breakdown mechanism is the enormous increase in the injection of electrons from the metal electrode into the insulator at very high fields as a result of field-assisted emission.[8] It has been proposed that insulation breakdown under short durations in some thin polymer films is due to tunneling injection.

Thermal Breakdown   Finite conductivity of the insulation means that there is Joule heat $\sigma \mathcal{E}^2$ being released within the solid. Further, at high frequencies, the dielectric loss, $V^2 \omega \tan \delta$, becomes especially significant. For example, the work done by the external field in rotating the dipoles is transferred more frequently to random molecular collisions as heat as the frequency of the field increases. Both conduction and dielectric losses therefore generate heat within the dielectric. If this heat cannot be removed from the solid sufficiently quickly by thermal conduction (or by other means), then the temperature of the dielectric will increase. The increase in the temperature invariably increases the conductivity of an insulator. The increase in the conductivity then leads to more joule heating and hence further rises in the temperature and so on. If the heat cannot be conducted away to limit the temperature, then the result is a thermal runaway condition in which the temperature and the current increase until a discharge occurs through various sections of the solid. As a consequence of sample inhomogeneities, frequently thermal runaway is severe in certain parts of the solid that become hot spots and suffer local melting and physical and chemical erosion. Hot spots are those local regions or inhomogeneities where σ or ε_r'' is larger or where the thermal conductivity is poor to remove the heat generated. Local breakdown at various hot spots eventually leads to a conducting channel connecting the opposite electrodes and hence to a dielectric breakdown. Since it takes time to raise the temperature of the dielectric, due to the heat capacity, this breakdown process has a marked thermal lag. The time to achieve thermal breakdown depends on the heat generated, and hence on $\mathcal{E}^2$. Conversely, this means that the dielectric strength $\mathcal{E}_{br}$ depends on the duration of application of the field. For example, at 70 °C, pyrex has $\mathcal{E}_{br}$ typically 9 MV cm^{-1} if the applied field duration is kept short, not more than 1 ms or so. If the field is kept for 30 seconds, then the breakdown field is only 2.5 MV cm^{-1}. Dielectric breakdown in various ceramics and glasses at high frequencies has been attributed directly to thermal breakdown. A characteristic feature of thermal breakdown is not only the thermal lag, the time dependence, but also the temperature dependence. Thermal breakdown is facilitated by increasing the temperature of the dielectric, which means that $\mathcal{E}_{br}$ decreases with temperature.

Electromechanical Breakdown and Electrofracture   A dielectric medium between oppositely charged electrodes experiences compressional forces because the opposite charges, $+Q$ and $-Q$, on the plates attract each other, as depicted in Figure 7.23.

---

[8] The emission of electrons by tunneling from an electrode in the presence of a large field was treated in Chapter 4 as Fowler–Nordheim field emission.

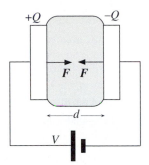

Figure 7.23   An exaggerated schematic illustration of a soft dielectric medium experiencing strong compressive forces due to the applied voltage.

As the voltage increases, so does the compressive load, and the dielectric becomes squeezed, or the thickness d gets smaller. At each stage, the increase in the compressive load is normally balanced by the elastic deformation of the insulation to a new smaller thickness. However, if the elastic modulus is sufficiently small, then compressive loads cannot be simply balanced by the elasticity of the solid, and there is a mechanical runaway for the following reasons. The decrease in d, due to the compressive load, leads to a higher field ($E = V/d$) and also to more charges on the electrodes ($Q = CV$, $C = \varepsilon_o \varepsilon_r A/d$). This is turn leads to a greater compressive load, which further decreases d, and so on, until the shear stresses within the insulation cause the insulation to flow plastically (for example, by viscous deformation). Eventually, the insulation breaks down. In addition, the increase in E as d gets smaller results in more joule (σE^2) and dielectric-loss heating ($\omega E^2 \tan \delta$) in the dielectric, which increases the temperature and hence lowers the elastic modulus and viscosity, thereby further deteriorating the mechanical stability. It is also possible for the field during the mechanical deformation of the dielectric to reach the thermal breakdown field, in which case the dielectric failure is not truly a mechanical breakdown mechanism though initiated by mechanical deformations. Another possibility is the initiation and growth of internal cracks (perhaps filamentary cracks) by internal stresses around inhomogeneous regions inside the dielectric. For example, an imperfection or a tiny cavity experiences shear stresses and also large local electric fields. Combined effects of both large shear stresses and large electric fields eventually lead to crack propagation and mechanical and, hence, dielectric failure. This type of process is sometimes called **electrofracture.** It is generally believed that certain thermoplastic polymers suffer from electromechanical dielectric breakdown, especially close to their softening temperatures. Polyethylene and polyisobutylene have been cited as examples.

Internal Discharges   These are partial discharges that take place in microstructural voids, cracks, or pores within the dielectric where the gas atmosphere (usually air) has lower dielectric strength. A porous ceramic, for example, would experience

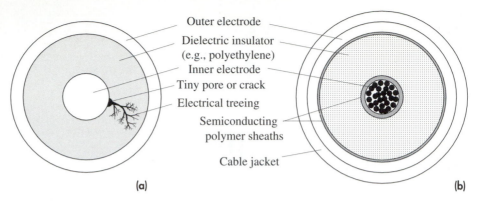

Figure 7.24

(a) A schematic illustration of electrical treeing breakdown in a high voltage coaxial cable that was initiated by a partial discharge in the void at the inner conductor–dielectric interface.
(b) A schemaic diagram of a typical high voltage coxial cable with semiconducting polymer layers around the inner conductor and around the outer surface of the dielectric.

partial discharges if the applied field is sufficiently large. The discharge current in a void, such as those in Figure 7.22b and c, can be easily sustained under ac conditions, which accounts for the severity of this type of breakdown mechanism under ac conditions. Initially, the pore size (or the number of pores) may be small and the partial discharge insignificant, but with time the partial discharge erodes the internal surfaces of the void. Partial discharges can locally melt the insulator and can easily cause chemical transformations. Eventually, and usually, an **electrical tree** type of discharge develops from a partial discharge that has been eroding the dielectric, as depicted in Figure 7.24a for a high voltage cable in which there is a tiny void at the interface between the dielectric and the inner conductor (generated perhaps by the differential thermal expansion of the electrode and polymeric insulation). The erosion of the dielectric by the partial discharge propagates like a branching tree. The "tree branches" are erosion channels—hollow filaments of various sizes—in which gaseous discharge takes place and forms a conducting channel during operation.

In the case of a coaxial high voltage cable in Figure 7.24a, the dielectric is usually a polymer, polyethylene (PE) being one of the most popular. The electric field is maximum at the surface of the inner conductor, which is the reason for the initiation of most electrical trees near this surface. Electrical treeing is substantially controlled by having semiconductive polymer layers or sheaths surrounding the inner conductor and the outer surface of the insulator, as shown in Figure 7.24b. For flexibility, the inner conductor is frequently multicored, or stranded, rather than solid. Due to the extrusion process used to draw the insulation, the semiconductive polymer sheaths are bonded to the insulation. There are therefore practically no microvoids at the interfaces between the insulator and the semiconducting sheath. Further, these semiconducting polymer sheaths are sufficiently conductive to become "part of the electrodes." Both the conductor and the adjacent semiconductor are roughly at the same voltage, which means that there is no breakdown in the semiconductor–conductor interfaces. There is normally an outer jacket (e.g., PVC) to protect the cable.

Insulation Aging   It is well recognized that during service, the properties of an insulating material become degraded and eventually dielectric breakdown occurs at a field below that predicted by experiments on fresh forms of the insulation. **Aging** is a term used to describe, in a general sense, the deterioration in the properties of the insulation. Aging therefore determines the useful life of the insulation. There are many factors that either directly or indirectly affect the properties and performance of an insulator in service. Even in the absence of an electric field, the insulation will experience physical and chemical aging whereby its physical and chemical properties change considerably. An insulation that is subjected to temperature and mechanical stress variations can develop structural defects, such as microcracks, which are quite damaging to the dielectric strength, as mentioned above. Irradiation by ionizing radiation such as x-rays, exposure to severe ambient conditions such as excessive humidity, ozone, and many other external conditions, through various chemical processes, deteriorate the chemical structure and properties of an insulator. This is generally much more severe for polymers than ceramics, but it is not practical to use a solid ceramic insulation in a coaxial power cable. Oxidation of a polymeric insulation with time is another form of chemical aging and is well-known to degrade the insulation performance. This is the reason for adding various antioxidants into semicrystalline polymers for use in insulation. The chemical aging processes are generally accelerated with temperature. In service, the insulation also experiences electrical aging as a result of the effects of the field on the properties of the insulation. For example, dc fields can disassociate and transport various ions in the structure and thereby slowly change the structure and properties of the insulation. Electrical trees develop as a result of electrical aging because, in service, the ac field gives rise to continual partial discharges in an internal or surface microcavity, which then erodes the region around it and slowly grows like a branching tree. In well-manufactured insulation systems, electrical treeing has been substantially reduced or eliminated from microvoids. A form of electrical aging that is currently in vogue is **water treeing,** which eventually leads to electrical treeing. The definition of a water tree, as viewed under an optical microscope, is a diffused bushy (or broccoli) type growth that consists of millions of microscopic voids (per mm^3) containing water or aqueous electrolyte. They invariably occur in moist environments and are relatively nonconducting, which means that they do not themselves lead to a direct discharge.

External Discharges   There are many examples where the surface of the insulation becomes contaminated by ambient conditions such as excessive moisture, deposition of pollutants, dirt, dust, salt spraying, and so forth. Eventually the contaminated surface develops sufficient conductance to allow discharge between the electrodes at a field below the normal breakdown strength of the insulator. This type of dielectric breakdown over the surface of the insulation is termed **surface tracking.**

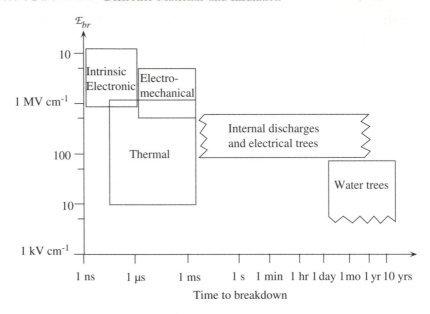

Figure 7.25   Time to breakdown and the field at breakdown, $\mathcal{E}_{br}$, are interrelated and depend on the mechanism that causes the insulation breakdown.
External discharges have been excluded.

SOURCE: Based on L. A. Dissado and J. C. Fothergill, *Electrical Degradation and Breakdown in Polymers* (United Kingdom: Peter Peregrinus Ltd. for IEE, 1992), p. 63.

It is apparent that there are a number of dielectric breakdown mechanisms and the one that causes eventual breakdown depends not only on the properties and quality of the material but also on the operating conditions, environmental factors being no less important. Figure 7.25 provides an illustrative diagram showing the relationship between the breakdown field and the time to breakdown. An insulation that can withstand large fields for a very short duration will break down at a lower field if the duration of the field increases. The breakdown mechanism is also likely to change from being intrinsic to being, perhaps, thermal. When insulation breakdown occurs in times beyond a few days, it is generally attributed to the degradation of the insulation, which eventually leads to a breakdown through, most probably, electrical treeing. It is also apparent that it is not possible to clearly identify a specific dielectric breakdown mechanism for a given material.

Example 7.10 | **DIELECTRIC BREAKDOWN IN A COAXIAL CABLE**   Consider the coaxial cable in Figure 7.26 with a and b defining the radii of the inner and outer conductors.

a. Using Gauss's law, find the capacitance of the coaxial cable.

b. What is the electric field at r from the center of the cable ($r > a$)? Where is the field maximum?

c. Consider two candidate materials for the dielectric insulation: cross-linked polyethylene, XLPE, and silicone rubber. Suppose that the inner conductor diameter

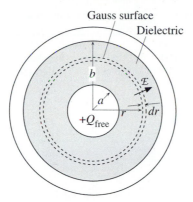

Figure 7.26 A schematic diagram for the calculation of the capacitance of a coaxial cable and the field at point r from the axis.

Consider an infinitessimally thin cylindrical shell of radius r and thickness dr in the dielectric and concentrically around the inner conductor. This surface is chosen as the Gauss surface. The voltage across the dielectric thickness dr is dV. The field $\mathcal{E} = -dV/dr$.

is 5 mm and the insulation thickness is also 5 mm. What is the voltage that will cause dielectric breakdown in each insulator?

d. What typical voltage will initiate a partial discharge in a small air pore (perhaps formed during mechanical and thermal stressing) at the inner conductor–insulator interface? Assume that the breakdown field for air at 1 atm and gap spacing around 0.1 mm is about 100 kV cm^{-1}.

Solution

Consider a cyclindrical shell of thickness dr of the dielectric as shown in Figure 7.26. Suppose that the voltage across the shell thickness is dV. Then the field $\mathcal{E}$ at r is $-dV/dr$ (this is the definition of $\mathcal{E}$). Suppose that Q_{free} is the free charge on the inner conductor. We take a Gauss surface that is a cylinder of radius r and concentric with the inner conductor as depicted in Figure 7.26. The surface area, A, of this cylinder is $2\pi rL$ where L is the length of the cable. The field at the surface, at distance r, is $\mathcal{E}$, which is normal to A and coming out of A. Then from Equation 7.35

$$\mathcal{E}(2\pi rL) = \frac{Q_{free}}{\varepsilon_o \varepsilon_r} \qquad [7.36]$$

Thus

$$-\frac{dV}{dr} = \frac{Q_{free}}{\varepsilon_o \varepsilon_r 2\pi rL}$$

This can be integrated from $r = a$, where the voltage is V, to b, where $V = 0$. Then

$$V = \frac{Q_{free}}{\varepsilon_o \varepsilon_r 2\pi L} \ln\left(\frac{b}{a}\right) \qquad [7.37]$$

Table 7.5   Dielectric insulation candidates for a coaxial cable

Dielectric	ε_r (60 Hz)	Strength (60 Hz) (kV cm^{-1})	C (60 Hz) (pF m^{-1})	Breakdown Voltage	Voltage for Partial Discharge in a Microvoid
XLPE	2.3	217	116	59.6 kV	11.9 kV
Silicone rubber	3.7	158	187	43.4 kV	7.4 kV

We can obtain the capacitance of the coaxial cable from $C_{coax} = Q_{free}/V$, which is

$$C_{coax} = \frac{\varepsilon_o \varepsilon_r 2\pi L}{\ln\left(\dfrac{b}{a}\right)}$$

[7.38]

The capacitance per unit length can be calculated using $a = 2.5$ mm and

$$b = a + \text{Thickness} = 7.5 \text{ mm}$$

and the appropriate dielectric constants, $\varepsilon_r = 2.3$ for XLPE and 3.7 for silicone rubber. The values are around 100–200 pF per meter, as listed in the fourth column in Table 7.5.

The electric field $\mathcal{E}$ follows directly when we substitute for Q_{free} from Equation 7.37 into Equation 7.36,

Field in a coaxial cable

$$\mathcal{E} = \frac{V}{r\ln\left(\dfrac{b}{a}\right)}$$

[7.39]

The above is valid for r from a to b (there is no field within the conductors). The field is maximum where $r = a$,

Maximum field in a coaxial cable

$$\mathcal{E}_{max} = \frac{V}{a\ln\left(\dfrac{b}{a}\right)}$$

[7.40]

The breakdown voltage V_{br} is reached when this maximum field, $\mathcal{E}_{max}$, reaches the dielectric strength or the breakdown field, $\mathcal{E}_{br}$

$$V_{br} = \mathcal{E}_{br} a \ln\left(\frac{b}{a}\right)$$

[7.41]

The breakdown voltages calculated from the above equation are listed in the fifth column in Table 7.5. Although the values are high, it must be remembered that, due to a number of other factors such as insulation aging, one cannot expect the cable to withstand these voltages forever.

If there is an air cavity or bubble at the inner conductor to dielectric surface, then the field in this gaseous space will be $\mathcal{E}_{air} \approx \varepsilon_r \mathcal{E}_{max}$, where $\mathcal{E}_{max}$ is the field at $r = a$. Air breakdown occurs when

$$\mathcal{E}_{air} = \mathcal{E}_{air\text{-}br} = 100 \text{ kV cm}^{-1}$$

at 1 atm and 25 °C for a 0.1 mm gap. Then $\mathcal{E}_{\max} \approx \mathcal{E}_{\text{air-}br}/\varepsilon_r$. The corresponding voltage from Equation 7.41 is

$$V_{\text{air-}br} \approx \frac{\mathcal{E}_{\text{air-}br}}{\varepsilon_r} a \ln\left(\frac{b}{a}\right)$$

The voltages for partial discharges for the two coaxial cables are shown in the sixth column of Table 7.5. It should be noted that these voltages will only give partial discharges contained within microvoids and will not normally lead to the immediate breakdown of the insulation. The partial discharges erode the cavities and also release vapor from the polymer that accumulates in the cavities. Thus, gaseous content and pressure in a cavity will change as the partial discharge continues. For example, the pressure buildup will increase the breakdown field and elevate the voltage for partial breakdown. Eventual degradation is likely to lead to electrical treeing.

We should also note that the actual field in the air cavity depends on the shape of the cavity, and the above treatment is only valid for a thin disk-like cavity lying perpendicular to the field (see Section 7.9, Additional Topics).

7.7   CAPACITOR DIELECTRIC MATERIALS

7.7.1   Typical Capacitor Constructions

The selection criteria of dielectric materials for capacitors depend on the capacitance value, frequency of application, maximum tolerable loss, and maximum working voltage, with size and cost being additional external constraints. Requirements for high voltage power capacitors are distinctly different than those used in small integrated circuits. Large capacitance values are more easily obtained at low frequencies because low frequency polarization mechanisms such as interfacial and dipolar polarization make a substantial contribution to the dielectric constant. At high frequencies, it becomes more difficult to achieve large capacitances and at the same time maintain acceptable low dielectric loss, inasmuch as the dielectric loss per unit volume is $\varepsilon_o \varepsilon_r' \omega E^2 \tan \delta$.

The bar-chart diagrams in Figures 7.27 and 7.28 provide some typical examples of dielectrics for a range of capacitance values and for a range of usable frequencies. For example, electrolytic dielectrics characteristically provide capacitances between one to thousands of microfarads, but their frequency response is typically limited to below 10 kHz. On the other hand, polymeric film capacitors typically have values less than 10 μF but a frequency response that is flat well into the gigahertz range.

We can understand the principles utilized in capacitor design from the capacitance of a parallel plate capacitor,

$$C = \frac{\varepsilon_o \varepsilon_r A}{d} \qquad\qquad \text{[7.42]}$$

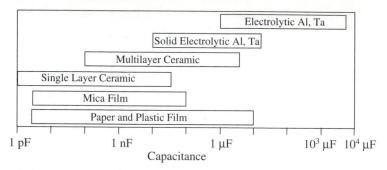

Figure 7.27   Examples of dielectrics that can be used for various capacitance values.

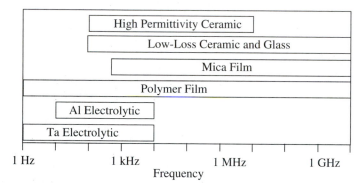

Figure 7.28   Examples of dielectrics that can be used in various frequency ranges.

where ε_r infers ε_r'. Large capacitances can be achieved by using high ε_r dielectrics, thin dielectrics, and large areas. There are various commercial ceramics, usually a mixture of various oxides or ferroelectric ceramics, that have high dielectric constants, ranging up to several thousands. These are typically called high-K (or high-κ), where K (or κ) stands for the relative permittivity. A ceramic dielectric with $\varepsilon_r = 10$, d of perhaps 10 μm, and an area of 1 cm^2 has a capacitance of 885 pF. Figure 7.29a shows a typical single-layer ceramic capacitor. The thin ceramic disk or plate has suitable metal electrodes, and the whole structure has been encapsulated in an epoxy by dipping it in a thermosetting resin. The epoxy coating prevents moisture from degrading the dielectric properties of the ceramic (increasing ε_r'' and the loss, tan δ). One way to increase the capacitance is to connect N number of these in parallel, and this is done in a space-efficient way by using the multilayer ceramic structure shown in Figure 7.29b. In this case there are N electroded dielectric layers. Each ceramic has offset metal electrodes that align with the opposite sides of the plate and make contact with the metal terminations on these sides. The result is N number of parallel plate capacitors. There is therefore an effective use of volume as the surface area of the component stays the same but

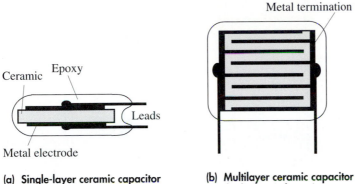

(a) **Single-layer ceramic capacitor**
(e.g. disk capacitors)

(b) **Multilayer ceramic capacitor**
(stacked ceramic layers)

Figure 7.29   Single- and multilayer dielectric capacitors.

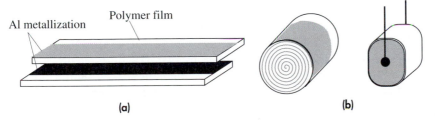

Figure 7.30   Two polymer tapes in (a), each with a metallized film electrode on the surface (offset from each other), can be rolled together (like a swiss roll) to obtain a polymer film capacitor as in (b).

As the two separate metal films are lined at oppose edges, electroding is done over the whole side surface.

the height increases to at least Nd. By using multilayer ceramic structures, capacitances up to a few hundred microfarads have been recently obtained.

Many wide-frequency-range capacitors utilize **polymeric thin films** for two reasons. Although ε_r is typically 2 to 3 (less than those for many ceramics), it is constant over a wide frequency range. The dielectric loss, $\varepsilon_o \varepsilon_r \omega E^2 \tan \delta$, becomes significant at high frequencies and polymers have low $\tan \delta$ values. Low ε_r values mean that one has to find a space-efficient way of constructing polymer film capacitors. One method is shown in Figure 7.30a and b for constructing a metallized film polymer capacitor. Two polymeric tapes have metallized electrodes (typically vacuum deposited Al) on one surface, leaving a margin on one side. These metal film electrodes have been offset in opposite directions so that they line up with the opposite sides of the tapes. The two tapes together are rolled up (like a swiss-roll cake) and the opposite sides are electroded using suitable conducting glues or other means. The concept is therefore similar to the multilayer ceramic capacitor except that the layers are rolled up to form a circular cross section. It is also possible to cut and stack the layers as in the multilayer ceramic construction.

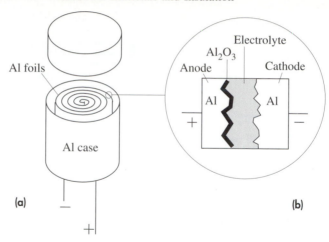

Figure 7.31   Aluminum electrolytic capacitor.

Electrolytic capacitors provide large values of capacitance while maintaining a tolerable size. There are various types of electrolytic capacitors. In aluminum electrolytic capacitors, the metal electrodes are two Al foils, typically 50–100 μm thick, that are separated by a porous paper medium soaked with a liquid electrolyte. The two foils together are wound into a cylindrical form and held within a cylindrical case, as shown in Figure 7.31a. Contrary to intuition, the paper-soaked electrolyte is not the dielectric. The dielectric medium is the thin alumina Al_2O_3 layer grown on the roughened surface of one of the foils, as shown in Figure 7.31b. This foil is then called the anode (+ terminal). Both Al foils are etched to obtain rough surfaces, which increases the surface area compared with smooth surfaces. The capacitor is called electrolytic because the Al_2O_3 layer is grown electrolytically on one of the foils and is typically 0.1 μm in thickness. This small thickness and the large surface area are responsible for the large capacitance. The electrolyte is conducting and serves to heal local minor breakdowns in the Al_2O_2 by an electrolytic reaction, provided that the anode has been positively biased. The capacitive behavior is due to the $Al/(Al_2O_3)/$ electrolyte structure. Furthermore, Al/Al_2O_3 contact is like a metal to p-type semiconductor contact and has rectifying properties. It must be reverse biased to prevent charge injection into the Al_2O_3 and hence conduction through the capacitor. Thus the Al must be connected to the positive terminal, which makes it the anode. A reverse-biased Al electrolytic capacitor is normally conducting.

Electrolytic capacitors using liquid electrolytes tend to dry up over a long period, which is a disadvantage. **Solid electrolyte tantalum capacitors** overcome the drying-up problem by using a solid electrolyte. The structure of a typical solid Ta capacitor is shown in Figure 7.32a and b. The anode (+ electrode) is a porous (sintered) Ta pellet that has the surface anodized to obtain a thin surface layer of tantalum pentoxide, Ta_2O_5, which is the dielectric medium (with $\varepsilon_r' = 28$). The Ta pellet with Ta_2O_5 is then coated with a thick solid electrolyte, in this case MnO_2. Subsequently, graphite and silver paste layers are applied. Leads are then attached

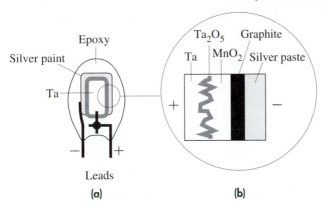

Figure 7.32   Solid electrolyte tantalum capacitor.
(a) A cross section without fine detail.
(b) An enlarged section through the Ta capacitor.

and the whole construction is molded into a resin chip. Solid tantalum capacitors are widely used in numerous electronics applications due to their small size, temperature and time stability, and high reliability.

7.7.2   Dielectrics: Comparison

The capacitance per unit volume, C_{vol}, which characterizes the **volume efficiency** of a dielectric, can be obtained by dividing C by Ad,

$$C_{vol} = \frac{\varepsilon_o \varepsilon_r}{d^2}$$ **[7.43]**

It is clear that large capacitances require high dielectric constants and thin dielectrics. We should note that d appears as d^2 so that the importance of d cannot be understated. Although mica has a higher ε_r than polymer films, the latter can be made quite thin, a few microns, which leads to a greater capacitance per unit volume. The reason that electrolytic aluminum capacitors can achieve large capacitance per unit volume is that d can be made very thin over a large surface area by using the liquid electrolyte to heal minor local dielectric breakdowns. Table 7.6 shows a selection of dielectric materials for capacitor applications and compares the volume efficiency, C_{vol}, based on a typical minimum thickness that a convenient process can handle. It is apparent that, compared with polymeric films, ceramics have substantial volume efficiency as a result of large dielectric constants (high-K ceramics) in some cases and as a consequence of a thin dielectric thickness in other cases (Al_2O_3).

Another engineering consideration in selecting a dielectric is the working voltage. Although d can be decreased to obtain large capacitances per unit volume, this also decreases the working voltage. The maximum voltage that can be applied

Table 7.6 Comparison of dielectric properties at 60 Hz for capacitor applications: typical values. (Assume $\eta = 2$, PS = polystyrene, PET = polyethyleneterephthalate. X7R is the name of a particular ceramic solid solution.)

	Polymer Film PS	Polymer Film PET	Mica	Ceramic Al_2O_3	Ceramic TiO_2	High-K Ceramic ($BaTiO_3$ based)
Name	Polystyrene	Polyester	Mica	Anodized alumina film	Polycrystalline titania	X7R
ε_r'	2.5	3.2	6.9	8.5	90	1,800
$\tan \delta$	3×10^{-4}	5×10^{-3}	5×10^{-4}	1×10^{-3}	4×10^{-4}	5×10^{-2}
$\mathcal{E}_{br}$ (kV cm^{-1})	200–250	150–200	1000	1000	50–100	100
d (typical minimum)	1–2 μm	1–2 μm	2.5 μm	0.1 μm	10 μm	10 μm
C_{vol} (μF cm^{-3})	22.1	28	10	7500	7.5	159.4
E_{vol} (mJ cm^{-3})	1.1	0.80	76	94	2.5	200
W_{vol} (W cm^{-3})	0.00025	0.0030	0.029	0.071	0.00075	7.5
Polarization	Electronic bond	Electronic bond and dipolar	Ionic	Ionic	Ionic	Large ionic displacement

to a capacitor depends on the breakdown field of the dielectric medium, $\mathcal{E}_{br}$, which itself is a highly variable quanity. A safe working voltage must be some safety factor η less than the breakdown voltage, $\mathcal{E}_{br}d$. Thus, if V_m is the maximum safe working voltage, then the maximum energy that can be stored per unit volume is given by

$$E_{vol} = \frac{1}{2}CV_m^2 \times \frac{1}{Ad} = \frac{\varepsilon_o \varepsilon_r'}{2\eta^2}\,\mathcal{E}_{br}^2 \qquad \text{[7.44]}$$

It is clear that both ε_r' and $\mathcal{E}_{br}$ of the dielectric are significant in determining the energy storage ability of the capacitor. Moreover, at the maximum working voltage, the rate of dielectric loss per unit volume in the capacitor becomes

$$W_{vol} = \frac{\mathcal{E}_{br}^2}{\eta^2}\,\omega\varepsilon_o \varepsilon_r' \tan\delta \qquad \text{[7.45]}$$

Those materials that have relatively higher $\tan\delta$ exhibit greater dielectric losses. Although dielectric losses may be small at 60 Hz, at high frequencies they become quite significant. Table 7.6 compares the energy storage efficiency, E_{vol}, and the relative rate of dielectric loss per unit volume, W_{vol}, for various dielectrics. It seems that ceramics have a better energy storage efficiency than polymers. High-K ceramics tend to have large $\tan\delta$ values and suffer from greater dielectric loss. Polystyrene has a particularly low $\tan\delta$ as the polarization mechanism is due to electronic bond polarization and the dielectric losses are the least. Indeed, polystyrene capacitors have found applications in high-quality audio electronics.

The temperature stability of a capacitor is determined by the temperature dependences of ε_r' and $\tan\delta$, which are controlled by the dominant polarization

Dipolar side group   Polymer chain

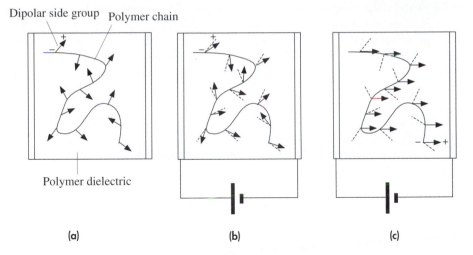

Polymer dielectric

(a)   (b)   (c)

Figure 7.33

(a) A polymer dielectric that has dipolar side groups attached to the polymer chains. With no applied field, the dipoles are randomly oriented.
(b) In the presence of an applied field, some very limited rotation enables dipolar polarization to take place.
(c) Near the softening temperature of the polymer, the molecular motions are rapid and there is also sufficient volume between chains for the dipoles to align with the field. The dipolar contribution to ε_r is substantial, even at high frequencies.

mechanism. For example, polar polymers have permanent dipole groups attached to the polymer chains as in polyethyleneterephthalate (PET). In the absence of an applied field, these dipoles are randomly oriented and also restricted in their rotations by neighboring chains, as depicted in Figure 7.33a. In the presence of an applied dc field, as in Figure 7.33b, some very limited rotation enables partial dipolar (orientational) polarization to take place. Typically, at room temperature, dipolar contribution to ε_r under ac conditions, however, is small because restricted and hindered rotation prevents the dipoles to closely follow to the ac field. Close to the softening temperature of the polymer, the molecular motions become easier and, further, there is more volume between chains for the dipoles to rotate. The dipolar side groups and polarized chains become capable of responding to the field. They can align with the field and also follow the field variations, as shown in Figure 7.33c. Dipolar contribution to ε_r is substantial even at high frequencies. Both ε_r' and tan δ therefore increase with temperature. Thus, polar polymers exhibit temperature dependent ε_r and tan δ, which reflect in the properties of the capacitor.

On the other hand, in nonpolar polymers such as polystyrene, the polarization is due to electronic bond polarization and ε_r and tan δ remain relatively constant. Thus polystyrene capacitors are more stable compared with PET (polyester) capacitors. The change in the capacitance with temperature is measured by the **temperature coefficient of capacitance** (TCC), which is defined as the fractional (or percentage) change in the capacitance per unit temperature change. The temperature controls not only ε_r but also the linear expansion of the dielectric, which changes the dimensions A and d. For example, polystyrene, polycarbonate, and mica capacitors are particularly stable with small TCC values. Plastic capacitors are typically limited to operations well below their melting temperatures, which is one of their main drawbacks. The specified operating temperature, for example, from $-55°$ C to $125°$ C, for many of the ceramic capacitors is often a limitation of the epoxy coating of the capacitor rather than the actual limitation of the ceramic material. In many capacitors, the working voltage has to be derated for operation at high temperatures and high frequencies because $\mathcal{E}_{br}$ decreases with ambient temperature and the frequency of the applied field. For example, a 1000 V dc polypropolyene capacitor will have a substantially lower ac working voltage, e.g., 100 V at 10 kHz.

Example 7.11

DIELECTRIC LOSS AND EQUIVALENT CIRCUIT OF A POLYESTER CAPACITOR AT 1 kHz   Figure 7.34 shows the temperature dependence of ε_r' and $\tan \delta$ for a polyester film. Calculate the equivalent circuit at 25 °C at 1 kHz for a 560 pF PET capacitor that uses a 0.5 micron thick polyester film. What happens to these values at 100 °C?

Solution

From Figure 7.34 at 25 °C, $\varepsilon_r' = 2.60$ and $\tan \delta \approx 0.002$. The capacitance C at 25 °C is given as 560 pF. The equivalent parallel conductance G_P, representing the dielectric loss, is given by

$$G_P = \frac{\omega A \varepsilon_o \varepsilon_r' \tan \delta}{d} = \omega C \tan \delta$$

Substituting

$$\omega = 2\pi f = 2000\pi,$$

and $\tan \delta = 0.002$, we get

$$G_P = (2000\pi)(560 \times 10^{-12})(0.002) = 7.04 \times 10^{-9} \frac{1}{\Omega}$$

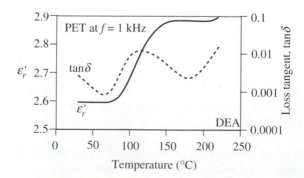

Figure 7.34   Real part of the dielectric constant, ε_r', and loss tangent, tan δ, at 1 kHz versus temperature for PET.

SOURCE: Data obtained by Kasap and Maeda (1995) using a dielectric analyzer (DEA).

This is equivalent to a resistance of 142 MΩ. The equivalent circuit is an ideal (lossless) capacitor of 560 pF in parallel with a 142 MΩ resistance (this resistance value decreases with the frequency).

At 100 °C, $\varepsilon_r' = 2.69$ and tan $\delta \approx 0.01$, so the new capacitance is

$$C_{100\,°C} = C_{25\,°C} \frac{\varepsilon_r(100\,°C)}{\varepsilon_r(25\,°C)} = 560\frac{2.69}{2.60} = 579 \text{ pF}$$

The equivalent parallel conductance at 100 °C is

$$G_P = (2000\pi)(579 \times 10^{-12})(0.01) = 3.64 \times 10^{-8}\frac{1}{\Omega}$$

This is equivalent to a resistance of 27.5 MΩ. The equivalent circuit is an ideal (lossless) capacitor of 579 pF in parallel with a 27.5 MΩ resistance.

7.8   PIEZOELECTRICITY, FERROELECTRICITY, AND PYROELECTRICITY PHENOMENOLOGY

7.8.1   Piezoelectricity

Certain crystals, for example, quartz (crystalline SiO_2) and $BaTiO_3$, become polarized when they are mechanically stressed. Charges appear on the surfaces of the crystal, as depicted in Figure 7.35a and b. Appearance of surface charges leads to a voltage difference between the two surfaces of the crystal. The same crystals also exhibit mechanical strain or distortion when they experience an electric field, as shown in Figure 7.35c and d. The direction of mechanical deformation (e.g., extension or compression) depends on the direction of the applied field, or the

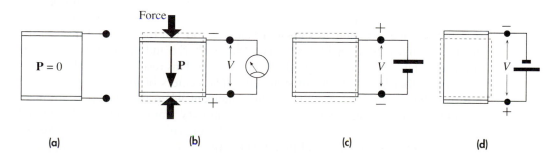

(a)   (b)   (c)   (d)

Figure 7.35   The piezoelectric effect
(a) A piezoelectric crystal with no applied stress or field.
(b) The crystal is strained by an applied force that induces polarization in the crystal and generates surface charges.
(c) An applied field causes the crystal to become strained. In this case the field compresses the crystal.
(d) The strain changes direction with the applied field and now the crystal is extended.

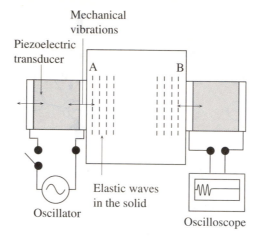

Figure 7.38   Piezoelectric transducers are widely used to generate ultrasonic waves in solids and also to detect such mechanical waves.

The transducer on the left is excited from an ac source and vibrates mechanically. These vibrations are coupled to the solid and generate elastic waves. When the waves reach the other end, they mechanically vibrate the transducer on the right, which converts the vibrations to an electrical signal.

They are used in many engineering applications that involve electromechanical conversions, as in ultrasonic transducers, microphones, accelerometers, and so forth. Piezoelectric transducers are widely used to generate ultrasonic waves in solids and also to detect such mechanical waves, as illustrated in Figure 7.38. The transducer is simply a piezoelectric crystal, for example, quartz, that is appropriately cut and electroded to generate the desired types of mechanical vibrations (e.g., longitudinal or transverse vibrations). The transducer on the left is attached to the surface A of the solid under examination, as shown in Figure 7.38. It is excited from an ac source, which means that it mechanically vibrates. These vibrations are coupled to the solid by a proper coupling medium (typically grease) and generate mechanical waves or elastic waves that propagate away from A. They are called **ultrasonic waves** as their frequencies are typically above the audible range. When the waves reach the other end, B, they mechanically vibrate the transducer attached to B, which converts the vibrations to an electrical signal that can readily be displayed on an oscilloscope. In this trivial example, one can easily measure the time it takes for elastic waves to travel in the solid from A to B and hence determine the ultrasonic velocity of the waves since the distance AB is known. From the ultrasonic velocity one can determine the elastic constants (Young's modulus) of the solid. Furthermore, if there are internal imperfections such as cracks in the solid, then they reflect or scatter the ultrasonic waves. These reflections can lead to echoes that can be detected by suitably located transducers.

Such ultrasonic testing methods are widely used for nondestructive evaluations of solids in mechanical engineering.

It is clear that an important engineering factor in the use of piezoelectric transducers is the efficiency of conversion between electrical and mechanical energies. The electromechanical conversion factor K is defined in terms of K^2 by

$$K^2 = \frac{\text{Output of mechanical energy}}{\text{Input of electrical energy}} \qquad \text{[7.48]}$$

or equivalently by

$$K^2 = \frac{\text{Output of electrical energy}}{\text{Input of mechanical energy}} \qquad \text{[7.49]}$$

Table 7.7 summarizes some typical piezoelectric materials with some applications. The so-called PZT ceramics are widely used in many piezoelectric applications. PZT stands for lead zirconate titanate and the ceramic is a solid solution of lead zirconate, $PbZrO_3$, and lead titanate, $PbTiO_3$, so that its composition is $PbTi_{1-x}Zr_xO_3$ where x is determined by the extent of the solid solution but typically is around 0.5. PZT piezoelectric components are manufactured by sintering, which is a characteristic ceramic manufacturing process in which PZT powders are placed in a mold and subjected to a pressure at high temperatures. During sintering the ceramic powders are fused through interdiffusion. The final properties depend not only on the composition of the solid solution but also on the manufacturing process, which controls the average grain size or polycrystallinity. Electrodes are deposited onto the final ceramic component, which is then poled by the application of a temporary electric field to induce it to become piezoelectric. **Poling** refers to the application of a temporary electric field, generally at an elevated temperature, to align the polarizations of various grains and thereby develop piezoelectric behavior.

Table 7.7   Piezoelectric materials and some typical values for d and K.

Crystal	d (m V^{-1})	K	Comment
Quartz (crystal SiO_2)	2.3×10^{-12}	0.1	Crystal oscillators, ultrasonic transducers, delay lines, filters
Rochelle salt ($NaKC_4H_4O_6 \cdot 4H_2O$)	350×10^{-12}	0.78	—
Barium titanate ($BaTiO_3$)	190×10^{-12}	0.49	Accelerometers
PZT, lead zirconate titanate ($PbTi_{1-x}Zr_xO_3$)	480×10^{-12}	0.72	Wide range of applications including earphones, microphones, spark generators (gas lighters, car ignition), displacement transducers, accelerometers
Polyvinylidene fluoride (PVDF)	18×10^{-12}	—	Must be poled; heated, put in an electric field and then cooled. Large area and inexpensive

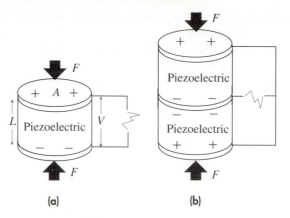

Figure 7.39   The piezoelectric spark generator.

Example 7.12 | **PIEZOELECTRIC SPARK GENERATOR**   The piezoelectric spark generator, as used in various applications, lighters, car ignitions, and so on, operates by stressing a piezoelectric crystal to generate a high voltage that is discharged through a spark gap in air, as schematically shown in Figure 7.39a. Consider a piezoelectric crystal (most likely a polycrystalline PZT ceramic cylinder as in Figure 7.39a) that has a Piezoelectric coefficient $d = 200 \times 10^{-12}$ m V^{-1}, Young's modulus of 70 GPa, a thickness of 10 mm, and a diameter of 3 mm. The spark gap is 1 mm and the breakdown of air with this gap is 5 kV mm^{-1}. What is the force required to spark the gap? Is this a realistic force?

Solution

The breakdown occurs when the field in the spark gap is 5 kV mm^{-1}. Since the spark gap is 1 mm, we need to generate a voltage of about 5 kV. The expression we need is the piezoelectric relationship between the field generated $\mathcal{E}$ and the induced strain S, or $S = d\mathcal{E}$, where d is the piezoelectric coefficient. The voltage V that appears across the piezoelectric crystal is given by $V = \mathcal{E}L$, where L is the length of the crystal. Then

$$S = d\left(\frac{V}{L}\right)$$

The strain S and stress T are related through $T = YS$, where Y is the elastic modulus (Young's modulus; the usual mechanical symbols for strain and stress are ε and σ). Thus

$$\frac{T}{Y} = \frac{dV}{L}$$

Further, the stress $T = F/A$, where F is applied force and A is the cross-sectional area. Thus,

$$F = \frac{AY\,dV}{L} = \frac{\pi(1.5 \times 10^{-3})^2(70 \times 10^9)(200 \times 10^{-12})(5000)}{(10 \times 10^{-3})} = 49.5 \text{ N}$$

This force can be applied by squeezing an appropriate lever arrangement by hand; it is the weight of about 50 apples. The energy in the spark depends on the amount of charge generated. This can be increased by using two piezoelectric crystals back to back, as in Figure 7.39b, which is a more practical arrangement for a spark generator.

7.8.2   Piezoelectricity; Quartz Oscillators and Filters

One of the most important applications of the piezoelectric quartz crystal in electronics is in the frequency control of oscillators and filters. Consider a suitably cut thin plate of a quartz crystal that has thin gold electrodes on the opposite faces. Suppose that we set up mechanical vibrations in the crystal by connecting the electrodes to an ac source, as in Figure 7.40a. It is possible to set up a mechanical resonance, or mechanical standing waves, in the crystal if the wavelength λ of the waves and the length L along which the waves are traveling satisfy the condition for standing waves:

$$n\left(\frac{1}{2}\lambda\right) = L \qquad\qquad [7.50]$$

where n is an integer.

The frequency of these mechanical vibrations, f_s, is given by $f_s = v/\lambda$, where v is the velocity of the waves in the medium and λ is the wavelength. These mechanical vibrations in quartz experience very small losses and therefore have a high quality factor Q, which means that resonance can only be set up if the frequency of the excitation, the electrical frequency, is close to f_s. Because of the coupling of energy between the electrical excitation and mechanical vibrations through the piezoelectric effect, mechanical vibrations appear like a series LCR circuit to the ac source, as shown in Figure 7.40b. This LCR series circuit has an impedance that is minimum at the mechanical resonant frequency f_s, given by

$$f_s = \frac{1}{2\pi\sqrt{LC}} \qquad\qquad [7.51]$$

In this series LCR circuit, L represents the mass of the transducer, C the stiffness, and R the losses or mechanical damping. Since the quartz crystal has electrodes at opposite faces, there is, in addition, the parallel plate capacitance, C_o,

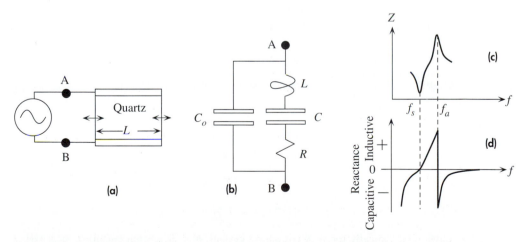

Figure 7.40   When a suitably cut quartz crystal with electrodes is excited by an ac voltage as in (a) it behaves as if it has the equivalent circuit in (b). (c) The frequency dependence of the magnitude of the impedance Z between A and B. (d) The reactance between A and B vs. frequency.

between the electrodes. Thus, the whole equivalent circuit is C_o in parallel with LCR, as in Figure 7.40b. As far as L is concerned, C_o and C are in series. There is a second resonant frequency, f_a, that is at a higher frequency and is due to L resonating with C and C_o in series,

$$f_a = \frac{1}{2\pi \sqrt{LC'}}$$

[7.52]

where

$$\frac{1}{C'} = \frac{1}{C_o} + \frac{1}{C}.$$

The impedance between the terminals of the quartz crystal has the frequency dependence shown in Figure 7.40c. The two frequencies, f_s and f_a, are called the series and parallel resonant frequencies. It is apparent that around f_a, the crystal behaves like a filter with a high Q value. If we were to examine the reactance of the crystal, whether it is behaving capacitively or inductively, we would find the behavior in Figure 7.40d, where positive reactance refers to an inductive and negative reactance to a capacitive behavior. Between f_s and f_a the crystal behaves inductively, and capacitively outside this range. Indeed, between f_s and f_a the response of the transducer is controlled by the mass of the crystal. This property has been utilized by electrical engineers in designing quartz oscillators.

In quartz oscillators, the crystal is invariably used in one of two modes. First, it can be used at f_s where it behaves as a resistance of R without any reactance. The circuit is designed so that oscillations can take place only when the crystal in the circuit exhibits no reactance or phase change—in other words, at f_s. Outside this frequency, the crystal introduces reactance or phase changes that do not lead to sustained oscillations. In a different mode of operation, the oscillator circuit is designed to make use of the **inductance** of the crystal just above f_s. Oscillations are maintained close to f_s because even very large changes in the inductance result in small changes in the frequency between f_s and f_a.

Example 7.13 | **THE QUARTZ CRYSTAL AND ITS EQUIVALENT CIRCUIT**   From the following equivalent definition of the coupling coefficient,

$$K^2 = \frac{\text{Mechanical energy stored}}{\text{Total energy stored}}$$

show that

$$K^2 = 1 - \frac{f_s^2}{f_a^2}$$

Given that typically for an X-cut quartz crystal, $K = 0.1$, what is f_a for $f_s = 1$ MHz? What is your conclusion?

Solution

C represents the mechanical mass where the mechanical energy is stored, whereas C_o is where the electrical energy is stored. If V is the applied voltage, then

$$K^2 = \frac{\text{Mechanical energy stored}}{\text{Total energy stored}} = \frac{\frac{1}{2}CV^2}{\frac{1}{2}CV^2 + \frac{1}{2}C_o V^2} = \frac{C}{C + C_o} = 1 - \frac{f_s^2}{f_a^2}$$

Rearranging the above equation, we find

$$f_a = \frac{f_s}{\sqrt{1 - K^2}} = \frac{1 \text{ MHz}}{\sqrt{1 - (0.1)^2}} = 1.005 \text{ MHz}$$

Thus, $f_a - f_s$ is only 5 kHz. The two frequencies f_s and f_a in Figure 7.40d are very close. An oscillator designed to oscillate at f_s, that is, at 1 MHz, therefore, cannot drift far (for example, a few kHz) because that would change the reactance enormously, which would upset the oscillation conditions.

QUARTZ CRYSTAL AND ITS INDUCTANCE   A typical 1 MHz quartz crystal has the following properties: | **Example 7.14**

$$f_s = 1 \text{ MHz} \qquad f_a = 1.0025 \text{ MHz} \qquad C_o = 5 \text{ pF} \qquad R = 20 \text{ }\Omega$$

What are C and L in the equivalent circuit of the crystal? What is the quality factor of the crystal, given that

$$Q = \frac{1}{2\pi f_s RC}$$

Solution

The expression for f_s is

$$f_s = \frac{1}{2\pi \sqrt{LC}}$$

From the expression for f_a, we have

$$f_a = \frac{1}{2\pi \sqrt{LC'}} = \frac{1}{2\pi \sqrt{L\dfrac{CC_o}{C + C_o}}}$$

Dividing f_a by f_s eliminates L, and we get

$$\frac{f_a}{f_s} = \sqrt{\frac{C + C_o}{C_o}}$$

so that C is

$$C = C_o\left[\left(\frac{f_a}{f_s}\right)^2 - 1\right] = (5 \text{ pF})(1.0025^2 - 1) = 0.025 \text{ pF}$$

Thus

$$L = \frac{1}{C(2\pi f_s)^2} = \frac{1}{0.025 \times 10^{-12}(2\pi\, 10^6)^2} = 1.01 \text{ H}$$

This is a substantial inductance, and the enormous increase in the inductive reactance above f_s is intuitively apparent. The quality factor

$$Q = \frac{1}{2\pi f_s RC} = 3.18 \times 10^5$$

is very large.

7.8.3   Ferroelectric and Pyroelectric Crystals

Certain crystals are permanently polarized even in the absence of an applied field. The crystal already possesses a finite polarization vector due to the separation of positive and negative charges in the crystal. These crystals are called **ferroelectric**.[10] Barium titanate ($BaTiO_3$) is probably the best cited example. Above approximately 130 °C, the crystal structure of $BaTiO_3$ has a cubic unit cell, as shown in Figure 7.41a. The centers of mass of the negative charges (O^{2-}) and the positive charges, Ba^{2+} and Ti^{4+}, coincide at the Ti^{4+} ion, as shown in Figure 7.41b. There is therefore no net polarization and $\mathbf{P} = 0$. Above 130 °C, therefore, the barium titanate crystal exhibits no permanent polarization and is not ferroelectric. However, below 130 °C, the structure of barium titanate is tetragonal, as shown in Figure 7.41c, in which the Ti^{4+} atom is not located at the center of mass of the negative charges. The crystal is therefore polarized by the separation of the centers of mass of the negative and positive charges. The crystal possesses a finite polarization vector $\mathbf{P}$ and is ferroelectric. The critical temperature above which ferroelec-

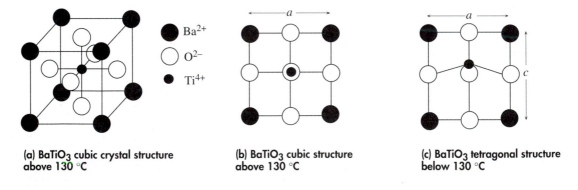

(a) $BaTiO_3$ cubic crystal structure above 130 °C

(b) $BaTiO_3$ cubic structure above 130 °C

(c) $BaTiO_3$ tetragonal structure below 130 °C

Figure 7.41 $BaTiO_3$ has different crystal structures above and below 130 °C that lead to different dielectric properties.

---

[10] In analogy with the ferromagnetic crystals that already possess magnetization.

tric property is lost, in this case 130 °C, is called the **Curie temperature** (T_C). Below the Curie temperature, the whole crystal becomes spontaneously polarized. The onset of spontaneous polarization is accompanied by the distortion of the crystal structure, as shown by the change from Figure 7.41b to Figure 7.41c. The spontaneous displacement of the Ti^{4+} ion below the Curie temperature elongates the cubic structure, which becomes tetragonal. It is important to emphasize that we have only described an observation and not the reasons for the spontaneous polarization of the whole crystal. The development of the permanent dipole moment below the Curie temperature involves long-range interactions between the ions outside the simple unit cell pictured in Figure 7.41. The energy of the crystal is lower when the Ti^{4+} ion in each unit cell is slightly displaced along the c direction, as in Figure 7.41c, which generates a dipole moment in each unit cell. The interaction energy of these dipoles when all are aligned in the same direction lowers the energy of the whole crystal. It should be mentioned that the distortion of the crystal that takes place when spontaneous polarization occurs just below T_C is very small relative to the dimensions of the unit cell. For $BaTiO_3$, for example, c/a is 1.01 and the displacement of the Ti^{4+} ion from the center is only 0.012 nm, compared with $a = 0.4$ nm.

An important and technologically useful characteristic of a ferroelectric crystal is its ability to be poled. Above 130 °C there is no permanent polarization in the crystal. If we apply a temporary field $\mathcal{E}$ and let the crystal cool to below 130 °C, we can induce the spontaneous polarization **P** to develop along the field direction. In other words, we would define the c axis by imposing a temporary external field. This process is called poling. The c axis is the polar axis along which **P** develops. It is also called the **ferroelectric axis.** Since below the Curie temperature the ferroelectric crystal already has a permanent polarization, it is not possible to use the expression

$$P = \varepsilon_o(\varepsilon_r - 1)\mathcal{E}$$

to define a relative permittivity. Suppose that we use a ferroelectric crystal as a dielectric medium between two parallel plates. Since any change ΔP normal to the plates changes the stored charge, what is of significance to the observer is the change in the polarization. We can appreciate this by noting that $C = Q/V$ is not a good definition of capacitance if there are already charges on the plates, even in the absence of voltage.[11] We then prefer a definition of C based on $\Delta Q/\Delta V$ where ΔQ is the change in stored charge due to a change ΔV in the voltage. Similarly, we define the relative permittivity ε_r in this case in terms of the change ΔP in P induced by $\Delta \mathcal{E}$ in the field $\mathcal{E}$,

$$\Delta P = \varepsilon_o(\varepsilon_r - 1)\, \Delta \mathcal{E}$$

An applied field along the a axis can displace the Ti^{3+} ion more easily than that along the c axis, and experiments show that $\varepsilon_r \approx 4100$ along a is much greater than $\varepsilon_r \approx 160$ along c. Because of their large dielectric constants, ferroelectric ceramics are used as high-K dielectrics in capacitors.

---

[11] A finite Q on the plates of a capacitor when $V = 0$ implies an infinite capacitance, $C = \infty$. However, $C = dQ/dV$ definition avoids this infinity.

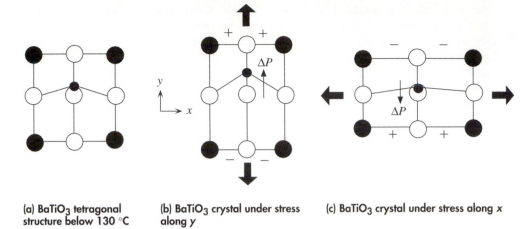

(a) BaTiO$_3$ tetragonal structure below 130 °C

(b) BaTiO$_3$ crystal under stress along y

(c) BaTiO$_3$ crystal under stress along x

Figure 7.42   Piezoelectric properties of BaTiO$_3$ below its Curie temperature.

All ferroelectric crystals are also piezoelectric, but the reverse is not true: not all piezoelectric crystals are ferroelectric. When a stress along y is applied to the BaTiO$_3$ crystal in Figure 7.42a, the crystal is stretched along y, as a result of which the Ti^{4+} atom becomes displaced, as shown in Figure 7.42b. There is, however, no shift in the center of mass of the negative charges, which means that there is a change, ΔP, in the polarization vector along y. Thus, the applied stress induces a change in the polarization, which is a piezoelectric effect. If the stress is along x, as illustrated in Figure 7.42c, then the change in the polarization is along y. In both cases, ΔP is proportional to the stress, which is a characteristic of the piezoelectric effect.

The barium titanate crystal in Figure 7.41 is also said to be pyroelectric because when the temperature increases, the crystal expands and the relative distances of ions change. The Ti^{4+} ion becomes shifted, which results in a change in the polarization. Thus, a temperature change δT induces a change δP in the polarization of the crystal. This is called **pyroelectricity,** which is illustrated in Figure 7.43. The magnitude of this effect is quantized by the pyroelectric coefficient p, which is defined by

Pyroelectric coefficient

$$p = \frac{dP}{dT}$$ [7.53]

A few typical pyroelectric crystals and their pyroelectric coefficients are listed in Table 7.8. Very small temperature changes, even in thousandths of degrees, in the material can develop voltages that can be readily measured. For example, for a PZT-type pyroelectric ceramic in Table 7.8, taking $\delta T = 10^{-3}$ K   and $p \approx 380 \times 10^{-6}$, we find $\delta P = 3.8 \times 10^{-7}$ C m^{-2}. From

$$\delta P = \varepsilon_o(\varepsilon_r - 1) \, \delta \mathcal{E}$$

with $\varepsilon_r = 290$, we find

$$\delta \mathcal{E} = 148 \text{ V m}^{-1}$$

Temperature change = δT

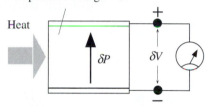

Figure 7.43   The heat absorbed by the crystal increases the temperature by δT, which induces a change δP in the polarization.

This is the pyroelectric effect. The change δP gives rise to a change δV in the voltage that can be measured.

Table 7.8   Some pyroelectric (and also ferroelectric) crystals and typical properties

Material	ε_r'	tan δ	Pyroelectric Coefficient ($\times 10^{-6}$ C m^{-2} K^{-1})	Curie Temperature (°C)
BaTiO$_3$	4100 $\perp$ polar axis; 160 //polar axis	7×10^{-3}	20	130
PZT modified for pyroelectric	290	2.7×10^{-3}	380	230
PVDF, polymer	12	0.01	27	80

If the distance between the faces of the ceramic where the charges are developed is 0.1 mm, then

$$\delta V = 0.0148 \text{ V} \qquad \text{or} \qquad 15 \text{ mV}$$

which can be readily measured. Pyroelectric crystals are widely used as infrared detectors. Any infrared radiation that can raise the temperature of the crystal even by a thousandth of a degree can be detected. For example, many intruder alarms use pyroelectric detectors because as the human or animal intruder passes by the view of detector, the infrared radiation from the warm body raises the temperature of the pyroelectric detector, which generates a voltage that actuates an alarm.

Figure 7.44 shows a simplified schematic circuit for a pyroelectric radiation detector. The detecting element, labeled A, is actually a thin crystal or ceramic (or even a polymer) of a pyroelectric material that has electrodes on opposite faces. Pyroelectric materials are also piezoelectric and therefore also sensitive to stresses. Thus, pressure fluctuations, for example, vibrations from the detector mount or sound waves, interfere with the response of the detector to radiation alone. These can be compensated for by having a second dummy detector, B, that has a reflecting coating and is subjected to the same vibrations (air and mount), as depicted in Figure 7.44. Thus, there are two elements in the detector, one with an absorbing surface, detecting element A, and the other with a reflecting surface, compensating element B. Stress fluctuations give rise to the same piezoelectric voltage in both,

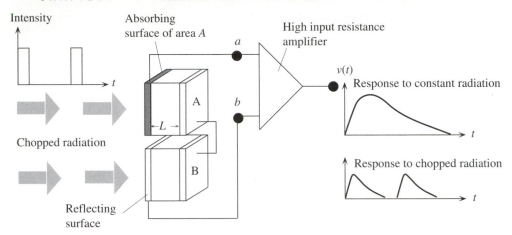

Figure 7.44   The pyroelectric detector

Radiation is absorbed in the detecting element, A, which generates a pyroelectric voltage that is measured by the amplifier. The second element, B, has a reflecting electrode and does not absorb the radiation. It is a dummy element that compensates for the piezoelectric effects. Piezoelectric effects generate equal voltages in both A and B, which cancel each other across a and b, the input of the amplifier.

which then cancel each other between a and b at the imput of the amplifier. When radiation is incident, then only the detecting element absorbs the radiation, becomes warmer, and hence generates a pyroelectric voltage. This voltage appears directly across a and b. As the incident radiation warms the detecting element and increases its temperature, the pyroelectric voltage increases with time. Eventually the temperature reaches a steady-state value determined by heat losses from the element. We therefore expect the pyroelectric voltage to reach a constant value as well. However, the problem is that a constant pyroelectric voltage cannot be sustained because the surface charges slowly become neutralized or leak away. The constant radiation is therefore normally chopped to subject the detector to periodic bursts of radition, as shown in the figure. The pyroelectric voltage is then a changing function of time, which is readily measured and related to the power in the incident radiation.

Example 7.15   **A PYROELECTRIC RADIATION DETECTOR**   Consider the radiation detector in Figure 7.44. Suppose that the radiation is chopped so that the radiation is passed to the detector for a time Δt seconds every τ seconds, where $\Delta t \ll \tau$. If Δt is sufficiently small, then the temperature rise ΔT is small and hence the heat losses are negligible during Δt. Using the heat capacity to find the temperature change during Δt, relate the magnitude of the voltage, ΔV, to the incident radiation intensity, $\mathcal{I}$. What is your conclusion?

Consider a PZT-type pyroelectric material with a density of about 7 g cm^{-3} and a specific heat capacity of about 380 J K^{-1} kg^{-1}. If $\Delta t = 0.2$ second and the minimum voltage that can be detected above the background noise is 1 mV, what is the minimum radiation intensity that can be measured?

Solution

Suppose that the radiation of intensity $\mathcal{I}$ is received during a time interval Δt and delivers an amount of energy ΔH to the pyroelectric detector. This energy ΔH, in the absence of any heat losses, increases the temperature by ΔT. If c is the specific heat capacity (heat capacity per unit mass) and ρ is the density,

$$\Delta H = (AL\rho)c\,\Delta T,$$

where A is the surface area and L the thickness of the detector. The change in the polarization, ΔP, is

$$\Delta P = p\,\Delta T = \frac{p\,\Delta H}{AL\rho c}$$

The change in the surface charge ΔQ is

$$\Delta Q = A\,\Delta P = \frac{p\,\Delta H}{L\rho c}$$

This change in the surface charge gives a voltage change, ΔV, across the electrodes of the detector,

$$\Delta V = \frac{\Delta Q}{C} = \frac{p\,\Delta H}{L\rho c} \times \frac{L}{\varepsilon_o \varepsilon_r A} = \frac{p\,\Delta H}{A\rho c \varepsilon_o \varepsilon_r}$$

The absorbed energy (heat), ΔH, during Δt depends on the intensity of incident radiation. Incident intensity, I, is the energy arriving per unit area per unit time. In time Δt, I delivers an energy $\Delta H = IA\,\Delta t$. Substituting for ΔH in the expression for ΔV, we find

$$\Delta V = \frac{pI\,\Delta t}{\rho c \varepsilon_r \varepsilon_o} = \left(\frac{p}{\rho c \varepsilon_r \varepsilon_o}\right) I\,\Delta t \qquad\qquad \text{[7.54]}$$

The parameters in the parentheses are material properties and reflect the "goodness" of the pyroelectric material for the application. We should emphasize that in deriving Equation 7.54 we did not consider any heat losses that will prevent the rise of the temperature indefinitely. If Δt is short, then the temperature change will be small and heat losses negligible.

For a PZT-type pyroelectric, we can take $p = 380 \times 10^{-6}$ C m^{-2} K^{-1}, $\varepsilon_r = 290$, $c = 380$ J K^{-1} kg^{-1}, and $\rho = 7 \times 10^3$ kg m^{-3}, and then from Equation 7.54 with $\Delta V = 0.001$ V and $\Delta t = 0.2$ s, we have

$$I = \left(\frac{p}{\rho c \varepsilon_o \varepsilon_r}\right)^{-1} \frac{\Delta V}{\Delta t} = \left(\frac{380 \times 10^{-6}}{(7000)(380)(290)(8.85 \times 10^{-12})}\right)^{-1} \frac{0.001}{0.2}$$

$$= 0.090 \text{ W m}^{-2} \quad \text{or} \quad 9\ \mu\text{W cm}^{-2}$$

ADDITIONAL TOPICS

7.9   ELECTRIC DISPLACEMENT AND DEPOLARIZATION FIELD

Electric Displacement (D) and Free Charges   Consider a parallel plate capacitor with free space between the plates, as shown in Figure 7.45a, which has been charged to a voltage V_o by connecting it to a battery of voltage V_o. The battery has been suddenly removed, which has left the free positive and negative charges, Q_{free}, on

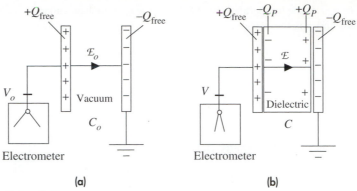

Figure 7.45

(a) Parallel plate capacitor with free space between plates that has been charged to a voltage V_o. There is no battery to maintain the voltage constant across the capacitor. The electrometer measures the voltage difference across the plates and, in principle, does not affect the measurement.

(b) After the insertion of the dielectric, the voltage difference is V, less than V_o, and the field in the dielectric is $\mathcal{E}$ less than $\mathcal{E}_o$.

the plates. These charges are free in the sense that they can be conducted away. An ideal electrometer (with no leakage current) measures the total charge on the positive plate (or voltage of the positive plate with respect to the negative plate). The voltage across the plates is V_o and the capacitance is C_o. The field in the free space between the plates is

$$\mathcal{E}_o = \frac{Q_{\text{free}}}{\varepsilon_o A} = \frac{V_o}{d} \tag{7.55}$$

where d is the separation of the plates.

When we insert a dielectric to fit between the plates, the field polarizes the dielectric and polarization charges $-Q_P$ and $+Q_P$ appear on the left and right surfaces of the dielectric, as shown in Figure 7.45b. As there is no battery to supply more free charges, the net charge on the left plate (positive plate) becomes $Q_{\text{free}} - Q_P$. Similarly the net negative charge on the right plate becomes $-Q_{\text{free}} + Q_P$. The field inside the dielectric is no longer $\mathcal{E}_o$ but less because induced polarization charges have the opposite polarity to the original free charges and the net charge on each plate has been reduced. The new field can be found by applying Gauss's law. Consider a Gauss surface just enclosing the left plate and the surface region of the dielectric with its negative polarization charges, as shown in Figure 7.46. Then Gauss's law gives

$$\oint_{\text{Surface}} \varepsilon_o \mathcal{E}\, dA = Q_{\text{total}} = Q_{\text{free}} - Q_P \tag{7.56}$$

where A is the plate area (same as dielectric surface area) and we take the field $\mathcal{E}$ to be normal to the surface area dA, as indicated in Figure 7.46. If the polarization charge is dQ_P over a small surface area dA of the dielectric, then the polarization

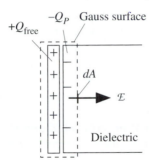

Figure 7.46   A Gauss surface just around the left plate and within the dielectric, encompassing both $+Q_{\text{free}}$ and $-Q_P$.

charge density σ_P at this point is defined as

$$\sigma_P = \frac{dQ_P}{dA}$$

For uniform polarization, the charge distribution is Q_P/A, as we have used previously. Since $\sigma_P = P$, where P is the polarization vector, we can write

$$P = \frac{dQ_P}{dA}$$

and therefore express Q_P as

$$Q_P = \oint_{\text{Surface}} P \, dA \qquad\qquad \text{[7.57]}$$

We can now substitute for Q_P in Equation 7.56 and take this term to the left-hand side to add the two surface integrals. The right-hand side is left with only Q_{free}. Thus,

$$\oint_{\text{Surface}} (\varepsilon_o \mathcal{E} + P) \, dA = Q_{\text{free}} \qquad\qquad \text{[7.58]}$$

What is important here is that the surface integration of the quantity $\varepsilon_o \mathcal{E} + P$ is always equal to the total free charges on the surface. Whatever the dielectric material, this integral is always Q_{free}. It becomes convenient to define $\varepsilon_o \mathcal{E} + P$ as a usable quantity, called the **electric displacement** and denoted as D, that is,

$$D = \varepsilon_o \mathcal{E} + P \qquad\qquad \text{[7.59]} \qquad \textit{Definition of electric displacement}$$

Then, Gauss's law in terms of free charges alone in Equation 7.58 becomes

$$\oint_{\text{Surface}} D \, dA = Q_{\text{free}} \qquad\qquad \text{[7.60]} \qquad \textit{Gauss's law for free charges}$$

In equation 7.60 we take D to be normal to the surface area dA as in the case of $\mathcal{E}$ in Gauss's law. Equation 7.60 provides a convenient way to calculate the electric displacement D, from which one should be able to determine the field. We should note that, in general, $\mathcal{E}$ is a vector and so is P, so the definition in Equation 7.59 is strictly in terms of vectors. Inasmuch as the electric displacement depends only on free charges, as a vector it starts at negative free charges and finishes on positive free charges.

Equation 7.60 for D defines it in terms of $\mathcal{E}$ and P, but we can express D in terms of the field $\mathcal{E}$ in the dielectric alone. The polarization P and $\mathcal{E}$ are related by the definition of the relative permittivity, ε_r,

$$P = \varepsilon_o(\varepsilon_r - 1)\mathcal{E}$$

Substituting for P in Equation 7.59 and rearranging, we find that D is simply given by

Electric displacement and the field

$$D = \varepsilon_o \varepsilon_r \mathcal{E} \qquad [7.61]$$

We should note that this simple equation applies in an isotropic medium where the field along one direction, for example, x, does not generate polarization along a different direction, for example, y. In those cases, Equation 7.61 takes a tensor form whose mathematics is beyond the scope of this book.

We can now apply Equation 7.60 for a Gauss surface surrounding the left plate,

$$D = \frac{Q_{\text{free}}}{A} = \varepsilon_o \mathcal{E}_o \qquad [7.62]$$

where we used Equation 7.55 to replace Q_{free}. Thus D does not change when we insert the dielectric because the same free charges are still on the plates (they cannot be conducted away anywhere). The new field $\mathcal{E}$ between the plates after the insertion of the dielectric is

$$\mathcal{E} = \frac{1}{\varepsilon_o \varepsilon_r} D = \frac{1}{\varepsilon_r} \mathcal{E}_o \qquad [7.63]$$

The original field is reduced by the polarization of the dielectric. We should recall that the field does *not* change in the case where the parallel plate capacitor is connected to a battery that keeps the voltage constant across the plates and supplies additional free charges (ΔQ_{free}) to make up for the induced opposite-polarity polarization charges.

Gauss's law in Equation 7.60 in terms of D and the enclosed free charges, Q_{free}, can also be written in terms of the field $\mathcal{E}$, but including the relative permittivity, because D and $\mathcal{E}$ are related by Equation 7.61. Using Equation 7.61, Equation 7.60 becomes

$$\oint_{\text{Surface}} \varepsilon_o \varepsilon_r \mathcal{E} \, dA = Q_{\text{free}}$$

For an isotropic medium where ε_r is the same everywhere,

$$\oint_{\text{Surface}} \mathcal{E} \, dA = \frac{Q_{\text{free}}}{\varepsilon_o \varepsilon_r} \qquad \text{[7.64]}$$

Gauss's law in an isotropic dielectric

As before, $\mathcal{E}$ in the surface integral is taken as normal to dA everywhere, Equation 7.64 is a convenient way of evaluating the field from the free charges alone, given the dielectric constant of the medium.

The Depolarizing Field   We can view the field $\mathcal{E}$ as arising from two electric fields: that due to the free charges, $\mathcal{E}_o$, and that due to the polarization charges, denoted as $\mathcal{E}_{\text{dep}}$. These two fields are indicated in Figure 7.47. $\mathcal{E}_o$ is called the **applied field** as it is due to the free charges that have been put on the plates. It starts and ends at free charges on the plates. The field due to polarization charges starts and ends at these bound charges and is in the *opposite* direction to the $\mathcal{E}_o$. Although $\mathcal{E}_o$ polarizes the molecules of the medium, $\mathcal{E}_{\text{dep}}$, being in the opposite direction, tries to depolarize the medium. It is called the **depolarizing field** (and hence the subscript). Thus the field inside the medium is

$$\mathcal{E} = \mathcal{E}_o - \mathcal{E}_{\text{dep}} \qquad \text{[7.65]}$$

The depolarizing field depends on the amount of polarization since it is determined by $+Q_P$ and $-Q_P$. For the dielectric plate in Figure 7.47, we know the field $\mathcal{E}$ is $\mathcal{E}_o/\varepsilon_r$, so we can eliminate $\mathcal{E}_o$ in Equation 7.65 and relate $\mathcal{E}_{\text{dep}}$ directly to $\mathcal{E}$,

$$\mathcal{E}_{\text{dep}} = \mathcal{E}(\varepsilon_r - 1)$$

However, the polarization P is related to the field $\mathcal{E}$ by

$$P = \varepsilon_o(\varepsilon_r - 1)\mathcal{E}$$

which means that the depolarization field is

$$\mathcal{E}_{\text{dep}} = \frac{1}{\varepsilon_o} P \qquad \text{[7.66]}$$

Depolarizing field in a dielectric plate

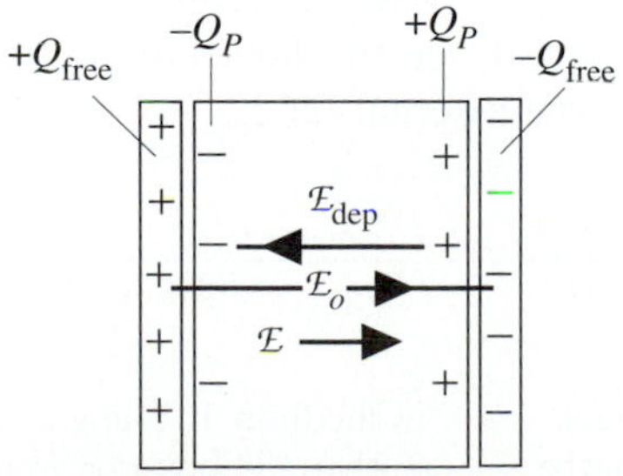

Figure 7.47   The field inside the dielectric can be considered to be the sum of the field due to the free charges (Q_{free}) and a field due to the polarization of the dielectric, called the depolarization field.

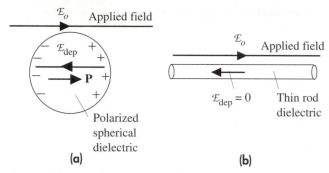

Figure 7.48
(a) Polarization and the depolarizing field in a spherical-shaped dielectric placed in an applied field.
(b) Depolarization field in a thin rod placed in an applied field is nearly zero.

As we expected, the depolarizing field is proportional to the polarization, P. We should emphasize that $\mathcal{E}_{\text{dep}}$ is in the *opposite direction* to $\mathcal{E}$ and P and Equation 7.65 is for magnitudes only. If we write it as a vector equation, then we must introduce a negative sign to give $\mathcal{E}_{\text{dep}}$ a direction opposite to that of P. Moreover, the above relationship is special to the dielectric plate geometry in Figure 7.47. In general, the depolarizing field is still proportional to the polarization, as in Equation 7.66, but it is given by

Depolarizing field in a dielectric

$$\mathcal{E}_{\text{dep}} = \frac{N_{\text{dep}}}{\varepsilon_o}P \qquad [7.67]$$

where N_{dep} is a numerical factor called the **depolarization factor.** It takes into account the shape of the dielectric and the variation in the polarization within the medium. For a dielectric plate placed perpendicularly to an external field, $N_{\text{dep}} = 1$, as we found above in Equation 7.66. For the spherical dielectric medium as in Figure 7.48a, $N_{\text{dep}} = 1/3$. For a long thin dielectric rod placed with its axis along the applied field, as in Figure 7.48b, $N_{\text{dep}} \approx 0$ and becomes exactly zero as the diameter shrinks to zero. N_{dep} is always between 0 and 1. If we know N_{dep}, we can determine the field inside the dielectric, for example, in a small spherical cavity within an insulation given the external field.

IMPORTANT TERMS

Boundary conditions relate the normal and tangential components of the electric field next to the boundary. The tangential component must be continuous through the boundary. Suppose that $\mathcal{E}_{n1}$ is the normal component of the field in medium 1 at the boundary and ε_{r1} is the relative permittivity

in medium 1. Using a similar notation for medium 2, then the boundary condition is $\varepsilon_{r1}\mathcal{E}_{n1} = \varepsilon_{r2}\mathcal{E}_{n2}$.

Clausius–Mossotti equation relates the dielectric constant (ε_r), a macroscopic property, to the polarizability (α), a microscopic property.

Complex relative permittivity $(\varepsilon_r' + j\varepsilon_r'')$ has a real part (ε_r') that determines the charge storage ability and an imaginary part (ε_r'') that determines the energy losses in the material as a result of the polarization mechanism. The real part determines the capacitance through $C = \varepsilon_o \varepsilon_r' A/d$ and the imaginary part determines the electric power dissipation per unit volume as heat by $\mathcal{E}^2 \omega \varepsilon_o \varepsilon_r''$.

Corona discharge is a local discharge in a gaseous atmosphere where the field is suffi-ciently high to cause dielectric breakdown, for example, by avalanche ionization.

Curie temperature T_C is the temperature above which ferroelectricity disappears, that is, the spontaneous polarization of the crystal is lost.

Debye equations attempt to describe the frequency response of the complex relative permittivity, $\varepsilon_r' + j\varepsilon_r''$, of a dipolar medium through the use of a single relaxation time τ to describe the sluggishness of the dipoles driven by the external ac field.

Dielectric is a material in which energy can be stored by the polarization of the molecules. It is a material that increases the capacitance or charge storage ability of a capacitor. Ideally, it is a nonconductor of electrical charge so that an applied field does not cause a flow of charge but instead relative displacement of opposite charges and hence polarization of the medium.

Dielectric loss is the electrical energy lost as heat in the polarization process in the presence of an applied ac field. The energy is absorbed from the ac voltage and converted to heat during the polarization of the molecules. It should not be confused with conduction loss $\sigma \mathcal{E}^2$ or V^2/R.

Dielectric strength is the maximum field $(\mathcal{E}_{br})$ that can be sustained in a dielectric beyond which dielectric breakdown ensues; that is, there is a large conduction current through the dielectric shorting the plates.

Dipolar (orientational) polarization arises when randomly oriented polar molecules in a dielectric are rotated and aligned by the application of a field so as to give rise to a net average dipole moment per molecule. In the absence of the field, the dipoles (polar molecules) are randomly oriented and there is no average dipole moment per molecule. In the presence of the field, the dipoles are rotated, some partially and some fully, to align with the field and hence give rise to a net dipole moment per molecule.

Dipolar relaxation equation describes the time response of the induced dipole moment per molecule in a dipolar material in the presence of a time-dependent applied field. The response of the dipoles depends on their relaxation time, which is the mean time required to dissipate the stored electrostatic energy in the dipole alignment to heat through lattice vibrations or molecular collisions.

Dipole relaxation (dielectric resonance) occurs when the frequency of the applied ac field is such that there is maximum energy transfer from the ac voltage source to heat in the dielectric through the alternate polarization and depolarization of the molecules by the ac field. The stored electrostatic energy is dissipated through molecular collisions and lattice vibrations (in solids). The peak occurs when the angular frequency of the ac field is the reciprocal of the relaxation time.

Electric dipole moment exists when a positive charge $+Q$ is separated from a negative charge $-Q$. Even though the net charge is zero, there is nonetheless an electric dipole moment, $\mathbf{p}$, given by $\mathbf{p} = Q\mathbf{x}$ where $\mathbf{x}$ is the distance vector from $-Q$ to $+Q$. Just as two charges exert a coulomb force on each other, two dipoles also exert a force on each other that depends on the magnitudes of the dipoles, their separation, and orientation.

Electric susceptibility (χ_e) is a material quantity that measures the extent of polarization in the material per unit field. It relates the amount of polarization P at a point in the dielectric to the field $\mathcal{E}$ at that point via $P = \chi_e \varepsilon_o \mathcal{E}$. If ε_r is the relative permittivity, then $\chi_e = \varepsilon_r - 1$. Vacuum has no electric susceptibility.

Electromechanical breakdown and electrofracture is a breakdown process that directly or indirectly involves electric field–induced mechanical weakening, for example, crack propagation, or mechanical deformation that eventually leads to dielectric breakdown.

Electronic bond polarization is the displacement of valence electrons in the bonds in covalent solids (e.g., Ge, Si). It is a collective displacement of the electrons in the bonds with respect to the positive nuclei.

Electronic polarization is the displacement of the electron cloud of an atom with respect to the positive nucleus. Its contribution to the relative permittivity of a solid is usually small.

External discharges are discharges or shorting currents over the surface of the insulator when the conductance of the surface increases as a result of surface contamination, for example, excessive moisture, deposition of pollutants, dirt, dust, salt spraying, and so forth. Eventually the contaminated surface develops sufficient conductance to allow discharge between the electrodes at a field below the normal breakdown strength of the insulator. Dielectric breakdown over the surface of an insulation is termed **surface tracking.**

Ferroelectricity is the occurrence of spontaneous polarization in certain crystals such as barium titanate ($BaTiO_3$). Ferroelectric crystals have a permanent polarization **P** as a result of spontaneous polarization. The direction of **P** can be defined by the application of an external field.

Gauss's law is a fundamental law of physics that relates the surface integral of the electric field over a closed (hypothetical) surface to the sum of all the charges enclosed within the surface. If $\mathcal{E}_n$ is the field normal to a small surface area dA and Q_{total} is the enclosed total charge, then over the whole closed surface $\varepsilon_o \oint \mathcal{E}_n \, dA = Q_{\text{total}}$.

Induced polarization is the polarization of a molecule as a result of its placement in an electric field. The induced polarization is along the direction of the field. If the molecule is already polar, then induced polarization is the additional polarization that arises due to the applied field alone and it is directed along the field.

Insulation aging is a term used to describe the physical and chemical deterioration in the properties of the insulation so that its dielectric breakdown characteristics worsen with time. Aging therefore determines the useful life of the insulation.

Interfacial polarization occurs whenever there is an accumulation of charge at an interface between two materials or between two regions within a material. Grain boundaries and electrodes are regions where charges generally accumulate and give rise to this type of polarization.

Internal discharges are partial discharges that take place in microstructural voids, cracks, or pores within the dielectric where the gas atmosphere (usually air) has lower dielectric strength. A porous ceramic, for example, would experience partial discharges if the field is sufficiently large. Initially, the pore size (or the number of pores) may be small and the partial discharge insignificant, but with time the partial discharge erodes the internal surfaces of the void. Eventually (and usually) an *electrical tree* type of discharge develops from a partial discharge that has been eroding the dielectric. The erosion of the dielectric by the partial discharge propagates like a branching tree. The "tree branches" are erosion channels, filaments of various sizes, in which gaseous discharge takes place and forms a conducting channel during operation.

Intrinsic breakdown or electronic breakdown commonly involves the avalanche multiplication of electrons (and holes in solids) by impact ionization in the presence of high electric fields. The large number of free carriers generated by the avalanche of impact ionizations leads to a runaway current between the electrodes and hence to insulation breakdown.

Ionic polarization is the relative displacement of oppositely charged ions in an ionic crystal that results in the polarization of the whole material. Typically, ionic polarization is important in ionic crystals below the infrared wavelengths.

Local field ($\mathcal{E}_{\text{loc}}$) is the true field experienced by a molecule in a dielectric that arises from the free charges on the plates and all the induced dipoles surrounding the molecule. The true field at a molecule is not simply the applied field (V/d) because of the field of the neighboring induced dipoles.

Loss tangent or tan δ is the ratio of the imaginary part to the real part, $\varepsilon_r'' / \varepsilon_r'$. The angle δ is the phase angle between the capacitive current and the total current. If there is no dielectric loss, then the two currents are the same and $\delta = 0$.

Partial discharge occurs when only a local region of the dielectric is exhibiting discharge, so that the discharge does not directly connect the two electrodes.

Piezoelectric material has a noncentrosymmetric crystal structure that leads to the generation of a polarization vector P, or charges on the crystal surfaces, upon the application of a mechanical stress. When strained, a piezoelectric crystal develops an internal field and therefore exhibits a voltage difference between two of its faces.

PLZT, lead lanthanum zirconate titanate, is a PZT-type material with lanthanum occupying the Pb site.

Polarizability (α) is the ability of an atom or molecule to become polarized in the presence of an electric field. It is induced polarization in the molecule per unit field along the field.

Polarization is the separation of positive and negative charges in a system so that there is a net electric dipole moment per unit volume.

Polarization vector (P) measures the extent of polarization in a unit volume of dielectric matter. It is the vector sum of dielectric dipoles per unit volume. If **p** is the average dipole moment per molecule and n is the number of molecules per unit volume, then $\mathbf{P} = n\mathbf{p}$. In a polarized dielectric matter (e.g., in an electric field), the bound surface charge density, σ_p, due to polarization is equal to the normal component of **P** at that point, $\sigma_p = P_{normal}$.

Poling is the application of a temporary electric field to a piezoelectric (or ferroelectric) material, generally at an elevated temperature, to align the polarizations of various grains and thereby develop piezoelectric behavior.

Pyroelectric material is a polar dielectric (such as barium titanate) in which a temperature change ΔT induces a proportional change ΔP in the polarization, that is, $\Delta P = p\,\Delta T$, where p is the pyroelectric coefficient of the crystal.

PZT is a general acronym for the lead zirconate titanate ($PbZrO_3$-$PbTiO_3$ or $PbTi_{0.48}Zr_{0.52}O_3$) family of crystals.

Q-factor or quality factor for an impedance is the ratio of its reactance to its resistance. The Q-factor of a capacitor is X_c/R_p where $X_c = 1/\omega C$ and R_p is the equivalent parallel resistance that represents the dielectric and conduction losses. The Q-factor of a resonant circuit measures the circuit's peak response at the resonant frequency and also its band-width. The greater the Q, the higher the peak response and the narrower the bandwidth. For a series RLC resonant circuit,

$$Q = \frac{\omega_o L}{R} = \frac{1}{\omega_o CR}$$

where ω_o is the resonant angular frequency, $\omega_o = 1/\sqrt{LC}$. The width of the resonant response curve between half-power points is $\Delta\omega = \omega_o/Q$.

Relative permittivity (ε_r) or the dielectric constant of a dielectric is the fractional increase in the stored charge per unit voltage on the capacitor plates due to the presence of the dielectric between the plates (the whole space between the plates is assumed to be filled). Alternatively, we can define it as the fractional increase in the capacitance of a capacitor when the insulation between the plates is changed from vacuum to a dielectric material, keeping the geometry the same.

Relaxation time (τ) is a characteristic time that determines the sluggishness of the dipole response to an applied field. It is the mean time for the dipole to lose its alignment with the field due to its random interactions with the other molecules through molecular collisions, lattice vibrations, and so forth.

Surface tracking is an *external dielectric breakdown* that occurs over the surface of the insulation.

Temperature coefficient of capacitance (TCC) is the fractional change in the capacitance per unit temperature change.

Thermal breakdown is a breakdown process that involves thermal runaway, which leads to a runaway current or discharge between the electrodes. If the heat generated by dielectric loss, due to ε_r'', or joule heating, due to finite σ, cannot be removed sufficiently rapidly, then the temperature of the dielectric rises, which increases the conductivity and the dielectric loss. The increases in ε_r'' and σ

lead to more heat generation and a further rise in the temperature, so that thermal runaway ensues, followed by either a large shorting current or local thermal decomposition of the insulation accompanied by a partial discharge in this region.

Transducer is a device that converts electrical energy into another form of usable energy or vice versa, for example, piezoelectric transducers convert electrical energy to mechanical energy and vice versa.

QUESTIONS AND PROBLEMS

7.1   Relative permittivity and polarizability

a.   Show that the local field is given by

$$\mathcal{E}_{\text{loc}} = \mathcal{E}\left(\frac{\varepsilon_r + 2}{3}\right)$$

b.   Amorphous selenium (a-Se) is a high resistivity semiconductor that has a density of approximately 4.3 g cm^{-3} and an atomic number and mass of 34 and 78.96. Its relative permittivity at 1 kHz has been measured to be 6.7. Calculate the relative magnitude of the local field in a-Se. Calculate the polarizability per Se atom in the structure. What type of polarization is this? How will ε_r depend on the frequency?

c.   If the electronic polarizability of an isolated atom is given by

$$\alpha_e \approx 4\pi \varepsilon_o r_o^3$$

where r_o is the radius of the atom, then calculate the electronic polarizability of an isolated Se atom, which has $r_o = 0.12$ nm, and compare your result with that for an atom in a-Se. Why is there a difference?

7.2   Relative permittivity, bond strength, bandgap and refractive index
Diamond, silicon, and germanium are covalent solids with the same crystal structure. Their relative permittivities are shown in Table 7.9.

a.   Explain why ε_r increases from diamond to germanium.

b.   Calculate the polarizability per atom in each crystal and then plot polarizability against the elastic modulus Y (Young's modulus). Should there be a correlation?

Table 7.9   Properties of diamond, Si, and Ge

	ε_r	M_{at}	Density (g cm^{-3})	α_e	Y (GPa)	E_g (eV)	n
Diamond	5.8	12	3.52		827	5.5	2.42
Si	11.9	28.09	2.33		190	1.12	3.45
Ge	16	72.61	5.32		75.8	0.67	4.09

c. Plot the polarizability from part *b* against the bandgap energy, E_g. Is there a relationship?

d. Show that the refractive index *n* is $\sqrt{\varepsilon_r}$. When does this relationship hold and when does it fail?

e. Would your conclusions apply to ionic crystals such as NaCl?

7.3 Ionic and electronic polarization   Consider a CsCl crystal that has the CsCl unit cell crystal structure (one Cs^+–Cl^- pair per unit cell) with a lattice parameter (*a*) of 0.412 nm. The electronic polarizability of Cs^+ and Cl^- ions are 3.35×10^{-40} F m² and 3.40×10^{-40} F m² respectively, and the mean ionic polarizability per ion pair is 6×10^{-40} F m². What is the low frequency dielectric constant and that at optical frequencies?

7.4 Electronic polarizability and KCl   KCl has the NaCl crystal structure with a lattice parameter of 0.629 nm. Calculate the relative permittivity of a KCl crystal at optical frequencies given that the electronic polarizability of K^+ is 1.264×10^{-40} F m² and that of Cl^- is 3.408×10^{-40} F m². How does this compare with the measured value of 2.19?

7.5 Equivalent circuit of a polyester capacitor   Consider a 1 nF polyester capacitor that has a polymer (PET) film thickness of 1 μm. Calculate the equivalent circuit of this capacitor at 50 °C and at 120 °C for operation at 1 kHz. What is your conclusion?

7.6 Dielectric loss per unit capacitance   Consider the three dielectric materials listed in Table 7.10 with the real and imaginary dielectric constants, ε_r' and ε_r''. At a given voltage, which dielectric will have the lowest power dissipation per unit capacitance at 1 kHz and at an operating temperature of 50 °C? Is this also true at 120 °C?

***7.7 TCC of a polyester capacitor**   Consider the parallel plate capacitor equation

$$C = \frac{\varepsilon_o \varepsilon_r x y}{z}$$

where ε_r is the relative permittivity (or ε_r'), *x* and *y* are the side lengths of the dielectric so that *xy* is the area *A*, and *z* is the thickness of the dielectric.

Table 7.10   Dielectric properties of three insulators at 1 kHz

Material	T = 50 °C		T = 120 °C	
	ε_r'	ε_r''	ε_r'	ε_r''
Polycarbonate	2.47	0.003	2.535	0.003
PET	2.58	0.003	2.75	0.027
PEEK	2.24	0.003	2.25	0.003

| Data taken using a DEA by Nomura and Kasap (1995).

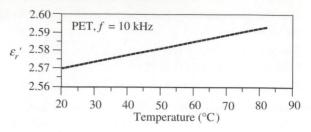

Figure 7.49   Temperature dependence of ε'_r at 10 kHz

The quantities ε_r, x, y, and z change with temperature. By differentiating this equation with respect to temperature, show that the temperature coefficient of capacitance (TCC) is

$$\text{TCC} = \frac{1}{C}\frac{dC}{dT} = \frac{1}{\varepsilon_r}\frac{d\varepsilon_r}{dT} + \lambda$$

where λ is the linear expansion coefficient (thermal coefficient of linear expansion) defined by

$$\lambda = \frac{1}{L}\frac{dL}{dT}$$

where L stands for any length of the material (x, y, or z). Assume that the dielectric is isotropic, λ is the same in all directions. Using ε'_r versus T behavior in Figure 7.49 and taking $\lambda = 50 \times 10^{-6}$ K^{-1} as a typical value for polymers, predict the TCC at room temperature and at 10 kHz.

7.8   **Dielectric breakdown of gases and Paschen curves**   Dielectric breakdown in gases typically involves the avalanche ionization of the gas molecules by energetic electrons accelerated by the applied field. The mean free path between collisions must be sufficiently long to allow the electrons to gain sufficient energy from the field to impact-ionize the gas molecules. The breakdown voltage, V_{br}, between two electrodes depends on the distance, d, between the electrodes as well as the gas pressure, P, as shown in Figure 7.50. V_{br} versus Pd plots are called **Paschen curves.** We consider gaseous insulation, air and SF$_6$, in an HV switch.

a.   What is the breakdown voltage between two electrodes of a switch separated by a 5 mm gap with air and SF$_6$ as gaseous insulation at 1 atm?

b.   What are the breakdown voltages in the two cases when the pressure is 10 times greater? What is your conclusion?

c.   At what pressure is the breakdown voltage minimum?

d.   What air gap spacing, d, at 1 atm gives the minimum breakdown voltage?

e.   What would be the reasons for preferring gaseous insulation over liquid or solid insulation?

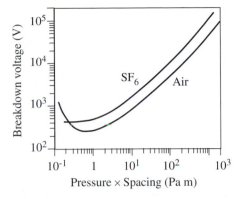

Figure 7.50   Breakdown voltage versus
(pressure × electrode spacing) (Paschen curves)

Table 7.11   Comparison of dielectric properties at 60 Hz (typical values)

	Polymer Film PET	**Ceramic TiO$_2$**	**High-K Ceramic (BaTiO$_3$ based)**
Name	Polyester	Polycrystalline titania	X7R
ε_r'	3.2	90	1800
tan δ	5×10^{-3}	4×10^{-4}	5×10^{-2}
$\mathcal{E}_{br}$(kV cm^{-1})	150	50	100
Typical minimum thickness	1–2 μm	10 μm	10 μm

*7.9   Capacitor design   Consider a nonpolarized 100 nF capacitor design at 60 Hz operation. Note that there are three candidate dielectrics, as listed in Table 7.11.

a.   Calculate the volume of the 100 nF capacitor for each dielectric, given that they are to be used under low voltages and each dielectric has its minimum fabrication thickness. Which one has the smallest volume?

b.   How is the volume affected if the capacitor is to be used at a 500 V application and the maximum field in the dielectric must be a factor of 2 less than the dielectric strength? Which one has the smallest volume?

c.   At a 500 V application, what is the power dissipated in each capacitor at 60 Hz operation? Which one has the lowest dissipation?

*7.10   Dielectric breakdown in a coaxial cable   Consider a coaxial underwater high voltage cable as in Figure 7.51a. The current flowing through the inner conductor generates heat, which has to flow through the dielectric

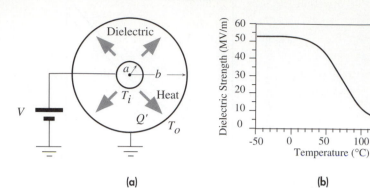

Figure 7.51

(a) The joule heat generated in the core conductor flows outward radially through the dielectric material.
(b) Typical temperature dependence of the dielectric strength of a polyethylene-based polymeric insulation.

insulation to the outer conductor where it will be carried away by conduction and convection. We will assume that steady state has been reached and the inner conductor is carrying a dc current I. Heat generated per unit second, $Q' = dQ/dt$, by joule heating of the inner conductor is

$$Q' = RI^2 = \frac{\rho L I^2}{\pi a^2} \qquad \text{[7.68]}$$

where ρ is the resistivity, a the radius of the conductor, and L the cable length.

This heat flows radially out from the inner conductor through the dielectric insulator to the outer conductor, then to the ambient. This heat flow is by thermal conduction through the dielectric. The rate of heat flow Q' depends on the temperature difference, $T_i - T_o$, between the inner and outer conductors; on the sample geometry (a, b, and L); and on the thermal conductivity κ of the dielectric. From elementary thermal conduction theory, this is given by

$$Q' = (T_i - T_o)\frac{2\pi\kappa L}{\ln\left(\dfrac{b}{a}\right)} \qquad \text{[7.69]}$$

The inner core temperature, T_i, rises until, in the steady state, the rate of joule heat generation by the electric current in Equation 7.68 is just removed by the rate of thermal conduction through the dielectric insulation, given by Equation 7.69.

a.   Show that the inner conductor temperature is

$$T_i = T_o + \frac{\rho I^2}{2\pi^2 a^2 \kappa} \ln\left(\frac{b}{a}\right) \qquad \text{[7.70]}$$

b.   The breakdown occurs at the maximum field point, which is at $r = a$, just outside the inner conductor, and is given by (see Example 7.10)

$$\mathcal{E}_{max} = \frac{V}{a \ln\left(\dfrac{a}{b}\right)} \qquad \text{[7.71]}$$

The dielectric breakdown occurs when $\mathcal{E}_{max}$ reaches the dielectric strength $\mathcal{E}_{br}$. However the dielectric strength $\mathcal{E}_{br}$ for many polymeric insulation materials depends on the temperature, and generally it decreases with temperature, as shown for a typical example in Figure 7.51b. If the load current, I, increases, then more heat, Q', is generated per second and this leads to a higher inner core temperature, T_i, by virtue of Equation 7.70. The increase in T_i with I eventually lowers $\mathcal{E}_{br}$ so much that it becomes equal to $\mathcal{E}_{max}$ and the insulation breaks down (thermal breakdown). Suppose that a certain coaxial cable has an aluminum inner conductor of diameter 10 mm and resistivity 27 nΩ m. The insulation is 3 mm thick and is a polyethylene-based polymer whose long-term dc dielectric strength is shown in Figure 7.51b. Suppose that the cable is carrying a voltage of 40 kV and the outer shield temperature is the ambient temperature, 25 °C. Given that the thermal conductivity of the polymer is about 0.3 W K^{-1} m^{-1}, at what dc current will the cable fail? See Table 2.1, p. 113 for ρ_0 and α_0.

7.11   Piezoelectricity   Consider a quartz crystal and a PZT ceramic filter both designed for operation at $f_s = 1$ MHz. What is the bandwidth of each? Given Young's modulus (Y), density (ρ) for each, and that the filter is a disk with electrodes and is oscillating radially, what is the diameter of the disk for each material? For quartz, $Y = 80$ GPa and $\rho = 2.65$ g cm^{-3}. For PZT, $Y = 70$ GPa and $\rho = 7.7$ g cm^{-3}. Assume that the velocity of mechanical oscillations in the crystal is $v = \sqrt{Y/\rho}$ and the wavelength $\lambda = v/f_s$. Consider only the fundamental mode ($n = 1$).

7.12   Piezoelectricity and the piezoelectric bender

a.   Consider using a piezoelectric material in an application as a mechanical positioner where the displacements are expected to be small (as in a scanning tunneling microscope). For the piezoelectric plate shown in Figure 7.52a, we will take $L = 20$ mm, $W = 10$ mm, and D (thickness) $= 0.25$ mm. Under an applied voltage of V, the plate changes length, width, and thickness according to the piezoelectric coefficients d_{ij}, relating the applied field along i to the resulting strain along j.

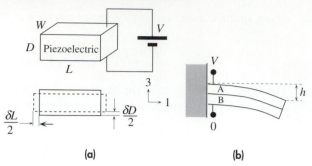

Figure 7.52

(a) A mechanical positioner using a piezoelectric plate under an applied voltage of V.
(b) A cantilever-type piezoelectric bender. An applied voltage bends the cantilever.

Suppose we define direction 3 along the thickness D and direction 1 along the length L, as shown in the figure. Show that the changes in the thickness and length are

$$\delta D = d_{33}\, V$$

$$\delta L = \left(\frac{L}{D}\right) d_{31}\, V$$

Given $d_{33} \approx 500 \times 10^{-12}$ m V^{-1} and $d_{31} \approx -250 \times 10^{-12}$ m V^{-1}, calculate the changes in the length and thickness for an applied voltage of 100 V. What is your conclusion?

b. Consider two oppositely poled and joined ceramic plates, A and B, forming a bimorph, as shown in Figure 7.52b. This piezoelectric bimorph is mounted as a cantilever: one end is fixed and the other end is free to move. Oppositely poled means that the electric field elongates A and contracts B, and the two relative motions *bend* the plate. The displacement h of the tip of the cantilever is given by

$$h = \frac{3}{2} d_{31} \left(\frac{L}{D}\right)^{2} V$$

What is the deflection of the cantilever for an applied voltage of 100 V? What is your conclusion?

7.13 Piezoelectricity   The wavelength, λ, of mechanical oscillations in a piezoelectric slab satisfies

$$n\left(\frac{1}{2}\lambda\right) = L$$

where n is an integer, L is the length of the slab along which mechanical oscillations are set up, and the wavelength λ is determined by the frequency

Table 7.12   Properties of PZT and PVDF

	ε_r'	Pyroelectric Coefficient ($\times 10^{-6}$ C m^{-2} K^{-1})	Density (g cm^{-3})	Heat Capacity (J K^{-1} g^{-1})
PZT	290	380	7.7	0.3
PVDF	12	27	1.76	1.3

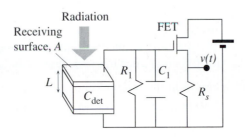

Figure 7.53   A pyroelectric detector with an FET voltage follower circuit.

f and velocity v of the waves. The ultrasonic wave velocity, v, depends on Young's modulus as

$$v = \left[\frac{Y}{\rho}\right]^{1/2}$$

where ρ is the density. For quartz, $Y = 80$ GPa and $\rho = 2.65$ g cm^{-3}. Considering the fundamental mode ($n = 1$), what are practical dimensions for crystal oscillators operating at 1 kHz and 1 MHz?

7.14   Pyroelectric detectors   Consider two different radiation detectors using PZT and PVDF as pyroelectric materials whose properties are summarized in Table 7.12. The receiving area is 4 mm^2. The thicknesses of the PZT ceramic and the PVDF polymer film are 0.1 mm and 0.005 mm respectively. In both cases the incident radiation is chopped periodically to allow the radiation to pass for a duration of 0.05 second.

a.   Calculate the magnitude of the output voltage for each detector if both receive a radiation of intensity 10 μW cm^{-2}. What is the corresponding current in the circuit? In practice, what would limit the magnitude of the output voltage?

b.   What is the minimum detectable radiation intensity if the minimum detectable signal voltage is 10 nV?

7.15   Spark generator design   Design a PZT piezoelectric spark generator that provides a 50 μJ spark in an air gap of 0.5 mm from a force of 70 N. The design will need to specify the dimensions of the crystal and also ε_r.

*7.16   Pyroelectric detectors   Consider a typical pyroelectric radiation detector circuit as shown in Figure 7.53. The FET circuit acts as a voltage follower (source follower). The resistance R_1 represents the input resistance of the

FET in parallel with a bias resistance that is usually inserted between the gate and source. C_1 is the overall input capacitance of the FET including any stray capacitance but excluding the capacitance of the pyroelectric detector. Suppose that the incident radiation intensity is constant and equal to I. Emissivity η of a surface characterizes what fraction of the incident radiation that is absorbed. ηI is the energy absorbed per unit area per unit time. Some of the absorbed energy will increase the temperature of the detector and some of it will be lost to surroundings by thermal conduction and convection. Let the detector receiving area be A, thickness be L, density be ρ, and specific heat capacity (heat capacity per unit mass) be c. The heat losses will be proportional to the temperature difference between the detector temperature, T, and the ambient temperature, T_o, as well as the surface area A (much greater than L). Energy balance requires that

> Rate of increase in the internal energy (heat content) of the detector
> = Rate of energy absorption − Rate of heat losses

that is,

$$(AL\rho)c\frac{dT}{dt} = A\eta I - KA(T - T_o)$$

where K is a constant of proportionality that represents the heat losses and hence depends on the thermal conductivity κ. If the heat loss involves pure thermal condition from the detector surface to the detector base (detector mount), then $K = \kappa/L$. In practice, this is generally not the case and $K = \kappa/L$ is an oversimplification.

a. Show that the temperature of the detector rises exponentially as

$$T = T_o + \frac{\eta I}{K}\left[1 - \exp\left(-\frac{t}{\tau_{th}}\right)\right]$$

where τ_{th} is a **thermal time constant** defined by $\tau_{th} = L\rho c/K$. Further show that for very small K, the above equation simplifies to

$$T = T_o + \frac{\eta I}{L\rho c}t$$

b. Show that temperature change dT in time dt leads to a pyroelectric current, i_p, given by

$$i_p = Ap\frac{dT}{dt} = \frac{Ap\eta I}{L\rho c}\exp\left(-\frac{t}{\tau_{th}}\right)$$

where p is the pyroelectric coefficient. What is the initial current?

c. The voltage across the FET and hence the output voltage $v(t)$ is given by

$$v(t) = V_o\left[\exp\left(-\frac{t}{\tau_{th}}\right) - \exp\left(-\frac{t}{\tau_{el}}\right)\right]$$

where V_o is a constant and τ_{el} is the **electrical time constant** given by $R_1 C_t$, where C_t, total capacitance, is $(C_1 + C_{det})$, where C_{det} is the capacitance of the detector. Consider a particular PZT pyroelectric detector with an area 1 mm^2, and thickness 0.05 mm. Suppose that this PZT has $\varepsilon_r = 250$, $\rho = 7.7$ g cm^{-3}, $c = 0.3$ J K^{-1} g^{-1}, and $\kappa = 1.5$ W K^{-1} m^{-1}. The detector is connected to an FET circuit that has $R_1 = 10$ MΩ and $C_1 = 3$ pF. Taking the thermal conduction loss constant K as κ/L, and $\eta = 1$, calculate τ_{th} and τ_{el}. Sketch schematically the output voltage. What is your conclusion?

8

Magnetic Properties and Superconductivity

Many electrical engineering devices such as inductors, transformers, rotating machines, and ferrite antennas are based on utilizing the magnetic properties of materials. There are many instances where permanent magnets are also used either on their own or as part of a device such as a rotating machine or a loud speaker. The majority of engineering devices make use of the ferromagnetic and ferrimagnetic properties, which are therefore treated in much more detail than other magnetic properties such as diamagnetism and paramagnetism. Although superconductivity involves the vanishing of the resistivity of a conductor at low temperatures and is normally explained within quantum mechanics, we treat the subject in this chapter because all superconductors are perfect diamagnets and, further, they have present or potential uses that involve magnetic fields. The advent of high-T_c superconductivity, discovered in 1986 by Georg Bednorz and Alex Müller at IBM Research Laboratories in Zürich, is undoubtedly one of the most significant discoveries over the last 50 years, as popularized in various magazines. High-T_c superconductors are already finding applications in such devices as superconducting solenoids, sensitive magnetometers, and high-Q microwave filters.

8.1 MAGNETIZATION OF MATTER

8.1.1 Magnetic Dipole Moment

Magnetic properties of materials involve concepts based on the magnetic dipole moment. Consider a current loop, as shown in Figure 8.1, where the circulating current is I. This may, for example, be a coil carrying a current. For simplicity we will assume that the current loop lies within a single plane. The area enclosed by the current is A. Suppose that $\mathbf{u}_n$ is a unit vector coming out from the area A. The

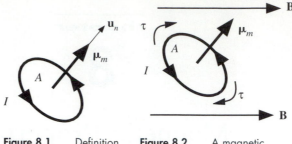

Figure 8.1   Definition of a magnetic dipole moment

Figure 8.2   A magnetic dipole moment in an external field experiences a torque.

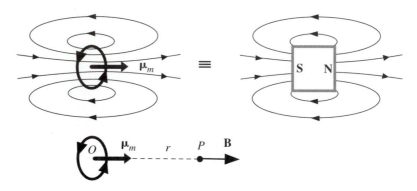

Figure 8.3   A magnetic dipole moment creates a magnetic field just like a bar magnet. The field **B** depends on $\boldsymbol{\mu}_m$.

direction of $\mathbf{u}_n$ is such that looking along it, the current circulates clockwise. Then the **magnetic dipole moment** or simply the **magnetic moment** $\boldsymbol{\mu}_m$, is defined by[1]

Definition of magnetic moment

$$\boldsymbol{\mu}_m = IA\mathbf{u}_n \qquad [8.1]$$

When a magnetic moment is placed in a magnetic field, it experiences a torque that tries to rotate the magnetic moment to align its axis with the magnetic field, as depicted in Figure 8.2. Moreover, since a magnetic moment is a current loop, it gives rise to a magnetic field **B** around it, as shown in Figure 8.3, which is similar to the magnetic field around a bar magnet. We can find the field **B** from the current I and its geometry, which are treated in various physics textbooks. For example, the field **B** at a point P at a distance r along the axis of the coil from the center, as shown in Figure 8.3, is directly proportional to the magnitude of the magnetic moment but inversely proportional to r^3, that is $\mathbf{B} \propto \boldsymbol{\mu}_m/r^3$.

---

[1] The symbol $\boldsymbol{\mu}$ for the magnetic dipole moment should not be confused with the permeability. Absolute and relative permeabilities will be denoted by μ_o and μ_r.

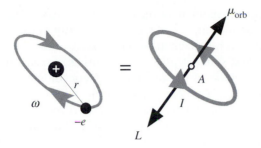

Figure 8.4   An orbiting electron is equivalent to a
magnetic dipole moment μ_{orb}.

8.1.2   Atomic Magnetic Moments

An orbiting electron in an atom behaves much like a current loop and has a
magnetic dipole moment associated with it, called the **orbital magnetic moment**
(μ_{orb}), as illustrated in Figure 8.4. If ω is the angular frequency of the electron, then
the current I due to the orbiting electron is

$$I = \text{Charge flowing per unit time} = \frac{-e}{\text{Period}} = \frac{-e\omega}{2\pi}$$

If r is the radius of the orbit, then the magnetic dipole moment is

$$\mu_{orb} = I(\pi r^2) = \frac{-e\omega r^2}{2}$$

But the velocity, v, of the electron is ωr and its orbital angular momentum is

$$L = (m_e v)r = m_e \omega r^2$$

Using this in μ_{orb}, we get

$$\mu_{orb} = -\frac{e}{2m_e} L \qquad \text{[8.2]}$$

*Orbital magnetic
moment of the
electron*

We see that the magnetic moment is proportional to the orbital angular momentum through a factor that has the charge to mass ratio of the electron. The
numerical factor, in this case $e/2m_e$, relating the angular momentum to the magnetic moment, is called the **gyromagnetic ratio**. The negative sign in Equation 8.2
indicates that μ_{orb} is in the opposite direction to L and is due to the negative charge
of the electron.

The electron also has an intrinsic angular momentum, S, that is, spin. The spin
of the electron has a **spin magnetic moment,** denoted by μ_{spin}, but the relationship
between μ_{spin} and S is not the same as that in Equation 8.2. The gyromagnetic ratio
is a factor of 2 greater,

$$\mu_{spin} = -\frac{e}{m_e} S \qquad \text{[8.3]}$$

*Spin magnetic
moment of the
electron*

The overall magnetic moment of the electron consists of $\boldsymbol{\mu}_{\text{orb}}$ and $\boldsymbol{\mu}_{\text{spin}}$ appropriately added. We cannot simply add them numerically as they are vector quantities. Furthermore, the overall magnetic moment $\boldsymbol{\mu}_{\text{atom}}$ of the atom itself depends on the orbital motions and spins of *all* the electrons. Electrons in closed subshells, however, do not contribute to the overall magnetic moment because for every electron with a given **L** (or **S**), there is another one with an opposite **L** (or **S**). The reason is that the direction of **L** is space quantized by m_ℓ and all negative and positive values of m_ℓ are occupied in a closed shell. Similarly, there are as many electrons spinning up as there are spinning down, so there is no net electron spin in a closed shell and no net $\boldsymbol{\mu}_{\text{spin}}$. Thus, only **unfilled subshells** contribute to the overall magnetic moment of an atom.

Consider an atom that has closed inner shells and a single electron in an *s* orbital ($\ell = 0$). This means that the orbital magnetic moment is zero and the atom has a magnetic moment due to the spin of the electron alone, $\boldsymbol{\mu}_{\text{atom}} = \boldsymbol{\mu}_{\text{spin}}$. In the presence of an external magnetic field along the *z* direction, the magnetic moment cannot simply rotate and align with the field because quantum mechanics requires the spin angular momentum to be space quantized, that is, S_z (the component of **S** along *z*) must be $m_s\hbar$ where $m_s = \pm 1/2$ is the spin magnetic quantum number. The torque experienced by the spinning electron causes the spin magnetic moment to precess about the external magnetic field, as shown in Figure 8.5. This precession is such that $S_z = -(1/2)\hbar$ and leads to an average magnetic moment μ_z along the field given by Equation 8.3 with S_z, that is,

$$\mu_z = -\frac{e}{m_e} S_z = -\frac{e}{m_e}(m_s\hbar) = \frac{e\hbar}{2m_e} = \beta \qquad [8.4]$$

The quantity $\beta = e\hbar/2m_e$ is called the **Bohr magneton** and has the value 9.27×10^{-24} A m^2 or J T^{-1}.

Thus, the spin of a single electron has a magnetic moment of one Bohr magneton along the field.

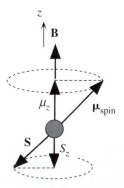

Figure 8.5   The spin magnetic moment precesses about an external magnetic field along *z* and has a value μ_z along *z*.

8.1.3   Magnetization Vector M

Consider a tightly wound long solenoid, ideally infinitely long, with free space (or vacuum) as the medium inside the solenoid, as shown in Figure 8.6a. The magnetic field inside the solenoid is denoted by $\mathbf{B}_o$ to specifically identify this field as in free space. This field depends on the current I through the solenoid wire and the number of turns per unit length n and is given by[2]

$$B_o = \mu_o n I = \mu_o I' \qquad \text{[8.5]}$$

where I' is the current per unit length of the solenoid, that is, $I' = nI$, and μ_o is the absolute permeability of free space in henries per meter, H m^{-1}.

   If we now place a cylindrical material medium to fill the inside of this solenoid, as in Figure 8.6b, we find that the magnetic field has changed. The new magnetic field in the presence of a medium is denoted as $\mathbf{B}$. We will take $\mathbf{B}_o$ to be the applied magnetic field into which the material medium is placed.

   Each atom of the material responds to the applied field, $\mathbf{B}_o$, and develops, or acquires, a net magnetic moment, $\boldsymbol{\mu}_m$, along the applied field. We can view each magnetic moment $\boldsymbol{\mu}_m$ as the result of the precession of each atomic magnetic moment about $\mathbf{B}_o$. The medium therefore develops a net magnetic moment along the field and becomes **magnetized.** The magnetic vector $\mathbf{M}$ describes the extent of magnetization of the medium. $\mathbf{M}$ is defined as the **magnetic dipole moment per unit volume.** Suppose that there are N atoms in a small volume ΔV and each atom, i, has a magnetic moment $\boldsymbol{\mu}_{mi}$ (where $i = 1$ to N). Then $\mathbf{M}$ is defined by

$$\mathbf{M} = \frac{1}{\Delta V} \sum_{i=1}^{N} \boldsymbol{\mu}_{mi} = n_{at}\boldsymbol{\mu}_{av} \qquad \text{[8.6]} \qquad \textit{Magnetization vector}$$

where n_{at} is the number of atoms per unit volume and $\boldsymbol{\mu}_{av}$ is the average magnetic moment per atom. We can assume that each atom acquires a magnetic moment $\boldsymbol{\mu}_{av}$

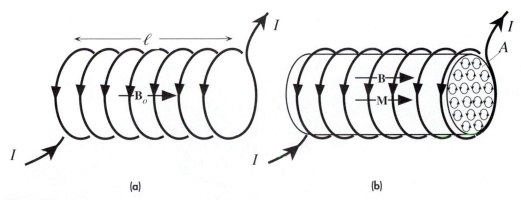

(a)   **(b)**

Figure 8.6
(a) Consider a long solenoid. With free space as the medium inside, the magnetic field is $\mathbf{B}_o$.
(b) A material medium inserted into the solenoid develops a magnetization $\mathbf{M}$.

| [2] The proof of this comes out from Ampere's law and can be found in any textbook of electromagnetism.

along $\mathbf{B}_o$. Each of these magnetic moments along $\mathbf{B}_o$ can be viewed as an elementary current loop at the atomic scale, as schematically depicted in Figure 8.6b. These elementary current loops are due to electronic currents within the atom and arise from both orbital and spin motions of the electrons. Each current loop has its current plane normal to $\mathbf{B}_o$.

Consider a cross section of the magnetized medium, as in Figure 8.7. All the elementary current loops in this plane have the current circulation in the same direction inasmuch as each atom acquires the same magnetic moment, $\boldsymbol{\mu}_{av}$. All neighboring loops in the bulk have adjacent currents in opposite directions that cancel each other, as apparent in Figure 8.7. Thus, there are no net bulk currents, or internal currents, within the bulk of the material. However, the currents at the surface in the surface loops cannot be canceled and this leads to a net surface current, as depicted in Figure 8.7. The surface currents are induced by the magnetization of the medium by the applied magnetic field and therefore depend on the magnetization M of the specimen.

From the definition of M, the total magnetic moment of the cylindrical specimen is

$$\text{Total magnetic moment} = M\,(\text{Volume}) = MA\ell$$

Suppose that the magnetization current on the surface per unit length of the specimen is I_m. Then the total circulating surface current is $I_m\ell$ and the total magnetic moment of the specimen, by definition, is

$$\text{Total magnetic moment} = (\text{Total current}) \times (\text{Cross-sectional area}) = I_m\ell A$$

Equating the two total magnetic moments, we find

Magnetization and surface currents

$$M = I_m \tag{8.7}$$

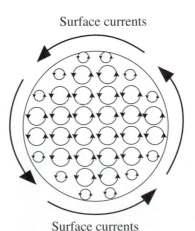

Surface currents

Surface currents

Figure 8.7   Elementary current loops result in surface currents.

There is no internal current, as adjacent currents on neighboring loops are in opposite directions.

We derived this for a particular sample geometry, a cylindrical specimen, in which **M** is along the axis of the cylindrical specimen and I_m flows in a plane perpendicular to **M**. The relationship, however, is more general, as derived in more advanced texts. It should be emphasized that the magnetization current I_m is not due to the flow of free charges carriers, as in a current carrying copper wire, but due to localized electronic currents within the atoms of the solid at the surface. Equation 8.7 states that we can represent the magnetization of a medium by a surface current per unit length, that is, I_m and equal to M.

8.1.4   Magnetizing Field or Magnetic Field Intensity, H

The magnetized specimen in Figure 8.6b placed inside the solenoid develops magnetization currents on the surface. It therefore behaves like a solenoid. We can now regard the solenoid with medium inside, as depicted in Figure 8.8. The magnetic field within the medium now arises from not only the conduction current per unit length I' in the solenoid wires but also from the magnetization current I_m on the surface. The magnetic field B inside the solenoid is now given by the usual solenoid expression but with a current that includes both I' and I_m, as shown in Figure 8.8:

$$B = \mu_o(I' + I_m) = B_o + \mu_o M$$

This relationship is generally valid and can be written in vector form as

$$\mathbf{B} = \mathbf{B}_o + \mu_o\mathbf{M} \qquad [8.8]$$

Magnetic field in a magnetized medium

The field at a point inside a magnetized material is the sum of the applied field, $\mathbf{B}_o$, and a contribution from the magnetization **M** of the material. The magnetization arises from the application of $\mathbf{B}_o$ due to the current of free carriers in the solenoid wires, called the **conduction current,** which we can externally adjust. It becomes useful to introduce a vector field that represents the effect of the external or conduction current alone. In general, $\mathbf{B} - \mu_o\mathbf{M}$ at a point is the contribution of

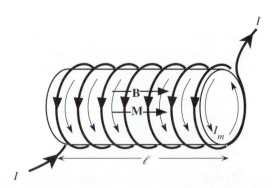

Figure 8.8   The field **B** in the material inside the solenoid is due to the conduction current I through the wires and the magnetization current I_m on the surface of the magnetized medium, or $\mathbf{B} = \mathbf{B}_o + \mu_o\mathbf{M}$.

the external currents alone to the magnetic field at that point inside the material that we called $\mathbf{B}_o$. $\mathbf{B} - \mu_o\mathbf{M}$ represents a magnetizing field because it is the field of the external currents that magnetize the material. The **magnetizing field H** is defined as

Definition of the magnetizing field

$$\mathbf{H} = \frac{1}{\mu_o}\mathbf{B} - \mathbf{M} \qquad [8.9]$$

or

$$\mathbf{H} = \frac{1}{\mu_o}\mathbf{B}_o$$

The magnetizing field is also known as the **magnetic field intensity** and is measured in A m^{-1}. The reason for the division by μ_o is that the resulting vector field $\mathbf{H}$ becomes simply related to the external conduction currents (through Ampere's law). Since in the solenoid $\mathbf{B}_o$ is $\mu_o nI$, we see that the magnetizing field in a solenoid is

$$H = nI = \text{Total conduction current per unit length} \qquad [8.10]$$

It is generally helpful to imagine $\mathbf{H}$ as the *cause* and $\mathbf{B}$ as the *effect*. The cause $\mathbf{H}$ depends only on the external conduction currents, whereas the effect $\mathbf{B}$ depends on the magnetization $\mathbf{M}$ of matter.

8.1.5 Magnetic Permeability and Magnetic Susceptibility

Suppose that at a point P in a material, the magnetic field is $\mathbf{B}$ and the magnetizing field is $\mathbf{H}$. We let $\mathbf{B}_o$ be the magnetic field at P in the absence of any material (i.e., in free space). The magnetic permeability of the medium at P is defined as the magnetic field per unit magnetizing field,

Definition of magnetic permeability

$$\mu = \frac{B}{H} \qquad [8.11]$$

It relates the effect, B, to the cause, H, at the same point P inside a material. In simple qualitative terms, μ represents to what extent a medium is permeable by magnetic fields. Relative permeability, μ_r, of a medium is the fractional increase in the magnetic field with respect to the field in free space when a material medium is introduced. For example, suppose that the field in a solenoid with free space in it is B_0 but with material inserted it is B. Then μ_r is defined by

Definition of relative permeability

$$\mu_r = \frac{B}{B_o} = \frac{B}{\mu_o H} \qquad [8.12]$$

From Equations 8.11 and 8.12, clearly,

$$\mu = \mu_o\mu_r$$

The magnetization $\mathbf{M}$ produced in a material depends on the net magnetic field $\mathbf{B}$. It would be natural to proceed as in dielectrics by relating $\mathbf{M}$ to $\mathbf{B}$ analogously to relating P (polarization) to $\mathcal{E}$ (electric field). However, for historic reasons, $\mathbf{M}$ is related to $\mathbf{H}$, the magnetizing field. Suppose that the medium is isotropic (same properties in all directions), then magnetic susceptibility, χ_m, of the medium is defined simply by

$$\mathbf{M} = \chi_m \mathbf{H} \qquad [8.13]$$

Definition of magnetic susceptibility

This relationship is not obeyed by all magnetic materials. For example, as we will see later, ferromagnetic materials do not obey Equation 8.12. Since the magnetic field

$$\mathbf{B} = \mu_o(\mathbf{H} + \mathbf{M})$$

we have

$$B = \mu_o H + \mu_o M = \mu_o H + \mu_o \chi_m H = \mu_o(1 + \chi_m)H$$

and

$$\mu_r = 1 + \chi_m \qquad [8.14]$$

Relative permeability and susceptibility

Table 8.1 Magnetic quantities and their units

Magnetic Quantity	Symbol	Definition	Units	Comment
Magnetic field; Magnetic induction	**B**	$\mathbf{F} = q\mathbf{v} \times \mathbf{B}$	T = tesla = webers m^{-2}	Produced by moving charges or currents, acts on moving charges or currents.
Magnetic flux	Φ	$\Delta\Phi = B_{\text{normal}}\,\Delta A$	Wb = weber	$\Delta\Phi$ is flux through ΔA and B_{normal} is normal to ΔA. Total flux through any closed surface is zero.
Magnetic dipole moment	$\boldsymbol{\mu}_m$	$\mu_m = IA$	A m^2	Experiences a torque in **B** and a net force in a non-uniform **B**.
Bohr magneton	β	$\beta = e\hbar/2m_e$	A m^2 or J T^{-1}	Magnetic moment due to the spin of the electron. $\beta = 9.27 \times 10^{-24}$ A m^2
Magnetization vector	**M**	Magnetic moment per unit volume	A m^{-1}	Net magnetic moment in a material per unit volume.
Magnetizing field; magnetic field intensity	**H**	$\mathbf{H} = \mathbf{B}/\mu_o - \mathbf{M}$	A m^{-1}	**H** is due to external conduction currents only and is the cause of **B** in a material.
Magnetic susceptibility	χ_m	$\mathbf{M} = \chi_m \mathbf{H}$	None	Relates the magnetization of a material to the magnetizing field **H**.
Absolute permeability	μ_o	$c = [\varepsilon_o \mu_o]^{-1/2}$	H m^{-1} = Wb m^{-1} A^{-1}	A fundamental constant in magnetism. In free space, $\mu_o = B/H$.
Relative permeability	μ_r	$\mu_r = B/\mu_o H$.	None	
Magnetic permeability	μ	$\mu = \mu_o \mu_r$	H m^{-1}	Not to be confused with magnetic moment.
Inductance	L	$L = \Phi_{\text{total}}/I$	H (henries)	Total flux threaded per unit current.
Magnetostatic energy density	E_{vol}	$dE_{\text{vol}} = H\,dB$	J m^{-3}	dE_{vol} is the energy required per unit volume in changing B by dB.

The presence of a magnetizable material is conveniently accounted for by using the relative permeability μ_r, or $(1 + \chi_m)$, to simply multiply μ_o. Alternatively, one can simply replace μ_o with $\mu = \mu_o\mu_r$. For example, the inductance of the solenoid with a magnetic medium inside increases by a factor of μ_r.

Table 8.1 provides a summary of various important magnetic quantities, their definitions, and units.

Example 8.1 | **AMPERE'S LAW AND THE INDUCTANCE OF A TOROIDAL COIL**   Ampere's law provides a relationship between the conduction current I and the magnetic field intensity H threading this current. The conduction current I is the current due to the flow of free charge carriers through a conductor and not due to the magnetization of any medium. Consider an arbitrary closed path, C, around a conductor carrying a current I, as shown in Figure 8.9. The tangential component of $\mathbf{H}$ to the curve C at point P is H_t. If dl is an infinitesimally small path length of C at P, as shown in Figure 8.9, then the summation of $H_t\,dl$ around the path C gives the conduction current enclosed within C. This is **Ampere's law,**

Ampere's law

$$\oint_C H_t\,dl = I \qquad\qquad [8.15]$$

Consider the toroidal coil with N turns shown in Figure 8.10. First assume that the toroid core is air ($\mu_r \approx 1$). Suppose that the current through the coils is I. By symmetry, the magnetic field intensity H inside the toroidal core is the same everywhere and is directed along the circumference. Suppose that l is the length of the mean circumference C. The current is linked N times by the circumference C, so that Equation 8.15 is

$$\oint_C H_t\,dl = H\ell = NI$$

or

$$H = \frac{NI}{\ell}$$

The magnetic field B_o with air as core material is then simply

$$B_o = \mu_o H = \frac{\mu_o NI}{\ell}$$

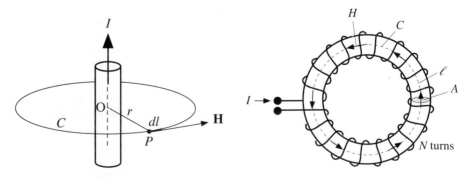

Figure 8.9   Ampere's circuital law

Figure 8.10   A toroidal coil with N turns

When the toroidal coil has a magnetic medium with relative permeability μ_r, the magnetic field intensity is still H because the conduction current I has not changed. But the magnetic field B is now different than B_o and is given by

$$B = \mu_o\mu_r H = \frac{\mu_o\mu_r NI}{\ell}$$

If A is the cross-sectional area of the toroid, then the total flux Φ through the core is BA or $\mu_o\mu_r NAI/\ell$. The current I in Figure 8.10 threads the flux N times. The inductance L of the toroidal coil, by definition, is then

$$L = \frac{\text{Total flux threaded}}{\text{Current}} = \frac{N\Phi}{I} = \frac{\mu_o\mu_r N^2 A}{\ell}$$

Having a magnetic material as the toroid core increases the inductance by a factor of μ_r in the same way a dielectric materials increases the capacitance by a factor of ε_r.

MAGNETOSTATIC ENERGY PER UNIT VOLUME   Consider a toroidal coil with N turns that is energized from a voltage supply through a rheostat, as shown in Figure 8.11. The core of the toroid may be any material. Suppose that by adjusting the rheostat we increase the current, i, supplied to the coil. The current i produces magnetic flux Φ in the core, which is BA, where B is the magnetic field and A is the cross-sectional area. We can now use Ampere's law for H to relate the current i to H, as in Example 8.1. If ℓ is the mean circumference, then

$$H\ell = Ni \qquad \qquad \textbf{[8.16]}$$

The changing current means that the flux is also changing (both increasing). We know from Faraday's law that a changing flux that threads a circuit generates a voltage v in that circuit given by the rate of change of total threaded flux, or $N\Phi$. Lenz's law makes the polarity of the induced voltage oppose the applied voltage. Suppose that in a time interval δt seconds, the magnetic field within the core changes by δB, then $\delta\Phi = A\delta B$ and

$$v = \frac{\delta(\text{Total flux threaded})}{\delta t} = \frac{N\delta\Phi}{\delta t} = NA\frac{\delta B}{\delta t} \qquad \textbf{[8.17]}$$

The battery has to supply the current i against this induced voltage, v, which means that it has to do electrical work iv every second. In other words, the battery has to do work $iv\,\delta t$ in a time interval δt to supply the necessary current to increase the magnetic field by δB.

Example 8.2

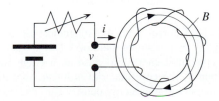

Figure 8.11   Energy required to magnetize a toroidal coil

The electrical enegy, δE, that is input into the coil in time δt is then, using Equations 8.16 and 8.17,

$$\delta E = iv \; \delta t = \left(\frac{H\ell}{N}\right)\left(NA\frac{\delta B}{\delta t}\right) \delta t = (A\ell)H \; \delta B$$

This energy, δE, is the work done in increasing the field in the core by δB. The volume of the toroid is $A\ell$. Therefore, the total energy or work required per unit volume to increase the magnetic field from an initial value B_1 to a final value B_2 in the toroid is

Work done per unit volume during magnetization

$$E_{\text{vol}} = \int_{B_1}^{B_1} H \; dB \tag{8.18}$$

where the integration limits are determined by the initial and final magnetic fields. This is the expression for calculating the **energy density** (energy per unit volume) required to change the field from B_1 to B_2. It should be emphasized that Equation 8.18 is valid for *any medium*. We conclude that an incremental energy density of $dE_{\text{vol}} = H \; dB$ is required to increase the magnetic field by dB at a point in any medium including free space.

We can now consider a core material that we can represent by a *constant* relative permeability μ_r. This means we can exclude those materials that do not have a linear relationship between B and H, such as ferromagnetic and ferrimagnetic materials, which we will discuss later. If the core is free space or air, then $\mu_r = 1$.

Suppose that we increase the current in Figure 8.11 from zero to some final value, I, so that the magnetic field changes from zero to some final value B. Since the medium has a constant relative permeability μ_r, we can write

$$B = \mu_r\mu_o H$$

and use this in Equation 8.18 to integrate and find the energy per unit volume needed to establish the field B or field intensity H

Energy density of a magnetic field

$$E_{\text{vol}} = \frac{1}{2}\mu_r\mu_o H^2 = \frac{B^2}{2\mu_r\mu_o} \tag{8.19}$$

This is the energy absorbed from the battery per unit volume of core medium to establish the magnetic field. This energy is stored in the magnetic field and is called **magnetostatic energy density.** It is a form of magnetic potential energy. If we were to suddenly remove the battery and short those terminals, the current will continue to flow for a short while (determined by L/R) and do external work in heating the resistor. This external work comes from the stored energy in the magnetic field. If the medium is free space, or air, then the energy density is

Magnetostatic energy density in free space

$$E_{\text{vol}}(\text{air}) = \frac{1}{2}\mu_o H^2 = \frac{B^2}{2\mu_o}$$

A magnetic field of 2 T corresponds to a magnetostatic energy density of 1.6 MJ m^{-3} or 1.6 J cm^{-3}. The energy in a magnetic field of 2 T in a 1 cm^3 volume (size of a thimble) has the work ability (potential energy) to raise an average-sized apple by five feet. We should note that as long as the core material is linear, that is, μ_r is independent of the magnetic field itself, magnetostatic energy density can also be written as

Magnetostatic energy in a linear magnetic medium

$$E_{\text{vol}} = \frac{1}{2}H \; B \tag{8.20}$$

8.2   MAGNETIC MATERIAL CLASSIFICATIONS

In general, magnetic materials are classified into five distinct groups: diamagnetic, paramagnetic, ferromagnetic, antiferromagnetic, and ferrimagnetic. Table 8.2 provides a summary of the magnetic properties of these classes of materials.

8.2.1   Diamagnetism

Typical diamagnetic materials have a magnetic susceptibility that is negative and small. For example, the silicon crystal is diamagnetic with $\chi_m = -5.2 \times 10^{-6}$. The relative permeability of diamagnetic materials is slightly less than unity. When a diamagnetic substance such as a silicon crystal is placed in a magnetic field, the

Table 8.2   Classification of magnetic materials

Type	χ_m (typical values)	χ_m versus T	Comment and Examples
Diamagnetic	Negative and small (-10^{-6})	T independent	Atoms of the material have closed shells. Organic materials, e.g., many polymers; covalent solids, e.g., Si, Ge, diamond; some ionic solids, e.g., alkalihalides; some metals, e.g., Cu, Ag, Au.
	Negative and large (-1)	Below a critical temperature	Superconductors
Paramagnetic	Positive and small $(10^{-5}-10^{-4})$	Independent of T	Due to the alignment of spins of conduction electrons. Alkali and transition metals.
	Positive and small (10^{-5})	Curie or Curie–Weiss law, $\chi_m = C/(T - T_C)$	Materials in which the constituent atoms have a permanent magnetic moment, e.g., gaseous and liquid oxygen; ferromagnets (Fe), antiferromagnets (Cr), and ferrimagnets (Fe_3O_4) at high temperatures.
Ferromagnetic	Positive and very large.	Ferromagnetic above and paramagnetic below the Curie temperature	May possess a large permanent magnetization even in the absence of an applied field. Some transition and rare earth metals, Fe, Co, Ni, Gd, Dy.
Antiferromagnetic	Positive and small	Antiferromagnetic above and paramagnetic below the Néel temperature	Mainly salts and oxides of transition metals, e.g., MnO, NiO, MnF_2, and some transition metals, α–Cr, Mn.
Ferrimagnetic	Positive and very large	Ferrimagnetic above and paramagnetic below the Curie temperature	May possess a large permanent magnetization even in the absence of an applied field. Ferrites.

magnetization vector **M** in the material is in the *opposite* direction to the applied field $\mu_o\mathbf{H}$ and the resulting field **B** within the material is less than $\mu_o\mathbf{H}$. The negative susceptibility can be interpreted as the diamagnetic substance trying to expel the applied field from the material. When a diamagnetic specimen is placed in a nonuniform magnetic field, the magnetization **M** of the material is in the opposite direction to **B** and the specimen experiences a net force toward smaller fields, as depicted in Figure 8.12. A substance exhibits diamagnetism whenever the constituent atoms in the material have closed subshells and shells. This means that each constituent atom has no permanent magnetic moment in the absence of an applied field. Covalent crystals and many ionic crystals are typical diamagnetic materials because the constituent atoms have no unfilled subshells. Superconductors, as we will discuss later, are perfect diamagnets with $\chi_m = -1$ and totally expel the applied field from the material.

8.2.2   Paramagnetism

Paramagnetic materials have a small positive megnetic susceptibility. For example, oxygen gas is paramagnetic with $\chi_m = 2.1 \times 10^{-6}$ at atmospheric pressure and room temperature. Each oxygen molecule has a net magnetic dipole moment, $\boldsymbol{\mu}_{\mathrm{mol}}$. In the absence of an applied field, these molecular moments are randomly oriented due to the random collisions of the molecules, as depicted in Figure 8.13a. The magnetization of the gas is zero. In the presence of an applied field, the molecular magnetic moments take various alignments with the field, as illustrated in Figure 8.13b. The degree of alignment of $\boldsymbol{\mu}_{\mathrm{mol}}$ with the applied field and hence magnetization **M** increases with the strength of the applied field $\mu_o\mathbf{H}$. Magnetization M typically decreases with increasing temperature because at higher temperatures there are more molecular collisions, which destroy the alignments of molecular magnetic moments with the applied field. When a paramagnetic substance is

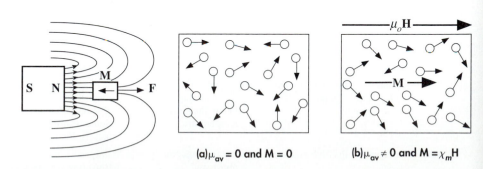

(a)$\mu_{av} = 0$ and $M = 0$ (b)$\mu_{av} \neq 0$ and $M = \chi_m\mathbf{H}$

Figure 8.12   A diamagnetic material placed in a nonuniform magnetic field experiences a force toward smaller fields. This repels the diamagnetic material away from a permanent magnet.

Figure 8.13
(a) In a paramagnetic material, each individual atom possesses a permanent magnetic moment, but due to thermal agitation there is no average moment per atom and **M** = 0.
(b) In the presence of an applied field, individual magnetic moments take alignments along the applied field and **M** is finite and along **B**.

placed in a nonuniform magnetic field, the induced magnetization, **M**, is along **B** and there is a net force toward greater fields. For example, when liquid oxygen is poured close to a strong magnet, as depicted in Figure 8.14, the liquid becomes attracted to the magnet.

Many metals are also paramagnetic, such as magnesium with $\chi_m = 1.2 \times 10^{-5}$. The origin of paramagnetism (called **Pauli spin paramagnetism**) in these metals is due to the alignment of the majority of spins of conduction electrons with the field.

8.2.3   Ferromagnetism

Ferromagnetic materials such as iron can possess large permanent magnetizations even in the absence of an applied magnetic field. The magnetic susceptibility χ_m is typically positive and very large (even infinite) and, further, depends on the applied field intensity. The relationship between the magnetization **M** and the applied magnetic field, μ_o**H**, is highly nonlinear. At sufficiently high fields, the magnetization **M** of the ferromagnet saturates. The origin of ferromagnetism is the quantum mechanical exchange interaction (discussed later) between the constituent atoms that results in regions of the material possessing permanent magnetization. Figure 8.15 depicts a region of the Fe crystal, called a **magnetic domain,** that has a net magnetization vector **M** due to the alignment of the magnetic moments of all Fe atoms in this region. This crystal domain has **magnetic ordering** as all the atomic magnetic moments have been aligned parallel to each other. Ferromagnetism occurs below a critical temperature called the Curie temperature, T_C. At temperatures above T_C, ferromagnetism is lost and the material becomes paramagnetic.

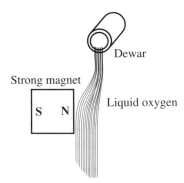

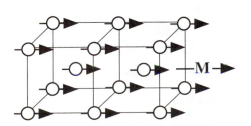

Figure 8.14   A paramagnetic material placed in a nonuniform magnetic field experiences a force toward greater fields. This attracts the paramagnetic material (e.g., liquid oxygen) toward a permanent magnet.

Figure 8.15   In a magnetized region of a ferromagnetic material such as iron, all the magnetic moments are spontaneously aligned in the same direction. There is a strong magnetization vector **M** even in the absence of an applied field.

spatial distributions (recall that m_ℓ determines the orientation of an orbit). Different m_ℓ values result in a smaller coulombic repulsion energy between the electrons compared with the case where the electrons have opposite spins (different m_s), where they would be in the same orbital (sample m_ℓ), that is, in the same spatial region. It is apparent that even though the interaction energy between the electrons has nothing to do with magnetic forces, it does depend nonetheless on the orientations of their spins (m_s), or on their spin magnetic moments, and it is less when the spins are parallel. Two electrons parallel their spins not because of the direct magnetic interaction between the spin magnetic moments but because of the **Pauli exclusion principle** and the **electrostatic interaction energy.** Together they constitute what is known as an **exchange interaction,** which forces two electrons to take m_s and m_ℓ values that result in the minimum of electrostatic energy. In an atom, the exchange interaction therefore forces two electrons to take the same m_s but different m_ℓ if this can be done within the Pauli exclusion principle. This is the reason an isolated Fe atom has four unpaired spins in the 3d subshell.

In the crystal, of course, the outer electrons are no longer strictly confined to their parent Fe atoms, particularly the 4s electrons. The electrons now have wavefunctions that belong to the whole solid. Something like Hund's rule also operates at the crystal level for Fe, Co, and Ni. If two 3d electrons parallel their spins and occupy different wavefunctions (and hence different negative charge distributions), the resulting mutual coulombic repulsion between them and also with all the other electrons and the attraction to the positive Fe ions result in an overall reduction of potential energy. This reduction in energy is again due to the exchange interaction and is a direct consequence of the Pauli exclusion principle and the coulombic forces. Thus, the majority of 3d electrons spontaneously parallel their spins without the need for the application of an external magnetic field. The number of electrons that actually parallel their spins depends on the strength of the exchange interaction, and for the iron crystal this turns out to be about 2.2 electrons per atom. Since typically the wavefunctions of the 3d electrons in the whole iron crystal show localization around the iron ions, some people prefer to view the 3d electrons as spending the majority of their time around Fe atoms, which explains the reason for drawing the magnetized iron crystal as in Figure 8.15

It may be thought that all solids should follow the example of Fe and become spontaneously ferromagnetic since paralleling spins would result in different spatial distributions of negative charge and probably a reduction in the electrostatic energy, but this is not generally the case at all. We know that, in the case of covalent bonding, the electrons have the lowest energy when the two electrons spin in opposite directions. In covalent bonding in molecules, the exchange interaction does not reduce the energy. Making the electron spins parallel leads to spatial negative charge distributions that result in a net mutual electrostatic repulsion between the positive nuclei.

In the simples case, for two atoms only, the exchange energy depends on the interatomic separation between two interacting atoms and the relative spins of the two outer electrons (labeled as 1 and 2). From quantum mechanics, the exchange interaction can be represented in terms of an exchange energy, E_{ex}, as

$$E_{ex} = -2J_e \mathbf{S}_1 \cdot \mathbf{S}_2 \qquad \textbf{[8.21]}$$

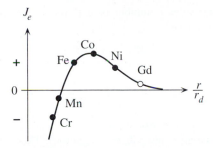

Figure 8.20   The exchange integral as a function of r/r_d, where r is the interatomic distance and r_d the radius of the d orbit (or the average d subshell radius).

Cr to Ni are transition metals. For Gd, the x axis is r/r_f, where r_f is the radius of the f orbit.

where $\mathbf{S}_1$ and $\mathbf{S}_2$ are the spin angular momenta of the two electrons and J_e is a numerical quantity called the **exchange integral** that involves integrating the wavefunctions with the various potential energy interaction terms. It therefore depends on the electrostatic interactions and hence on the interatomic distance. For the majority of solids, J_e is negative, so the exchange energy is negative if $\mathbf{S}_1$ and $\mathbf{S}_2$ are in the opposite directions, that is, the spins are antiparallel (as we found in covalent bonding). This is the antiferromagnetic state. For Fe, Co, and Ni, however, J_e is positive. E_{ex} is then negative if $\mathbf{S}_1$ and $\mathbf{S}_2$ are parallel. Spins of the 3d electrons on the Fe atoms therefore spontaneously align in the same direction to reduce the exchange energy. This spontaneous magnetization is the phenomenon of ferromagnetism. Figure 8.20 illustrates how J_e changes with the ratio of interatomic separation to the radius of the 3d subshell (r/r_d). For the transition metals Fe, Co, and Ni, the r/r_d is such that J_e is positive.[3] In all other cases, it is negative and does not produce ferromagnetic behavior. It should be mentioned that Mn, which is not ferromagnetic, can be alloyed with other elements to increase r/r_d and hence endow ferromagnetism in the alloy.

SATURATION MAGNETIZATION IN IRON   The maximum magnetization, called **saturation magnetization**, M_{sat}, in iron is about 1.75×10^6 A m^{-1}. This corresponds to all possible net spins aligning parallel to each other. Calculate the effective number of Bohr magnetons per atom that would give M_{sat}, given that the density and relative atomic mass of iron are 7.86 g cm^{-3} and 55.85 respectively.

Example 8.3

Solution

The number of Fe atoms per unit volume is

$$n_{\text{at}} = \frac{\rho N_A}{M_{\text{at}}} = \frac{(7.86 \times 10^3 \text{ kg m}^{-3})(6.022 \times 10^{23} \text{ mol}^{-1})}{(55.85 \times 10^{-3} \text{ kg mol}^{-1})}$$

$$= 8.48 \times 10^{28} \text{ atoms m}^{-3}$$

---

[3] According to H. P. Myers, *Introductory Solid State Physics* (London: Taylor and Francis Ltd., 1990), p. 375, there have been no theoretical calculations of the exchange integral J_e for any real magnetic substance.

If each Fe atom contributes x number of net spins, then since each net spin has a magnetic moment of β, we have,

$$M_{\text{sat}} = n_{\text{at}}(x\beta)$$

so that

$$x = \frac{M_{\text{sat}}}{n_{\text{at}}\beta} = \frac{(1.75 \times 10^6)}{(8.48 \times 10^{28})(9.27 \times 10^{-24})} \approx 2.2$$

In the solid, each Fe atom contributes only 2.2 Bohr magnetons to the magnetization even though the isolated Fe atom has 4 Bohr magnetons. There is no orbital contribution to the magnetic moment per atom in the solid because all the outer electrons, 3d and 4s electrons, can be viewed as belonging to the whole crystal, or being in an energy band, rather than orbiting individual atoms. A 3d electron is attracted by various Fe ions in the crystal and therefore does not experience a central force, in contrast to the 3d electron in the isolated Fe atom that orbits the nucleus. The orbital momentum in the crystal is said to be quenched.

We should note that when the magnetization is saturated, all atomic magnetic moments are aligned. The resulting magnetic field within the iron specimen in the absence of an applied magnetizing field ($H = 0$) is

$$B_{\text{sat}} = \mu_o M_{\text{sat}} = 2.2 \text{ T}$$

8.4   SATURATION MAGNETIZATION AND CURIE TEMPERATURE

The maximum magnetization in a ferromagnet when all the atomic magnetic moments have been aligned as much as possible is called the saturation magnetization, M_{sat}. In the iron crystal, for example, this corresponds to each Fe atom with an effective spin magnetic moment of 2.2 Bohr magnetons aligning in the same direction to give a magnetic field $\mu_o M_{\text{sat}}$ or 2.2 Teslas. As we increase the temperature, lattice vibrations become more energetic, which leads to a frequent disruption of the alignments of the spins. The spins cannot align perfectly with each other as the temperature increases due to lattice vibrations randomly agitating the individual spins. When an energetic lattice vibration passes through a spin site, the energy in the vibration may be sufficient to disorientate the spin of the atom. The ferromagnetic behavior disappears at a critical temperature called the **Curie temperature,** denoted by T_C, when the thermal energy of lattice vibrations in the crystal can overcome the potential energy of the exchange interaction and hence destroy the spin alignments. Above the Curie temperature, the crystal behaves as if it were paramagnetic. The saturation magnetization, M_{sat}, therefore decreases from its maximum value, $M_{\text{sat}}(0)$, at absolute zero of temperature to zero at the Curie temperature. Figure 8.21 shows the dependence of M_{sat} on the temperature when M_{sat} has been normalized to $M_{\text{sat}}(0)$ and the temperature is the reduced temperature, that is T/T_C. At $T/T_C = 1$, $M_{\text{sat}} = 0$. When plotted in this way, the ferromagnets cobalt and nickel follow closely the observed behavior for iron. We should note that since for iron $T_C = 1043$ K, at room temperature, $T/T_C = 0.29$ and M_{sat} is very close to its value at $M_{\text{sat}}(0)$.

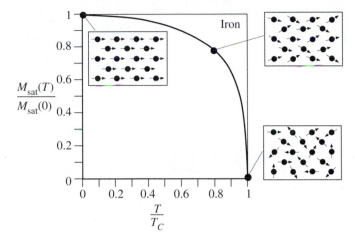

Figure 8.21   Normalized saturated magnetization versus reduced temperature, T/T_C, where T_C is the Curie temperature (1043 K).

Table 8.3   Properties of the ferromagnets Fe, Co, Ni, and Gd

	Fe	**Co**	**Ni**	**Gd**
Crystal structure	BCC	HCP	FCC	HCP
Bohr magnetons per atom	2.22	1.72	0.60	7.1
$M_{sat}(0)(\text{MA m}^{-1})$	1.75	1.45	0.50	2.0
$B_{sat} = \mu_o M_{sat}(T)$	2.2	1.82	0.64	2.5
T_C	770 °C	1127 °C	358 °C	16 °C
	1043 K	1400 K	631 K	289 K

Since at the Curie temperature, the thermal energy, of the order of kT_C, is sufficient to overcome the energy of the exchange interaction, E_{ex}, that aligns the spins, we can take kT_C as an order of magnitude estimate of E_{ex}. For iron, E_{ex} is ~0.09 eV and for cobalt this is ~0.1 eV.

Table 8.3 summarizes some of the important properties of the ferromagnets Fe, Co, Ni, and Gd (rare earth metal).

8.5   MAGNETIC DOMAINS: FERROMAGNETIC MATERIALS

8.5.1   Magnetic Domains

A single crystal of iron does not necessarily possess a net permanent magnetization in the absence of an applied field. If a magnetized piece of iron is heated to a temperature above its Curie temperature and then allowed to cool in the absence

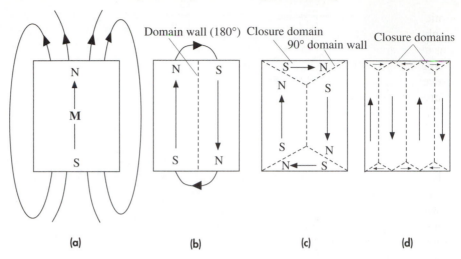

(a)   (b)   (c)   (d)

Figure 8.22

(a) Magnetized bar of ferromagnet in which there is only one domain and hence an external magnetic field.

(b) Formation of two domains with opposite magnetizations reduces the external field. There are, however, field lines at the ends.

(c) Domains of closure fitting at the ends eliminate the external fields at the ends.

(d) A specimen with several domains and closure domains. There is no external magnetic field and the specimen appears unmagnetized.

of a magnetic field, it will possess no net magnetization. The reason for the absence of net magnetization is due to the formation of magnetic domains that effectively cancel each other, as discussed below. A **magnetic domain** is a region of the crystal in which all the spin magnetic moments are aligned to produce a magnetic moment in one direction only.

Figure 8.22a shows a single crystal of iron that has a permanent magnetization as a result of ferromagnetism (aligning of all atomic spins). The crystal is like a bar of magnet with magnetic field lines around it. As we know, there is potential energy (*PE*), called **magnetostatic energy,** stored in a magnetic field and we can reduce this energy in the external field by dividing the crystal into two domains where the magnetizations are in the opposite directions, as shown in Figure 8.22b. The external magnetic field lines are reduced and there is now less potential energy stored in the magnetic field. There are only field lines at the ends. This arrangement is energetically favorable because the magnetostatic energy has been reduced by decreasing the external field lines. However, there is now a boundary, called a **domain wall** (or **Bloch wall**), between the two domains where the magnetization changes from one direction to the opposite direction and hence the atomic spins do, also. It requires energy to rotate the atomic spin through 180° with respect to its neighbor because the exchange energy favors aligning neighboring atomic spins (0°). The wall in Figure 8.22b is a 180° wall inasmuch as the magnetization through the wall is rotated by 180°. It is apparent that the wall region where the neighboring atomic spins change their relative direction (or orientation) from one domain to the neighboring one has higher *PE* than the bulk of the domain, where all the atomic spins are aligned. As we will show below, the domain wall is not simply one atomic spacing but has a finite thickness, which for iron is typically of the order of 0.1 μm, or several hundred atomic spacings. The excess energy in the wall increases with the area of the wall.

The magnetostatic energy associated with the field lines at the ends in Figure 8.22b can be further reduced by eliminating these external field lines by closing the ends with sideway domains with magnetizations at 90°, as shown in Figure 8.22c. These end domains are **closure domains** and have walls that are 90° walls. The magnetization is rotated through 90° through the wall. Although we have reduced the magnetostatic energy, we have increased the potential energy in the walls by adding additional walls. The creation of magnetic domains continues (spontaneously) until the potential energy reduction in creating an additional domain is the same as the increase in creating an additional wall. The specimen then possesses minimum potential energy and is in equilibrium with no net magnetization. Figure 8.22d shows a specimen with several domains and no net magnetization. The sizes, shapes, and distributions of domains depend on a number of factors, including the size and shape of the whole specimen. For iron particles of dimensions less than of the order of 0.01 μm, the increase in the potential energy in creating a domain wall is too costly and these particles are single domains and hence always magnetized.

The magnetization of each domain is normally along one of preferred directions in which the atomic spin alignments are easiest (the exchange interaction is the strongest). For iron, the magnetization is easiest along any one of six <100> directions (along cube edges), which are called **easy directions.** The domains have magnetizations along these easy directions. The magnetization of the crystal along an applied field occurs, in principle, by the growth of domains with magnetizations (or components of **M**) along the applied field (**H**), as illustrated in Figure 8.23a and b. For simplicity, the magnetizing field is taken along an easy direction. The Bloch wall between the domains A and B migrates toward the right, which enlarges the domain A and shrinks domain B, with the net result that the crystal has an effective magnetization **M** along **H**. The migration of the Bloch wall is caused by the spins in the wall, and also spins in section B adjacent to the wall, being gradually rotated by the applied field (they experience a torque). The magnetization process therefore involves the motions of Bloch walls in the crystal.

8.5.2   Magnetocrystalline Anisotropy

Ferromagnetic crystals characteristically exhibit magnetic anisotropy, which means that the magnetic properties are different along different crystal directions. In the case of iron (BCC), the spins in a domain are most easily aligned in any of the six [100] type directions, collectively labeled as <100>, and correspond to the six edges of the cubic unit cell. The exchange interactions are such that spin magnetic moments are most easily aligned with each other if they all point in one of the six

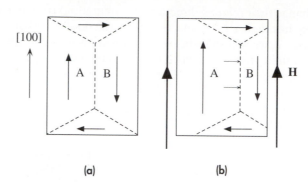

[100]

A B

A B **H**

(a)

(b)

Figure 8.23

(a) An unmagnetized crystal of iron in the absence of an applied magnetic field. Domains A and B are the same size and have opposite magnetizations.

(b) When an external magnetic field is applied, the domain wall migrates into domain B, which enlarges A and shrinks B. The result is that the specimen now acquires net magnetization.

<100> directions. Thus <100> directions in the iron crystal constitute the easy directions for magnetization. When a magnetizing field **H** along a [100] direction is applied, as illustrated in Figure 8.23a and b, domain walls migrate to allow those domains (e.g., A) with magnetizations along **H** to grow at the expense of those domains (e.g., B) with magnetizations opposing **H**. The observed M versus H behavior is shown in Figure 8.24. Magnetization rapidly increases and saturates with applied field less than 0.01 T.

On the other hand, if we want to magnetize the crystal along the [111] direction by applying a field along this direction, then we have to apply a stronger field than that along [100]. This is clearly shown in Figure 8.24, where the resulting magnetization along [111] is smaller than that along [100] for the same magnitude of applied field. Indeed, saturation is reached at an applied field that is about a factor of four greater than that along [100]. The [111] direction in the iron crystal is consequently known as the **hard direction.** The M versus H behavior along [100], [110], and [111] directions in an iron crystal and the associated anisotropy are shown in Figure 8.24.

When an external field is applied along the diagonal direction OD in Figure 8.24, initially all those domains with **M** along OA, OB, and OC, that is, those with magnetization components along OD, grow by consuming those with **M** in the wrong direction and eventually take over the whole specimen. This is an easy process (similar to the process along [100]) and requires small fields and represents the processes from 0 to P on the magnetization curve for [111] in Figure 8.24. However, from P onwards, the magnetizations in the domains have to be rotated away from their easy directions, that is, from OA, OB, and OC toward OD. This process consumes substantial energy and hence needs much stronger applied fields.

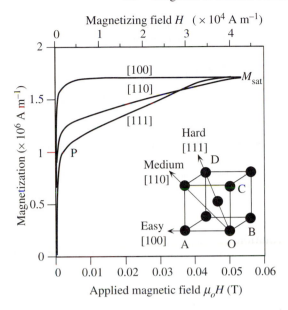

Figure 8.24   Magnetocrystalline anisotropy in a single iron crystal.

M versus H depends on the crystal direction and is easiest along [100] and hardest along [111].

It is apparent that the magnetization of the crystal along [100] needs the least energy, whereas that along [111] consumes the greatest energy. The excess energy required to magnetize a unit volume of a crystal in a particular direction with respect to that in the easy direction is called the **anisotropy energy** and is denoted by K. For iron, the anisotropy energy is zero for [100] and largest for the [111] direction, about 40 kJ m^{-3} or 2.9×10^{-5} eV per atom. For cobalt, which has the HCP crystal structure, the anisotropy energy is at least an order of magnitude greater.

8.5.3   Domain Walls

We recall that the spin magnetic moments rotate across a domain wall. We mentioned that the wall is not simply one atomic spacing wide, as this would mean two neighboring spins being at 180° to each other and hence possessing excessive exchange interaction. A schematic illustration of the structure of a typical 180° Bloch wall, between two domains A and B, is depicted in Figure 8.25. It can be seen that the neighboring spin magnetic moments are rotated gradually, and over several hundred atomic spacings the magnetic moment reaches a rotation of 180°. Exchange forces between neighboring atomic spins favor very little relative rotation. Had it been left to exchange forces alone, relative rotation of neighboring

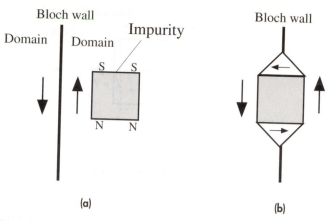

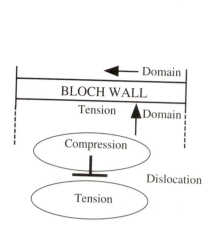

Figure 8.27   Stress and strain distributions around a dislocation and near a domain wall.

Figure 8.28   Interaction of a Bloch wall with a nonmagnetic (no permanent magnetization) inclusion.

(a) The inclusion becomes magnetized and there is magnetostatic energy.

(b) This arrangement has lower potential energy and is thus favorable.

wall that has a tensile strain on the side of the dislocation. If the wall gets close to the dislocation, the tensile and compressive strains cancel, which results in an unstrained lattice and hence a lower strain energy. This energetically favorable arrangement keeps the domain boundary close to the dislocation. It now takes greater magnetic field to snap away the boundary from the dislocation. Domain walls also interact with nonmagnetic impurities and inclusions. For example, an inclusion that finds itself in a domain becomes magnetized and develops south and north poles, as shown in Figure 8.28a. If the domain wall were to intersect the inclusion and if there were to be two neighboring domains around the inclusion, as in Figure 8.28b, then the magnetostatic energy would be lowered—energetically a favorable event. This reduction in magnetostatic potential energy means that it now takes greater force to move past the domain wall, as if the wall were "pinned" by the impurity.

The motion of a domain wall in a crystal is therefore not smooth but rather jerky. The wall becomes pinned somewhere by a defect or an impurity and then needs a greater applied field to break free. Once it snaps off, the domain wall is moved until it is attracted by another type of imperfection, where it is held until the field increases further to snap it away again. Each time the domain wall is snapped loose, lattice vibrations are generated, which means loss of energy as heat. The whole domain wall motion is nonreversible and involves energy losses as heat to the crystal.

8.5.6   Polycrystalline Materials and the *M* versus *H* Behavior

The majority of the magnetic materials used in engineering are polycrystalline and therefore have a microstructure that consists of many grains of various sizes and orientations depending on the preparation and thermal history of the component.

Small grain with a single domain

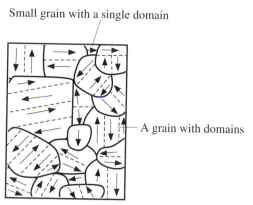

A grain with domains

Figure 8.29   Schematic illustration of magnetic domains in the grains of an unmagnetized polycrystalline iron sample. Very small grains have single domains.

In an unmagnetized polycrystalline sample, each crystal grain will possess domains, as depicted in Figure 8.29. The domain structure in each grain will depend on the size and shape of the grain and, to some extent, on the magnetizations in neighboring grains. Although very small grains, perhaps smaller than 0.1 μm, may be single domains, in most cases the majority of the grains will have many domains. Overall, the structure will possess no net magnetization, provided that it was not previously subjected to an applied magnetic field. We can assume that the component was heated to a temperature above the Curie point and then allowed to cool to room temperature without an applied field.

Suppose that we start applying a very small external magnetic field ($\mu_o H$) along some direction, which we can arbitrarily label as $+x$. The domain walls within various grains begin to move small distances, and favorably oriented domains (those with a component of M along $+x$) grow a little larger at the expense of those pointing away from the field, as indicated by point a in Figure 8.30. The domain walls that are pinned by imperfections tend to bow out. There is a very small but net magnetization along the field, as indicated by the Oa region in the magnetization versus magnetizing field (M versus H) behavior in Figure 8.30. As we increase the magnetizing field, the domain motions extend larger distances, as shown for point b in Figure 8.30, and walls encounter various obstacles such as crystal imperfections, impurities, second phases, and so on, which tend to attract the walls and thereby hinder their motions. A domain wall that is stuck (or pinned) at an imperfection at a given field cannot move until the field increases sufficiently to provide the necessary force to snap the wall free, which then suddenly surges forward to the next obstacle. As a wall suddenly snaps free and shoots forward to the next obstacle, essentially two causes lead to heat generation. Sudden changes in the lattice distortion, due to magnetostriction, create lattice waves that carry off some of the energy. Sudden changes in the magnetization induce eddy currents that dissipate energy via joule heating (domains have a finite electrical resistance). These processes involve energy conversion to heat and are irreversible. Sudden

the *B–H* loops are so small that they end up at the origin when *H* reaches zero. The demagnetization process in Figure 8.34 is commonly known as **deperming.** Undesirable magnetization of various magnetic devices such as recording heads is typically removed by this deperming process (for example, a demagnetizing gun brought close to a magnetized recording head implements deperming by applying a cycled *H* with decreasing magnitude).

Example 8.4

ENERGY DISSIPATED PER UNIT VOLUME AND THE HYSTERESIS LOOP   Consider a toroidal coil with an iron core that is energized from a voltage supply through a rheostat, as shown in Figure 8.11. Suppose that by adjusting the rheostat we can adjust the current *i* supplied to the coil and hence the magnetizing field *H* in the core material. *H* and *i* are simply related by Ampere's law. However, the magnetic field *B* in the core is determined by the *B–H* characteristics of the core material. From electromagnetism (see Example 8.2), we know that the battery has to do work, dE_{vol}, per unit volume of core material to increase the magnetic field by *dB*, where

$$dE_{vol} = H\,dB$$

so that the total energy or work involved per unit volume in changing the magnetic field from an initial value B_1 to a final value B_2 in the core is

Work done per unit volume during magnetization

$$E_{vol} = \int_{B_1}^{B_2} H\,dB \qquad \text{[8.22]}$$

where the integration limits are determined by the initial and final magnetic fields.

Equation 8.22 corresponds to the area between the *B–H* curve and the *B* axis between B_1 and B_2. Suppose that we take the iron core in the toroid from point *P* on the hysteresis curve to *Q*, as shown in Figure 8.35. This is a magnetization process for which energy is put into the sample. The work done per unit volume from *P* to *Q* is the area *PQRS*, shown as hatched. On returning the sample to the same initial magnetization (same magnetic field *B* as we had at *P*), taking it from *Q* to *S*, energy is returned from the core into the electrical circuit. This energy per unit volume is the area *QRS*, shown as gray, and is less than *PQRS* during magnetization. The difference is the energy dissipated in the sample as heat (moving domain walls and so on) and corresponds to the hysteresis loop area *PQS*. Over one full cycle, the energy dissipated per unit volume is the total hysteresis loop area.

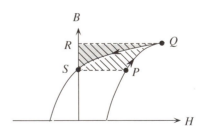

Figure 8.35   The area between the *B–H* curve and the *B* axis is the energy absorbed per unit volume in magnetization or released during demagnetization.

The hysteresis loop and hence the energy dissipated per unit volume per cycle depend not only on the core material but also on the magnitude of the magnetic field (B_m), as apparent in Figure 8.32. For example, for magnetic steels used in transformer cores, the hysteresis **power loss,** P_h, per unit volume of core is empirically expressed in terms of the maximum magnetic field B_m and the ac frequency f as[4]

$$P_h = K f B_m^n \qquad\qquad \text{[8.23]}$$

Hysteresis power loss per m^3

where K is a constant that depends on the core material (typically, $K = 150.7$), f is the ac frequency (Hz), B_m is the maximum magnetic field (T) in the core (assumed to be in the range $0.1-1.5$ T), and $n = 1.6$. According to Equation 8.23, the hysteresis loss can be decreased by operating the transformer with a reduced magnetic field.

8.6   SOFT AND HARD MAGNETIC MATERIALS

8.6.1   Definitions

Based on their $B-H$ behavior, engineering materials are typically classified into soft and hard magnetic materials. Their typical $B-H$ hysteresis curves are shown in Figure 8.36. Soft magnetic materials are easy to magnetize and demagnetize and hence require relatively low magnetic field intensities. Put differently, their $B-H$ loops are narrow, as shown in Figure 8.36. The hysteresis loop has a small area, so the hysteresis power loss per cycle is small. Soft magnetic materials are typically suitable for applications where repeated cycles of magnetization and demagnetization are involved, as in electric motors, transformers, and inductors, where the

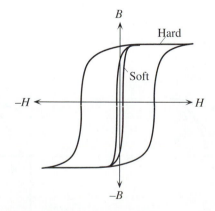

Figure 8.36   Soft and hard magnetic materials

---

[4] This is the power engineers Steinmetz equation for commercial magnetic steels. It has been applied not only to silicon irons (Fe + few percent Si) but also to a wide range of magnetic materials.

magnetic field varies cyclically. These applications also require low hysteresis losses, or small hysteresis loop area. Electromagnetic relays that have to be turned on and off require the relay iron to be magnetized and demagnetized and therefore need soft magnetic materials.

Hard magnetic materials, on the other hand, are difficult to magnetize and demagnetize and hence require relatively large magnetic field intensities, as apparent in Figure 8.36. Their B–H curves are broad and almost rectangular. They possess relatively large coercivities, which means that they need large applied fields to be demagnetized. The coercive field for hard materials can be millions of times greater than those for soft magnetic materials. Their characteristics make hard magnetic materials useful as permanent magnets in a variety of applications. It is also clear that the magnetization can be switched from one very persistent direction to another very persistent direction, from $+B_r$ to $-B_r$, by a suitably large magnetizing field intensity. As the coercivity is strong, both the states, $+B_r$ and $-B_r$, persist until a suitable (large) magnetic field intensity switches the field from one direction to the other. It is apparent that hard magnetic materials can also be used in magnetic storage of digital data, where the states $+B_r$ and $-B_r$ can be made to represent 1 and 0 (or vice versa).

8.6.2   Initial and Maximum Permeability

It is useful to characterize the magnetization of a material by a relative permeability, μ_r, since this simplifies magnetic calculations. For example, inductance calculations become straightforward if one could represent the magnetic material by μ_r alone. But it is clear from Figure 8.37a that

$$\mu_r = \frac{B}{\mu_o H}$$

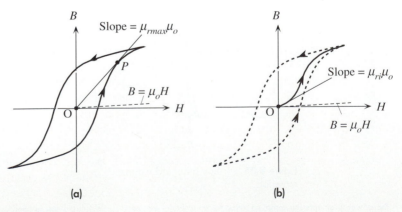

Figure 8.37   Definitions of (a) maximum permeability and (b) initial permeability

is not even approximately constant because it depends on the applied field and the magnetic history of the sample. Nonetheless, we still find it useful to specify a relative permeability to compare various materials and even use it in various calculations. The definition $\mu_r = B/(\mu_o H)$ represents the slope of the straight line from the origin, O, to the point P, as shown in Figure 8.37a. This is a maximum when the line becomes a tangent to the $B-H$ curve at P, as in the figure. Any other line from O to the $B-H$ curve that is not a tangent does not yield a maximum relative permeability (the mathematical proof is left to the reader, though the argument is intuitively acceptable from the figure). The **maximum relative permeability,** as defined in Figure 8.37a, is denoted by μ_{rmax} and serves as a useful magnetic parameter. The point P in Figure 8.37a that defines the maximum permeability corresponds to what is called the "knee" of the $B-H$ curve. Many transformers are designed to operate with the maximum magnetic field in the core reaching this knee point. For pure iron, μ_{rmax} is less than 10^4, but for certain soft magnetic materials such as supermalloys (a nickel–iron alloy), the values of μ_{rmax} can be as high as 10^6.

Initial relative permeability, denoted as μ_{ri}, represents the initial slope of the initial B versus H curve as the material is first magnetized from an unmagnetized state, as illustrated in Figure 8.37b. This definition is useful for soft magnetic materials that are employed at very low magnetic fields (e.g., small signals in electronics and communications engineering). In practice, weak applied magnetic fields where μ_{ri} is useful are typically less than 10^{-4} T. In contrast, μ_{rmax} is useful when the magnetic field in the material is not far removed from saturation. Initial relative permeability of a magnetically soft material can vary by orders of magnitude. For example, μ_{ri} for iron is 150, whereas for a supernumetal-200, a commercial alloy of nickel and iron, it is about 2×10^5.

8.7 SOFT MAGNETIC MATERIALS: EXAMPLES AND USES

Table 8.4 identifies what properties are desirable in soft magnetic materials and also lists some typical examples with various applications. An *ideal* soft magnetic material would have zero coercivity (H_c), a very large saturation magnetization (B_{sat}), zero remanent magnetization (B_r), zero hysteresis loss, and very large μ_{rmax} and μ_{ri}. A number of example materials, from pure iron to ferrites, which are ferrimagnetic, are listed in Table 8.4. Pure iron, although soft, is normally not used in electrical machines (except in a few specific relay-type applications) because its good conductivity allows large eddy currents to be induced under varying fields. Induced eddy currents in the iron lead to joule losses (RI^2), which are undesirable. Addition of a few percentages of silicon to iron (silicon–iron), known typically as silicon–steels, increases the resistivity and hence reduces the eddy current losses. Silicon–iron is widely used in power transformers and electrical machinery.

The nickle–iron alloys with compositions around 77%Ni–23%Fe constitute an important class of soft magnetic materials with low coercivity, low hysteresis

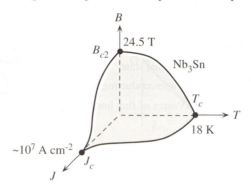

Figure 8.51 The critical surface for a niobium–tin alloy, which is a Type II superconductor

with preparation conditions. For example, in Y–Ba–Cu–O, J_c may be greater than 10^7 A cm^{-2} in some carefully prepared thin films and single crystals but around 10^3–10^6 A cm^{-2} in some of the polycrystalline bulk material (e.g., sintered bulk samples). In Nb$_3$Sn, used in superconducting solenoid magnets, on the other hand, J_c is close to 10^7 A cm^{-2} at near 0 K.

The critical current density is important in engineering because it limits the total current that can be passed through a superconducting wire or a device. The limits of superconductivity are therefore defined by the critical temperature T_c, critical magnetic field B_c (or B_{c2}), and critical current density J_c. These constitute a surface in a three-dimensional plot, as shown in Figure 8.51, which separates the superconducting state from the normal state. Any operating point (T_1, B_1, J_1) inside this surface is in the superconducting state. When the cuprate ceramic superconductors were first discovered, their J_c values were too low to allow immediate significant applications in engineering. Their synthesis over the last 10 years has advanced to a level that we can now benefit from large critical currents and fields. Over the same temperature range, ceramic cuprate superconductors now easily outperform the traditional superconductors. There are already a number of applications of these high T_c superconductors in the commercial market.

Example 8.7 | **SUPERCONDUCTING SOLENOIDS**[7]   Superconducting solenoid magnets can produce very large magnetic fields up to ~15 T or so, whereas the magnetic fields available from a ferromagnetic core solenoid is limited to ~2 T. High field magnets used in magnetic resonance imaging are based on superconducting solenoids wound using a superconducting wire. They are operated around 4 K with expensive liquid helium as the cryogen. These superconducting wires are typically Nb$_3$Sn or NbTi alloy filaments embedded in a copper matrix. A very large current, several hundred amperes, is passed through the solenoid winding to obtain the necessary high magnetic fields. There is, of course, no joule heating

[7] Designing a superconducting solenoid is by no means trivial, and the enthusiastic student is referred to a very readable description given by James D. Doss, *Engineer's Guide to High Temperature Superconductivity* (New York: John Wiley & Sons, 1989), Ch 4. Photographs and descriptions of catastrophic failure in high field solenoids can be found in an article by G. Broebinger, A. Passner, and J. Bevk, "Building World-Record Magnets" in *Scientific American*, June 1995, pp. 59–66.

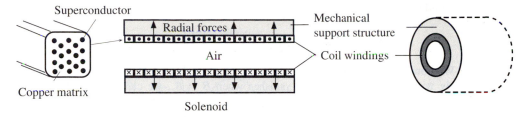

Figure 8.52   A solenoid carrying a current experiences radial forces pushing the coil apart and axial forces compressing the coil.

once the current is flowing in the superconducting state. The main problem is the large forces and hence stresses in the coil due to large currents. Two wires carrying currents in the opposite direction repel each other, and the force is proportional to I^2. Thus the magnetic forces between the wires of the coil give rise to outward radial forces trying to "blow open" the solenoid, as depicted in Figure 8.52. The forces between neighboring wires are attractive and hence give rise to compressional forces squeezing the solenoid axially. The solenoid has to have a proper mechanical support structure around it to prevent mechanical fracture and failure due to large forces between the windings. The copper matrix serves as mechanical support to cushion against the stresses as well as a good thermal conductor in the event that superconductivity is inadvertently lost during operation.

Suppose that we have a superconducting solenoid that is 10 cm in diameter and 1 m in length and has 500 turns of Nb$_3$Sn wire, whose critical field, B_c, at 4.2 K (liquid He temperature) is about 20 T and critical current density, J_c, is 3×10^6 A cm^{-2}. What is the current necessary to set up a field of 5 T at the center of a solenoid? What is the approximate energy stored in the solenoid? Assume that the critical current density decreases linearly with the applied field. Further, assume also that the field across the diameter of the solenoid is approximately uniform (field at the windings is the same as that at the center).

Solution

We can assume that we have a long solenoid, that is, length (100 cm) $\gg$ diameter (10 cm). The field at the center of a long solenoid is given by

$$B = \frac{\mu_o N I}{\ell}$$

so that the current necessary for $B = 5$ T is

$$I = \frac{B\ell}{\mu_o N} = \frac{(5)(1)}{(4\pi \times 10^{-7})(500)} = 7958 \text{ A} \qquad \text{or} \qquad 7.96 \text{ kA}$$

As the coil is 1 m and there are 500 turns, the coil wire radius must be 1 mm. If all the cross section of the wire were of superconducting medium, then the corresponding current density would be

$$J_{\text{wire}} = \frac{I}{\pi r^2} = \frac{7958}{\pi (0.001)^2} = 2.5 \times 10^9 \text{ A m}^{-2} \qquad \text{or} \qquad 2.5 \times 10^5 \text{ A cm}^{-2}$$

The actual current density through the superconductors will be greater than this as the wires are embedded in a metal matrix. Suppose that 20% by cross-sectional area (and hence as volume percentage) is the superconductor; then the actual current density through the

generally ferrite heads are preferred since ferrites are insulators and suffer no eddy current losses. Ferrites however have low saturation magnetizations and require magnetic storage media of low coercivity. The main problem in ferrite recording heads is that the corners of the poles at the air gap become saturated first. Once saturated, the field around the gap is not proportional to the input current signal, and this degrades the quality of recording. This is overcome by coating the pole faces with a high magnetization metal alloy such as Sendust, or, more recently, a magnetic amorphous metal (e.g., CoZrNb), as depicted in Figure 8.55. Since the magnetic metal alloy is only at the tips of the head, the eddy current losses are still small. This type of head where the poles of the ferrite core have a metal coating is called a metal-in-gap (MIG) head and is widely used in various recording applications. The gap distance itself also influences the extent of the fringing field around it and hence the field penetrating into the magnetic tape. The smaller the gap, the greater the fringing. The necessary fringing fields for proper recording on a tape require gap sizes around 1 μm or less.

More recently, recording head devices have been fabricated using thin films of various ferromagnetic metals or ferrite alloys that have sufficiently small eddy current losses to be useable at high frequencies. A highly simplified illustration of the principle of a thin film head is shown in Figure 8.56. The head is manufactured by using typical thin film deposition techniques, such as sputtering of the metal film in a vacuum chamber, photolithography, or some other method. The magentic core is in the form of a thin film whose thickness is a few microns and whose width is about the same as the tape. The gap at the end of the core has the same width as the core, but its spacing is very small (e.g., 0.25 μm) and generates the necessary fringing field. A spiral-type coil made by depositing a nonmagnetic metal thin film threads the core. The magnetic core is like a U-shaped core that is threaded by the

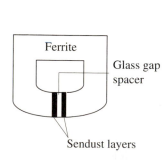

Figure 8.55   A simplified schematic illustration of a MIG (metal-in-gap) head.

The ferrite core has the poles coated with a ferromagnetic soft metal to enhance the head performance.

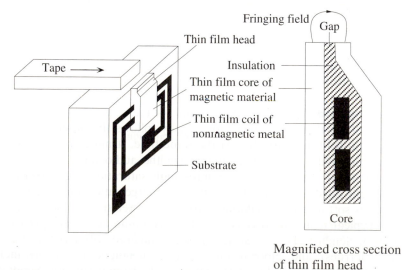

Figure 8.56   A highly simplified schematic illustration of the principle of a thin film head.

metal strips of the coil. If the core is a metallic material, the coil metal is appropriately insulated from it by thin films of insulation.

Magnetic Storage Media Materials   The properties of magnetic storage media such as magnetic tapes, floppy disks, and hard disks used in various magnetic recording applications (audio, video, digital, etc.) must be such that they are able to retain the spatial magnetization pattern written on them after they have passed the recording head. This requires high remanent magnetization, M_r. High remanent magnetization is also important in the reading process because the magnetic flux that induces voltages in the read head depends on this remanent magnetization, given a particular speed of motion under the read head. Thus the read operation requires media with high M_r.

Further, it should be difficult to undesirably erase the magnetic information on the tape by demagnetizing it under stray fields, and this requires high coercivity, H_c. A strong magnet passed over a floppy disk can destroy the information stored in it. The coercivity therefore determines the stability of the recording. The coercivity cannot be too high, however, as this would prevent the writing operation, that is, magnetization, under the recording head. One therefore has to find a compromise that allows the information to be written and at the same time retained without ease of demagnetization.

The above two requirements, high M_r and medium-to-high H_c, lead to a choice of medium to hard magnetic materials as magnetic storage media. Typical flexible storage media (e.g., audio or video tapes) use particulate coatings on flexible polymeric sheets or tapes, as pictured schematically in Figure 8.57. Elongated particles of various magnetic materials are magnetically hard due to a combination of two factors. First, these particles tend to be single domains and are hard due to the magnetocrystalline anisotropy energy. Second, they are also elongated, have greater length to width ratio (aspect ratio), which means they are also hard due to shape anisotropy; they prefer to be magnetized along the length.

Typical particulate matter used in coatings are $\gamma\text{-Fe}_2\text{O}_3$, Co-modified $\gamma\text{-Fe}_2\text{O}_3$ or Co($\gamma\text{-Fe}_2\text{O}_3$), CrO_2, and metallic particles (Fe), as summarized in Table 8.7. The overall magnetic properties of the particulate coating depend not only on the properties of the individual particles (which are hard) but also on the concentration

Thin particulate coating (5–15 µm)

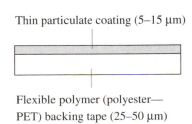

Flexible polymer (polyester—
PET) backing tape (25–50 µm)

Figure 8.57   A magnetic tape is typically a magnetic coating on a flexible polymer (e.g., PET) sheet in the form of a tape.

Soft magnetic materials characteristically have high saturation magnetizations (B_{sat}) but low saturation magnetizing fields (H_{sat}) and low coercivities (H_c) so they can be magnetized and demagnetized easily. They have tall and narrow B–H hysteresis loops.

Superconductivity is a phenomenon in which a substance loses all resistance to current flow (acquires zero resistivity) and also exhibits the Meissner effect (becomes a perfect diamagnet).

Type I superconductors have a single critical field (B_c) above which the superconducting behavior is totally lost.

Type II superconductors have a lower (B_{c1}) and an upper (B_{c2}) critical field. Below B_{c1}, the substance is in the superconducting phase with Meissner effect; all magnetic flux is excluded from the interior. Between B_{c1} and B_{c2}, magnetic flux lines pierce through local filamen-tary regions of the superconductor, which behave normally. Above B_{c2}, the superconductor reverts to normal behavior.

QUESTIONS AND PROBLEMS

8.1   Inductance of a long solenoid   Consider the very long (ideally infinitely long) solenoid shown in Figure 8.61. If r is the radius of the core and ℓ is the length of the solenoid, then $\ell \gg r$. The total number of turns is N and the number of turns per unit length is $n = N/\ell$. The current through the coil wires is I. Apply Ampere's law around C, which is the rectangular circuit PQRS, and show that

$$B \approx \mu_o \mu_r n I$$

Further, show that the inductance is

$$L \approx \mu_o \mu_r n^2 V_{core}$$

where V_{core} is the volume of the core. How would you increase the inductance of a long solenoid?

What is the approximate inductance of an air-cored solenoid with a diameter of 1 cm, length of 20 cm, and 500 turns? What is the magnetic field inside the solenoid and the energy stored in the whole solenoid when the current is 1 A? What happens to these values if the core medium has a relative permeability μ_r of 1000?

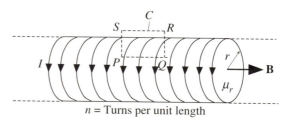

n = Turns per unit length

Figure 8.61

8.2 **Magnetization** Consider a long solenoid with a core that is an iron alloy (see Problem 8.1 for the relevant formulae). Suppose that the diameter of the solenoid is 2 cm and the length of the solenoid is 20 cm. The number of turns on the solenoid is 200. The current is increased until the core is magnetized to saturation at about $I = 2$ A and the saturated magnetic field is 1.5 T.

    a. What is the magnetic field intensity at the center of the solenoid and the applied magnetic field, $\mu_o H$, for saturation?

    b. What is the saturation magnetization M_{sat} of this iron alloy?

    c. What is the total magnetization current on the surface of the magnetized iron alloy specimen?

    d. If we were to remove the iron-alloy core and attempt to obtain the same magnetic field of 1.5 T inside the solenoid, how much current would we need? Is there a practical way of doing this?

8.3 **Pauli spin paramagnetism** Paramagnetism in metals depends on the number of conduction electrons that can flip their spins and align with the applied magnetic field. These electrons are near the Fermi level, E_F, and their number is determined by the density of states $g(E_F)$ at E_F. Since each electron has a spin magnetic moment of β, paramagnetic susceptibility can be shown to be given by

$$\chi_{\text{para}} \approx \mu_o \beta^2 g(E_F)$$

where the density of states is given by Equation 4.10. The Fermi energy of calcium, E_F, is 4.68 eV. Evaluate the paramagnetic susceptibility of calcium and compare with the experimental value of 1.9×10^{-5}.

8.4 **Ferromagnetism and the exchange interaction** Consider dysprosium (Dy), which is a rare earth metal with a density of 8.54 g cm^{-3} and atomic mass of 162.50 g mol^{-1}. The isolated atom has the electron structure [Xe] $4f^{10}6s^2$. What is the spin magnetic moment in the isolated atom in terms of number of Bohr magnetons? If the saturation magnetization of Dy near absolute zero of temperature is 2.4 MA m^{-1}, what is the effective number of spins per atom in the ferromagnetic state? How does this compare with the number of spins in the isolated atom? What is the order of magnitude for the exchange interaction in eV per atom in Dy if the Curie temperature is 85 K?

*8.5 **Toroidal inductor and radio engineers toroidal inductance equation**

    a. Consider a toroidal coil (Figure 8.10) whose mean circumference is ℓ and that has N tightly wound turns around it. Suppose that the diameter of the core is $2a$ and $\ell \gg 2a$. By applying Ampere's law, show that if the current through the coil is I, then the magnetic field in the core is

$$B = \frac{\mu_o \mu_r N I}{\ell} \qquad [8.30]$$

where μ_r is the relative permeability of the medium. Why do you need $\ell \gg a$ for this to be valid? Does this equation remain valid if the core cross section is not circular but rectangular, $a \times b$, and $\ell \gg a$ and b?

b. Show that the inductance of the toroidal coil is

$$L = \frac{\mu_o \mu_r N^2 A}{\ell} \qquad \text{[8.31]}$$

where A is the cross-sectional area of the core.

c. Consider a toroidal inductor used in electronics that has a ferrite core size FT-37, that is, round but with a rectangular cross section. The outer diameter is 0.375 in (9.52 mm), the inner diameter is 0.187 in (4.75 mm), and the height of the core is 0.125 in (3.175 mm). The initial relative permeability of the ferrite core is 2000, which corresponds to a ferrite called the 77 Mix. If the inductor has 50 turns, then using Equation 8.31, calculate the approximate inductance of the coil.

d. Radio engineers use the following equation to calculate the inductances of toroidal coils,

$$L \text{ (mH)} = \frac{A_L N^2}{10^6} \qquad \text{[8.32]}$$

where L is the inductance in millihenries (mH) and A_L is an inductance parameter, called an **inductance index,** that characterizes the core of the inductor. A_L is supplied by the manufacturers or ferrite cores and is typically quoted as millihenries (mH) per 1000 turns. In using Equation 8.32, one simply substitutes the numerial value of A_L to find L in millihenries. For the FT-37 ferrite toroid with the 77 Mix as the ferrite core, A_L is specified as 884 mH/1000 turns. What is the inductance of the toroidal inductor in part c from the radio engineers equation in Equation 8.32? What is the percentage difference in values calculated by Equations 8.32 and 8.31? What is your conclusion? [*Comment:* The agreement is not always this close.]

*8.6   A toroidal inductor

a. Equations 8.31 and 8.32 allow the inductance of a toroidal coil in electronics to be calculated. Equation 8.32 is the equation that is used in practice. Consider a toroidal inductor used in electronics that has a ferrite core of size FT-23 that is round but with a rectangular cross section. The outer diameter is 0.230 in (5.842 mm), the inner diameter is 0.120 in (3.05 mm), and the height of the core is 0.06 in (1.5 mm). The ferrite core is a 43-Mix that has an initial relative permeability of 850 and a maximum relative permeability of 3000. The inductance index for this 43-Mix ferrite core of size FT-23 is $A_L = 188$ (mH/1000 turns). If the inductor has 25 turns, then using Equations 8.31 and 8.32, calculate the inductance of the coil under small-signal conditions and comment on the two values.

b.   The saturation field, B_{sat}, of the 43-Mix ferrite is 0.2750 T. What will be typical dc currents that will saturate the ferrite core (an estimate calculation is required)? It is not unusual to find such an inductor in an electronic circuit also carrying a dc current. Will your calculation of the inductance remain valid in these circumstances?

c.   Suppose that the above toroidal inductor is in the vicinity of a very strong magnet that saturates the magnetic field inside the ferrite core. What will be the inductance of the coil?

*8.7   The transformer

a.   Consider the transformer shown in Figure 8.62a whose primary is excited by an ac voltage of frequency f. The current flowing into the primary coil sets up a magnetic flux in the transformer core. By virtue of Faraday's law of induction and Lenz's law, the flux generated in the core is the flux necessary to induce a voltage nearly equal and opposite to the applied voltage. Thus,

$$v = \frac{d(\text{Total flux linked})}{dt} = \frac{NAdB}{dt}$$

where A is the cross-sectional area, assumed constant, and N is the number of turns in the primary. Show that if V_{rms} is the rms voltage at the primary ($V_{max} = V_{rms}\sqrt{2}$) and B_m is the maximum magnetic field in the core, then

$$V_{rms} = 4.44\, NAfB_m \qquad \textbf{[8.33]}$$

Transformers are typically operated with B_m at the "knee" of the $B–H$ curve, which corresponds roughly to maximum permeability. For transformer irons, $B_m \approx 0.12$ T. Taking $V_{rms} = 120$ V and a transformer core with A = 10 cm × 10 cm, what should N be for the primary winding? If the secondary winding is to generate 240 V, what should be the number of turns for the secondary coil?

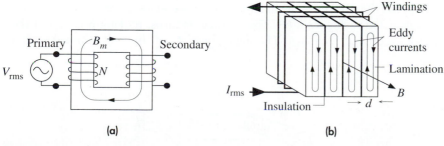

(a)   (b)

Figure 8.62
(a) A transformer with N turns in the primary.
(b) Laminated core reduces eddy current losses.

b. The transformer core will exhibit hysteresis and eddy current losses. The hysteresis loss per unit second, as power loss in watts, is given by

$$P_h = KfB_m^n V_{core} \qquad \text{[8.34]}$$

where $K = 150.7$, f is the ac frequency (Hz), B_m is the maximum magnetic field (T) in the core (assumed to be in the range 0.2–1.5 T), $n = 1.6$, and V_{core} is the volume of the core. The eddy current losses are reduced by laminating the transformer core, as shown in Figure 8.62b. The eddy current loss is given by

$$P_e = 1.65 \; f^2 B_m^2 \left(\frac{d^2}{\rho}\right) V_{core} \qquad \text{[8.35]}$$

where d is the thickness of the laminated iron sheet in meters (Figure 8.62b) and ρ is its resistivity (Ω m).

Suppose that the transformer core has a volume of 0.0108 m³ (corresponds to a mean circumference of 1.08 m). If the core is laminated into sheets of thickness 1 mm and the resistivity of the transformer iron is 6×10^{-7} Ωm, calculate both the hysteresis and eddy current losses and comment on their relative magnitudes. How would you reduce each loss?

8.8 **Losses in a magnetic recording head**   Consider eddy current losses in a permalloy magnetic head for audio recording up to 10 kHz. We will use the following equation for the eddy current losses,

$$P_e = 1.65 \; f^2 B_m^2 \left(\frac{d^2}{\rho}\right) V_{core}$$

where V_{core} is the volume of the core. Consider a magnetic head weighing 30 grams and made from a permalloy with density 8.8 g cm⁻³ and resistivity 6×10^{-7} Ω m. The head is to operate at B_m of 0.5 T. If the eddy current losses are not to exceed 1 mW, estimate the thickness of laminations needed. How would you achieve this?

*8.9 **Design of a ferrite antenna for an AM receiver**   We consider an AM radio receiver that is to operate over the frequency range 530–1600 kHz. Suppose that the receiving antenna is to be a coil with a ferrite rod as core, as depicted in Figure 8.63. The coil has N turns, its length is ℓ, and the cross-sectional area is A. The inductance, L, of this coil is tuned with a

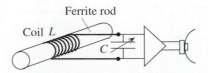

Figure 8.63   A ferrite antenna of an AM receiver

variable capacitor C. The maximum value of C is 265 pF, which with L should correspond to tuning in the lowest frequency at 530 kHz. The coil with the ferrite core receives the EM waves, and the magnetic field of the EM wave permeates the ferrite core and induces a voltage across the coil. This voltage is detected by a sensitive amplifier, and in subsequent electronics it is suitably demodulated. The coil with the ferrite core therefore acts as the antenna of the receiver (ferrite antenna). We will try to find a suitable design for the ferrite coil by carrying out approximate calculations—in practice some trial and error experimentation would also be necessary. We will assume that the inductance of a finite solenoid is

$$L = \frac{\gamma \mu_{ri} \mu_o A N^2}{\ell} \qquad \text{[8.36]}$$

Inductance of a solenoid

where A is the cross-sectional area of the core, ℓ is the coil length, N is the number of turns, and γ is a geometric factor that accounts for the solenoid coil being of finite length. Assume $\gamma \approx 0.75$. The resonant frequency f of an LC circuit is given by

$$f = \frac{1}{2\pi(LC)^{1/2}} \qquad \text{[8.37]}$$

a. If d is the diameter of the enameled wire to be used as the coil winding, then the length $\ell \approx Nd$. If we use an enameled wire of diameter 1 mm, what is the number of coil turns, N, we need for a ferrite rod given its diameter is 1 cm and given initial relative permeability is 100?

b. Suppose that the magnetic field intensity H of the signal in free space is varying sinusoidally, that is

$$H = H_m \sin(2\pi ft) \qquad \text{[8.38]}$$

where H_m is the maximum magnetic field intensity. H is related to the electric field $\mathcal{E}$ at the point by $H = \mathcal{E}/Z_{\text{space}}$, where Z_{space} is the impedance of free space given by 377 Ω. Show that the induced voltage at the antenna coil is

$$V_m = \frac{\mathcal{E}_m d}{2\pi \, 377 C f \gamma} \qquad \text{[8.39]}$$

Induced voltage across a ferrite antenna

where f is the frequency of the AM wave and $\mathcal{E}_m$ is the electric field intensity of the AM station at the receiver point. Suppose that the electric field of a local AM station at the receiver is 10 mV m^{-1}. What is the voltage induced across the ferrite antenna and can this voltage be detected by an amplifier? What is your conclusion? Would you use a ferrite rod antenna at short wave frequencies (higher than ~5 MHz)?

*8.10 **A permanent magnet with an air gap**   The magnetic field energy in the gap of a permanent magnet is available to do work. Suppose that B_m and B_g are the magnetic field in the magnet and the gap, H_m and H_g are the field intensities in the magnet and the gap, and V_m and V_g are the volumes of the magnet and gap; show that, in terms of magnitudes,

$$B_g H_g V_g \approx B_m H_m V_m$$

What is the significance of this result?

8.11   A permanent magnet with an air gap

a.   Show that the maximum energy stored in the air gap of a permanent magnet can be written very roughly as

$$E_{gap} \approx \frac{1}{8} B_r H_c V_m$$

where V_m is the volume of the magnet, which is much greater than that of the gap, B_r is the remanent magnetic field, and H_c is the coercivity of the magnet.

b.   Using Table 8.5, compare the $(BH)_{max}$ with the product $[(1/2)H_c]$ $[(1/2)B_r]$ and comment on the closeness of agreement.

c.   Calculate the energy in the gap of a rare earth cobalt magnet that has a volume of 0.1 m³. Give an example of typical work (e.g., raising so many apples, each 100 grams, by so many meters) that could be done if all this energy could be converted to mechanical work.

8.12   Weight, cost, and energy of a permanent magnet with an air gap   For a certain application, an energy of 1 kJ is required in the gap of a permanent magnet. There are three candidates, as shown in Table 8.9. Which material will give the lightest magnet? Which will give the cheapest magnet?

8.13   Superconductivity and critical current density   Consider two superconducting wires, niobium (Type I) and Nb₃Sn (Type II), each 1 mm in thickness. The magnetic field on the surface of a current-carrying conductor is given by

$$B = \frac{\mu_o I}{2\pi r}$$

Table 8.9   Three permanent magnet candidates

Magnet	$(BH)_{max}$ (kJ m⁻³)	Density (g cm⁻³)	Yesterday's Relative Price (per unit mass)
Alnico	50	7.3	1
Rare earth	200	8.2	2
Ferrite	30	4.8	0.5

a. Assuming that Nb wire loses its superconductivity when the field at the surface reaches the critical field (0.2 T), calculate the maximum current and hence the critical current density that can be passed through the Nb wire near absolute zero of temperature.

b. Calculate the maximum current and critical current density for the Nb_3Sn wire using the same assumption as in part *a* but taking the critical field to be the upper critical field, B_{c2}, which is 24.5 at 0 K. How does your calculation of J_c compare with the critical density of about 10^{11} A m^{-2} for Nb_3Sn at 0 K.

*8.14 **Enterprising engineers in Northwest Territories building a superconducting inductor** A current-carrying inductor has energy stored in its magnetic field that can be converted to electrical work. A group of enterprising engineers and scientists living in Resolute in Northwest Territories (Canada) have decided to build a toroidal inductor to store energy so that this energy can be used to supply a small community of 10 houses each consuming on average 3 kW of energy during the night (6 months). They have discovered a superconductor (Type II) that has a $B_{c2} = 100$ T and a critical current density of $J_c = 5 \times 10^{10}$ A m^{-2} at night temperatures (it is obviously a novel high-T_c superconductor of some sort). Their superconducting wire has a diameter of 5 mm and is available in any desirable length. All the wiring in the community is done by superconductors except where energy needs to be converted to other forms (mechanical, heat, etc.). They have decided on the following design specification for their toroid:

The mean diameter, D_{toroid}, of toroid, (1/2) (Outside diameter + Inside diameter), is 10 times longer than the core diameter, D_{core}. The field inside the toroid is therefore reasonably uniform to within 10%.

The maximum operating magnetic field in the core is 35 T. Fields larger than this can result in mechanical fracture and failure.

Assume that J_c decreases linearly with the magnetic field and that the mechanical engineers in the group can take care of the forces trying to blow open the toroid by building a proper support structure.

Find the size of the toroid (mean diameter and circumference), the number of turns and the length of the superconducting wire they need, the current in the coil, and whether this current is sufficiently below the critical current at that field. Is it feasible?

8.15 **Magnetic storage media**

a. Consider the storage of video information (FM signal) on a video tape. Suppose that the maximum signal frequency to be recorded as aspatial magnetic pattern is 10 MHz. The heads helically scan the tape, and the relative velocity of the tape to head is about 10 m s^{-1}. What is the minimum spatial wavelength of the stored magnetic pattern (information) on the tape?

b. Suppose that the speed of an audio cassette tape in a cassette player is 5 cm s^{-1}. If the maximum frequency that needs to be recorded is 20 kHz, what is the minimum spatial wavelength on the tape?

  c. Discuss the advantages and disadvantages of longitudinal and perpendicular recording technologies. What do you think is the ultimate limit to the size of a magnetic domain? What would be the corresponding estimate of storage density?

A

Major Symbols and Abbreviations

A	area; cross-sectional area; amplification
a	lattice parameter; acceleration; amplitude of vibrations; half-channel thickness in a JFET (Ch. 6)
a (subscript)	acceptor, e.g., N_a = acceptor concentration (m^{-3})
ac	alternating current
a_o	Bohr radius (0.0529 nm)
A_V, A_P	voltage amplification, power amplification
APF	atomic packing factor
B, B	magnetic field vector (T), magnetic field
B	frequency bandwidth; base terminal of a BJT
B_c	critical magnetic field
B_m	maximum magnetic field
B_o, B_e	Richardson-Dushman constant, effective Richardson-Dushman constant
BC	base collector
BCC	body-centered cubic
BE	base emitter
BJT	bipolar junction transistor
C	capacitance; composition; the Nordheim coefficient (Ω m); collector terminal of a BJT
c	speed of light (3×10^8 m s^{-1}); specific heat capacity (J K^{-1} kg^{-1})
C_{dep}	depletion layer capacitance
C_m	molar heat capacity (J K^{-1} mol^{-1})
C_{diff}	diffusion (storage) capacitance of a forward-biased pn junction
CAD	computer aided design
CB	conduction band; common base
CE	common emitter
CMOS	complementary MOS
CN	coordination number
CRT	cathode-ray tube
CVD	chemical vapor deposition
D	diffusion coefficient (m^2 s^{-1}); thickness; drain terminal of an FET; electric displacement (C m^{-2})
d	density (kg m^{-3}); distance; separation of the atomic planes in a crystal, separation of capacitor plates; piezoelectric coefficient
d (subscript)	donor, e.g., N_d = donor concentration (m^{-3})
dc	direct current
d_{ij}	piezoelectric coefficients
DEA	dielectric analyzer; dielectric analysis
DRAM	dynamic random access memory

E	energy; emitter terminal of a bipolar transistor
E_a, E_d	acceptor and donor energy levels
E_c, E_v	conduction band edge, valence band edge
E_{ex}	exchange interaction energy
E_F, E_{FO}	Fermi energy, Fermi energy at 0 K
E_g	band gap energy
E_{mag}	magnetic energy
$\mathcal{E}$	electric field (V m^{-1})
$\mathcal{E}_{br}$	dielectric strength or break down field (V m^{-1})
$\mathcal{E}_{loc}$	local electric field
e	electronic charge (1.60218 $\times$ 10^{-19} C)
e	Napierian base, 2.71828. . .
e (subscript)	electron, e.g., μ_e = electron drift mobility
eff (subscript)	effective, e.g., μ_{eff} = effective drift mobility
EB	emitter base
EHP	electron hole pair
EM	electromagnetic
EMF (emf)	electromagnetic force (V)
F	force (N); function
f	frequency; function
$f(E)$	Fermi-Dirac function
F_{ext}, F_{int}	external force (due to an applied field) or $F_{external}$, internal force (due to ions in the crystal) or $F_{internal}$
FCC	face-centered cubic
FET	field effect transistor
G	rate of generation; gate of an FET
G_{ph}	rate of photogeneration
G_p	parallel conductance (Ω^{-1})
$g(E)$	density of states
g	conductance; transconductance (A/V)
g_d	incremental conductance (A/V)
g_m	mutual transconductance (A/V)
$\mathbf{H}, H$	magnetic field intensity (strength) or magnetizing field (A m^{-1})
h	Planck's constant (6.6261 $\times$ 10^{-34} J s)
$\hbar$	Planck's constant divided by 2π (1.0546 $\times$ 10^{-34} J s)
h (subscript)	hole, e.g., μ_h = hole drift mobility
h_{FE}, h_{fe}	dc current gain, small signal (ac) current gain in the common emitter configuration
HCP	hexagonal close-packed
HF	high frequency
I	electric current (A); moment of inertia (kg m^2) (Ch.1)
$\mathcal{I}$	light intensity (W m^{-2})
I_{br}	breakdown current
$I_B I_C, I_E$	base, collector, and emitter currents in a BJT
i	instantaneous current (A); small signal (ac) current, $i = \delta I$
i (subscript)	intrinsic, e.g., n_i = intrinsic concentration
i_b, i_c, i_e	small signal base, collector, and emitter currents ($\delta I_B, \delta I_C, \delta I_E$) in a BJT
IC	integrated circuit
J	current density (A m^{-2})
$\mathbf{J}$	total angular momentum vector
j	imaginary constant: $\sqrt{-1}$
J_c	critical current density (A m^{-2})
JFET	junction FET

K	spring constant (Ch. 1); dielectric constant (Ch. 7)
k	Boltzmann constant ($k = R/N_A = 1.3807 \times 10^{-38}$ J K^{-1}); wavenumber ($k = 2\pi/\lambda$), wavevector (m^{-1})
KE	kinetic energy
$\mathbf{L}$	total orbital angular momentum
L	length; inductance
ℓ	length; mean free path (ℓ_e = "effective" mean free path); orbital angular momentum quantum number
L_{ch}	channel length in an FET
L_e, L_h	electron and hole diffusion lengths
ℓ_n, ℓ_p	lengths of the n and p regions outside depletion region in a pn junction
$\ln(x)$	natural logarithm of x
LCAO	linear combination of atomic orbitals
$\mathbf{M}, M$	magnetization vector (A m^{-1}), magnetization (A m^{-1})
M	multiplication in avalanche effect
M_{at}	relative atomic mass; atomic mass; "atomic weight" (grams per mole)
M_r	remanent or residual magnetization (A m^{-1})
M_{sat}	saturation magnetization (A m^{-1})
m	mass (kg)
m_e	mass of the electron in free space (9.10939×10^{-31} kg)
m_e^*	effective mass of the electron in a crystal
m_h^*	effective mass of a hole in a crystal
m_ℓ	magnetic quantum number
m_s	spin magnetic quantum number
MOS (MOST)	metal-oxide-semiconductor (transistor)
MOSFET	metal-oxide-semiconductor FET
N	number of atoms or molecules; number of atoms per unit volume (m^{-3})(Ch. 7); number of turns on a coil (Ch. 8)
n	electron concentration (number per unit volume); atomic concentration; principal quantum number; integer number
n^+	heavily doped n-region
n_{at}	number of atoms per unit volume
N_c, N_v	effective density of states at the conduction and valence band edges (m^{-3})
N_d, N_d^+	donor and ionized donor concentrations (m^{-3})
n_i	intrinsic concentration (m^{-3})
n_{no}, p_{po}	equilibrium majority carrier concentrations (m^{-3})
n_{po}, p_{no}	equilibrium minority carrier concentrations (m^{-3})
N_S	concentration of electron scattering centers
n_v	velocity density function; vacancy concentration (m^{-3})
P	probability; pressure (Pa); power (W) or power loss (W)
$\mathbf{p}, p$	electric dipole moment (C m)
p	hole concentration (m^{-3}); momentum (kg m s^{-1}); pyroelectric coefficient (Ch. 7)
p^+	heavily doped p-region
$\mathbf{p}_{av}$	average dipole moment per molecule
p_e	electron momentum (kg m s^{-1})
PE	potential energy
$p_{induced}$	induced pole moment (C m)
p_o	permanent dipole moment (C m)
PET	polyester, polyethylene terephtalate
PZT	lead zirconate titanate
Q	charge (C); heat (J); quality factor
Q'	rate of heat flow (W)
q	charge (C)

R	gas constant ($N_a k = 8.31457$ J mol^{-1} K^{-1}); resistance; radius; reflection coefficient (Ch. 3); rate of recombination (Ch. 5)
r	position vector
r	radial distance; radius; interatomic separation; resistance per unit length
R_H	Hall coefficient (m^3 C^{-1})
r_o	bond length, equilibrium separation
rms	root mean square
S	total spin momentum, intrinsic angular momentum
S	cross-sectional area of a scattering center; Seebeck coefficient, thermoelectric power (V m^{-1}); source terminal of an FET
S_{band}	number of states per unit volume in the band
S_j	strain along direction j
SCL	space charge layer
t	time (s); thickness (m)
T	temperature in Kelvin; transmission coefficient
tanδ	loss tangent
T_C	Curie temperature
T_c	critical temperature (K)
T_j	mechanical stress along direction j (Pa)
TC	thermocouple
TCC	temperature coefficient of capacitance (K^{-1})
TCR	temperature coefficient of resistivity (K^{-1})
U	total internal energy of 1 mole
u	mean speed (of electron) (m s^{-1})
V	voltage; volume; PE function of the electron, $PE(x)$
V_{br}	breakdown voltage
V_o	built-in voltage
V_P	pinch-off voltage
V_r	reverse bias voltage
v	velocity (m s^{-1}); instantaneous voltage (V)
$\overline{v^2}$	mean square velocity; mean square voltage
v_{dx}	drift velocity in the x direction
v_e, v_{rms}	effective velocity or rms velocity of the electron
v_F	Fermi speed
v_g	group velocity
v_{th}	thermal velocity
VB	valence band
VLSI	very large scale integration
W	width; width of depletion layer with applied voltage; dielectric loss
W_o	width of depletion region with no applied voltage
W_n, W_p	width of depletion region on the n side and on the p side with no applied voltage
X	atomic fraction
Y	admittance (Ω^{-1}); Young's modulus (Pa)
Z	impedance (Ω); atomic number, number of electrons in the atom
α	polarizability; temperature coefficient of resistivity (K^{-1}); absorption coefficient (m^{-1}); gain or current transfer ratio from emitter to collector of a BJT
β	current gain I_C/I_B of a BJT; Bohr magneton (9.2732×10^{-24} J T^{-1})
β_S	Schottky coefficient
γ	emitter injection efficiency (Ch. 6); gyromagnetic ratio (Ch. 8)
Γ, Γ_{ph}	flux (m^{-2} s^{-1}), photon flux (photons m^{-2} s^{-1})
δ	small change; skin depth (Ch. 2); loss angle (Ch. 7)
Δ	change, excess (e.g., Δn = excess electron concentration)

∇^2	$\partial^2/\partial x^2 + \partial^2/\partial y^2 + \partial^2/\partial z^2$
ε	$\varepsilon_o\varepsilon_r$, permittivity of a medium (C V^{-1} m^{-1} or F m^{-1}); elastic strain
ε_o	permittivity of free space or absolute permittivity (8.8542 $\times$ 10^{-12} C V^{-1} m^{-1} or F m^{-1})
ε_r	relative permittivity or dielectric constant
η	efficiency; quantum efficiency; ideality factor; emissivity
θ	angle; an angular spherical coordinate; thermal resistance
κ	thermal conductivity (W m^{-1} K^{-1}); dielectric constant
λ	wavelength (m); thermal coefficient of linear expansion (K^{-1}); characteristic length (Ch. 8)
μ	$\mu_o\mu_r$, magnetic permeability (H m^{-1}); chemical potential (Ch. 5)
μ_o	absolute permeability (4π $\times$ 10^{-7} H m^{-1})
μ_r	relative permeability
μ_m	magnetic dipole moment (A m^2)
μ_d	drift mobility (m^2 V^{-1} s^{-1})
μ_h, μ_e	hole drift mobility, electron drift mobility (m^2 V^{-1} s^{-1})
v	frequency (Hz); Poisson's ratio
π	pi, 3.14159. . .
Π	Peltier coefficient (V)
ρ	resistivity (Ω m); density (kg m^{-3}); charge density (C m^{-3})
ρ_E	energy density (J m^{-3})
ρ_{net}	net space charge density (C m^{-3})
ρJ^2	joule heating per unit volume (W m^{-3})
σ	electrical conductivity (Ω^{-1} m^{-1}); surface concentration of charge (C m^{-2}) (Ch. 7)
σ_P	polarization charge density (C m^{-2})
σ_o	free surface charge density (C m^{-2})
σ_S	Stefan's constant (5.670 $\times$ 10^{-8} W m^{-2} K^{-4})
τ	time constant; mean electron scattering time; relaxation time; torque (N m)
τ_g	mean time to generate an electron-hole pair
ϕ	angle; an angular spherical coordinate
Φ	work function (J or eV), magnetic flux (Wb)
Φ_m	metal work function (J or eV)
Φ_n	energy required to remove an electron from an n-type semiconductor (J or eV)
χ	volume fraction; electron affinity; susceptibility (χ_e is electrical; χ_m is magnetic)
$\Psi(x, t)$	total wavefunction
$\psi(x)$	spatial dependence of the electron wavefunction under steady-state conditions
$\psi_k(x)$	Bloch wavefunction, electron wavefunction in a crystal
ψ_{hyb}	hybrid orbital
ω	angular frequency (2πv); oscillation frequency (rad s^{-1})

B

Constants and Useful Information

Physical Constants

Atomic mass unit	amu	1.66054×10^{-27} kg
Avogadro's number	N_A	6.02214×10^{23} mol^{-1}
Bohr magneton	β	9.2732×10^{-24} J T^{-1}
Electron mass in free space	m_e	9.10939×10^{-31} kg
Electron charge	e	1.60218×10^{-19} C
Gas constant	R	8.31451 J K^{-1} mole^{-1} or m^3 Pa K^{-1} mol^{-1}
Gravitational constant	G	6.6726×10^{-11} N m^2 kg^{-2}
Boltzmann constant	k	1.3806×10^{-23} J K^{-1} = 8.6174×10^{-5} eV K^{-1}
Permeability of vacuum or absolute permeability	μ_o	$4\pi \times 10^{-7}$ H m^{-1} (or Wb A^{-1} m^{-1})
Permittivity of vacuum or absolute permittivity	ε_o	8.8542×10^{-12} F m^{-1}
Planck's constant	h	6.626×10^{-34} J s = 4.135×10^{-15} eV s
Planck's constant / 2π	$\hbar$	1.055×10^{-34} J s = 6.582×10^{-16} eV s
Proton rest mass	m_p	1.67262×10^{-27} kg
Rydberg constant	R_∞	1.0937×10^{-7} m^{-1}
Speed of light	c	2.9979×10^8 m s^{-1}
Stefan's constant	σ_S	5.670×10^{-8} W m^{-2} K^{-4}

Useful Information

Acceleration due to gravity at 45° latitude	g	9.81 m s^{-2}
kT at $T = 293$ K (20 °C)	kT	0.02525 eV
kT at $T = 300$ K (27 °C)	kT	0.02585 eV
Bohr radius	a_o	0.0529 nm
1 ångstrom	Å	10^{-10} m
1 micron	μm	10^{-6} m

1 eV = 1.6022×10^{-19} J

1 kJ mol^{-1} = 0.010363 eV atom^{-1}

1 atm. = 1.013×10^5 Pa

C

Elements to Uranium

Element	Symbol	Z	Atomic mass (g mol^{-1})	Electron Structure	Density (g cm^{-3}) (*At 0 °C, 1 atm)
Hydrogen	H	1	1.008	$1s^1$	0.00009*
Helium	He	2	4.003	$1s^2$	0.00018*
Lithium	Li	3	6.941	$[He]2s^1$	0.53
Beryllium	Be	4	9.012	$[He]2s^2$	1.85
Boron	B	5	10.81	$[He]2s^2p^1$	2.54
Carbon	C	6	12.01	$[He]2s^2p^2$	2.62
Nitrogen	N	7	14.007	$[He]2s^2p^3$	0.00125*
Oxygen	O	8	15.99	$[He]2s^2p^4$	0.00143*
Fluorine	F	9	18.99	$[He]2s^2p^5$	0.00170*
Neon	Ne	10	20.18	$[He]2s^2p^6$	0.00090*
Sodium	Na	11	22.99	$[Ne]3s^1$	0.97
Magnesium	Mg	12	24.30	$[Ne]3s^2$	1.74
Aluminum	Al	13	26.96	$[Ne]3s^2p^1$	2.70
Silicon	Si	14	28.09	$[Ne]3s^2p^2$	2.33
Phosphorus	P	15	30.97	$[Ne]3s^2p^3$	1.82
Sulfur	S	16	32.06	$[Ne]3s^2p^4$	2.07
Chlorine	Cl	17	35.45	$[Ne]3s^2p^5$	0.0032*
Argon	Ar	18	39.95	$[Ne]3s^2p^6$	0.0018*
Potassium	K	19	39.09	$[Ar]4s^1$	0.86
Calcium	Ca	20	40.08	$[Ar]4s^2$	1.55
Scandium	Sc	21	44.96	$[Ar]3d^14s^2$	3.0
Titanium	Ti	22	47.88	$[Ar]3d^24s^2$	4.50
Vanadium	V	23	50.94	$[Ar]3d^34s^2$	5.8
Chromium	Cr	24	51.99	$[Ar]3d^54s^1$	7.19
Manganese	Mn	25	54.93	$[Ar]3d^54s^2$	7.43
Iron	Fe	26	55.85	$[Ar]3d^64s^2$	7.86
Cobalt	Co	27	58.93	$[Ar]3d^74s^2$	8.90
Nickel	Ni	28	58.70	$[Ar]3d^84s^2$	8.90
Copper	Cu	29	63.54	$[Ar]3d^{10}4s^1$	8.96
Zinc	Zn	30	65.38	$[Ar]3d^{10}4s^2$	7.14
Gallium	Ga	31	69.72	$[Ar]3d^{10}4s^2p^1$	5.91
Germanium	Ge	32	72.61	$[Ar]3d^{10}4s^2p^2$	5.32
Arsenic	As	33	74.92	$[Ar]3d^{10}4s^2p^3$	5.72
Selenium	Se	34	78.96	$[Ar]3d^{10}4s^2p^4$	4.80
Bromine	Br	35	79.90	$[Ar]3d^{10}4s^2p^5$	3.12
Krypton	Kr	36	83.80	$[Ar]3d^{10}4s^2p^6$	3.74
Rubidium	Rb	37	85.47	$[Kr]5s^1$	1.53
Strontium	Sr	38	87.62	$[Kr]5s^2$	2.6

Element	Symbol	Z	Atomic mass (g mol^{-1})	Electron Structure	Density (g cm^{-3}) (*At 0 °C, 1 atm)
Yttrium	Y	39	88.90	[Kr]4d^{1}5s^2	4.5
Zirconium	Zr	40	91.22	[Kr]4d^{2}5s^2	6.49
Niobium	Nb	41	92.91	[Kr]4d^{4}5s^1	8.55
Molybdenum	Mo	42	95.94	[Kr]4d^{5}5s^1	10.2
Technetium	Tc	43	(97.91)	[Kr]4d^{5}5s^2	11.5
Ruthenium	Ru	44	101.07	[Kr]4d^{7}5s^1	12.2
Rhodium	Rh	45	102.91	[Kr]4d^{8}5s^1	12.4
Palladium	Pd	46	106.42	[Kr]4d^{10}5s^0	12.0
Silver	Ag	47	107.87	[Kr]4d^{10}5s^1	10.5
Cadmium	Cd	48	112.41	[Kr]4d^{10}5s^2	8.65
Indium	In	49	114.82	[Kr]4d^{10}5s^{2}p^1	7.31
Tin	Sn	50	118.71	[Kr]4d^{10}5s^{2}p^2	7.30
Antimony	Sb	51	121.75	[Kr]4d^{10}5s^{2}p^3	6.62
Tellurium	Te	52	127.60	[Kr]4d^{10}5s^{2}p^4	6.24
Iodine	I	53	126.91	[Kr]4d^{10}5s^{2}p^5	4.92
Xenon	Xe	54	131.29	[Kr]4d^{10}5s^{2}p^6	0.0059*
Cesium	Cs	55	132.90	[Xe]6s^1	1.87
Barium	Ba	56	137.33	[Xe]6s^2	3.78
Lanthanum	La	57	138.91	[Xe]5d^{1}6s^2	6.7
Cerium	Ce	58	140.12	[Xe]4f^{1}5d^{1}6s^2	6.78
Praseodymium	Pr	59	140.91	[Xe]4f^{3}6s^2	6.77
Neodymium	Nd	60	144.24	[Xe]4f^{4}6s^2	7.00
Promethium	Pm	61	(145)	[Xe]4f^{5}6s^2	6.475
Samarium	Sm	62	150.4	[Xe]4f^{6}6s^2	7.54
Europium	Eu	63	151.97	[Xe]4f^{7}6s^2	5.26
Gadolinium	Gd	64	157.25	[Xe]4f^{7}5d^{1}6s^2	7.89
Terbium	Tb	65	158.92	[Xe]4f^{9}6s^2	8.27
Dysprosium	Dy	66	162.50	[Xe]4f^{10}6s^2	8.54
Holmium	Ho	67	164.93	[Xe]4f^{11}6s^2	8.80
Erbium	Er	68	167.26	[Xe]4f^{12}6s^2	9.05
Thulium	Tm	69	168.93	[Xe]4f^{13}6s^2	9.33
Ytterbium	Yb	70	173.04	[Xe]4f^{14}6s^2	6.98
Lutetium	Lu	71	174.97	[Xe]4f^{14}5d^{1}6s^2	9.84
Hafnium	Hf	72	178.49	[Xe]4f^{14}5d^{2}6s^2	13.1
Tantalum	Ta	73	180.94	[Xe]4f^{14}5d^{3}6s^2	16.6
Tungsten	W	74	183.85	[Xe]4f^{14}5d^{4}6s^2	19.3
Rhenium	Re	75	186.20	[Xe]4f^{14}5d^{5}6s^2	21.0
Osmium	Os	76	190.2	[Xe]4f^{14}5d^{6}6s^2	22.6*
Iridium	Ir	77	192.22	[Xe]4f^{14}5d^{7}6s^2	22.5
Platinum	Pt	78	195.08	[Xe]4f^{14}5d^{9}6s^1	21.4
Gold	Au	79	196.97	[Xe]4f^{14}5d^{10}6s^1	19.3
Mercury	Hg	80	200.59	[Xe]4f^{14}5d^{10}6s^2	13.55
Thallium	Tl	81	204.38	[Xe]4f^{14}5d^{10}6s^{2}p^1	11.85
Lead	Pb	82	207.2	[Xe]4f^{14}5d^{10}6s^{2}p^2	11.34
Bismuth	Bi	83	208.98	[Xe]4f^{14}5d^{10}6s^{2}p^3	9.8
Polonium	Po	84	(209)	[Xe]4f^{14}5d^{10}6s^{2}p^4	9.4
Astatine	At	85	(210)	[Xe]4f^{14}5d^{10}6s^{2}p^5	—
Radon	Rn	86	(222)	[Xe]4f^{14}5d^{10}6s^{2}p^6	0.0099
Francium	Fr	87	(223)	[Rn]7s^1	—
Radium	Ra	88	226.02	[Rn]7s^2	5
Actinium	Ac	89	227.02	[Rn]6d^{1}7s^2	10.07
Thorium	Th	90	232.04	[Rn]6d^{2}7s^2	11.7
Protactinium	Pa	91	(231.03)	[Rn]5f^{2}6d^{1}7s^2	15.4
Uranium	U	92	(238.05)	[Rn]5f^{3}6d^{1}7s^2	19.07

INDEX

Q